# Population Genetics of Bacteria

A TRIBUTE TO

**Thomas S. Whittam**

# Population Genetics of Bacteria

## A TRIBUTE TO
## Thomas S. Whittam

EDITED BY

**SETH T. WALK**
*Department of Internal Medicine, University of Michigan Health System, Ann Arbor, MI 48109*

**PETER C. H. FENG**
*Division of Microbiology, U.S. Food and Drug Administration, College Park, MD 20740*

ASM PRESS WASHINGTON, DC

American Society for Microbiology
1752 N Street, N.W.
Washington, DC 20036-2904

**Library of Congress Cataloging-in-Publication Data**

Population genetics of bacteria : a tribute to Thomas S. Whittam / edited by Seth T. Walk, Peter C. H. Feng.
p. ; cm.
Includes bibliographical references and index.
ISBN 978-1-55581-535-6
1. Bacterial genetics. 2. Escherichia coli. 3. Population genetics. I. Walk, Seth T. II. Feng, Peter C. H. III. Whittam, Thomas S. IV. American Society for Microbiology.
[DNLM: 1. Whittam, Thomas S. 2. Bacteria—genetics—Festschrift. 3. Escherichia coli—genetics—Festschrift. 4. Genetics, Population—methods—Festschrift. QW 51]

QH434.P67 2011
579.3'42—dc22

2011001596



10 9 8 7 6 5 4 3 2 1

Address editorial correspondence to: ASM Press, 1752 N St., N.W., Washington, DC 20036-2904, U.S.A.

Send orders to: ASM Press, P.O. Box 605, Herndon, VA 20172, U.S.A.
Phone: 800-546-2416; 703-661-1593
Fax: 703-661-1501
Email: Books@asmusa.org
Online: estore.asm.org

# DEDICATION

"A teacher affects eternity; he can never tell where his influence stops."
—American historian and philosopher Henry Adams

Whoever they were and wherever they were headed—whether eager undergraduates in search of all that science has to offer, opportunistic graduate students in pursuit of their dreams, postdoctorates looking to gain experience and the almighty publication, or any of the above individuals at the very end of their rope—Tom Whittam made an indelible mark on all, with plain words of wisdom, humor, and actions that set an example. At some point in time, every member of the Whittam lab was witness to Tom's humility in the face of success, kindness in the face of cruelty, and civility in the face of rejection. His mentees were taught that professionalism is not attained solely by the acquisition of knowledge—it also requires a deep commitment to personal development. Tom's sound character and respect for others helped develop the very best in his students, because he rarely, if ever, advised to "Do as I say, not as I do." More than this, Tom was strong enough emotionally to take a personal interest in those who sought his guidance. His willingness to become part of the mentee's experience, not simply a facilitator of it, created a bond that inspired and one that will continue to ripple outward through his students and his students' students.

The Whittam family took pride in our (i.e., the Whittamites') accomplishments and shared the pain of our failures. They too allowed us to share in their loss, for which we are grateful and by which we have been eternally affected. This book is dedicated with our gratitude to the Whittam family.

# CONTENTS

# CONTRIBUTORS

**Galeb S. Abu-Ali** • Center for Food Safety and Applied Nutrition, U.S. Food and Drug Administration, 8301 Muirkirk Road, Laurel, MD 20708

**Mark Achtman** • Department of Microbiology & Environmental Research Institute, University College Cork, Cork, Ireland

**Elizabeth W. Alm** • Department of Biology, Central Michigan University, 157 Brooks Hall, Mount Pleasant, MI 48859

**Andrea Ammon** • European Centre for Disease Prevention and Control (ECDC), 17183 Stockholm, Sweden

**Teresa M. Bergholz** • Department of Food Science, 405 Stocking Hall, Cornell University, Ithaca, NY 14853

**Thomas E. Besser** • Department of Veterinary Microbiology and Pathology, School for Global Animal Health, Washington State University, Pullman, WA 99164

**Martina Bielaszewska** • Institute for Hygiene, University of Münster, 48149 Münster, Germany

**Alyssa C. Bumbaugh** • Department of Biology, Shippensburg University, Shippensburg, PA 17257

**Isabel Chinen** • Servicio Fisiopatogenia, Instituto Nacional de Enfermedades Infecciosas—ANLIS "Dr. C. G. Malbrán," Ciudad Autónoma de Buenos Aires, Argentina

**Frederick M. Cohan** • Department of Biology, Wesleyan University, Middletown, CT 06459-0170

**Margaret A. Davis** • Department of Veterinary Microbiology and Pathology, School for Global Animal Health, Washington State University, Pullman, WA 99164

**Edward G. Dudley** • Department of Food Science, Pennsylvania State University, University Park, PA 16802

**Peter C. H. Feng** • Division of Microbiology, U.S. Food and Drug Administration, College Park, MD 20740

**Lucía Galli** • Servicio Fisiopatogenia, Instituto Nacional de Enfermedades Infecciosas—ANLIS "Dr. C. G. Malbrán," Ciudad Autónoma de Buenos Aires, Argentina

**David M. Gordon** • Research School of Biology, Australian National University, Building 116, Gould Wing, Canberra, ACT 0200, Australia

**James R. Johnson** • VA Medical Center and University of Minnesota, Minneapolis, MN 55417

**Helge Karch** • Institute for Hygiene, University of Münster, 48149 Münster, Germany

**Chih-Horng Kuo** • Institute of Plant and Microbial Biology, Academia Sinica, Taipei 11529, Taiwan

**David W. Lacher** • Division of Molecular Biology, Office of Applied Research and Safety Assessment, Center for Food Safety and Applied Nutrition, U.S. Food and Drug Administration, Laurel, MD 20708

**Shana R. Leopold** • Departments of Pediatrics and Molecular Microbiology, Washington University School of Medicine, St. Louis, MO 63105

**Bruce R. Levin** • Department of Biology, Emory University, Atlanta, GA 30322

**Amanda L. Lloyd** • Department of Microbiology and Immunology, University of Michigan Medical School, Ann Arbor, MI 48109

**Martin C. J. Maiden** • Department of Zoology, University of Oxford, South Parks Road, Oxford, OX1 3PS, United Kingdom

**Shannon D. Manning** • Microbiology and Molecular Genetics, Michigan State University, 165 Food Safety and Toxicology Building, E. Lansing, MI 48824

**Marcelo Masana** • Instituto Tecnología de Alimentos, Centro de Investigación de Agroindustria, Instituto Nacional de Tecnología Agropecuaria (INTA), Pcia. de Buenos Aires, Argentina

**Angela R. Melton-Celsa** • Department of Microbiology and Immunology, Uniformed Services University of the Health Sciences, Bethesda, MD 20814

**Elizabeth Miliwebsky** • Servicio Fisiopatogenia, Instituto Nacional de Enfermedades Infecciosas—ANLIS "Dr. C. G. Malbrán," Ciudad Autónoma de Buenos Aires, Argentina

**Harry L. T. Mobley** • Department of Microbiology and Immunology, University of Michigan Medical School, Ann Arbor, MI 48109

**Steven R. Monday** • Division of Microbiology, U.S. Food and Drug Administration, College Park, MD 20740

**Alison D. O'Brien** • Department of Microbiology and Immunology, Uniformed Services University of the Health Sciences, Bethesda, MD 20814

**Howard Ochman** • Department of Ecology and Evolutionary Biology, Yale University, New Haven, CT 06520

**Lindsey M. Ouellette** • Microbiology and Molecular Genetics, Michigan State University, 165 Food Safety and Toxicology Building, E. Lansing, MI 48824

**Weihong Qi** • Functional Genomics Center Zurich Uni, ETH Zurich, Winterthurerstrasse 190, Y32 H66, CH-8057 Zurich, Swizerland

**Rahul Raghavan** • Department of Ecology and Evolutionary Biology, Yale University, New Haven, CT 06520

**David A. Rasko** • Department of Microbiology and Immunology, University of Maryland School of Medicine, Baltimore, MD 21201

**Horacio A. Repetto** • Servicio de Pediatría, Hospital Nacional "Prof. Alejandro Posadas," Departamento de Pediatría, Facultad de Medicina, Universidad de Buenos Aires, Pcia. de Buenos Aires, Argentina

**James T. Riordan** • Department of Cell Biology, Microbiology and Molecular Biology, University of South Florida, 4202 East Fowler Avenue, BSF 217, Tampa, FL 33063

**Marta Rivas** • Servicio Fisiopatogenia, Instituto Nacional de Enfermedades Infecciosas—ANLIS "Dr. C. G. Malbrán," Ciudad Autónoma de Buenos Aires, Argentina

**A. Cody Springman** • Microbiology and Molecular Genetics, Michigan State University, 165 Food Safety and Toxicology Building, E. Lansing, MI 48824

**Hans Steinsland** • Department of Biomedicine, University of Bergen, Jonas Lies Vei 91, N-5009 Bergen, Norway

**Phillip I. Tarr** • Departments of Pediatrics and Molecular Microbiology, Washington University School of Medicine, St. Louis, MO 63105

**Louise D. Teel** • Department of Microbiology and Immunology, Uniformed Services University of the Health Sciences, Bethesda, MD 20814

**Jillian A. Tietjen** • University of Michigan Genetic Counseling Department, A909 Buhl, Ann Arbor, MI 48109

**Sivapriya Kailasan Vanaja** • Molecular Genetics and Microbiology, University of Massachusetts Medical School, 55 Lake Ave. North, S6-221, Worcester, MA 01655

**Seth T. Walk** • Department of Internal Medicine, Division of Infectious Diseases, University of Michigan Health System, 4618 Medical Sciences Building II, 1150 West Medical Center Dr., Ann Arbor, MI 48109

**Lukas Wick** • Biosynth AG, Rietlistrasse 4, 9422 Staad, Switzerland

# FOREWORD: THE ACCIDENTAL ECOLOGIST

One August day in 1976, Tom Whittam appeared at Arizona as if out of thin air. No one expected him. No one had prepared for him. There was not even the money to fund his teaching assistantship! He was the accidental ecologist, the phantom 13th floor in a superstitious high-rise.

You see, we were an inchoate department, cobbled together out of scraps, and suffering in the steadfast conviction of our provost that ecology and evolutionary biology was not a science. Those were fat days in academia, however, and somehow we managed to wrangle a few new faculty positions from the system. I instructed the graduate secretary that a particular four of the graduate students of these new faculty were to be admitted to the University of Arizona without delay or review. One was Tom Whitham, today a distinguished professor at Northern Arizona University, but then an unknown, a promising young student of Jim Brown.

Later in the semester, unknown to any faculty member, we received the application of one Thomas Whittam to our department. It should have gone to our graduate admissions committee. But the secretary, never the world's best proofreader, failed to notice that an "h" in the surname had given way to a second "t." And she also forgot that she had already processed *Whitham*'s application. She dutifully admitted Thomas *Whittam.*

Our departmental records were clear. We admitted 12 students that year. Surprise! Thirteen showed up.

I felt personally responsible. There was every likelihood that we were on the hook, legally. Besides, we needed the teaching help and the guy was qualified. So we scrounged the money to support him and I opened my lab to him.

On paper, you could not make a prima facie case that Tom Whittam would have been a shoo-in for admission. I had to let him know where he stood. We two had a solemn meeting. "Tom," I confessed, "you are an accident. We will honor our commitment to you for this year but you will have to prove yourself or there won't be a second."

Very harsh.

I found out much later that Tom's world came crashing in on him at that meeting. The joy of a new beginning, the optimism of a good student were replaced by foreboding and fear of coming up short. It was cruel and unusual teaching. But Tom never showed it. "I understand," he said. He left the room and went forth to work his heart out.

There is another harsh reality to describe about that year and, in fact, Tom's entire graduate career. Until then, my graduate students had been my colleagues and partners in my own research projects. But that model, which works for a lot of us, was not working for me. I had decided to forsake it for a model much more like that of my mentor, Robert MacArthur. I might cooperate with a student on a research paper but it would not be any that formed the core of the student's own research. Students would have to produce their own research questions and develop them into dissertations.

Tom was the perfect guinea pig. The new model meant a smaller investment in him, and, as I was not convinced he would succeed, I would be able to minimize what I risked.

Very, very harsh.

Because of Tom, I never changed to any other mentorship model. Tom lit up the department with scholarship, collegiality, and academic success. By the end of the year, we began to question the criteria we had used to assemble the group of 12 new students. How could we change them to get more students like number 13?

Soon, Tom took to the field to study two interesting but obscure species of mouse, both members of the genus *Onychomys*. *Onychomys* spp. have the common name "grasshopper mice" because they are miniature carnivores most of whose diet consists of large arthropods. I have caught 10 or 15 *Onychomys* in a lucky hot spot, but catching even one individual is unusual compared to the abundant heteromyids and more omnivorous cricetine rodents. Despite their scarcity, Tom wondered about their behavior and the interactions that might exist between them. But his career as a behavioral/community ecologist was over almost as soon as it began.

His preliminary reading took him to some papers that discussed *Onychomys* taxonomy. There he learned that a third species had been discovered not far from Tucson. I have to tell you that members of the genus *Onychomys* all look pretty much alike and, given the great intensity of mammal study in our part of the world, this third one had, no doubt, been collected many times over the previous century. But to discriminate the third species required looking inside the nuclei of their cells; the key was the karyotypes.

I did not realize it at first, but that cryptic third species of *Onychomys* swept Tom into the vortex of genetics. His dissertation, "Variation in Gene Frequencies in Rodent Populations: the Roles of Selective and Nonselective Evolutionary Forces," was pure ecological genetics. It was also a tour de force in meta-analysis. The patterns that Tom found depended on his carefully combining many other genetic studies to tease out the general truths. Mostly he was able to show the strong influence of genetic drift on gene frequencies.

Finding that his mentor was a virtual idiot in the field, Tom learned to reach out to the better informed, which is always a good strategy and an excellent habit to foster. Notable especially was his lifelong association with Michael Gaines, whom he met when Mike was spending some time in my lab. By the time Tom was ready for a postdoctoral appointment, he did not hesitate to abandon mammals entirely and head off to work with much smaller organisms.

My new model of graduate mentorship had worked perfectly—I was left in Tom's intellectual dust!

Since Tom's time I have learned that my model is no assurance of success. Many others have crashed and burned when allowed to follow their own star. Tom soared. Tom's success, although not unique, is uncommon, a sign of discipline, curiosity, and creativity all layered onto a base of high intelligence. But the latter is easy to recognize. The rest does not emerge until one gives it a chance. I am so proud of what Tom accomplished in his all too brief time.

MICHAEL L. ROSENZWEIG

# PREFACE

Thomas S. Whittam was born in Newton, PA, in 1954. He began his scientific career at Franklin & Marshall, a small liberal arts college in Lancaster, PA, and graduated with a Bachelor of Arts in Biology in 1976. After graduation, Tom moved with his wife, Beth, to Tucson, AZ, and began his doctoral studies at the University of Arizona, studying Evolutionary Biology with the eminent ecologist Michael Rosenzweig. Tom excelled in research that addressed a number of ecological and evolutionary hypotheses concerning rodent populations and seabird communities. He received an Outstanding Dissertation Award at the completion of his Ph.D. education and published four papers between 1980 and 1981 on this research. These were his last manuscripts on such (macro)organisms.

In 1981, Tom moved his family, which included a son, Benjamin (b. 1980), to Rochester, NY, where he began postdoctoral training in the laboratory of the renowned population geneticist Robert Selander. Dr. Selander's group was studying *Escherichia coli* and had substantial genetic data already collected when Tom arrived. Having mastered a variety of statistical genetics techniques, Tom made rapid progress in describing the evolutionary dynamics of *E. coli* populations. In 1982, he presented his research on the "Multilocus genetic structure of *E. coli*" at an international workshop in Bethesda, MD, that was sponsored by the National Institutes of Health (NIH) Fogarty Center. By 1983, just 2 years after beginning his postdoctoral work, he had already published a book chapter and four articles including a paper in the *Proceedings of the National Academy of Sciences USA*. That last paper described an evolutionary framework for naturally occurring *E. coli* populations (the abundance, distribution, and fate of genetic polymorphisms) and foreshadowed a career of outstanding scientific contributions to the field of bacterial population genetics.

Tom accepted his first academic position in 1983 at Rutgers University and moved his family, including a new daughter, Elizabeth (b. 1983 in Rochester, NY), to Piscataway, NJ. He continued his research and taught courses in Environmental Science, Genetics, and Vertebrate Zoology. The stay in New Jersey was short, however, and in 1985 Tom accepted an assistant professorship at the Pennsylvania State University and moved his family to State College, PA. He was awarded his first research grant in 1986 (an institutional Biomedical Research Support Grant) to study the "Clonal nature of pathogenic *E. coli* associated with outbreaks of hemorrhagic colitis." Tom used this grant to propel his research into the public health arena, a move that would gain him global recognition in years to come as *the* expert on the evolution of *E. coli* O157:H7. Tom and Beth had their second son and third child, Andrew (b. 1987), and raised their family in State College until they moved to Okemos, MI, in 2001, when Tom accepted a Hannah Distinguished Professorship at Michigan State University.

Tom made numerous scientific discoveries that had global impact. Of special note, he developed an internationally recognized and utilized collection of pathogenic *E. coli* strains, called the STEC Center, which has been critical to the research of clinician scientists, evolutionary biologists, basic microbiologists, and population geneticists. The STEC Center houses well over 12,000 pathogenic and nonpathogenic strains, most with metadata (e.g., contributor, year of isolation, source, and genetic characterizations). These strains are cryopreserved and have been shipped to investigators at no cost through supporting funds from NIH. The STEC Center is still maintained at Michigan State University (www.shigatox.net) and represents a physical legacy of Tom's lifetime achievement in infectious disease research.

Over the course of his 25-year research career, Tom was awarded 19 competitive grants and government contracts from NIH, the National Science Foundation, the United States Department of Agriculture, and the Pennsylvania Department of Agriculture, totaling in excess of $14.5 million in funding. He was invited to give 98 seminars and presentations at colleges, universities, and national and international scientific meetings. At the time of this writing, Tom's work totals over 70 published abstracts from scientific meetings and 175 authored and coauthored articles in peer-reviewed journals. According to the ISI Web of Knowledge, 22 of Tom's articles have been cited more than 100 times. Tom supervised the education of 10 undergraduate honors, 9 masters, and 13 doctoral students, and mentored 10 postdoctoral fellows. Dozens of undergraduate researchers gained training and experience in Tom's laboratory, and thousands more were educated in courses that he taught (e.g., Genetics, Advanced Genetics, Evolution of Infectious Diseases, and Introductory Biology). Tom was dedicated to the universities that employed him and served on numerous search, executive, scientific, and other advisory committees.

Tom's love for his family was paramount, but his second love was science. He used to jest proudly, "I can't believe anyone would pay me to do this!," meaning he loved his job and what he was doing. He inspired his collaborators and peers—and science benefited. He gave students the chance they needed to prove and improve themselves—and science benefited. He placed research discoveries before politics and personal ambitions—and science benefited.

Just as Tom was in the midst of a productive and successful scientific and professorial career, he was diagnosed with brain cancer. Tom passed away on December 5, 2008, at 54 years old. His death was a tragic loss, not only for family, friends, and his many collaborators and students, but also for science. This book is a tribute to a great scientist and educator and, in most cases, to a great friend. The book is not meant to be a comprehensive nor a complete historical narrative of Tom's work. Instead, it is a compilation of research projects and ideas that were influenced by Tom and his science. The book is focused on bacterial population genetics and its application to pathogenic *E. coli*. The centerpiece of this focus is the story of O157:H7, which comprises a significant portion of the book. We hope that the reader will gain an understanding and appreciation of these contributions to the field as well as an understanding of how the population genetic methodologies that Tom used to study pathogenic *E. coli* and O157:H7 now serve as a model to investigate other pathogenic bacterial populations.

*Population Genetics of Bacteria: a Tribute to Thomas S. Whittam*
Edited by S. T. Walk and P. C. H. Feng


*Chapter 1*

# Introduction: a Personal Homage to Tom Whittam

***Mark Achtman***

I interacted with Tom Whittam on various occasions, including some correspondence at a time in the 1980s when we overlapped in our work on *Escherichia coli* (1, 4, 5, 26). I was extremely impressed by his great expertise, erudition, and professionalism in statistical methods and evolutionary biology, and studied his masterly review article on *E. coli* population genetics (25) as well as other publications on multiple occasions. We discussed a number of scientific topics at the ASM general meeting in Salt Lake City in 2002, and Tom invited me to present my work at a Keystone Conference in Colorado in 2009. Unfortunately, he was already so sick by that time that he could not attend, and I don't remember having met him again after 2002.

One particular interaction with Tom that remains vivid in my memory was at a small conference in Italy in 1998 that was organized by Jan Klein, where Tom informed me that he had been one of the reviewers for a recent manuscript on *Helicobacter pylori* that I coauthored with Sebastian Suerbaum, John Maynard Smith, Noel Smith, and others (23). His review had suggested the use of the program Reticulate (11) to display homoplasies, i.e., mutations that existed along multiple independent branches of a phylogenetic tree. After we followed Tom's suggestion and included the output from Reticulate analyses in the revised version, the paper was accepted by *Proceedings of the National Academy of Sciences of the USA*, and published later that year. What particularly impressed me about that interaction was the review, whose style of writing was not only scholarly but extremely gentle, and not in the slightest abrasive. In subsequent publications on *H. pylori*, we again used Reticulate (3), as well as Happlot, a Fortran program that Tom provided us to graphically depict polymorphic nucleotides (9). In fact, we were still using Happlot as recently as 2009 (21). Thus, Tom had a particularly important role in the beginnings of

***Mark Achtman*** • Department of Microbiology and Environmental Research Institute, University College Cork, Cork, Ireland.

my research on *H. pylori* and was partially responsible for the contents of our seminal article (23), an article which had already been cited more than 300 times by 2010.

Tom's insights into phylogenetics in the late 1980s continued to influence how we calculated the ages of clonal lineages until very recently. His calculations used the assumption that a universal clock rate (18), which was calibrated by the genetic distance between *E. coli* and *Salmonella enterica* serovar Typhimurium and an estimated time of separation of 160 million years, could be applied to date the coalescent of a clonal lineage. This clock rate was used to calculate that the *Mycobacterium tuberculosis* complex evolved 20,000 years ago (22) and that the *E. coli* O157:H7 lineage evolved from O55:H7 some 40,000 years ago (29). Indeed, when faced with the problem of dating the age of *Yersinia pestis* (7), a genetically monomorphic organism, I partially copied this approach in widely cited publications on *Y. pestis* in 1999 (7) and 2004 (6). This approach has also been used extensively for clonal lineages in various organisms by multiple other groups (2), including myself (6, 7, 16, 20, 28), and has become a simple paradigm for calculating their ages. Unfortunately, the bacterial mutational clock rate is not invariant (15, 17), and the concept of using clock rates based on fixed single nucleotide polymorphisms between species to calculate the accumulation of genetic diversity within a species has been criticized (12). Indeed, clock rates for the accumulation of diversity within a species can vary by several orders of magnitude between different species (15), and are very likely to be considerably more rapid over the short than the long term (8, 10). Tom was beginning to take account of such considerations in his more recent publications (14).

I would have welcomed an opportunity to collaborate with Tom. I consider him to have been a professional phylogeneticist, whereas my knowledge of population genetics was acquired autodidactically and suffers from multiple gaps. I think we would have complemented each other well. However, an opportunity never arose because we worked on different organisms and used different approaches, except for our brief overlaps with *E. coli* in the 1980s (4, 26) and again in the last decade (19, 27), when we both established multilocus sequence typing databases for *E. coli* (13, 27). These brief overlaps have resulted in independent publications that are difficult if not impossible to correlate with each other, with a few exceptions where the same isolates were tested by both methods (24, 27). I regret this situation, because I am convinced that I could have learned from him had we found an opportunity to collaborate. Possibly the reverse is also true.

Tom was one of the few early founders of modern bacterial population genetics and has left a lasting impression on the field, particularly so on those who met him. I continue to grieve for his untimely death and feel that my scientific area of interest has been impoverished as a result. Bacterial population genetics has had only a few leading figures, among whom Tom was one of the most prominent. I also miss him as a human being because of his gentle persona, and wish I could emulate his behavior more often.

## REFERENCES


1. **Achtman, M.** 1985. Clonal groups and virulence factors among *Escherichia coli* K1 strains, p. 65–74. *In* T. K. Korhonen, E. A. Dawes, and P. H. Mäkelä (ed.), *Enterobacterial Surface Antigens: Methods for Molecular Characterisation.* Elsevier Science Publishers, Amsterdam, The Netherlands.
2. **Achtman, M.** 2008. Evolution, population structure, and phylogeography of genetically monomorphic bacterial pathogens. *Annu. Rev. Microbiol.* **62:**53–70.
3. **Achtman, M., T. Azuma, D. E. Berg, Y. Ito, G. Morelli, Z.-J. Pan, S. Suerbaum, S. Thompson, A. van der Ende, and L. J. van Doorn.** 1999. Recombination and clonal groupings within *Helicobacter pylori* from different geographical regions. *Mol. Microbiol.* **32:**459–470.
4. **Achtman, M., M. Heuzenroeder, B. Kusecek, H. Ochman, D. Caugant, R. K. Selander, V. Väisanen-Rhen, T. K. Korhonen, S. Stuart, F. Ørskov, and I. Ørskov.** 1986. Clonal analysis of *Escherichia coli* O2:K1 isolated from diseased humans and animals. *Infect. Immun.* **51:**268–276.
5. **Achtman, M., A. Mercer, B. Kusecek, A. Pohl, M. Heuzenroeder, W. Aaronson, A. Sutton, and R. P. Silver.** 1983. Six widespread bacterial clones among *Escherichia coli* K1 isolates. *Infect. Immun.* **39:**315–335.
6. **Achtman, M., G. Morelli, P. Zhu, T. Wirth, I. Diehl, A. J. Vogler, D. M. Wagner, C. J. Allender, W. R. Easterday, V. Chenal-Francisque, P. Worsham, N. R. Thomson, J. Parkhill, L. E. Lindler, E. Carniel, and P. Keim.** 2004. Microevolution and history of the plague bacillus, *Yersinia pestis. Proc. Natl. Acad. Sci. USA* **101:**17837–17842.
7. **Achtman, M., K. Zurth, G. Morelli, G. Torrea, A. Guiyoule, and E. Carniel.** 1999. *Yersinia pestis*, the cause of plague, is a recently emerged clone of *Yersinia pseudotuberculosis. Proc. Natl. Acad. Sci. USA* **96:**14043–14048.
8. **Drummond, A. J., S. Y. Ho, M. J. Phillips, and A. Rambaut.** 2006. Relaxed phylogenetics and dating with confidence. *PLoS Biol.* **4:**e88.
9. **Falush, D., C. Kraft, N. S. Taylor, P. Correa, J. G. Fox, M. Achtman, and S. Suerbaum.** 2001. Recombination and mutation during long-term gastric colonization by *Helicobacter pylori*: estimates of clock rates, recombination size, and minimal age. *Proc. Natl. Acad. Sci. USA* **98:**15056–15061.
10. **Ho, S. Y., and G. Larson.** 2006. Molecular clocks: when times are a-changin'. *Trends Genet.* **22:**79–83.
11. **Jakobsen, I. B., and S. Easteal.** 1996. A program for calculating and displaying compatibility matrices as an aid in determining reticulate evolution in molecular sequences. *Comput. Appl. Biosci.* **12:**291–295.
12. **Kryazhimskiy, S., and J. B. Plotkin.** 2008. The population genetics of dN/dS. *PLoS Genet.* **4:**e1000304.
13. **Lacher, D. W., H. Steinsland, T. E. Blank, M. S. Donnenberg, and T. S. Whittam.** 2007. Molecular evolution of typical enteropathogenic *Escherichia coli*: clonal analysis by multilocus sequence typing and virulence gene allelic profiling. *J. Bacteriol.* **189:**342–350.
14. **Leopold, S. R., V. Magrini, N. J. Holt, N. Shaikh, E. R. Mardis, J. Cagno, Y. Ogura, A. Iguchi, T. Hayashi, A. Mellmann, H. Karch, T. E. Besser, S. A. Sawyer, T. S. Whittam, and P. I. Tarr.** 2009. A precise reconstruction of the emergence and constrained radiations of *Escherichia coli* O157 portrayed by backbone concatenomic analysis. *Proc. Natl. Acad. Sci. USA* **106:**8713–8718.
15. **Morelli, G., X. Didelot, B. Kusecek, S. Schwarz, D. Falush, C. Bahlawane, S. Suerbaum, and M. Achtman.** 2010. Microevolution of *Helicobacter pylori* during prolonged infection of single hosts and within families. *PLoS Genet.* **6:**e1001036.
16. **Nübel, U., P. Roumagnac, M. Feldkamp, J. H. Song, K. S. Ko, Y. C. Huang, G. Coombs, M. Ip, H. Westh, R. Skov, M. J. Struelens, R. V. Goering, B. Strommenger, A. Weller, W. Witte, and M. Achtman.** 2008. Frequent emergence and limited geographic dispersal of methicillin-resistant *Staphylococcus aureus. Proc. Natl. Acad. Sci. USA* **105:**14130–14135.

17. **Ochman, H., S. Elwyn, and N. A. Moran.** 1999. Calibrating bacterial evolution. *Proc. Natl. Acad. Sci. USA* **96:**12638–12643.
18. **Ochman, H., and A. C. Wilson.** 1987. Evolution in bacteria: evidence for a universal substitution rate in cellular genomes. *J. Mol. Evol.* **26:**74–86.
19. **Reid, S. D., C. J. Herbelin, A. C. Bumbaugh, R. K. Selander, and T. S. Whittam.** 2000. Parallel evolution of virulence in pathogenic *Escherichia coli*. *Nature* **406:**64–67.
20. **Roumagnac, P., F.-X. Weill, C. Dolecek, S. Baker, S. Brisse, N. T. Chinh, T. A. Le, C. J. Acosta, J. Farrar, G. Dougan, and M. Achtman.** 2006. Evolutionary history of *Salmonella* Typhi. *Science* **314:**1301–1304.
21. **Schwarz, S., G. Morelli, B. Kusecek, A. Manica, F. Balloux, R. J. Owen, D. Y. Graham, S. van der Merwe, M. Achtman, and S. Suerbaum.** 2008. Horizontal vs. familial transmission of *Helicobacter pylori*. *PLoS Pathog.* **4:**e1000180.
22. **Sreevatsan, S., X. Pan, K. Stockbauer, N. D. Connell, B. N. Kreiswirth, T. S. Whittam, and J. M. Musser.** 1997. Restricted structural gene polymorphism in the *Mycobacterium tuberculosis* complex indicates evolutionarily recent global dissemination. *Proc. Natl. Acad. Sci. USA* **94:**9869–9874.
23. **Suerbaum, S., J. Maynard Smith, K. Bapumia, G. Morelli, N. H. Smith, E. Kunstmann, I. Dyrek, and M. Achtman.** 1998. Free recombination within *Helicobacter pylori*. *Proc. Natl. Acad. Sci. USA* **95:**12619–12624.
24. **Walk, S. T., E. W. Alm, D. M. Gordon, J. L. Ram, G. A. Toranzos, J. M. Tiedje, and T. S. Whittam.** 2009. Cryptic lineages of the genus *Escherichia*. *Appl. Environ. Microbiol.* **75:**6534–6544.
25. **Whittam, T. S.** 1996. Genetic variation and evolutionary processes in natural populations of *Escherichia coli*, p. 2708–2720. *In* R. Curtiss III, J. L. Ingraham, E. C. C. Lin, K. B. Low, B. Magasanik, W. S. Reznikoff, M. Riley, M. Schaechter, and H. E. Umbarger (ed.), Escherichia coli *and* Salmonella, vol. 2. ASM Press, Washington, DC.
26. **Whittam, T. S., and R. A. Wilson.** 1988. Genetic relationships among pathogenic strains of avian *Escherichia coli*. *Infect. Immun.* **56:**2458–2466.
27. **Wirth, T., D. Falush, R. Lan, F. Colles, P. Mensa, L. H. Wieler, H. Karch, P. R. Reeves, M. C. Maiden, H. Ochman, and M. Achtman.** 2006. Sex and virulence in *Escherichia coli*: an evolutionary perspective. *Mol. Microbiol.* **60:**1136–1151.
28. **Wirth, T., G. Morelli, B. Kusecek, A. van Belkum, C. van der Schee, A. Meyer, and M. Achtman.** 2007. The rise and spread of a new pathogen: seroresistant *Moraxella catarrhalis*. *Genome Res.* **17:**1647–1656.
29. **Zhang, W., W. Qi, T. J. Albert, A. S. Motiwala, D. Alland, E. K. Hyytia-Trees, E. M. Ribot, P. I. Fields, T. S. Whittam, and B. Swaminathan.** 2006. Probing genomic diversity and evolution of *Escherichia coli* O157 by single-nucleotide polymorphisms. *Genome Res.* **16:**757–767.

*Section I*

# BACTERIAL EVOLUTION AND THEORY

*Population Genetics of Bacteria: a Tribute to Thomas S. Whittam*
Edited by S. T. Walk and P. C. H. Feng


*Chapter 2*

# Population Geneticists Discover Bacteria and Their Genetic/Molecular Epidemiology

***Bruce R. Levin***

In this personal, opinionated, occasionally nostalgic, and doubtless egoistic rant, I consider the origins and early history of the enterprise to which Tom Whittam made his awesome contributions. While we all know what those contributions are, we almost certainly differ in what we call this enterprise: population genetics, genetic epidemiology, evolutionary genetics, molecular epidemiology, molecular evolution, molecular phylogeny, or molecular forensics. To paraphrase Andre Lwoff (15), we have to be molecular. Who is not? Tom's research included projects that fit all of these descriptions and more. By all standards, the breadth as well the importance, utility, and quality of Tom's contributions are impressive.

Tom was not the first *p,q*-trained and -inclined population and evolutionary geneticist to work with bacteria, but when he began to do so in the early 1980s he was a member of a small minority. As Carl Bergstrom and I once quipped, "The genetical theory of evolution was developed by sexually reproducing eukaryotes for sexually reproducing eukaryotes" (13). Indeed, shortly before I published my first research with bacteria (11), Theodosius Dobzhansky admonished me when I told him I gave up *Drosophila* to work on *Escherichia coli*, "Zat is not an organism."[1] Among the enlightened minority working with these "nonorganisms," Tom was one of the first to apply population genetics procedures and theory to the genetic epidemiology of bacteria from clinical and other natural sources.

## THE NEUTRAL GENE HYPOTHESIS AND THE CLONE CONCEPT

In my interpretation, the immediate antecedent of Tom's research with bacteria, as well as that of Bob Selander, Dominique Caugant, and Howard Ochman (eukaryote-trained biologists all), was the neutral gene controversy that dominated population genetics of the disco era, in particular an article by Roger Milkman (17). Using a collection of 829 isolates of *E. coli* from diverse sources

***Bruce R. Levin*** • Department of Biology, Emory University, Atlanta, GA 30322.

and cellulose acetate electrophoresis, Roger estimated the frequency of different electrophoretic variants for five housekeeping proteins. He interpreted the results of his study as evidence against the neutral gene hypothesis. He argued that for a species with as large a genetically effective population size as he assumed *E. coli* must have, in accord with the neutral gene hypothesis (i) the effective number of alleles would be vastly greater than that he estimated and (ii) the distribution of motility classes (alleles) would be different than that he observed. Central to and implicit in his interpretation was the assumption that *E. coli* was in linkage equilibrium; variation among isolates from the same individual was great enough to "suggest that recombination occurs regularly within hosts."

Was Milkman's 1973 study of enzyme variation in *E. coli* the definitive, "no excuse" test of the neutral gene hypothesis he proposed it would be (18)? It may well have seemed that way to the pan-selectionist,[2] and even the most astute neutralists like Motoo Kimura didn't counter it.[3] It wasn't until 1980 that an "excuse" for Milkman's study was provided in a report by Bob Selander and me (25).

The origins of that paper were a talk I gave at the University of Rochester. As I recall, the subject my presentation was the population dynamics of bacteriophage and plasmids. After I completed my talk, Bob—who was already well-known for his extensive and first-rate enzyme electrophoretic studies of genetic variation, evolution, and the genetic structure of natural populations of a number of species of eukaryotes—asked if he could come to my lab at University of Massachusetts Amherst to learn how to culture *E. coli*. Bob's two days in our lab was the start of a collaboration that included a number of visits to each other's labs over the next three or so years and five coauthored articles. On one of those visits, I met Tom for the first time. He had just arrived in Rochester, I assume expecting to study genetic variation in mice (see chapter 4).

At the time, both Bob and I had technical and theoretical issues with Roger's 1973 report, which we addressed in our 1980 "excuse" paper (25). In that study, we (really Bob and his assistants) used starch gel electrophoresis, which had greater resolving power than the cellulose acetate method Roger employed, and examined 20 rather than 5 enzymes. Most importantly, we then recorded the electrophoretic profiles separately for each clone, rather than for each enzyme, as did Roger. Of the 109 clones of *E. coli* we examined, 90 were included in Roger's study and the remaining 19 were clones I isolated from the feces of 16 infants and 2 adults in Massachusetts. In that study we also examined the enzyme electrophoretic profile for the three primary laboratory strains of *E. coli*, single clones of *E. coli* B and C and 24 clones of *E. coli* K-12.

Our estimate of the genetic diversity of *E. coli* from this sample, $H = 0.47$, exceeded what we calculated from Roger's data by about factor of 2. More importantly, our data clearly showed that the genetic structure of *E. coli* was very different from that of the oversexed, complete linkage-equilibrium eukaryotes assumed in the models upon which Roger's interpretations were based. Despite the relatively large amount of variability we observed for the 20 enzymes examined, there were only 98 distinct electrophoretic types in the 109 wild strains. Isolates with identical allozymes for all 20 proteins appeared in strains from two pairs of infants from Massachusetts and adults in Iowa, a giraffe from a zoo in Iowa, and a feral sheep in California. Moreover, a clone isolated from an infant in Massachusetts was identical at 20 enzymes to those expressed by *E. coli* K-12, although the strains were quite different in the plasmids they carried.

Although Bob Selander was hardly averse to the use of expletives, words like "eureka" weren't part of his repertoire. I, however, very distinctly recall a eureka feeling as we sat in my study in Amherst as Bob passed a ruler through a printout of the enzyme data and saw the identity and near identity of the allozyme profiles of independent isolates. In our second collaborative article (6), we provided evidence that was inconsistent with Roger Milkman's assertion of recombination occurring regularly between *E. coli* in mammalian hosts, at least not within me as the human host of this study.

## NOT REALLY NEWS

From a population genetics perspective, the result of our *E. coli* enzyme variation study and its interpretation had sufficient "man bites dog" appeal to appear in *Science* (a condition for publication then as well as now). On the other hand, neither the considerable variability nor the clonal population genetic structure we reported would have been surprising to microbiologists studying the epidemiology of bacteria from clinical and natural sources. This was certainly so for the late Fritz and Ida Ørskov, from the Serum Institute in Copenhagen. For some 40 years they and their predecessors used serotyping to study the epidemiology of *E. coli* (9, 20, 24). Although there were 164 O antigens, ~100 K antigens, and 56 H antigens (more than 900,000 possible combinations), isolates of the same O:K:H serotype appeared commonly from geographically and temporally different sources, and specific O:K:H serotypes were associated with different symptomatic infections.

Similar observations were made for a number of different species of bacteria with serological as well as other phenotypic markers, like phage resistance patterns (phage typing), repertoires of fermentation capabilities (biotyping), and the distribution of plasmids carried (plasmid typing). While direct sequencing of DNA was not yet in their toolbox, by the early 1980s investigators studying the genetic epidemiology of bacteria were also beginning to use various kinds of restriction endonuclease cutting procedures for these epidemiological studies. For a superb perspective on genetic epidemiology before population geneticists entered the biz, see the summary of a "workshop on the clone concept in the epidemiology, taxonomy, and evolution of the Enterobacteriaceae and other bacteria" (22). This 1982 workshop that Fritz and Ida Ørskov organized at the NIH Fogarty Center was, I believe, Tom's debut into the world of microbial genetic epidemiology. Tom gave a superb talk at that meeting presenting the results of his first bacterial population genetics article (26).

The Ørskovs were also well aware that recombination was a rare event in *E. coli*. They had done a study of the frequency and rate of homologous gene recombination with 199 different O:K:H *E. coli* serotypes as donors and an *E. coli* K-12 recipient (21). Were they population and evolutionary geneticists and not as genteel as they were, they would have probably set Roger straight for his assertion about *E. coli*'s proclivity for sex.[4]

## POPULATION GENETICS BECOMES GENETIC/MOLECULAR EPIDEMIOLOGY

In retrospect, I believe the single most important consequence of our early-1980s studies of genetic variation in *E. coli* was not direct genetic evidence for the clonal structure of *E. coli*, which the enzyme data provided, but rather the introduction of population genetics theory and approaches to the genetic epidemiology of bacteria. Perhaps as significant was the recruitment of *p,q*-trained population geneticists to this enterprise. Too freaking much!: would you believe evolutionary biologists could do something of value for the world beyond the precious realm of academe?

Although our 1980 *Science* paper may have made Motoo Kimura feel better about *E. coli* and the neutral gene hypothesis, neither this report nor subsequent studies on the population genetics of bacteria resolved the controversy, which seems to have appropriately faded away. What we did add to the genetic and now the molecular epidemiology of bacteria were explanations for why bacteria retain a clonal structure in the face of recurrent mutation and recombination (12, 16). And, thanks to Tom and others, population geneticists provided the theory and tools to analyze and interpret the genetic/molecular epidemiological data. Multilocus enzyme electrophoresis (MLEE), the main population genetics tool of the 1970s and early 1980s (a bit of an art form with superb practitioners like Bob and Dominique Caugant as well as Howard Ochman and Tom), also had a considerable virtue over the purely phenotypic methods microbiologists were using in epidemiological studies at the time. It made it possible to quantify the genetic (phylogenetic, if you prefer) relationships and distances between clones and within and between populations. To a large extent, the trees grown in the MLEE days of the last century have retained their shape into this century and I expect will do so beyond. What MLEE data and multilocus sequence typing data failed to detect was the considerable variation in size of the genomes of *E. coli* and other bacteria (see chapter 4) and the fundamental role of horizontal gene transfer as a source of variation for evolution in bacteria. Also missed by these multi-but-not-that-many-locus typing procedures were the movements of insertion sequences and other more rapid changes occurring in bacterial genomes, which are important for forensic considerations and probably adaptive evolution as well.

## WHAT A GOOD MEETING CAN DO

As part of my preparation for writing this essay, I reread Fritz and Ida Ørskov's summary of their 1982 clone concept workshop (22), a literary excursion that convinced me of the seminal role that a good, interdisciplinary meetings can play. Save for the forestry (trees), the anlagen of much of what is now the molecular epidemiology of bacteria were displayed at that meeting. In my interpretation, directly and indirectly, that workshop had a great deal of influence on the population and evolutionary geneticists whose work later became central to our understanding of the genetic epidemiology and population and evolutionary genetics of bacteria. I know Bob Selander, who was at the meeting, was listening, from a comment he made to a speaker: "If you fucking people really did genetics rather than just make believe you were . . ." While I don't know how much influence that workshop had on the awesome genetic/molecular epidemiological studies Bob and his students and collaborators did in the years that immediately followed, I am sure it had some (see, for example, references 1, 5, and 19). From the conversations he and I had at the workshop and after, and by this collaboration criterion (4, 23, 27), I know this was also the case for Tom. I also believe that meeting convinced the more enlightened molecular biologists and epidemiologically and clinically oriented microbiologists of the considerable value of the technology and theory of academic population genetics, if not the well-mannered nature of its practitioners.

**Notes**

[1]Although at the time, few card-carrying population geneticists and evolutionary biologists showed much interest in bacteria when they were not ill with some infection, a study with bacteria was responsible for one of the most significant contributions to the genetic theory of evolution. The first evidence for the preadaptive nature of mutations, that "Natural selection is the editor, rather than the composer, of the genetic message" (10), was experiments with *E. coli* (14). From his remark to me, which I later learned to be no more than a good-natured tease, it may appear that Dobzhansky was unaware of the evolutionary implications of Luria and Delbruck's experiments and other studies on the genetics of bacteria. That was not the case; see p. 87–88 in reference 8. It was certainly not the case for Jim Crow; not only was he very attuned to these studies with bacteria, but his student Jack Bennett provided the first evidence for preadaptive mutation in eukaryotes (3, 7). There was also an *E. coli* study by Kim Atwood and colleagues (2), which in a contemporary perspective would be seen as particularly classy experimental population genetics.

[2]Although I never really knew whether it was rhetorical posturing or sociological, at the time there were indeed pan-selectionists, despite the fact that the neutral gene and selection hypotheses were not mutually exclusive for the genome at large or any given gene. Needless to say, nobody was misguided enough to champion pan-neutrality, although some may have been accused of doing just that.

[3]In 1980 I received the first of my letters and beautiful Christmas cards from Motoo Kimura. While I can't say for sure, I always attributed that to the publication of Bob Selander's and my 1980 *Science* paper, or Motoo's learning about our results and interpretation before from a super-able mathematical population geneticist, the late Takeo Maruyama, who had visited our lab at about that time. Indeed, one can almost sense a tone of relief in a fine article Maruyama and Kimura wrote in 1980 about the genetic structure and effective population size of asexual haploids (16).

[4]Joshua Lederberg also knew how rare recombination is in *E. coli*. Had he worked with any of the other then prominent laboratory strains like *E. coli* B or C, or almost any wild strain, he would have gotten negative results in his 1946 experiments testing for recombination in this bacterium. The strain he used, *E. coli* K-12, was an oddball. It bore a plasmid, F, which was permanently derepressed for conjugative pili synthesis and a chromosome with insertion sequences homologous to those on the plasmid which enabled this conjugation-

encoding accessory element to integrate into the chromosome by Rec-mediated recombination. I once asked Josh how many other strains he would have looked at if he had gotten negative results with K-12. He told me, "One." Lederberg and Tatum's preparation to do these experiments, generating the amino acid and fermentation negative mutants, was a considerable task at the time. What a combination: serendipity and a prepared, brilliant mind!

**In Memoriam.** In addition to dedicating this chapter to the memory of Tom Whittam, I also dedicate it to the memories of Fritz and Ida Ørskov: all super scientists and kind, generous, and open-minded people whom we greatly miss.

**Acknowledgments.** I wish to thank Amy Kirby, Omar Cornejo, and Amoolya Singh for reading this manuscript and for making helpful suggestions and acknowledge my appreciation for their considerable diplomacy in that enterprise. I also want to thank Jim (James F.) Crow for insightful, always, historical information. This "rant" was prepared while I was funded by NIH Grant GM091875.

## REFERENCES

1. **Achtman, M., M. Heuzenroeder, B. Kusecek, H. Ochman, D. Caugant, R. K. Selander, V. Väisanen-Rhen, T. K. Korhonen, S. Stuart, F. Ørskov, and S. Ørskov.** 1986. Clonal analysis of *Escherichia coli* O2:K1 isolated from diseased humans and animals. *Infect. Immun.* **51:**268–276.
2. **Atwood, K. C., L. K. Schneider, and F. J. Ryan.** 1951. Periodic selection in *Escherichia coli. Proc. Natl. Acad. Sci. USA* **37:**146–155.
3. **Bennett, J.** 1960. A comparison of selective methods and a test of the pre-adaptation hypothesis. *Heredity* **15:**65–77.
4. **Beutin, L., I. Ørskov, F. Ørskov, S. Zimmermann, J. Prada, H. Gelderblom, R. Stephan, and T. S. Whittam.** 1990. Clonal diversity and virulence factors in strains of *Escherichia coli* of the classic enteropathogenic serogroup O114. *J. Infect. Dis.* **162:** 1329–1334.
5. **Caugant, D. A., B. R. Levin, I. Ørskov, F. Ørskov, C. Svanborg Eden, and R. K. Selander.** 1985. Genetic diversity in relation to serotype in *Escherichia coli. Infect. Immun.* **49:**407–413.
6. **Caugant, D. A., B. R. Levin, and R. K. Selander.** 1981. Genetic diversity and temporal variation in the *E. coli* population of a human host. *Genetics* **98:**467–490.
7. **Crow, J. F.** 1957. The genetics of insecticide resistance to chemicals. *Annu. Rev. Entomol.* **2:**227–246.
8. **Dobzhansky, T.** 1951. *Genetics and the Origin of the Species*, 3rd ed. Columbia University Press, New York, NY.
9. **Evans, D. J., Jr., D. G. Evans, H. L. DuPont, F. Ørskov, and I. Ørskov.** 1977. Patterns of loss of enterotoxigenicity by *Escherichia coli* isolated from adults with diarrhea: suggestive evidence for an interrelationship with serotype. *Infect. Immun.* **17:**105–111.
10. **King, J. L., and T. H. Jukes.** 1969. Non-Darwinian evolution. *Science* **164:**788–798.
11. **Levin, B. R.** 1972. Coexistence of two asexual strains on a single resource. *Science* **175:**1272–1274.
12. **Levin, B. R.** 1981. Periodic selection, infectious gene exchange and the genetic structure of *E. coli* populations. *Genetics* **99:**1–23.
13. **Levin, B. R., and C. T. Bergstrom.** 2000. Bacteria are different: observations, interpretations, speculations, and opinions about the mechanisms of adaptive evolution in prokaryotes. *Proc. Natl. Acad. Sci. USA* **97:**6981–6985.
14. **Luria, S. E., and M. Delbruck.** 1943. Mutations of bacteria from virus sensitivity to virus resistance. *Genetics* **28:**491–511.
15. **Lwoff, A.** 1966. The prophage and I, p. 88–99. *In* J. Cairns, G. S. Stent, and J. D. Watson (ed.), *Phage and the Origins of Molecular Biology.* Cold Spring Harbor Laboratory Press, Cold Spring Harbor, NY.

16. **Maruyama, T., and M. Kimura.** 1980. Genetic variability and effective population size when local extinction and recolonization of subpopulations are frequent. *Proc. Natl. Acad. Sci. USA* **77:**6710–6714.
17. **Milkman, R.** 1973. Electrophoretic variation in *Escherichia coli* from natural sources. *Science* **182:**1024–1026.
18. **Milkman, R.** 1972. How much room is left for non-Darwinian evolution? *Brookhaven Symp. Biol.* **23:**217–229.
19. **Musser, J. M., D. M. Granoff, P. E. Pattison, and R. K. Selander.** 1985. A population genetic framework for the study of invasive diseases caused by serotype b strains of *Haemophilus influenzae*. *Proc. Natl. Acad. Sci. USA* **82:**5078–5082.
20. **Ørskov, F., I. Ørskov, D. J. Evans, Jr., R. B. Sack, D. A. Sack, and T. Wadström.** 1976. Special *Escherichia coli* serotypes among enterotoxogenic strains from diarrhoea in adults and children. *Med. Microbiol. Immunol.* **162:**73–80.
21. **Ørskov, F., and I. Ørskov.** 1961. The fertility of *Escherichia coli* antigen test strains in crosses with K12. *Acta Pathol. Microbiol. Scand.* **51:**280–290.
22. **Ørskov, F., and I. Ørskov.** 1983. From the National Institutes of Health. Summary of a workshop on the clone concept in the epidemiology, taxonomy, and evolution of the Enterobacteriaceae and other bacteria. *J. Infect. Dis.* **148:**346–357.
23. **Ørskov, F., T. S. Whittam, A. Cravioto, and I. Ørskov.** 1990. Clonal relationships among classic enteropathogenic *Escherichia coli* (EPEC) belong to different O groups. *J. Infect. Dis.* **162:**76–81.
24. **Ørskov, I., and F. Ørskov.** 1977. Special O:K:H serotypes among enterotoxigenic *E. coli* strains from diarrhea in adults and children. Occurrence of the CF (colonization factor) antigen and of hemagglutinating abilities. *Med. Microbiol. Immunol.* **163:**99–110.
25. **Selander, R. K., and B. R. Levin.** 1980. Genetic diversity and structure in *Escherichia coli* populations. *Science* **210:**545–547.
26. **Whittam, T. S., H. Ochman, and R. K. Selander.** 1983. Multilocus genetic structure in natural populations of *Escherichia coli*. *Proc. Natl. Acad. Sci. USA* **80:**1751–1755.
27. **Whittam, T. S., M. L. Wolfe, I. K. Wachsmuth, F. Ørskov, I. Ørskov, and R. A. Wilson.** 1993. Clonal relationships among *Escherichia coli* strains that cause hemorrhagic colitis and infantile diarrhea. *Infect. Immun.* **61:**1619–1629.

*Population Genetics of Bacteria: a Tribute to Thomas S. Whittam*
Edited by S. T. Walk and P. C. H. Feng


## *Chapter 3*

# The Impact of Horizontal Genetic Exchange on Bacterial Population Structure: Insights from the Genera *Neisseria* and *Campylobacter*

*Martin C. J. Maiden*

Like many others, I first came to know of Tom through his work with Bob Selander's group at Pennsylvania State University, which was a strong influence in my development as a microbiologist. Later, I was fortunate enough to get to know him personally through interactions at scientific conferences, first at the Society for General Microbiology meeting on bacterial population structure that I coorganized in 1994 (4). Tom spent his professional life helping to develop the discipline of bacterial population biology, an area that, notwithstanding its relevance, especially to studies of pathogens, has been something of a Cinderella subject. I recall Tom once saying during a discussion—on population sampling, I think—"The trouble is that there are only a handful of people in the world interested in this problem, and three of them are in this room." While this was something of an exaggeration, the point of the relative neglect of population biology studies of bacteria at that time was well made. Tom did a great deal to promote and develop studies of this kind and to shed light on this fascinating and important area of microbiology; he was taken from us when he still had so much to contribute, and this chapter is respectfully dedicated to his memory.

## INTRODUCTION: MODELS OF BACTERIAL POPULATION STRUCTURE

Horizontal genetic exchange, long considered a rare process in bacterial populations, is now known to be widespread in both its occurrence and the extent to which it influences bacterial evolution. It is the dominant mechanism of genetic change for the majority of bacterial populations examined to date,

***Martin C. J. Maiden*** • Department of Zoology, University of Oxford, Oxford, OX1 3PS, United Kingdom.

and although most work has been done on pathogens, it is likely that genetic exchange is the major evolutionary process shaping populations throughout the prokaryotic world (46). Horizontal genetic exchange encompasses a wide range of processes, including the movement of insertion sequences and episomal elements (36), but here I shall concentrate on the impact of the sharing of genetic material within what is frequently referred to as the "core genome": those genes common to most or all of the members of a population and that perform essential metabolic functions (19). Specifically, I shall consider the impact and consequences that this process has on the evolution and maintenance of phenotypes within pathogen populations. I shall draw my examples from two very different bacterial genera, *Neisseria* and *Campylobacter*, which I have been studying over the last 20 years or so.

The concept of the clone is fundamental to the study of prokaryotic evolution and is the essential starting point for any discussion on horizontal genetic exchange. When a bacterium reproduces, it does so by binary fission, producing two identical daughter cells, and if this were the only process operating, the spread of diversity would be strictly "vertical," from mother to daughter. Genetic changes introduced by mutation accumulate successively, and particular mutations remain limited to the descendant of the cell in which they occurred. This inevitably gives a "clonal frame," a tree-like genealogy in which equivalent mutations only occur in different members of the same population if they share an ancestor or if, by chance, the same mutation has occurred independently in cells with different recent ancestry (giving rise to a homoplasy)—an eventuality that is rare and unlikely to be observed in the absence of positive selection that amplifies that mutation (54).

The high growth rates of the bacteria result in rapid population expansions and contractions, which, combined with the intimate contact that bacterial cells have with a potentially changeable environment, leads to a purging of diversity either by selection events (periodic selection) or by stochastic processes (bottlenecking). This process of mutation followed by diversity reduction will give rise to a clonal population structure composed of lineages, as envisaged by Levin (45). The clonal population structure provides the null model of bacterial population biology and has the merit of generating a number of testable predictions, including the following: (i) any genetic variation in the population will be nonrandomly distributed (linkage disequilibrium); (ii) the variation will be consistent with a tree-like phylogeny; and (iii) the same phylogeny will be recorded at all loci (the property of congruence) (26).

A number of bacterial populations conform to the clonal model, particularly the "single clone" monomorphic pathogens, which show strong clonal population structure and low diversity, probably as the result of relatively recent evolution into a novel pathogenic niche (1). Most other bacteria, however, show deviations at at least some loci, and some show very high levels of divergence, such that they have very little or indeed no discernible clonal frame. These deviations are a consequence of the acquisition of genetic material by horizontal genetic exchange, a process that can result in many polymorphisms being acquired by the host cell in a single genetic event (53).

A spectrum of population structure can be invoked that ranges from, at one extreme, those bacteria that undergo frequent horizontal gene transfer and therefore exhibit no clonal structure, to the other extreme, which is occupied by strictly clonal bacteria with no horizontal genetic exchange. The evidence accumulated to date suggests that the majority of the bacteria occupy a middle ground where their populations exhibit some clonal signal, due to patterns of recent clonal descent, which is disrupted to a variable degree by genetic reassortment caused by horizontal genetic exchange (66). While this model is coherent, fits the observed range of bacterial sequence diversity quite well, and accommodates a large number of different possible population structures, precise estimates of recombination rates and selection pressures are necessary to provide it with predictive and explanatory power (32).

## STUDYING BACTERIAL POPULATIONS

There are three elements required for a study of bacterial population biology: (i) obtaining a coherent population sample, (ii) characterizing the variation within the population, and (iii) analyzing that variation. Of these, the first is arguably the most challenging and the least frequently achieved satisfactorily. The reasons for this are many and include the size and diversity of bacterial populations, difficulties in defining bacterial populations precisely, the logistics of collecting samples, and the lack of adequate historical collections. This is an issue that urgently requires more attention, as the inferences that can be drawn from collections of bacterial specimens are usually constrained by the sample frame used. In particular, for bacteria that rarely cause disease, such as *Neisseria meningitidis* or *Campylobacter jejuni*, collections solely comprising isolates from invasive disease of humans will invariably underrepresent population diversity. Identifying the subsets of the population that cause disease is valuable for epidemiological studies and in investigating pathogenesis, but ultimately relies on understanding their place in the natural populations (15, 47).

Once a satisfactory sample for the population and question under investigation has been obtained, characterizing the genetic variation it contains is more straightforward and is becoming simpler and cheaper with the development of ever higher-capacity methods of genetic and phenotypic characterization (56). For the great majority of studies, nucleotide sequences have become the data of choice and developments in molecular biology techniques have removed the collection of such data as an obstacle: in principle it is now possible to access genetic variation at any chromosomal locus, or indeed all of them together. I shall return to the potential future impact of current developments in nucleotide sequence determination at the conclusion of this chapter, but suffice it to say here that it is no longer necessary for the design of population genetic analyses of bacteria to be constrained by considerations of the capacity and cost of nucleotide sequence determination.

Just as there is a range of diversity within bacterial populations and communities, diversity is unevenly distributed within bacterial genomes (8). Where there is a clonal frame, this is not too much of an issue for many population studies, as most of the variation will be congruent, even if one locus is more diverse than another due to selection or some kind of inherent instability. Frequent recombination, however, enables different parts of the genome to evolve independently. Unfortunately, the mixture of clonal signal and genetic reassortment found in many bacterial populations means that it is not always clear if relationships among genetic variants are due to descent or selection (2). For each bacterial population investigated it is therefore prudent to examine at least several loci to account for recombination. This also permits formal tests of congruence (41), and the choice of loci that are of similar diversity and under comparable types and levels of selection. Which loci and the number of loci examined will depend on the biology of the organism under examination, but the most widely used variation to date is that present in housekeeping genes, which are found in all members of a given population and which are subject to stabilizing selection for conservation of metabolic function. This started with multilocus enzyme electrophoresis (MLEE) (61), a technique borrowed from studies of the population biology of animals and plants, and has continued with multilocus sequence typing (MLST), a method based on the same concepts that was developed for bacterial typing and has had broad application to bacterial population studies (48, 50).

Sequence data are reproducible, accurate, and definitive, as well as being amenable to a range of analytical approaches. Population genetics and phylogenetics provide a wealth of these, yet relatively few have been designed with the challenges of bacterial population biology in mind. Phylogenetic approaches assume no recombination among clades, while the majority of population genetic approaches have been designed to analyze populations that recombine freely. A few techniques have now been developed that simultaneously account for clonal descent and genetic exchange, notably ClonalFrame (18), but more developments in analytical techniques will be needed to advance our understanding of this area (19).

## *NEISSERIA* AND *CAMPYLOBACTER* AS MODELS OF SEMICLONAL BACTERIAL PATHOGENS

Biologically and medically speaking, there are few obvious parallels between members of the genus *Neisseria*, which are obligate inhabitants of human respiratory and urogenital mucosa, and *Campylobacter*, which colonize the intestines of a wide range of mammals and birds. They do share the property of being "accidental" pathogens—in that they cause human disease as a consequence of dysfunctional or accidental associations with humans—but the disease syndromes that they cause are very different.

*N. meningitidis* is an occasional cause of meningitis and septicemia, although it is normally a harmless inhabitant of the human nasopharynx—like its relative *Neisseria lactamica*, which is commonly carried by children but which never causes invasive disease (5). A third member of the genus, *Neisseria gonorrhoeae*, causes gonorrhea and occasionally disseminated disease (57). Campylobacters are not normally associated with humans, but are common intestinal inhabitants of birds and animals. They are the most common cause of bacterial gastroenteritis in many industrialized countries and are also common in developing countries. They were identified only recently and cause human disease when individuals are infected incidentally, mostly as a consequence of consuming contaminated food or drink (76). Around 90% of human disease is caused by *C. jejuni*, with most of the remainder caused by its close relative, *Campylobacter coli*.

*Neisseria* and *Campylobacter* populations are, however, similarly structured, both exhibiting extensive evidence of frequent horizontal genetic exchange. The two genera provide model systems for the investigation of two different problems in relating phenotypes of interest to humans to pathogen biology. In the case of *Neisseria*, the phenotype of interest is the proclivity of certain genotypes to cause invasive disease, which is apparently paradoxical in organisms that have evolved to colonize humans asymptomatically. In the case of *Campylobacter*, the question relates to the genetic bases of the association of particular genotypes with different animal species. Population genetic studies have been instrumental in improving our understanding of the epidemiology of the diseases that these organisms cause and have illuminated important areas of their biology, which will be discussed later in this chapter. First, we shall consider the impact that studies of the population biology of these genera have had on our understanding of the bacterial species concept.

## HORIZONTAL GENETIC EXCHANGE AND THE BACTERIAL SPECIES PROBLEM

How bacterial species are defined, and even whether they actually exist (25), remains an area of active debate, and while pragmatic "microbiological" definitions exist, originally based on a variety of phenotypic characters, there is no accepted genetic definition that can be applied broadly. The application of molecular techniques has not produced an entirely satisfactory solution (67). Accepted systematic classification methods, such as DNA homology and 16S rRNA sequences, suggest that *N. meningitidis, N. gonorrhoeae*, and *N. lactamica* should be included in a single species (37); these approaches would also group *C. jejuni* and *C. coli* into a single species group. Recent studies have suggested, however, that the phenotypes that were originally used to define these microbiological groups as distinct species do reflect coherent clusters of genotypes and that the 16S rRNA and DNA homology species definitions are too broad for these bacteria. Horizontal genetic exchange has recently become central to discussion of bacterial species and speciation (32, 34), although there remains much argument concerning the nature of bacterial species and the means whereby they arise and are maintained (16, 25, 32).

Interest in interspecific genetic exchange in *Neisseria* was stimulated by observations that showed the movement of penicillin-binding proteins among bacteria assigned to different species groups (65). These were followed by observations that recombination also occurs in antigens and, against expectation at the time, in genes not so obviously under positive selection, the housekeeping genes (51, 77). As data have accumulated for additional loci, a more complete picture has emerged. The three *Neisseria* species do represent three coherent groups that exchange DNA infrequently (6). In the case of the gonococcus this can be explained by "geographic" isolation within the human host. The gonococcus is the only member of the genus regularly isolated from the human urogenital tract, and it is rarely isolated from the natural habitat of most *Neisseria*, the human nasopharynx (69). The gonococcus is an interesting case, as gonococci frequently exchange DNA although they are reproductively isolated from their closest relatives within the genus. Consequently, although the diversity present in gonococcal genes is limited, in terms of the number of polymorphisms present, because the species is relatively young and there has been little time for the accumulation of mutations, the number of alleles is inflated by extensive reassortment of those polymorphisms that have occurred by mutation since speciation. In contrast to the highly clonal monomorphic pathogens such as *Yersinia pestis* (28) that are apparently entirely asexual, the gonococcus is a sexual clone (6).

Geographic isolation within the human host cannot account for the lack of genetic exchange between *N. meningitidis* and *N. lactamica* that has resulted in their emergence as distinct entities. Although the meningococcus is more frequently isolated in adult humans and *N. lactamica* in children (5), both are regularly isolated from the same individuals and can exchange DNA (30). Indeed, at some loci they share a common gene pool, either by recent genetic exchange or as a consequence of descent from a common ancestral population (7). Reports of frequent persistent hybrids at housekeeping genes between these two groups such that they represent "fuzzy species" (39) have not been sustained by subsequent analyses, and the meningococcus and *N. lactamica* are coherent, distinct species that share few polymorphisms and with very low gene flow between them at housekeeping loci (6). There is some evidence that hybrids of the two species are less fit than nonhybrids, at least at some loci, and this provides a possible mechanism for the emergence and persistence of these as two separate groups (78). Therefore it is possible that selection against hybrid genotypes is playing a part in the maintenance of these species groups.

Analysis of concatenated housekeeping genes from *C. jejuni* and *C. coli* demonstrates that they, like the *Neisseria* species, form coherent and separate groups (21); however, there are some differences between these organisms. Unexpectedly for a transformable bacterium, *C. coli* isolates are grouped into three distinct clades, while the more diverse *C. jejuni* forms a single group. The three *C. coli* clades are phylogenetically robust groups apparently evolving independently, suggesting that there is a barrier to genetic exchange among them (63). If this is the case, they can be regarded as incipient species—closely related groups that are currently diverging. A further difference between *Campylobacter* and *Neisseria* is that hybrid genotypes between *C. jejuni* and *C. coli* do exist and are persisting long enough to be observed (63). Intriguingly, the great majority of genetic exchange is the movement of housekeeping genes from *C. jejuni* to *C. coli* clade 1, with 18% of the gene flow moving in this direction. This level of genetic exchange presents a paradox: *C. coli* clade 1 could not have emerged as a distinct group with this level of genetic exchange. The rate of genetic exchange must therefore have changed between these groups (i.e., the groups were genetically isolated but barriers to exchange have recently been reduced or removed). If the currently observed rate of genetic exchange is maintained over time, clade 1 will not be maintained as a distinct entity from *C. jejuni*. Hybridization has been seen in other bacteria, for example, between the biotypes of *Vibrio vulnificus* (10) and in the emergence of *Salmonella enterica* serovars Typhi and Paratyphi, where rates of genetic exchange also appear to have changed over time (17).

It is interesting to speculate what may have caused these changes in gene flow, and some hints of the process involved are to be found in the reasons why *Campylobacter* is of interest to public health authorities. As *C. jejuni* and *C. coli* are the major cause of bacterial gastroenteritis in many countries, much effort has been expended in establishing where humans acquire their *Campylobacter* infections. A range of evidence, including attribution studies that use MLST data, demonstrates that 60 to 80% of genotypes from cases of human disease are similar to those present in retail poultry meat, with most of the rest attributable to those present in ovine meat products (58, 62, 72). In other words, humans mostly get disease from farm animals, especially chickens, via contaminated food products. The worldwide production of chickens as a source of food protein has dramatically increased in the last 50 years, and large commercial flocks of fast-growing broiler chickens are frequently infected with both *C. jejuni* and *C. coli*. It is possible that intensive chicken production has generated a novel niche and resulted in changes in selection pressures that these bacteria are exposed to. Either *C. coli* clade 1 populations are now exposed to more *C. jejuni* DNA than previously or selection against hybrids has been relaxed. It is interesting to note that members of *C. coli* clade 1 are regularly isolated from food animals and human disease, while *C. coli* clades 2 and 3, which do not appear to be participating in this gene flow, are isolated mainly from other sources (63). Further research is necessary to investigate this, but the most recent estimates of the dates at which *C. jejuni* and *C. coli* diverged are consistent with these events coinciding with the development of agriculture (71).

Taken together, the multilocus data from *Neisseria* and *Campylobacter* are consistent with the idea that barriers to genetic exchange play a central role in the emergence and maintenance of the distinct phenotypic and genotypic groups currently classified as microbiological species (34). This does not, however, satisfactorily explain what these barriers are, which will require further investigation. The evidence accumulated to date suggests the possibility that selection against hybrid types is important in the maintenance of species groups under normal circumstances, but that this can break down as a consequence of environmental change.

## STRUCTURE WITHIN MICROBIOLOGICAL SPECIES GROUPS

As discussed above, recombination is very common within the microbiological species *N. meningitidis* and *C. coli* (23, 24, 41). Similar high rates of intraspecific recombination have been observed in numerous other bacteria, as diverse as *Streptococcus pneumoniae, Haemophilus influenzae* (31), and *Wolbachia pipientis* (3). The gastric pathogen *Helicobacter pylori* represents an extreme example wherein this recombination is sufficient to erase all clonal signal from the population (29). In many other species, including our two current examples, bacterial populations show structuring, with evidence for some clonal signal notwithstanding high rates of recombination. In any population sample of both *N. meningitidis* and *C. jejuni*, some genotypes are much more common than others and appear to persist longer than others; this leads to the paradox of how stable genotypes can emerge and persist in recombining populations (11).

Population structure can be a consequence of purely neutral or stochastic forces. In any genealogical analysis the more recent events are more likely to be identified than the more distant ones, and this can lead to an overrepresentation of recent clusters (44). Similarly, clusters could be a consequence of short-term spread of genotypes (54). Both of these explanations have to be rejected before invoking other forces such as adaptation and selection in the generation of clusters. If, however, selection for particular phenotypes does play a role in population structuring, with clusters of genotypes occupying peaks in a "fitness landscape" (74), then an examination of the relationship of the population structure with phenotype will be an invaluable source of inference to the biology of the bacteria in question.

Several analyses have demonstrated that the structure present in meningococcal populations deviates significantly from that expected under purely neutral processes. Studies of carried meningococci in the Czech Republic, for example (42), showed the profound influence of recombination on the population structure of this bacterium. Further, estimated rates of recombination and recombination fragment sizes determined that the number of sequence types observed in the population was significantly lower than expected under neutrality (43). Similarly, a genealogical reconstruction of the relationships of 75 meningococci at 20 housekeeping loci with a coalescent neutral model showed significant deviations of the number of internal branches from the coalescent expectation (20). An analysis of the "microepidemic" model, which formalized the epidemic clone concept (54), also rejected neutral explanations for the observed structure but was able to explain the frequency of clusters within carried populations of meningococci at a single point in time (33). This model cannot, however, account for the observed persistence of this structure over time (11). The persistence of meningococcal clonal complexes that are associated with particular phenotypes suggests the operation of selective processes rather than neutrality, and long-term persistence of types over time and during global spread within a recombining population is consistent with models that envisage competition among multilocus genotypes that differ slightly in their transmission efficiency. Such models can also explain the emergence of virulence in only the most transmissible variants, and are consistent with the observed structure and dynamics of meningococcal populations (11).

*C. jejuni* populations are also structured into clonal complexes, which can be detected by a variety of means (22). In this case, the clonal complexes appear to be strongly associated with particular host animals (55), with most of the clonal complexes identified to date associated with farm animals and human disease, but it appears that there is a very large, relatively poorly sampled population present in wild birds and animals (35). Some of the larger clonal complexes, in terms of numbers of isolates and sequence types observed to date, may be specifically adapted to the farm environment and contain subgroups that are adapted to particular species of farm animals (55). At least one environmental clonal complex, which rarely causes particularly severe disease, has been elevated to a subspecies level (*C. jejuni doylei*)—but this is likely to be due to its unusual and dramatic phenotype of infecting humans, rather than reflecting particular genetic isolation and true subspecies status (59).

## PRACTICAL CONSEQUENCES OF POPULATION STRUCTURING IN *NEISSERIA* AND *CAMPYLOBACTER*

Both *N. meningitidis* and *C. jejuni* present important public health problems that population studies have helped to address. This is a consequence of defined subgroups of genotypes within their populations being associated with phenotypes that are important in public health terms, while others are essentially never observed to cause human health problems. Meningococcal populations are characterized by a wide variety of genotypes, but most of these are only rarely represented in isolate collections. Among the common and persistent genotypes, a subset of fewer than a dozen so-called hyperinvasive lineages have caused the majority of meningococcal disease reported in the last 60 years or so (13). These lineages, first identified by MLEE studies (14), correspond to the clonal complexes discussed above (52). The phenotypic differences among meningococcal clonal complexes are dramatic, with the frequency of given clonal complexes in asymptomatic carriage and invasive disease varying by as much as 2 orders of magnitude (75). In addition to being somewhat unexpectedly associated with the invasive phenotype, meningococcal clonal complexes, although based on supposedly neutral variation, are also associated with particular repertoires of antigens, or antigenic types (68). A number of genetic elements that encode putative virulence determinants are also associated with particular clonal complexes, which may contribute to their increased association with disease (9, 40).

From the perspective of disease control, the association of meningococcal antigenic types with clonal complexes is fortuitous as it renders the clonal complexes potentially vulnerable to vaccines directed against the antigens with which they are associated (68). This association contributed greatly to the success of the meningococcal C conjugate vaccine introduced in the United Kingdom in 1999. The ST-11 clonal complex, which was responsible for the increase of meningococcal disease in Europe at that time, exhibited an unusually high level of expression of serogroup C capsule. As the vaccine was effective in preventing asymptomatic carriage (49), there was a dramatic reduction in transmission, which resulted in the virtual elimination of the disease even though the immunization schedule originally introduced was ineffective in long-term protection of the most at-risk age group, those under 1 year at vaccination (64).

It may also be possible to exploit the association of hyperinvasive lineages with antigenic types in the formulation of novel vaccines against serogroup B meningococci, for which no polysaccharide vaccine is likely to become available in the foreseeable future (68). The observed stable associations of clonal complexes with antigenic types in recombining pathogens are unexpected and to a degree paradoxical, as one might naively anticipate maximum diversity to be favored by selection during pathogen spread, but there is a growing body of work on strain theory that shows that selection imposed at the population level by host immune responses can explain this behavior (38) and perhaps also the emergence of hyperinvasive lineages themselves (11).

Despite the magnitude of the public health problem that they cause, analysis of pathogenic *Campylobacter* was hindered for many years by the lack of appropriate typing schemes (70). The application of MLST to these organisms (22, 23), therefore, had a more profound effect on the field than the adoption of MLST for *Neisseria*, for example, as in the latter case the hyperinvasive lineages had already been identified by MLEE and were in many cases associated, at least to an extent, with serological properties (12). In the case of *Campylobacter*, MLEE studies were limited and the serotyping typing schemes poorly predictive of the underlying genetic structure (22).

In large part thanks to the application of MLST, we now have a more complete appreciation of the structure and interrelationships of *C. coli* and *C. jejuni* populations. *C. jejuni* comprises a number of clonal complexes, with no or very few phylogenetic relationships among them, and only a subset of these clonal complexes is associated with the important phenotype of being isolated from cases of human disease. This is almost certainly a consequence of these clonal complexes being prevalent in certain human food sources, and this in turn is a consequence of these genotypes being associated with particular human food animals (22). These studies have enabled the attribution of cases from human disease to infection sources with increasing accuracy, enabling ever more informed public health intervention (58, 62, 71). These groups can potentially be exploited to identify those genetic determinants that drive host association, which must be the biological bases of these observations.

## FUTURE PROSPECTS FOR THE ANALYSIS OF SEMICLONAL PATHOGENS

MLST studies have substantially advanced our understanding of the population biology and evolution of many pathogenic bacteria and an increasing number of bacteria that are not pathogens of humans (48). The clonal complex description is appropriate to many of these organisms and, for these bacteria, has replaced the clonal population as the predominant paradigm. An advantage of the clone complex concept is that it can accommodate varying degrees of clonal signal in bacterial populations, although the mechanisms whereby these arise and are maintained remain incompletely understood. *N. meningitidis* and *C. jejuni*, for example, exhibit apparently stable clonal complexes even though most of the variation within them is generated by recombination with other clonal complexes. This contrasts with *Staphylococcus aureus* clonal complexes, for example, which seem to evolve predominantly by a clonal process (27). The multilocus approach is also an efficient way of identifying clonal lineages within populations, such as *Escherichia coli* (73) and the *Bacillus cereus* group (60).

MLST data have been enhanced by the addition of sequence or other data from loci under a variety of selection pressures, such as those encoding virulence determinants, antigens, and resistance to antibiotics. Collectively such studies can be considered to be "first-generation population genomics"—analyses that index and assess variation from around the chromosome, and under a variety of selection pressures, from population samples. While it has been possible to investigate thousands of samples in this way, the number of loci included is constrained to a maximum of 20 or so used to date, and most use around 7 loci—a number that has proved sufficient to identify phenotypically distinct genotypes for many bacteria, including hyperinvasive lineages of the meningococcus and host-associated types of *C. jejuni*. These studies provide a basis for future population genomics studies that are becoming possible with the increasing availability of parallel sequencing technology. The extent to which assembled, complete, and fully annotated sequences will become available is unclear at the time of writing, but it is reasonable to assume that studies that index variation in the majority of the genome will be technically feasible in the very near future (56). The examination of genome-wide variation across many isolates will be challenging and, given that selection and other evolutionary forces will not operate evenly across the whole genome, will not be appropriate in many cases. In many cases it will be necessary to adopt an MLST-like, hierarchical approach, analyzing loci grouped to reflect the phenotypic characteristics that they encode. Nevertheless, the prospects of the future are mouth-watering for those intrigued by the population biology and evolution of the bacteria.

Three elements need to be put in place to exploit these opportunities. First, for the great majority of bacteria, it will be necessary to assemble isolate collections that are significantly larger, better characterized phenotypically, and more representative of natural populations than those currently available. Second, large reference sets of nucleotide and peptide sequence data will have to be assembled and curated. These will need to include links to appropriate phenotypic information and incorporate usable, widely understood, and broadly accepted nomenclatures. The studies based on MLST have demonstrated the value of this approach, both in organizing and summarizing complex data sets and in enabling the hierarchical analysis of genome variation. Finally, new databases, together with novel analytical and computational approaches, will be necessary to exploit these data. Given the size and complexity of the data sets envisaged, it will be essential that these are robust, efficient, and widely available via the Internet.

As these issues are addressed, it will become feasible to unravel the various forces that structure populations of recombining semiclonal bacteria. It is possible, and perhaps likely, that population genomic analyses will reveal complex interrelationships within populations of genomes, with variation among genes encoding one set of cellular functions exhibiting a different structure to that seen in genes encoding other functions. Understanding which elements of this structuring are due to chance, descent, and selection (2) will require careful analysis but will move us toward an era where the barriers between reductionist investigations of cellular function and evolutionary studies are removed, enabling us to discern with increasing clarity the interrelationships of bacterial phenotypes and genotypes and how these evolved.

**Acknowledgments.** I would like to express my gratitude and appreciation to all past and current members of my research group, who have been instrumental in realizing the work described here. We are also fortunate to have benefited from much productive and enjoyable collaboration with many wonderful colleagues. Since 1997 I have been supported by a Senior Fellowship from the Wellcome Trust, without which much of this work would not have been possible.

## REFERENCES

1. **Achtman, M.** 2008. Evolution, population structure, and phylogeography of genetically monomorphic bacterial pathogens. *Annu. Rev. Microbiol.* **62:**53–70.
2. **Andam, C. P., D. Williams, and J. P. Gogarten.** 2010. Biased gene transfer mimics patterns created through shared ancestry. *Proc. Natl. Acad. Sci. USA* **107:**10679–10684.
3. **Baldo, L., J. C. Dunning Hotopp, K. A. Jolley, S. R. Bordenstein, S. A. Biber, R. R. Choudhury, C. Hayashi, M. C. Maiden, H. Tettelin, and J. H. Werren.** 2006. Multilocus sequence typing system for the endosymbiont *Wolbachia pipientis*. *Appl. Environ. Microbiol.* **72:**7098–7110.
4. **Baumberg, S., J. P. W. Young, E. M. H. Wellington, and J. R. Saunders (ed.).** 1995. *Population Genetics of Bacteria.* Cambridge University Press, Cambridge, United Kingdom.
5. **Bennett, J. S., D. T. Griffiths, N. D. McCarthy, K. L. Sleeman, K. A. Jolley, D. W. Crook, and M. C. Maiden.** 2005. Genetic diversity and carriage dynamics of *Neisseria lactamica* in infants. *Infect. Immun.* **73:**2424–2432.
6. **Bennett, J. S., K. A. Jolley, P. F. Sparling, N. J. Saunders, C. A. Hart, I. M. Feavers, and M. C. Maiden.** 2007. Species status of *Neisseria gonorrhoeae*: evolutionary and epidemiological inferences from multilocus sequence typing. *BMC Biol.* **5:**35.
7. **Bennett, J. S., E. A. Thompson, P. Kriz, K. A. Jolley, and M. C. Maiden.** 2009. A common gene pool for the *Neisseria* FetA antigen. *Int. J. Med. Microbiol.* **299:**133–139.
8. **Bentley, S. D., and J. Parkhill.** 2004. Comparative genomic structure of prokaryotes. *Annu. Rev. Genet.* **38:**771–792.
9. **Bille, E., J. R. Zahar, A. Perrin, S. Morelle, P. Kriz, K. A. Jolley, M. C. Maiden, C. Dervin, X. Nassif, and C. R. Tinsley.** 2005. A chromosomally integrated bacteriophage in invasive meningococci. *J. Exp. Med.* **201:**1905–1913.
10. **Bisharat, N., D. I. Cohen, R. M. Harding, D. Falush, D. W. Crook, T. Peto, and M. C. Maiden.** 2005. Hybrid *Vibrio vulnificus*. *Emerg. Infect. Dis.* **11:**30–35.
11. **Buckee, C. O., K. Jolley, M. Recker, B. Penman, P. Kriz, S. Gupta, and M. C. Maiden.** 2008. Role of selection in the emergence of lineages and the evolution of virulence in *Neisseria meningitidis*. *Proc. Natl. Acad. Sci. USA* **105:**15082–15087.
12. **Caugant, D. A.** 1998. Population genetics and molecular epidemiology of *Neisseria meningitidis*. *APMIS* **106:**505–525.

13. **Caugant, D. A., and M. C. Maiden.** 2009. Meningococcal carriage and disease—population biology and evolution. *Vaccine* **27**(Suppl. 2)**:**B64–B70.
14. **Caugant, D. A., L. F. Mocca, C. E. Frasch, L. O. Frøholm, W. D. Zollinger, and R. K. Selander.** 1987. Genetic structure of *Neisseria meningitidis* populations in relation to serogroup, serotype, and outer membrane protein pattern. *J. Bacteriol.* **169:**2781–2792.
15. **Cody, A. J., F. M. Colles, S. K. Sheppard, and M. C. J. Maiden.** 2010. Where does *Campylobacter* come from? A molecular odyssey. *Adv. Exp. Med. Biol.* **659:**47–56.
16. **Cohan, F. M., and A. F. Koeppel.** 2008. The origins of ecological diversity in prokaryotes. *Curr. Biol.* **18:**R1024–R1034.
17. **Didelot, X., M. Achtman, J. Parkhill, N. R. Thomson, and D. Falush.** 2007. A bimodal pattern of relatedness between the *Salmonella* Paratyphi A and Typhi genomes: convergence or divergence by homologous recombination? *Genome Res.* **17:**61–68.
18. **Didelot, X., and D. Falush.** 2007. Inference of bacterial microevolution using multilocus sequence data. *Genetics* **175:**1251–1266.
19. **Didelot, X., and M. C. Maiden.** 2010. Impact of recombination on bacterial evolution. *Trends Microbiol.* **18:**315–322.
20. **Didelot, X., R. Urwin, M. C. Maiden, and D. Falush.** 2009. Genealogical typing of *Neisseria meningitidis. Microbiology* **155:**3176–3186.
21. **Dingle, K. E., F. M. Colles, D. Falush, and M. C. Maiden.** 2005. Sequence typing and comparison of population biology of *Campylobacter coli* and *Campylobacter jejuni. J. Clin. Microbiol.* **43:**340–347.
22. **Dingle, K. E., F. M. Colles, R. Ure, J. Wagenaar, B. Duim, F. J. Bolton, A. J. Fox, D. R. A. Wareing, and M. C. J. Maiden.** 2002. Molecular characterisation of *Campylobacter jejuni* clones: a rational basis for epidemiological investigations. *Emerg. Infect. Dis.* **8:**949–955.
23. **Dingle, K. E., F. M. Colles, D. R. A. Wareing, R. Ure, A. J. Fox, F. J. Bolton, H. J. Bootsma, R. J. L. Willems, R. Urwin, and M. C. J. Maiden.** 2001. Multilocus sequence typing system for *Campylobacter jejuni. J. Clin. Microbiol.* **39:**14–23.
24. **Dingle, K. E., and M. C. Maiden.** 2005. Population genetics of *Campylobacter jejuni*, p. 43–58. *In* J. M. Ketley and M. E. Konkel (ed.), Campylobacter *Molecular and Cellular Biology.* Horizon Bioscience, Wymondham, United Kingdom.
25. **Doolittle, W. F.** 2008. Microbial evolution: stalking the wild bacterial species. *Curr. Biol.* **18:**R565–R567.
26. **Dykhuizen, D. E., D. S. Polin, J. J. Dunn, B. Wilske, V. Preac Mursic, R. J. Dattwyler, and B. J. Luft.** 1993. *Borrelia burgdorferi* is clonal: implications for taxonomy and vaccine development. *Proc. Natl. Acad. Sci. USA* **90:**10163–10167.
27. **Enright, M. C., D. A. Robinson, G. Randle, E. J. Feil, H. Grundmann, and B. G. Spratt.** 2002. The evolutionary history of methicillin-resistant *Staphylococcus aureus* (MRSA). *Proc. Natl. Acad. Sci. USA* **99:**7687–7692.
28. **Eppinger, M., P. L. Worsham, M. P. Nikolich, D. R. Riley, Y. Sebastian, S. Mou, M. Achtman, L. E. Lindler, and J. Ravel.** 2010. Genome sequence of the deep-rooted *Yersinia pestis* strain Angola reveals new insights into the evolution and pangenome of the plague bacterium. *J. Bacteriol.* **192:**1685–1699.
29. **Falush, D., C. Kraft, N. S. Taylor, P. Correa, J. G. Fox, M. Achtman, and S. Suerbaum.** 2001. Recombination and mutation during long-term gastric colonization by *Helicobacter pylori*: estimates of clock rates, recombination size, and minimal age. *Proc. Natl. Acad. Sci. USA* **98:**15056–15061.
30. **Feil, E., G. Carpenter, and B. G. Spratt.** 1995. Electrophoretic variation in adenylate kinase of *Neisseria meningitidis* is due to inter- and intraspecies recombination. *Proc. Natl. Acad. Sci. USA* **92:**10535–10539.
31. **Feil, E. J., E. C. Holmes, D. E. Bessen, M. S. Chan, N. P. Day, M. C. Enright, R. Goldstein, D. W. Hood, A. Kalia, C. E. Moore, J. Zhou, and B. G. Spratt.** 2001. Recombination within natural populations of pathogenic bacteria: short-term empirical

estimates and long-term phylogenetic consequences. *Proc. Natl. Acad. Sci. USA* **98:**182–187.

32. **Fraser, C., E. J. Alm, M. F. Polz, B. G. Spratt, and W. P. Hanage.** 2009. The bacterial species challenge: making sense of genetic and ecological diversity. *Science* **323:**741–746.
33. **Fraser, C., W. P. Hanage, and B. G. Spratt.** 2005. Neutral microepidemic evolution of bacterial pathogens. *Proc. Natl. Acad. Sci. USA* **102:**1968–1973.
34. **Fraser, C., W. P. Hanage, and B. G. Spratt.** 2007. Recombination and the nature of bacterial speciation. *Science* **315:**476–480.
35. **French, N., M. Barrigas, P. Brown, P. Ribiero, N. Williams, H. Leatherbarrow, R. Birtles, E. Bolton, P. Fearnhead, and A. Fox.** 2005. Spatial epidemiology and natural population structure of *Campylobacter jejuni* colonizing a farmland ecosystem. *Environ. Microbiol.* **7:**1116–1126.
36. **Frost, L. S., R. Leplae, A. O. Summers, and A. Toussaint.** 2005. Mobile genetic elements: the agents of open source evolution. *Nat. Rev. Microbiol.* **3:**722–732.
37. **Guibourdenche, M., M. Y. Popoff, and J. Y. Riou.** 1986. Deoxyribonucleic acid relatedness among *Neisseria gonorrhoeae, N. meningitidis, N. lactamica, N. cinerea* and "*Neisseria polysaccharea.*" *Ann. Inst. Pasteur Microbiol.* **137B:**177–185.
38. **Gupta, S., M. C. J. Maiden, I. M. Feavers, S. Nee, R. M. May, and R. M. Anderson.** 1996. The maintenance of strain structure in populations of recombining infectious agents. *Nat. Med.* **2:**437–442.
39. **Hanage, W. P., C. Fraser, and B. G. Spratt.** 2005. Fuzzy species among recombinogenic bacteria. *BMC Biol.* **3:**6.
40. **Harrison, O. B., N. J. Evans, J. M. Blair, H. S. Grimes, C. R. Tinsley, X. Nassif, P. Kriz, R. Ure, S. J. Gray, J. P. Derrick, M. C. Maiden, and I. M. Feavers.** 2009. Epidemiological evidence for the role of the hemoglobin receptor, HmbR, in meningococcal virulence. *J. Infect. Dis.* **200:**94–98.
41. **Holmes, E. C., R. Urwin, and M. C. J. Maiden.** 1999. The influence of recombination on the population structure and evolution of the human pathogen *Neisseria meningitidis*. *Mol. Biol. Evol.* **16:**741–749.
42. **Jolley, K. A., J. Kalmusova, E. J. Feil, S. Gupta, M. Musilek, P. Kriz, and M. C. Maiden.** 2000. Carried meningococci in the Czech Republic: a diverse recombining population. *J. Clin. Microbiol.* **38:**4492–4498.
43. **Jolley, K. A., D. J. Wilson, P. Kriz, G. McVean, and M. C. Maiden.** 2005. The influence of mutation, recombination, population history, and selection on patterns of genetic diversity in *Neisseria meningitidis*. *Mol. Biol. Evol.* **22:**562–569.
44. **Kingman, J. F.** 2000. Origins of the coalescent: 1974–1982. *Genetics* **156:**1461–1463.
45. **Levin, B. R.** 1981. Periodic selection, infectious gene exchange and the genetic structure of *E. coli* populations. *Genetics* **99:**1–23.
46. **Lorenz, M. G., and W. Wackernagel.** 1994. Bacterial gene transfer by natural genetic transformation in the environment. *Microbiol. Rev.* **58:**563–602.
47. **Maiden, M. C.** 2004. Dynamics of bacterial carriage and disease: lessons from the meningococcus. *Adv. Exp. Med. Biol.* **549:**23–29.
48. **Maiden, M. C.** 2006. Multilocus sequence typing of bacteria. *Annu. Rev. Microbiol.* **60:**561–588.
49. **Maiden, M. C., A. B. Ibarz-Pavon, R. Urwin, S. J. Gray, N. J. Andrews, S. C. Clarke, A. M. Walker, M. R. Evans, J. S. Kroll, K. R. Neal, D. A. Ala'aldeen, D. W. Crook, K. Cann, S. Harrison, R. Cunningham, D. Baxter, E. Kaczmarski, J. Maclennan, J. C. Cameron, and J. M. Stuart.** 2008. Impact of meningococcal serogroup C conjugate vaccines on carriage and herd immunity. *J. Infect. Dis.* **197:**737–743.
50. **Maiden, M. C. J.** 1998. The impact of molecular techniques on the study of meningococcal disease, p. 265–291. *In* N. Woodford and A. P. Johnson (ed.), *Molecular Bacteriology: Protocols and Clinical Applications.* Humana Press, Totowa, NJ.
51. **Maiden, M. C. J.** 1993. Population genetics of a transformable bacterium: the influence of horizontal genetical exchange on the biology of *Neisseria meningitidis*. *FEMS Microbiol. Lett.* **112:**243–250.

52. **Maiden, M. C. J., J. A. Bygraves, E. Feil, G. Morelli, J. E. Russell, R. Urwin, Q. Zhang, J. Zhou, K. Zurth, D. A. Caugant, I. M. Feavers, M. Achtman, and B. G. Spratt.** 1998. Multilocus sequence typing: a portable approach to the identification of clones within populations of pathogenic microorganisms. *Proc. Natl. Acad. Sci. USA* **95:**3140–3145.
53. **Maynard Smith, J., C. G. Dowson, and B. G. Spratt.** 1991. Localized sex in bacteria. *Nature* **349:**29–31.
54. **Maynard Smith, J., N. H. Smith, M. O'Rourke, and B. G. Spratt.** 1993. How clonal are bacteria? *Proc. Natl. Acad. Sci. USA* **90:**4384–4388.
55. **McCarthy, N. D., F. M. Colles, K. E. Dingle, M. C. Bagnall, G. Manning, M. C. Maiden, and D. Falush.** 2007. Host-associated genetic import in *Campylobacter jejuni*. *Emerg. Infect. Dis.* **13:**267–272.
56. **Medini, D., D. Serruto, J. Parkhill, D. A. Relman, C. Donati, R. Moxon, S. Falkow, and R. Rappuoli.** 2008. Microbiology in the post-genomic era. *Nat. Rev. Microbiol.* **6:** 419–430.
57. **Morse, S. A., and J. S. Knapp.** 1992. The genus *Neisseria*, p. 2495–2559. *In* A. Balows, H. G. Trüper, M. Dworkin, W. Harder, and K.-H. Schleifer (ed.), *The Prokaryotes*, 2nd ed. Springer-Verlag, New York, NY.
58. **Mullner, P., S. E. Spencer, D. J. Wilson, G. Jones, A. D. Noble, A. C. Midwinter, J. M. Collins-Emerson, P. Carter, S. Hathaway, and N. P. French.** 2009. Assigning the source of human campylobacteriosis in New Zealand: a comparative genetic and epidemiological approach. *Infect. Genet. Evol.* **9:**1311–1319.
59. **Parker, C. T., W. G. Miller, S. T. Horn, and A. J. Lastovica.** 2007. Common genomic features of *Campylobacter jejuni* subsp. *doylei* strains distinguish them from *C. jejuni* subsp. *jejuni*. *BMC Microbiol.* **7:**50.
60. **Priest, F. G., M. Barker, L. W. Baillie, E. C. Holmes, and M. C. Maiden.** 2004. Population structure and evolution of the *Bacillus cereus* group. *J. Bacteriol.* **186:**7959–7970.
61. **Selander, R. K., D. A. Caugant, H. Ochman, J. M. Musser, M. N. Gilmour, and T. S. Whittam.** 1986. Methods of multilocus enzyme electrophoresis for bacterial population genetics and systematics. *Appl. Environ. Microbiol.* **51:**837–884.
62. **Sheppard, S. K., J. F. Dallas, N. J. Strachan, M. MacRae, N. D. McCarthy, D. J. Wilson, F. J. Gormley, D. Falush, I. D. Ogden, M. C. Maiden, and K. J. Forbes.** 2009. *Campylobacter* genotyping to determine the source of human infection. *Clin. Infect. Dis.* **48:**1072–1078.
63. **Sheppard, S. K., N. D. McCarthy, D. Falush, and M. C. Maiden.** 2008. Convergence of *Campylobacter* species: implications for bacterial evolution. *Science* **320:**237–239.
64. **Snape, M. D., D. F. Kelly, B. Green, E. R. Moxon, R. Borrow, and A. J. Pollard.** 2005. Lack of serum bactericidal activity in preschool children two years after a single dose of serogroup C meningococcal polysaccharide-protein conjugate vaccine. *Pediatr. Infect. Dis. J.* **24:**128–131.
65. **Spratt, B. G.** 1988. Hybrid penicillin-binding proteins in penicillin-resistant strains of *Neisseria gonorrhoeae*. *Nature* **332:**173–176.
66. **Spratt, B. G., and M. C. J. Maiden.** 1999. Bacterial population genetics, evolution and epidemiology. *Philos. Trans. R. Soc. Lond. B Biol. Sci.* **354:**701–710.
67. **Stackebrandt, E., W. Frederiksen, G. M. Garrity, P. A. Grimont, P. Kämpfer, M. C. Maiden, X. Nesme, R. Rossello-Mora, J. Swings, H. G. Truper, L. Vauterin, A. C. Ward, and W. B. Whitman.** 2002. Report of the ad hoc committee for the re-evaluation of the species definition in bacteriology. *Int. J. Syst. Evol. Bacteriol.* **52:**1043–1047.
68. **Urwin, R., J. E. Russell, E. A. Thompson, E. C. Holmes, I. M. Feavers, and M. C. Maiden.** 2004. Distribution of surface protein variants among hyperinvasive meningococci: implications for vaccine design. *Infect. Immun.* **72:**5955–5962.
69. **Vázquez, J. A., L. de la Fuente, S. Berron, M. O'Rourke, N. H. Smith, J. Zhou, and B. G. Spratt.** 1993. Ecological separation and genetic isolation of *Neisseria gonorrheae* and *Neisseria meningitidis*. *Curr. Biol.* **3:**567–572.

70. **Wassenaar, T. M., and D. G. Newell.** 2000. Genotyping of *Campylobacter* species. *Appl. Environ. Microbiol.* **66:**1–9.
71. **Wilson, D. J., E. Gabriel, A. J. Leatherbarrow, J. Cheesbrough, S. Gee, E. Bolton, A. Fox, C. A. Hart, P. J. Diggle, and P. Fearnhead.** 2009. Rapid evolution and the importance of recombination to the gastroenteric pathogen *Campylobacter jejuni*. *Mol. Biol. Evol.* **26:**385–397.
72. **Wilson, D. J., E. Gabriel, A. J. H. Leatherbarrow, J. Cheesbrough, S. Gee, E. Bolton, A. Fox, P. Fearnhead, A. Hart, and P. J. Diggle.** 2008. Tracing the source of campylobacteriosis. *PLoS Genet.* **26:**e1000203.
73. **Wirth, T., D. Falush, R. Lan, F. Colles, P. Mensa, L. H. Wieler, H. Karch, P. R. Reeves, M. C. Maiden, H. Ochman, and M. Achtman.** 2006. Sex and virulence in *Escherichia coli*: an evolutionary perspective. *Mol. Microbiol.* **60:**1136–1151.
74. **Wright, S.** 1932. The roles of mutation, inbreeding, crossbreeding and selection in evolution, p. 356–366. *In* D. F. Jones (ed.), *Proceedings of the Sixth International Congress on Genetics*, vol. 1. Brooklyn Botanic Garden, Brooklyn, NY.
75. **Yazdankhah, S. P., P. Kriz, G. Tzanakaki, J. Kremastinou, J. Kalmusova, M. Musilek, T. Alvestad, K. A. Jolley, D. J. Wilson, N. D. McCarthy, D. A. Caugant, and M. C. Maiden.** 2004. Distribution of serogroups and genotypes among disease-associated and carried isolates of *Neisseria meningitidis* from the Czech Republic, Greece, and Norway. *J. Clin. Microbiol.* **42:**5146–5153.
76. **Young, K. T., L. M. Davis, and V. J. Dirita.** 2007. *Campylobacter jejuni*: molecular biology and pathogenesis. *Nat. Rev. Microbiol.* **5:**665–679.
77. **Zhou, J., and B. G. Spratt.** 1992. Sequence diversity within the *argF, fbp* and *recA* genes of natural isolates of *Neisseria meningitidis*: interspecies recombination within the *argF* gene. *Mol. Microbiol.* **6:**2135–2146.
78. **Zhu, P., A. van der Ende, D. Falush, N. Brieske, G. Morelli, B. Linz, T. Popovic, I. G. Schuurman, R. A. Adegbola, K. Zurth, S. Gagneux, A. E. Platonov, J. Y. Riou, D. A. Caugant, P. Nicolas, and M. Achtman.** 2001. Fit genotypes and escape variants of subgroup III *Neisseria meningitidis* during three pandemics of epidemic meningitis. *Proc. Natl. Acad. Sci. USA* **98:**5234–5239.

*Population Genetics of Bacteria: a Tribute to Thomas S. Whittam*
Edited by S. T. Walk and P. C. H. Feng


*Chapter 4*

# The Genomics of *Escherichia coli* and Beyond

***Chih-Horng Kuo, Howard Ochman, and Rahul Raghavan***

The actual reason why Tom Whittam switched from studying rodents to studying bacteria will never be known: like most life-altering decisions, it was probably serendipitous. He had selected to postdoc with an advisor best known for work on house mouse population genetics; but by the time Tom arrived, the lab had switched to studying *Escherichia coli* population structure (53) and was poised to examine a wide range of bacterial pathogens (52, 54). There were thousands of isolates already assayed for allelic variation at multiple loci, and this was the type of analytical challenge that Tom relished. He mocked the lab's practice of storing multilocus data on punch cards (although his preference for 10.5-inch open-reel magnetic tapes is equally laughable these days), but he was the only one in the lab who understood all those tests of recombination and selection, so his initial side project turned into a career.

By first applying multilocus enzyme electrophoresis (MLEE) and later multilocus sequence typing (MLST), Tom's research provided a comprehensive picture of the population-level variation within *E. coli* pathogens. A large number of his publications from that era sported titles that paired the word "genetic" or "clonal" with "structure" or "relationships," which gave the immediate sense that these contributions were going beyond the standard phenetic and epidemiologic treatments being offered at the time. A retrospective look at Tom's career indicates that he went from being a statistician to an experimentalist, a direction opposite to that traveled by most contemporary evolutionary geneticists, who by and large exchange the lab bench for a computer cubicle, possibly because the chairs are more comfortable. Of course, the primary motivation for the switch from *generating* sequence data to *analyzing* sequence data was the pace at which genomic information became available, and in this author-centric chapter, we focus on a few of the findings that emerged from the comparative genomics of *E. coli*.

***Chih-Horng Kuo, Howard Ochman, and Rahul Raghavan*** • Department of Ecology and Evolutionary Biology, University of Arizona, Tucson, AZ 85718.

## PATHOGENOMICS AND THE MOSAIC *E. COLI* GENOME

The initial decoding of the *E. coli* K-12 genome was an important milestone (5); however, a single lab-adapted, nonpathogenic strain does not provide a realistic view of the diversity within the species at large. *E. coli* is a versatile species that colonizes diverse ecosystems, and the genomes of *E. coli* strains are shaped by their habitats, a fact best reflected in the differing gene repertoires of strains that promote human diseases. To elucidate the evolution of pathogenic strains from nonpathogenic ancestors and to understand the genetic basis of their varying disease potential require a comparative genomics approach, and the opportunity to perform such analyses was realized when the genome sequences of pathogenic strains of *E. coli* became available.

Based on disease potential in humans, *E. coli* strains can be broadly classified as (i) harmless commensals, (ii) intestinal pathogens, and (iii) those capable of infecting extraintestinal sites. Moreover, lineages of *Shigella* have originated several times from ancestral strains of *E. coli* and should be considered as members of the species (19, 30, 44). Two O157:H7 intestinal pathogenic strains (enterohemorrhagic *E. coli*, or EHEC) were the first *E. coli* pathovars to be fully sequenced (20, 49). Prior work had already shown that *E. coli* strains could vary greatly in genome size (4, 7), but it was anticipated that pathogens would differ from commensals in the allelic variants of shared genes, by the expression of shared genes, or by the presence of a very limited number of virulence factors. However, there were substantial differences in the genome size and contents of K-12 MG1655 and the two O157:H7 strains. Both O157:H7 strains were larger than K-12 by approximately 1 Mbp but lacked about 0.5 Mbp of sequences found in K-12, resulting in nearly 1.4 Mbp (>25% of the genome!) of strain-specific sequences. The remarkable diversity in genome content could be largely attributed to the integration of numerous bacteriophages and to the acquisition en bloc of virulence gene arrays, i.e., pathogenicity islands (PAIs). In addition, many strain-specific genes, including virulence genes, arise through horizontal gene transfer events and are interspersed throughout the conserved genomic backbone, forming a mosaic structure to the *E. coli* and *Shigella* genomes (15). As expected, inspection of gene functions revealed that genes involved in essential cellular processes are located in conserved segments of *E. coli* genomes, whereas the strain-specific segments contain mostly genes that facilitate adaptation to particular environments.

One of the most important drivers of virulence in EHEC strains has been the locus of enterocyte effacement (LEE), a PAI also found in enteropathogenic *E. coli* (EPEC) and some distantly related enterics, such as *Hafnia alvei* and *Citrobacter rodentium* (37). The locus of enterocyte effacement encodes a type III secretion system (T3SS), its regulators, and some of its secreted effectors; however, several other effector genes controlled by this T3SS are encoded within other, independently acquired prophage loci (59). Having its T3SS effectors on prophages makes the *E. coli* genome highly dynamic by connecting it to the vast phage metagenome present in its environment, thereby providing an avenue for accelerated genome evolution and niche specialization.

The presence of Shiga toxin (Stx) distinguishes EHEC from EPEC (25), and the genes conferring this virulence factor are also supplied by prophages. The virulence potential of EHEC strains varies depending on the combination of *stx* genes present in a genome (47), again highlighting the impact of phage transmission and acquisition on bacterial pathogenesis. Although O157:H7 strains cause severe morbidity and mortality in humans, they typically infect cattle and other ungulates. Consequently, Stx and other virulence factors that render EHEC pathogenic to humans serve to enhance survival in their natural environments by aiding in bovine intestine colonization and by providing protection against bactivorous protozoa (57, 58). It has been posited that O157:H7 strains evolved from a nontoxigenic ancestor by stepwise acquisition of virulence traits, such as the Stx and the virulence plasmid pO157 (66). EHEC strains have multiple origins, as reflected by their divergent genomic structures; however, they display remarkable congruence in their virulence gene repertoires. The parallel evolution observed in EHEC is driven by the acquisition of mobile genetic elements via horizontal gene transfer—prophages constitute 16% of the O157 genome (1)—and their retention is due to selective pressures exerted in their natural environments (36, 45).

The mosaic nature of *E. coli* genomes was further confirmed with the elucidation of the genome sequence of uropathogenic (UPEC) strains of *E. coli* (64). The differences observed between UPEC strains and sequenced pathovar and commensal *E. coli* mostly involved horizontally acquired segments throughout the genome. Accordingly, most of the virulence genes dictating the EHEC or EIEC (enteroinvasive *E. coli*) lifestyle, such as T3SS and Stx, are absent in UPEC strains, but replaced by genes that promote infection of the urinary tract, such as fimbrial adhesins and phase-switch recombinases. Many of these virulence factors are encoded within five large PAIs (35), which have base compositions that deviate from the chromosome, are often inserted at the 3′ end of tRNA genes, and are flanked by direct repeats, all suggesting a prophage ancestry. The PAIs are capable of spontaneous excision from the chromosome by site-specific recombination between flanking direct repeats, making them amenable for horizontal gene transfer and leading to additional genome flexibility (23).

## GENE ACQUISITION AND GENOMIC DIVERSIFICATION

The sequencing of multiple *E. coli* strains has exposed the role of horizontal gene transfer in facilitating the evolution of pathogenic variants. Although mobile genetic elements acquired by independent gene transfer events are distributed throughout the genome, certain sites appear to serve as universal insertion targets (64). These "'hot spots'" can contain different genetic information in different genomes and correspond to regions of repeated insertions and deletions, reminiscent of ancient palimpsests that have been repeatedly erased and overwritten (13, 23). One such active region of genome evolution is the *mutS-rpoS* intergenic region (21). Both the *mutS* and *rpoS* genes are highly conserved among strains, but the intergenic region between them can vary from 9.8 kb in some EPEC strains to 6.9 kb in K-12 and some ExPEC (extraintestinal pathogenic *E. coli*) strains to 3.7 kb in O157:H7 strains (21, 51). The polymorphism in this genomic segment is likely related to the high frequency of mutations observed in both *mutS* and *rpoS* (16). Mutations in *mutS* enhance the ability to acquire foreign DNA and facilitate the emergence of pathogenic *E. coli* (65); however, for advantageous traits to persist, functional *mutS* needs to revert to its original state (26). Similarly, some environmental conditions drive the selection of *rpoS* mutants whereas others will restore functional *rpoS* (2, 14). This cycling of mutation and restoration via horizontal gene transfer of the flanking gene sequences makes the *rpoS-mutS* intergenic region highly dynamic and demonstrates the consequences of fluctuating selection pressures on the genome structure of *E. coli*. Many of the other hot spots for recombination on the *E. coli* genome encode pathogenicity determinants, such as the operon involved in O-antigen biosynthesis, fimbrial adhesin genes, and porin-encoding genes (61), revealing the connections among recombination, genomic plasticity, and pathogenic potential.

In addition to bacteriophages and PAIs, plasmids can enrich the pathogenic repertoire of *E. coli*. All sequenced *E. coli* pathogens harbor plasmids that have been implicated in some way in virulence (24). In enterotoxigenic *E. coli* (ETEC), enteroaggregative *E. coli* (EAEC), and EPEC, the fimbriae/pili and other adhesins that promote intestinal colonization are encoded on plasmids. In EIEC and *Shigella*, pINV contains genes that facilitate host-cell invasion, and ExPEC strains carry colicin-encoding plasmids. In addition, plasmids are known to carry drug resistance genes, iron acquisition genes, and ABC transport system genes (8). Despite the large number of virulence plasmids detected in *E. coli*, a vast majority of these plasmids belong to the same incompatibility group (IncF) and have similar genetic backbones decorated with strain-specific genes (24). Whereas each plasmid can have its own evolutionary history, the coevolution of plasmid and host genomes is evident from the extent of shared genes and the plasmid-dependent regulation of chromosomally encoded virulence genes (13). Plasmids often harbor large numbers of insertion (IS) elements, which can mediate integration into or excision from host chromosomes and the acquisition of novel genes, thereby providing plasmids with the ability to aid in the niche specialization of the *E. coli* host.

## OKAY, MUTATIONS ARE IMPORTANT, TOO

In addition to the substantial role that gene acquisition has played in the evolution and formation of pathogenic lineages, mutations that alter or abolish protein functions have also been implicated in the accelerated adaptation of *E. coli* to diverse environments (6, 9, 56). In many UPEC isolates, mutations in FimH—the adhesive subunit of type 1 fimbriae—have possibly hastened the shift in *E. coli* from being an intestinal commensal to being a urinary tract pathogen (55). Similarly, in *Shigella* and EIEC strains, production of cadaverine, a small polyamine that interferes with the action of enterotoxins, has been abolished by mutations in the *cad* locus, thereby increasing the fitness of the pathogen in its new intracellular niche (12).

## AS THE NUMBER OF GENOME SEQUENCES GROWS

Although the wide variation in genome among *E. coli* strains was already established well before the first *E. coli* genome sequence became available (3), the availability of completed genomes allows elucidation of the sources and effects of this variation. With rapid advancements in sequencing technologies, the number of completed prokaryotic genomes reached 1,000 by the end of 2009, which includes the sequences of 40 *E. coli* and *Shigella* strains. The genome size of sequenced strains ranges from 4.6 to 5.9 Mbp and the gene number ranges from 4,087 to 5,714 (Fig. 1). In addition to enumerating those genes present in a genome, the completed genome sequences also confirm the absence of genes in a particular strain; armed with this information, the comparative analysis of gene repertoires within a phylogenetic framework can shed light on the evolutionary dynamics of *E. coli* genomes.

Applying this approach, Hershberg et al. (22) compared several strains of *Shigella* to other pathogenic *E. coli* in order to understand the effects of niche specialization on genome evolution. Consistent with population genetics theory, the reduced population sizes of *Shigella*, as caused by their highly restricted host ranges, decrease the effectiveness of genome-wide selection against the accumulation of slightly deleterious mutations. As a consequence, *Shigella* strains have experienced higher rates of gene inactivation and loss during their evolutionary history and have genomes that are smaller than other *E. coli* strains.

To investigate the fate of newly emerged or acquired genes, van Passel et al. (62) examined the pattern of gene gain and loss in 13 sequenced strains of *E. coli* and *Shigella*. Following the designations of a previous study on lineage-specific genes (11), newly acquired genes were classified as those with distant

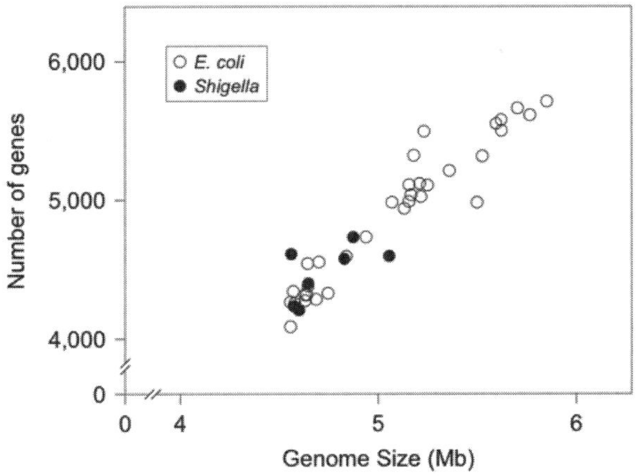


**Figure 1.** Association between genome size and gene number in sequenced strains of *E. coli* and *Shigella*.

homologs (termed "HOPs") or those with no homologs in current databases (termed "ORFans"). The differences in acquisition and retention rates between these two groups have provided several insights into the incorporation of novel components into *E. coli* genomes. For example, the high frequency of genes with distant homologs (HOPs) indicated that acquisition from distantly related bacteria is the major source of new sequences. However, the low retention rate of these acquired genes indicates that genes evolving in a different genetic background have contributed little to the long-term fitness of *E. coli*. In contrast, ORFan genes, whose source remains largely unknown, have a much higher retention rate, suggesting that they are more likely to confer adaptive benefit.

The most extensive study of *E. coli* comparative genomics to date includes 20 *E. coli/Shigella* strains and one *Escherichia fergusonii* strain (61). To ensure robust comparison, the authors reannotated all genomes by using the same methods to minimize artifacts stemming from individual genome sequencing efforts. This large-scale comparative study revealed the fluidity of *E. coli* genomes, as evident from the large number of gene gain and loss events occurring in all extant lineages. The high rate of gene gain and loss created a wide range of variation in the gene repertoires of *E. coli* strains, which was sufficiently large to discourage the use of gene contents as an indicator of divergence time within species. For example, among genome pairs that shared 75% of their genes, the average sequence divergence can range from 0.75 to 3%. Similarly, among genome pairs that have an average sequence divergence of 1.5%, the proportion of shared genes can vary from 60 to 90%.

This high level of diversity in gene repertoires raises questions about which genes define the characteristics of *E. coli*. The authors addressed this issue by analyzing the distribution of orthologous genes in all strains to define core genome (i.e., genes shared by the majority of examined strains) and pan-genome (i.e., genes present in any of the examined strains) (50). The set of genes in the core genome account for less than half of all genes in any *E. coli/Shigella* strain, and the number of pan-genome genes showed little sign of plateauing even with their sample of 20 strains, indicating that we are still far from capturing the full extent of genetic diversity in *E. coli*. This extreme level of gene diversity echoes a previous analysis of 573 sequenced bacterial genomes (31), which concluded that no upper bound can be established for the size of bacterial pan-genome.

## EVOLUTIONARY DYNAMICS OF BACTERIAL GENOMES

Bacterial genomes differ from those of eukaryotes in several aspects, such as their generally high coding densities (with 80 to 90% of the genome encoding functional genes) and the existence of a strong correlation between genome size and gene number (29). Therefore, changes in genome size in bacteria are directly associated with the gain or loss of genes, whereas differences in eukaryotic genome sizes need not reflect the amount of functional sequences. In addition to influencing the size and overall complexity of a genome, changes in gene content will also determine the metabolic capacity of a bacterial cell and the ecological niches that it can occupy.

The dominant process for introducing novel genes into a bacterial genome is horizontal gene transfer (18), which has been largely responsible for the genetic innovations present in *E. coli* genomes (11, 32, 50, 62). This is reinforced by the observation that most genes in bacterial genomes exist in single copy (34), suggesting that gene duplication plays a far less important role in generating new genes than it does in eukaryotes.

Because bacterial genomes never increase in size, the gain of genes through horizontal gene transfer must somehow be countered by gene loss. All genes that are not maintained by selection will accumulate mutations that eventually lead to gene inactivation (28); however, there are very few pseudogenes shared among closely related strains of *E. coli*, indicating that nonfunctional regions are rapidly purged from bacterial genomes (33). Although the exact mechanisms are not fully known, the mutational process in most bacterial genomes is strongly biased toward DNA loss, particularly in *E. coli* (27). This deletional bias is responsible for the rapid removal of pseudogenes and other nonfunctional regions, and is important in the maintenance of high coding density observed in bacterial genomes.

The high rate of gene loss in *Shigella* compared to *E. coli* suggests that the development of a host-dependent lifestyle is important to the process of genome evolution in bacteria (22, 61, 62). This link between host association and genome size has been documented across a wide range of bacterial lineages (43), and pathogens and symbionts that develop an obligate association with eukaryotic hosts have the most highly reduced genomes (42). One commonly held, albeit incorrect, hypothesis to explain this evolutionary trend is that positive selection for more efficient replication is driving the streamlining of pathogen genomes. However intuitive, this model is not supported by available data: replication rates are not correlated with genome size either among strains of *E. coli* (4) or across a wide range of bacterial species (17, 38).

## CASCADING EFFECTS OF POPULATION STRUCTURE ON BACTERIAL GENOME EVOLUTION

The process of genome reduction in host-associated bacteria is rooted in the effects of population-level processes on these lineages (29, 40). The dependence on eukaryotic hosts for survival and cell replication leads to reductions in effective population size, which in turn increases the level of genetic drift in the host-associated microbes. An elevated level of drift increases the fixation probability of slightly deleterious mutations that will begin to accumulate in nonessential genes, ultimately resulting in gene inactivation. The stable and rich environment provided by the host will relax purifying selection on genes involved in metabolic pathways that are vital to a free-living lifestyle but not essential for a symbiotic lifestyle, thus allowing many genes to become inactivated without large effects on fitness (39). In the early stage of the transition toward a smaller genome, high levels of gene decay can occur and create a large number of pseudogenes in the genome, as observed in *Mycobacterium leprae* (10) and *Sodalis glossinidius* (60). Initially, these newly created pseudogenes represent noncoding sequences, thereby lowering the overall gene density of the genome, but because the mutational process is biased toward DNA loss (27), these nonfunctional regions are eventually removed, even in the absence of selection for a compact genome.

Although the majority of a bacterial genome encodes functional genes, changes in nongenic regions also contribute to the evolution of genome size and complexity. For example, insertion sequences (i.e., IS elements) and other types of transposable sequences are important components of nongenic regions. Comparison between *E. coli* and *Shigella* strains revealed that the number of IS elements is much higher in *Shigella* (41, 46), an observation that parallels the pattern of pseudogene accumulation in host-associated pathogens due to the decreases in the efficacy of selection. In addition, the proliferation of IS elements can also promote genome rearrangements (48, 63). However, the large numbers of IS elements are eventually purged from the genome (39), presumably due to the strong deletional bias present in bacterial genomes. In summary, the genomes of *E. coli* and other bacteria have been influenced by numerous factors, but in the case of their overall size and complexity, nonadaptive processes, such as mutation and genetic drift, appear to be more important than selection.

## REFERENCES

1. **Asadulghani, M., Y. Ogura, T. Ooka, T. Itoh, A. Sawaguchi, A. Iguchi, K. Nakayama, and T. Hayashi.** 2009. The defective prophage pool of *Escherichia coli* O157: prophage-prophage interactions potentiate horizontal transfer of virulence determinants. *PLoS Pathog.* **5:**e1000408.
2. **Atlung, T., H. V. Nielsen, and F. G. Hansen.** 2002. Characterisation of the allelic variation in the *rpoS* gene in thirteen K12 and six other non-pathogenic *Escherichia coli* strains. *Mol. Genet. Genomics* **266:**873–881.
3. **Bergthorsson, U., and H. Ochman.** 1995. Heterogeneity of genome sizes among natural isolates of *Escherichia coli*. *J. Bacteriol.* **177:**5784–5789.
4. **Bergthorsson, U., and H. Ochman.** 1998. Distribution of chromosome length variation in natural isolates of *Escherichia coli*. *Mol. Biol. Evol.* **15:**6–16.
5. **Blattner, F. R., G. Plunkett III, C. A. Bloch, N. T. Perna, V. Burland, M. Riley, J. Collado-Vides, J. D. Glasner, C. K. Rode, G. F. Mayhew, J. Gregor, N. W. Davis, H. A. Kirkpatrick, M. A. Goeden, D. J. Rose, B. Mau, and Y. Shao.** 1997. The complete genome sequence of *Escherichia coli* K-12. *Science* **277:**1453–1462.
6. **Bono, J. L., J. E. Keen, M. L. Clawson, L. M. Durso, M. P. Heaton, and W. W. Laegreid.** 2007. Association of *Escherichia coli* O157:H7 *tir* polymorphisms with human infection. *BMC Infect. Dis.* **7:**98.
7. **Brenner, D. J., G. R. Fanning, F. J. Skerman, and S. Falkow.** 1972. Polynucleotide sequence divergence among strains of *Escherichia coli* and closely related organisms. *J. Bacteriol.* **109:**953–965.
8. **Brzuszkiewicz, E., G. Gottschalk, E. Ron, J. Hacker, and U. Dobrindt.** 2009. Adaptation of pathogenic *E. coli* to various niches: genome flexibility is the key. *Genome Dyn.* **6:**110–125.
9. **Chattopadhyay, S., S. J. Weissman, V. N. Minin, T. A. Russo, D. E. Dykhuizen, and E. V. Sokurenko.** 2009. High frequency of hotspot mutations in core genes of *Escher-*

*ichia coli* due to short-term positive selection. *Proc. Natl. Acad. Sci USA* **106:**12412–12417.

10. **Cole, S. T., K. Eiglmeier, J. Parkhill, K. D. James, N. R. Thomson, P. R. Wheeler, N. Honoré, T. Garnier, C. Churcher, D. Harris, K. Mungall, D. Basham, D. Brown, T. Chillingworth, R. Connor, R. M. Davies, K. Devlin, S. Duthoy, T. Feltwell, A. Fraser, N. Hamlin, S. Holroyd, T. Hornsby, K. Jagels, C. Lacroix, J. Maclean, S. Moule, L. Murphy, K. Oliver, M. A. Quail, M.-A. Rajandream, K. M. Rutherford, S. Rutter, K. Seeger, S. Simon, M. Simmonds, J. Skelton, R. Squares, S. Squares, K. Stevens, K. Taylor, S. Whitehead, J. R. Woodward, and B. G. Barrell.** 2000. Massive gene decay in the leprosy bacillus. *Nature* **409:**1007–1011.
11. **Daubin, V., and H. Ochman.** 2004. Bacterial genomes as new gene homes: the genealogy of ORFans in *E. coli*. *Genome Res.* **14:**1036–1042.
12. **Day, W. A., Jr., R. E. Fernández, and A. T. Maurelli.** 2001. Pathoadaptive mutations that enhance virulence: genetic organization of the *cadA* regions of *Shigella* spp. *Infect. Immun.* **69:**7471–7480.
13. **Dobrindt, U.** 2005. (Patho-)Genomics of *Escherichia coli*. *Int. J. Med. Microbiol.* **295:** 357–371.
14. **Dong, T., S. M. Chiang, C. Joyce, R. Yu, and H. E. Schellhorn.** 2009. Polymorphism and selection of *rpoS* in pathogenic *Escherichia coli*. *BMC Microbiol.* **9:**118.
15. **Escobar-Páramo, P., O. Clermont, A. B. Blanc-Potard, H. Bui, C. Le Bouguénec, and E. Denamur.** 2004. A specific genetic background is required for acquisition and expression of virulence factors in *Escherichia coli*. *Mol. Biol. Evol.* **21:**1085–1094.
16. **Ferenci, T.** 2003. What is driving the acquisition of *mutS* and *rpoS* polymorphisms in *Escherichia coli*? *Trends Microbiol.* **11:**457–461.
17. **Froula, J. L., and M. P. Francino.** 2007. Selection against spurious promoter motifs correlates with translational efficiency across bacteria. *PLoS One* **2:**e745.
18. **Gogarten., J. P., and J. P. Townsend.** 2005. Horizontal gene transfer, genome innovation and evolution. *Nat. Rev. Microbiol.* **3:**679–687.
19. **Hale, T. L.** 1991. Genetic basis of virulence in *Shigella* species. *Microbiol. Mol. Biol. Rev.* **55:**206–224.
20. **Hayashi, T., K. Makino, M. Ohnishi, K. Kurokawa, K. Ishii, K. Yokoyama, C. G. Han, E. Ohtsubo, K. Nakayama, T. Murata, M. Tanaka, T. Tobe, T. Iida, H. Takami, T. Honda, C. Sasakawa, N. Ogasawara, T. Yasunaga, S. Kuhara, T. Shiba, M. Hattori, and H. Shinagawa.** 2001. Complete genome sequence of enterohemorrhagic *Escherichia coli* O157:H7 and genomic comparison with a laboratory strain K-12. *DNA Res.* **8:**11–22.
21. **Herbelin, C. J., S. C. Chirillo, K. A. Melnick, and T. S. Whittam.** 2000. Gene conservation and loss in the *mutS-rpoS* genomic region of pathogenic *Escherichia coli*. *J. Bacteriol.* **182:**5381–5390.
22. **Hershberg, R., H. Tang, and D. A. Petrov.** 2007. Reduced selection leads to accelerated gene loss in *Shigella*. *Genome Biol.* **8:**R164.
23. **Hochhut, B., C. Wilde, G. Balling, B. Middendorf, U. Dobrindt, E. Brzuszkiewicz, G. Gottschalk, E. Carniel, and J. Hacker.** 2006. Role of pathogenicity island-associated integrases in the genome plasticity of uropathogenic *Escherichia coli* strain 536. *Mol. Microbiol.* **61:**584–595.
24. **Johnson, T. J., and L. K. Nolan.** 2009. Pathogenomics of the virulence plasmids of *Escherichia coli*. *Microbiol. Mol. Biol. Rev.* **73:**750–774.
25. **Kaper, J. B., J. P. Nataro, and H. L. Mobley.** 2004. Pathogenic *Escherichia coli*. *Nat. Rev. Microbiol.* **2:**123–140.
26. **Kotewicz, M. L., E. W. Brown, J. E. LeClerc, and T. A. Cebula.** 2003. Genomic variability among enteric pathogens: the case of the *mutS-rpoS* intergenic region. *Trends Microbiol.* **11:**2–6.
27. **Kuo, C.-H., and H. Ochman.** 2009. Deletional bias across the three domains of life. *Genome Biol. Evol.* **2009:**145–152.

28. **Kuo, C.-H., and H. Ochman.** 2009. The fate of new bacterial genes. *FEMS Microbiol. Rev.* **33:**38–43.
29. **Kuo, C.-H., N. A. Moran, and H. Ochman.** 2009. The consequences of genetic drift for bacterial genome complexity. *Genome Res.* **19:**1450–1454.
30. **Lan, R., and P. R. Reeves.** 2002. *Escherichia coli* in disguise: molecular origins of *Shigella. Microbes Infect.* **4:**1125–1132.
31. **Lapierre, P., and J. P. Gogarten.** 2009. Estimating the size of the bacterial pan-genome. *Trends Genet.* **25:**107–110.
32. **Lawrence, J. G., and H. Ochman.** 1998. Molecular archaeology of the *Escherichia coli* genome. *Proc. Natl. Acad. Sci. USA* **95:**9413–9417.
33. **Lerat, E., and H. Ochman.** 2004. Ψ-Φ: exploring the outer limits of bacterial pseudogenes. *Genome Res.* **14:**2273–2278.
34. **Lerat, E., V. Daubin, H. Ochman, and N. A. Moran.** 2005. Evolutionary origins of genomic repertoires in bacteria. *PLoS Biol.* **3:**807–814.
35. **Lloyd, A. L., T. A. Henderson, P. D. Vigil, and H. L. Mobley.** 2009. Genomic islands of uropathogenic *Escherichia coli* contribute to virulence. *J. Bacteriol.* **191:**3469–3481.
36. **Manning, S. D., A. S. Motiwala, A. C. Springman, W. Qi, D. W. Lacher, L. M. Ouellette, J. M. Mladonicky, P. Somsel, J. T. Rudrik, S. E. Dietrich, W. Zhang, B. Swaminathan, D. Alland, and T. S. Whittam.** 2008. Variation in virulence among clades of *Escherichia coli* O157:H7 associated with disease outbreaks. *Proc. Natl. Acad. Sci. USA* **105:**4868–4873.
37. **McDaniel, T. K., K. G. Jarvis, M. S. Donnenberg, and J. B. Kaper.** 1995. A genetic locus of enterocyte effacement conserved among diverse enterobacterial pathogens. *Proc. Natl. Acad. Sci. USA* **92:**1664–1668.
38. **Mira, A., H. Ochman, and N. A. Moran.** 2001. Deletional bias and the evolution of bacterial genomes. *Trends Genet.* **17:**589–596.
39. **Moran, N. A., and G. R. Plague.** 2004. Genomic changes following host restriction in bacteria. *Curr. Opin. Genet. Dev.* **14:**627–633.
40. **Novichkov, P. S., Y. I. Wolf, I. Dubchak, and E. V. Koonin.** 2009. Trends in prokaryotic evolution revealed by comparison of closely related bacterial and archaeal genomes. *J. Bacteriol.* **191:**65–73.
41. **Nyman, K., K. Nakamura, H. Ohtsubo, and E. Ohtsubo.** 1981. Distribution of the insertion sequence IS1 in gram-negative bacteria. *Nature* **289:**609–612.
42. **Ochman, H.** 2005. Genomes on the shrink. *Proc. Natl. Acad. Sci. USA* **95:**9413–9417.
43. **Ochman, H., and L. M. Davalos.** 2006. The nature and dynamics of bacterial genomes. *Science* **311:**1730–1733.
44. **Ochman, H., T. S. Whittam, D. A. Caugant, and R. K. Selander.** 1983. Enzyme polymorphism and genetic population structure in *Escherichia coli* and *Shigella. J. Gen. Microbiol.* **129:**2715–2726.
45. **Ogura, Y., T. Ooka, A. Iguchi, H. Toh, M. Asadulghani, K. Oshima, T. Kodama, H. Abe, K. Nakayama, K. Kurokawa, T. Tobe, M. Hattori, and T. Hayashi.** 2009. Comparative genomics reveal the mechanism of the parallel evolution of O157 and non-O157 enterohemorrhagic *Escherichia coli. Proc. Natl. Acad. Sci. USA* **106:**17939–17944.
46. **Ohtsubo, E., H. Ohtsubo, W. Doroszkiewicz, K. Nyman, D. Allen, and D. Davison.** 1984. An evolutionary analysis of iso-IS1 elements from *Escherichia coli* and *Shigella* strains. *J. Gen. Appl. Microbiol.* **30:**359–376.
47. **Orth, D., K. Grif, A. B. Khan, A. Naim, M. P. Dierich, and R. Würzner.** 2007. The Shiga toxin genotype rather than the amount of Shiga toxin or the cytotoxicity of Shiga toxin in vitro correlates with the appearance of the hemolytic uremic syndrome. *Diagn. Microbiol. Infect. Dis.* **59:**235–242.
48. **Parkhill, J., M. Sebaihia, A. Preston, L. D. Murphy, N. Thomson, D. E. Harris, M. T. G. Holden, C. M. Churcher, S. D. Bentley, K. L. Mungall, A. M. Cerdeño-Tárraga, L. Temple, K. James, B. Harris, M. A. Quail, M. Achtman, R. Atkin, S. Baker, D. Basham, N. Bason, I. Cherevach, T. Chillingworth, M. Collins, A. Cronin,

**P. Davis, J. Doggett, T. Feltwell, A. Goble, N. Hamlin, H. Hauser, S. Holroyd, K. Jagels, S. Leather, S. Moule, H. Norberczak, S. O'Neil, D. Ormond, C. Price, E. Rabbinowitsch, S. Rutter, M. Sanders, D. Saunders, K. Seeger, S. Sharp, M. Simmonds, J. Skelton, R. Squares, S. Squares, K. Stevens, L. Unwin, S. Whitehead, B. G. Barrell, and D. J. Maskell.** 2003. Comparative analysis of the genome sequences of *Bordetella pertussis, Bordetella parapertussis* and *Bordetella bronchiseptica. Nature Genet.* **35:**32–40.

49. **Perna, N. T., G. Plunkett III, V. Burland, B. Mau, J. D. Glasner, D. J. Rose, G. F. Mayhew, P. S. Evans, J. Gregor, H. A. Kirkpatrick, G. Pósfai, J. Hackett, S. Klink, A. Boutin, Y. Shao, L. Miller, E. J. Grotbeck, N. W. Davis, A. Lim, E. T. Dimalanta, K. D. Potamousis, J. Apodaca, T. S. Anantharaman, J. Lin, G. Yen, D. C. Schwartz, R. A. Welch, and F. R. Blattner.** 2001. Genome sequence of enterohaemorrhagic *Escherichia coli* O157:H7. *Nature* **409:**529–533.
50. **Rasko, D. A., M. J. Rosovitz, G. S. Myers, E. F. Mongodin, W. F. Fricke, P. Gajer, J. Crabtree, M. Sebaihia, N. R. Thomson, R. Chaudhuri, I. R. Henderson, V. Sperandio, and J. Ravel.** 2008. The pangenome structure of *Escherichia coli*: comparative genomic analysis of *E. coli* commensal and pathogenic isolates. *J. Bacteriol.* **190:**6881–6893.
51. **Rolland, K., N. Lambert-Zechovsky, B. Picard, and E. Denamur.** 1998. *Shigella* and enteroinvasive *Escherichia coli* strains are derived from distinct ancestral strains of *E. coli. Microbiology* **144**(Pt. 9):2667–2672.
52. **Selander, R. K., D. A. Caugant, H. Ochman, J. M. Musser, M. N. Gilmour, and T. S. Whittam.** 1986. Methods of multilocus enzyme electrophoresis for bacterial population genetics and systematics. *Appl. Environ. Microbiol.* **51:**873–884.
53. **Selander, R. K., and B. R. Levin.** 1980. Genetic diversity and structure in *Escherichia coli* populations. *Science* **210:**545–547.
54. **Selander, R. K., and J. M. Musser.** 1990. Population genetics of bacterial pathogenesis, p. 11–36. *In* J. R. Sokatch and L. N. Ornston (ed.), *The Bacteria: a Treatise on Structure and Function*, vol. XI, *Molecular Basis of Bacterial Pathogenesis*. Academic Press, Inc., New York, NY.
55. **Sokurenko, E. V., V. Chesnokova, D. E. Dykhuizen, I. Ofek, X. R. Wu, K. A. Krogfelt, C. Struve, M. A. Schembri, and D. L. Hasty.** 1998. Pathogenic adaptation of *Escherichia coli* by natural variation of FimH adhesin. *Proc Natl. Acad. Sci. USA* **95:** 8922–8926.
56. **Sokurenko, E. V., D. L. Hasty, and D. E. Dykhuizen.** 1999. Pathoadaptive mutations: gene loss and variation in bacterial pathogens. *Trends Microbiol.* **7:**191–195.
57. **Spears, K. J., A. J. Roe, and D. L. Gally.** 2006. A comparison of enteropathogenic and enterohaemorrhagic *Escherichia coli* pathogenesis. *FEMS Microbiol. Lett.* **255:**187–202.
58. **Steinberg, K. M., and B. R. Levin.** 2007. Grazing protozoa and the evolution of the *Escherichia coli* O157:H7 Shiga toxin-encoding prophage. *Proc. Biol. Sci.* **274:**1921–1929.
59. **Tobe, T., S. A. Beatson, H. Taniguchi, H. Abe, C. M. Bailey, A. Fivian, R. Younis, S. Matthews, O. Marches, G. Frankel, T. Hayashi, and M. J. Pallen.** 2006. An extensive repertoire of type III secretion effectors in *Escherichia coli* O157 and the role of lambdoid phages in their dissemination. *Proc. Natl. Acad. Sci. USA* **103:**14941–14946.
60. **Toh, H., B. L. Weiss, S. A. H. Perkin, A. Yamashita, K. Oshima, M. Hattori, and S. Aksoy.** 2006. Massive genome erosion and functional adaptations provide insights into the symbiotic lifestyle of *Sodalis glossinidius* in the tsetse host. *Genome Res.* **16:** 149–156.
61. **Touchon, M., C. Hoede, O. Tenaillon, V. Barbe, S. Baeriswyl, P. Bidet, E. Bingen, S. Bonacorsi, C. Bouchier, O. Bouvet, A. Calteau, H. Chiapello, O. Clermont, S. Cruveiller, A. Danchin, M. Diard, C. Dossat, M. El Karoui, E. Frapy, L. Garry, J. M. Ghigo, A. M. Gilles, J. Johnson, C. Le Bouguénec, M. Lescat, S. Mangenot,**

**V. Martinez-Jéhanne, I. Matic, X. Nassif, S. Oztas, M. A. Petit, C. Pichon, Z. Rouy, C. S. Ruf, D. Schneider, J. Tourret, B. Vacherie, D. Vallenet, C. Médigue, E. P. C. Rocha, and E. Denamur.** 2009. Organised genome dynamics in the *Escherichia coli* species results in highly diverse adaptive paths. *PLoS Genet.* **5:**e1000344.

62. **van Passel, M., P. R. Marri, and H. Ochman.** 2008. The emergence and fate of horizontally acquired genes in *Escherichia coli*. *PLoS Comput. Biol.* **4:**e1000059.
63. **Wei, J., M. B. Goldberg, V. Burland, M. M. Venkatesan, W. Deng, G. Fournier, G. F. Mayhew, G. Plunkett III, D. J. Rose, A. Darling, B. Mau, N. T. Perna, S. M. Payne, L. J. Runyen-Janecky, S. Zhou, D. C. Schwartz, and F. R. Blattner.** 2003. Complete genome sequence and comparative genomics of *Shigella flexneri* serotype 2a strain 2457T. *Infect. Immun.* **71:**2775–2786.
64. **Welch, R. A., V. Burland, G. Plunkett III, P. Redford, P. Roesch, D. Rasko, E. L. Buckles, S.-R. Liou, A. Boutin, J. Hackett, D. Stroud, G. F. Mayhew, D. J. Rose, S. Zhou, D. C. Schwartz, N. T. Perna, H. L. T. Mobley, M. S. Donnenberg, and F. R. Blattner.** 2002. Extensive mosaic structure revealed by the complete genome sequence of uropathogenic *Escherichia coli*. *Proc. Natl. Acad. Sci. USA* **99:**17020–17024.
65. **Whittam, T. S., S. D. Reid, and R. K. Selander.** 1998. Mutators and long-term molecular evolution of pathogenic *Escherichia coli* O157:H7. *Emerg. Infect. Dis.* **4:**615–617.
66. **Wick, L. M., W. Qi, D. W. Lacher, and T. S. Whittam.** 2005. Evolution of genomic content in the stepwise emergence of *Escherichia coli* O157:H7. *J. Bacteriol.* **187:**1783–1791.

*Population Genetics of Bacteria: a Tribute to Thomas S. Whittam*
Edited by S. T. Walk and P. C. H. Feng


*Chapter 5*

# Are Species Cohesive?— A View from Bacteriology

***Frederick M. Cohan***

Tom Whittam's research career created a productive synergy between the fields of evolutionary biology and public health. His insights into the evolutionary origin of enteric pathogens allowed him to contribute uniquely to public health efforts to characterize pathogen diversity. From my point of view as a speciologist, his studies of newly divergent lineages of pathogens provided important insights into the role of horizontal transfer in the origin of ecological diversity, the ecological dimensions by which bacterial lineages become irreversibly separate, and the order in which new lineages accumulate adaptations as they establish themselves in novel ecological niches. Tom's work provides a trove of inspiration for those like me who are fascinated by the origins of bacterial diversity.

Are there species in the world of bacteria? To bacterial systematists, there is no doubt that species exist and that it is useful to classify, name, and describe them (17, 73, 78). In bacterial systematics, species are defined as groups of close relatives that are separated from one another by large gaps in phenotypic and molecular characters (78). Charles Darwin would be quite comfortable with this concept of bacterial species, as he had proposed just this rationale for species demarcation in animals and plants—that species are closely related groups that can coexist as phenotypically distinct entities (22, 49, 75). Although bacterial systematists have argued about whether the species we recognize are too widely or too narrowly defined and the appropriateness of different criteria for species demarcation (29, 72, 74), it is clear that the named bacterial species hold Darwin's essential property of being groups of close relatives with gaps between them.

However, to some microbial ecologists (17, 67, 69, 82), as well as most systematists outside of microbiology (21, 23, 51, 53), species have usually been understood to be more than closely related groups separated by gaps. Largely

***Frederick M. Cohan*** • Department of Biology, Wesleyan University, Middletown, CT 06459-0170.

under the influence of Ernst Mayr (51, 53), Darwin's species concept has been rejected (49). Instead, species of animals and plants have come to be understood as cohesive groups, in the sense that some force of evolution constrains the diversity within species, while different species are not constrained from diverging. In highly sexual species, such as most animals and plants, the force constraining diversity within species is understood to be genetic exchange. In Mayr's biological species concept, speciation requires certain unusual circumstances that allow newly divergent populations to break free of cohesion by recurrent, high-frequency genetic exchange; speciation is therefore understood to be rare (51).

Many concepts of species have been proposed since Mayr's biological species concept, but they generally share certain essential, dynamic properties: that members of a species are subject to cohesion, species are invented only once, and different species are ecologically distinct and irreversibly separate (23). In what we might call Mayrian concepts of species, the essential dynamic properties of sexual animal and plant species have been extended to other groups where genetic exchange is rare or absent, such as the bacteria. The forces of cohesion within a bacterial species are thought to include periodic selection (17), which causes a genome-wide purging of diversity; genetic drift (42); and genetic exchange (27).

As Doolittle (25) has reviewed, some authors have argued that species in the bacterial world *must be* cohesive (14, 85). Doolittle has challenged advocates of dynamically defined, cohesive bacterial species to prove not just that such species do exist on occasion, but that they must generally exist. Here I will take up this challenge, but I will also address the possibility that while many bacteria reside within dynamically defined, cohesive species, others do not.

My own doubt about whether all bacteria belong to Mayrian, cohesive species stems in part from a related debate occurring now within the zoology community (24, 48, 64). James Mallet has challenged Mayr's model that species cohesion usually prevents the splitting of one species into two (48). Mallet explains that ordinary, adjacent populations within ordinary animal species are frequently poised to diverge into separate species. Provided that natural selection favors different adaptations in adjacent environments, populations may split into separate species despite genetic exchange between them. In short, "speciation is easy" (48).

I will explain how Mallet's hypothesis of easy speciation likely applies even more strongly to bacteria than to animals and plants, owing to intrinsic differences in the population dynamics of these major groups. Also, I will address what it means generally for any species to be cohesive. I will suggest that many animal and plant species are cohesive only in a very limited sense—that they are no more cohesive than are distantly related bacteria that occasionally exchange genes, and they are not cohesive in a way that prevents their populations from diverging as they adapt to different environments. Finally, I will make the case that speciologists from zoology and botany can learn a great deal from speciation in bacteria.

## COHESION IN MAYRIAN SPECIES

Under the influence of Mayr, neo-Darwinians have seen species as a unique and special taxonomic level (51, 52). Species are seen as the largest cohesive groups, such that divergence within a species is constrained by an active force, genetic exchange in the case of the highly sexual animals and plants, while divergence between species is not constrained. Thus, in sexual organisms, so long as populations can successfully interbreed and exchange genes, their divergence is understood to be constrained.

Before we discuss the existence and importance of cohesion, it will be important to consider the various ways that species may cohere. One form of cohesion may be termed "speciation-quashing cohesion," preventing one species from splitting into two (16). In the paradigm of the biological species concept, when nascent species lose their ability to exchange genes, they become free to diverge indefinitely and irreversibly. From the presumed difficulty of populations breaking loose to specieshood, improvements within a species lineage (anagenesis) are understood to occur much more frequently than speciation (cladogenesis).

A second mode of cohesion promotes homogeneity of populations that are ecologically different. In what we might call "niche-transcending cohesion," ecologically different populations across a range of organisms may share, through genetic exchange, genes that are adaptive across populations (16, 20).

Finally, "intrapopulation cohesion" promotes genetic identity among ecologically homogeneous individuals. Natural selection may promote genetic identity within such a population by favoring an adaptive mutation over other alleles, but within a highly sexual population of an animal or plant species, genetic exchange will purge the genetic diversity of only the locus under selection and linked loci (90). Genetic drift is likely the most important force of cohesion within ecologically homogeneous animal or plant populations.

The Mayrian concept of cohesive species may be extended to asexual and rarely sexual organisms, such as the bacteria (9, 55, 77). In rarely sexual organisms, natural selection provides a powerful constraint, genome-wide, on the accumulation of diversity within an ecologically homogeneous population, in a process known as periodic selection (4, 40, 43). Under rare recombination rates, such as observed in the bacteria (11, 54, 79), not only will natural selection favoring each adaptive mutation within a population cause the adaptive mutation to become fixed at 100%, but the selection will also cause nearly the entire genome to be purged of diversity. Rare recombination can prevent a total purging of diversity, but quantitative modeling has shown that bacterial recombination is insufficient to prevent purging of diversity, even in the most frequently recombining bacteria (10, 12; W. Hanage, personal communication).

We may then extend the Mayrian concept of cohesive species to bacteria by defining a species as an ecologically homogeneous group (called an "ecotype"), whose diversity is constrained recurrently by periodic selection or genetic drift (9, 17). Diversity within an ecotype may accumulate until it is purged by an adaptive mutation (or an adaptive recombination event), whereby one organism becomes superior in fitness to all others in the ecotype. In this model, intraecotype diversity is ephemeral, awaiting its collapse with the next periodic selection event. Diversity within an ecotype can also be constrained by genetic drift in bacteria with modest effective population sizes, such as pathogens and vertically transmitted mutualists, whose effective population size may approach that of their hosts (17, 42). Thus, bacteria, like the highly sexual animals and plants, are subject to intrapopulation cohesion.

My colleagues and I have previously proposed that speciation within bacteria occurs when a mutation or recombination event places a bacterium into a new ecological niche and creates a new ecotype (9, 17, 82). Then the diversity-purging events in the old ecotype cannot extend to the nascent ecotype, owing to the ecological differences (Fig. 1).

In one model of bacterial species and speciation known as the Stable Ecotype model (17), formation of new ecotypes occurs rarely, and ecotypes are long-lived, such that each ecotype undergoes many periodic selection events during its long lifetime. Each ecotype is constrained in its diversity owing to recurrent periodic selection events and/or to genetic drift; and divergence between different ecotypes is not limited by periodic selection or drift, owing to their ecological distinctness. Provided that an ecotype is founded by only one lineage, these ecotypes appear to hold one essential property of Mayrian species—that periodic selection and/or drift provides cohesion within ecotypes but not between them.

But are bacterial ecotypes irreversibly separate? While the adaptive divergence between ecotypes cannot be reversed by periodic selection, we must consider whether recombination between ecotypes might reverse their adaptive divergence. Clearly in the case of animals and plants, with obligate sexual exchange, populations must be sexually isolated to some extent to successfully diverge as ecologically distinct populations. This sexual isolation may involve genetically based changes in mating behavior, reduced fitness of hybrids, or perhaps most important, ecologically based isolation owing to residence in different microhabitats (21). Consider next whether recombination can likewise prevent adaptive divergence in bacteria, given profound differences in the nature of genetic exchange between bacteria and the highly sexual animals and plants.

Here I will briefly summarize the most important differences in genetic exchange (13). In bacteria, genetic exchange is unidirectional, where a usually small chromosomal segment transfers from a donor to a recipient. Also, bacterial recombination may involve homologous or heterologous transfer of genes. Recombination in bacteria occurs at extremely low rates, even in the most frequently recombining bacteria, ranging from less than the mutation rate to around 50 to 100 times the mutation rate, per gene segment (11, 13, 45, 79). Finally, bacterial recombination can occur across vastly more divergent organisms than is possible in animals and plants (38, 47).

Let us now consider whether recurrent recombination between bacterial ecotypes can hinder their adaptive divergence. I will begin with the potentially harmful effect of recombination of *niche-specifying genes*, those genes that promote adaptation of one population but are harmful when recombined into

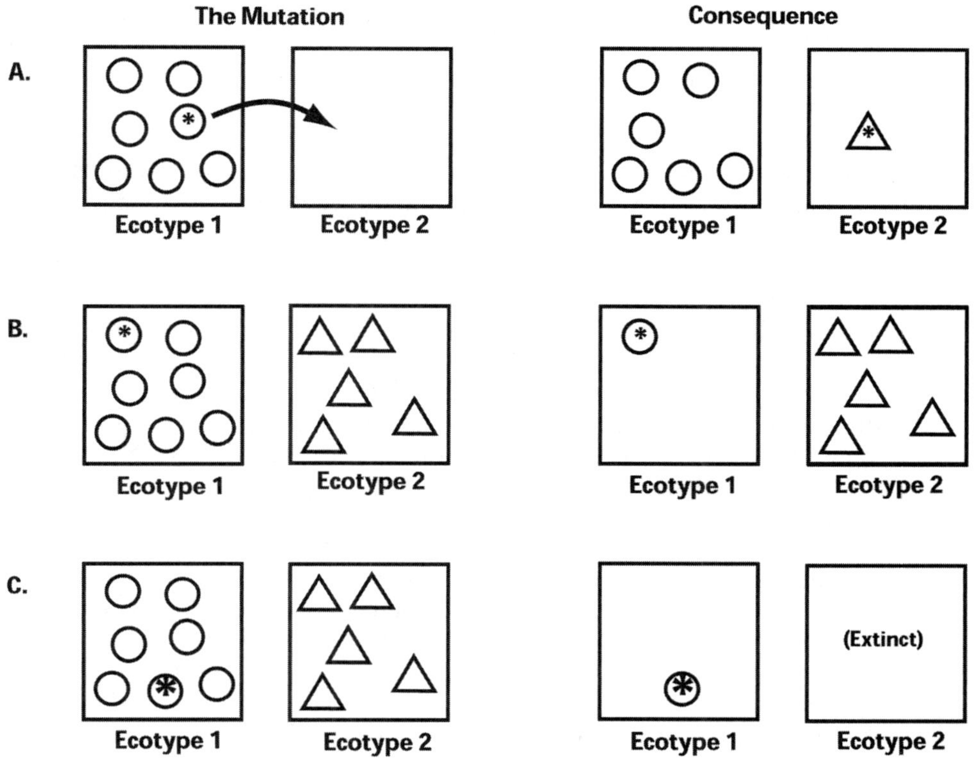


**Figure 1.** Three classes of mutation and recombination events that determine ecotype diversity in bacteria. The circles and triangles represent individuals within ecotypes 1 and 2, respectively; the asterisks represent adaptive mutations. (A) Niche-invasion mutations. Here a mutation changes the ecological niche of the cell, such that it can now escape periodic selection events in its former ecotype. This founds a new ecotype. (B) Periodic-selection mutations. These improve the fitness of an individual such that the mutant and its descendants outcompete all other cells within the ecotype; periodic selection events precipitated by these mutations generally do not affect the diversity within other ecotypes, owing to the differences in ecological niche. Periodic selection enhances the distinctness of ecotypes by purging the divergence within but not between ecotypes. (C) Speciation-quashing mutations in the Nano-Niche model. Even if two ecotypes have sustained a history of separate periodic selection events, an extraordinarily adaptive genotype may outcompete another ecotype to extinction. Competitive extinction of another ecotype (ecotype 2) is possible only if all of ecotype 2's resources are also used by ecotype 1. Used with permission from Landes Bioscience (10).

another population (16). In general, populations will successfully diverge in their adaptations in spite of recombination if the effect of selection is greater than that of recombination (9, 33, 48). More specifically, the equilibrium frequency of a recurrently recombined niche-specifying gene from another population is equal to the rate that the gene enters the population by recombination ($c_b$) divided by the selection intensity against the foreign allele ($s$), yielding $c_b/s$. Many adjacent animal populations have successfully diverged in adaptations in the face of recurrent recombination, at $c_b > 1\%$ per generation, presumably because $s >> c_b$ (48).

In the case of bacteria, it is straightforward to make the case that recurrent recombination of niche-specifying genes cannot prevent adaptive divergence (9, 16). Recombination among closest relatives within an ecotype ($c_w$) occurs extremely rarely (at about the mutation rate, $10^{-8}$ per gene segment per generation) (11, 54, 79). Then, even under the most conservative assumption of *no sexual isolation* between populations, recombination of niche-specifying genes is predicted to be inconsequential. That is, if recombination between ecotypes ($c_b$) were every bit as frequent as recombination within ecotypes ($c_b = c_w = 10^{-8}$), the equilibrium frequency of maladaptive niche-specifying genes imported from other populations would be extremely low (at $10^{-8}/s$). No one has quantified the fitness penalty ($s$) for incorporating a single niche-specifying gene from another ecotype, but a reasonable guess would be the selection advantage of a mutation contributing to adaptation in laboratory environments, with $s$ values ranging from 0.08 to 4.5% (39). It is then reasonable to assume that selection would keep the frequencies of maladaptive niche-specifying genes from other ecotypes extremely low, and that recombination cannot halt or reverse adaptive divergence between young ecotypes. Therefore, the evolution of sexual isolation between ecotypes is not a necessary milestone toward species formation. This is in contrast to the case for the highly sexual animals and plants, where between-population recombination must be reduced at least by residence in different microhabitats, if not through intrinsic sexual isolation (9, 17).

Recently, Sheppard and colleagues have claimed that recombination can prevent or even reverse the divergence among closely related bacterial species (66). They point to the case of two *Campylobacter* species that have recently been given an opportunity to exchange genes more frequently, owing to changes in animal husbandry that have put the species in close proximity. They note a recent history of increased recombination in certain genes used in multilocus sequence analysis, and claim that these two groups are "despeciating." However, it is important to consider that the genes included in multilocus sequence analysis have been chosen such that recombination between organisms is of little or no fitness or niche-specifying consequence (44). Thus, the genes seen to be recombining successfully between species may be considered to be niche neutral; these are not the niche-specifying genes whose recombination could potentially hinder the adaptive divergence of ecotypes. Recombination of niche-neutral genes may be a hindrance to our taxonomy (by making it difficult for us to assign an organism to its proper ecotype), but it is not a hindrance to the adaptive divergence of ecotypes.

We may conclude, then, that the critical threshold of divergence in ecotype formation is the ecological divergence that allows the new ecotype to escape the periodic selection events of its former ecotype. At this point, the new ecotype is free to accumulate additional niche-specifying genes that further distinguish it in ecology from related ecotypes. Furthermore, recombination between ecotypes cannot stall or reverse the adaptive divergence of ecotypes. To the extent that we are able to predict the future of divergence, ecotypes that are outside one another's periodic selection events appear to have the species-like quality of irreversible separateness. Thus, speciation-quashing cohesion postulated by Mayr, where recombination prevents one species from splitting into two irreversibly separate populations, does not appear to apply to bacteria.

We next move to the challenge of identifying ecotypes, or bacterial species that are subject to intrapopulation cohesion provided by periodic selection and/or drift.

## THE FAILURE OF SYSTEMATICS TO DISCOVER ECOLOGICALLY HOMOGENEOUS SPECIES OF BACTERIA

The discovery of dynamically defined, cohesive species would appear to be much easier for the familiar animals and plants than for bacteria. Systematists' demarcation of many animals and plants is informed by a familiarity with the morphological differences underlying their ecological divergence. Our knowledge of differences among animal species and our ability to classify animals based on these differences are likely to predate modern ecology and systematics, and may even be the result of eons of human evolution for ready distinction of organisms with life-or-death consequences for human hunters and gatherers (35, 92). Thus, zoological and botanical taxonomy may have benefited from (but ultimately may be distracted by) what has been called our species' "umwelt," the biological foundation for our species' way of seeing the natural world (92).

Bacteria are, of course, far outside our species' prehistorical umwelt, and the process of demarcating bacterial species has had to be invented wholly by microbiologists. Bacterial species taxonomy is also challenged by the difficulty of anticipating the traits most important for ecological divergence among closest species (17). The problem is that ecological divergence between bacterial species frequently results from acquisition of genes and operons through horizontal genetic transfer (16, 30, 59), and the genes imported by this mechanism and the donor sources of the genes are not known ahead of time. It is therefore difficult to predict the phenotypic characters that should be ecologically relevant in the divergence of new species (17).

It is easy to see the attraction that objective criteria held for midcentury bacterial systematists, beginning with numerical taxonomy (92). Numerical taxonomy attempts to find groups of similar organisms based on overall phenotypic similarity, without requiring knowledge of the most reliable or important traits. Beginning with Sneath's analysis of *Chromobacterium* species, numerical taxonomy was widely used in phenotypic demarcation of bacterial species from the 1950s through the 1970s (until molecular criteria became available). In principle, a numerical taxonomy approach based on a large set of random characters could have yielded the most recent products of speciation, the ecologically homogeneous and cohesive ecotypes.

However, there were some significant obstacles to reaching this goal through numerical taxonomy. First, identification of extremely newly divergent species by phenotypic analysis would be most likely successful if the phenotypic characters responsible for speciation were included in the analysis, but as I have noted, it is difficult to predict the characters underlying the ecological divergence driving speciation. Also, numerical taxonomy might have been handicapped by systematists' pragmatic choice of analyzing presence versus absence of metabolic characters. Each such difference is likely to reflect differences in ecologically significant traits. Thus, clusters of phenotypically similar but nonidentical organisms would likely contain ecologically heterogeneous populations, and defining species as such clusters would lump many long-coexisting, ecologically distinct groups within a single species.

Indeed, the species demarcated by numerical taxonomy are huge by any criterion, including the very characters used in numerical taxonomy (6). From the beginning, numerical taxonomists, as well as less quantitatively oriented taxonomists, could have demarcated smaller, more homogeneous phenotypic clusters. In Kämpfer and colleagues' *Streptomyces* study, for example, a similarity level of 80% for species demarcation was chosen, yet there were apparent clusters at higher levels of similarity (37). It is not clear why such huge phenotypic diversity within species was sought, but in the absence of an instinctive umwelt for describing bacteria, this level of within-species diversity was established artificially as our umwelt, and bacterial systematics has enforced this umwelt for decades (78).

As molecular approaches to species demarcation have become available, molecular criteria for demarcation have been calibrated to yield the species defined by the broad phenotypic clusters previously established (14). For example, whole-genome DNA-DNA hybridization, a measure of gene content shared between organisms, was the first molecular tool for species demarcation. A criterion of 70% annealing in DNA-DNA hybridization was set as the best general criterion to yield the established, recognized species (86). Given the technical and logistical difficulties of performing DNA-DNA hybridization (29), direct calculation of genome content shared has been suggested as a substitute for attaining the DNA-DNA value sought by systematics (31).

As DNA sequencing became available, systematists utilized the sequence of the small ribosomal subunit (16S rRNA) for species demarcation. A level of 3% divergence was originally set as a guide for declaring organisms to be in different species, but more recently a level of 1% divergence has been taken to be a reliable guide for demarcating the species originally classified by phenotypic clusters and then by DNA-DNA hybridization (72). We should note that the 16S levels used for demarcation, if applied to animals, would put all primates in the same species (74). Other molecular guides for species demarcation have been suggested, including the average nucleotide identity of all shared genes in the genome (at a level of 94%) (31).

There is no doubt that as more discerning and convenient molecular tools become available, bacterial systematists will be able to calibrate them to obtain the species fitting our newfound umwelt for bacterial species, as based originally on phenotypic clustering. If we knew these species to be ecologically homogeneous, and to be the most recent products of speciation, there would be nothing wrong with this. However, at least some of the recognized species of bacterial systematics are enormously diverse in their ecology. For example, the work of Tom Whittam and his colleagues has characterized the ecological diversity within *Escherichia coli* and within related species. They characterized lineages divergent in their preferred environments and ways of making a living, including commensalism, enteropathogenicity, and uropathogenicity, as well as in their degrees of virulence (50) and persistence outside of hosts (80, 81).

Other recognized pathogenic species have also shown ecological subdivision, with *Legionella pneumophila* containing 11 ecotypes, some of them distinguishable by host range differences and others by gene expression patterns (15). The lung pathogen *Mycobacterium tuberculosis* has several ecotypes distinguished by their associations with host mammalian species (69); similarly, there are lineages associated with different host ranges within the Lyme disease pathogen *Borrelia burgdorferi* (26).

Beyond the pathogens, the recognized marine heterotrophic species *Vibrio splendidus* has been found to include 15 sympatric phylogenetic groups that are distinct in their associations with substrate particle size and season (36). Another free-living heterotroph, *Bacillus simplex*, includes nine ecotypes varying in their associations with solar exposure in semiarid soils (41); a similar result has been obtained for *Bacillus subtilis*, with ecotypes distinguishable by solar exposure and soil texture associations (18). While systematists have long been untroubled by the ecological heterogeneity within the species they recognize, the heterogeneity of genome content, now explicitly seen through genome sequencing (87, 89), is leading to a growing sense among systematists that bacterial species demarcations should be reevaluated (29, 71).

My colleagues and I have previously argued that recognition of cohesive, ecologically homogeneous species by bacterial systematics would benefit microbiologists of many fields, including epidemiology, biotechnology, genomics, population genetics, and ecology; moreover, this could be done without upsetting the current systematics of species (17, 41). We next illustrate a general approach for identifying such cohesive and ecologically homogeneous ecotypes.

## ALGORITHMS FOR DISCOVERY OF BACTERIAL ECOTYPES

Several laboratories have recently developed universal, sequence-based approaches for identifying newly divergent bacterial ecotypes. These methods eschew the universal molecular cutoffs espoused by systematics; instead, they attempt to find appropriate criteria for demarcating ecotypes within a taxon of focus, taking into account the genes being analyzed. These each require a molecular survey of diversity among close relatives, based on molecular markers that are not necessarily responsible for the niche-specifying divergence among ecotypes (and usually are not). Clearly, the ability of any such sequence-based approach will be limited by the rate of speciation and the rate of evolution in the markers; so sequence-based approaches are best suited to finding the long-standing bacterial species that fit the Stable Ecotype model of bacterial speciation.

The algorithms Ecotype Simulation (ES) (41), BAPS (19), and GMYC (5) are designed to find long-lived, stable ecotypes using only sequence data, usually the nucleotide sequences of several protein-coding loci. Higher resolution of more newly divergent ecotypes could be obtained through analysis of many more genes, in the extreme the entire set of genes shared among genomes. These algorithms search for sequence clusters whose diversity appears to be purged recurrently. Because these algorithms do not utilize any data directly related to ecology (either phenotypes or habitats of isolation), the putative ecotypes hypothesized by these algorithms must be confirmed independently as ecologically distinct. On the plus side, the lack of ecological input allows these algorithms to demarcate ecotypes even when the ecological dimensions of divergence among ecotypes are beyond our imagination (16, 18). In contrast, the algorithm AdaptML requires information about the ecology of the organisms surveyed, in the form of an environmental description of their habitats of isolation (36). The AdaptML algorithm has the advantage that it can simultaneously identify ecotypes and confirm them as ecologically distinct, based on finding statistically significant differences in their habitat associations. The downside for AdaptML is that it can identify only those ecotypes that are divergent in habitat types anticipated by the investigators (16, 18).

Confirmation of ecotypes hypothesized by these algorithms requires two steps. The first and most straightforward part is demonstrating that the hypothesized ecotypes are ecologically distinct. Confirmation of ecological distinctness of hypothesized ecotypes is important because under some models of bacterial diversification, an ecotype may contain multiple, ecologically interchangeable sequence clusters, particularly when geography plays a role in the diversification of populations (17).

One approach to confirming the ecological distinctness of ecotypes is to test putative ecotypes for differences in their associations with contrasting microhabitats (18, 36, 41, 83). Also, ecotypes can be tested for physiological differences that might adapt the ecotypes to their favored microhabitats (2, 18, 68). Full-genome comparisons are expected to play a growing role in finding physiological differences between closely related ecotypes (91), either by identifying ecologically significant biochemical pathways that are not shared across ecotypes (7) or by identifying shared genes whose divergence has been accelerated by natural selection (70). These approaches have confirmed that the ecotypes hypothesized by ES and AdaptML are ecologically distinct (15, 18, 36, 41).

The more difficult, second part of confirming ecotypes is to demonstrate that the putative ecotypes identified by sequence analysis represent the most recent products of speciation. That is, each ecotype must be shown to be homogeneous in its ecology, and thereby potentially cohesive. Alternatively, ecotypes hypothesized by ES or AdaptML might contain within them different, ecologically distinct groups. In this case, the hypothesized ecotypes would not be cohesive because periodic selection and drift would be limited to the subset groups of organisms that represent the true, ecologically homogeneous ecotypes.

There are several models where an ecotype hypothesized by sequence-based approaches belies ecological homogeneity when there is actually ecological diversity hidden within it. One possibility is the Speedy Speciation model, a Mayrian model where each ecotype is cohesive, with recurrent purges of diversity; speciation is simply too rapid for the most recent products of speciation to be detected by the resolution provided by one to several loci (17). Perhaps in these cases, the resolution of whole-genome sequencing might discern the most newly divergent ecotypes.

We next consider a diversity of models of speciation that do not follow the Mayrian paradigm; in these cases, bacterial ecotypes do not represent units of intrapopulation cohesion. Under each of these non-Mayrian models, sequence-based approaches are unlikely to identify ecologically homogeneous ecotypes.

## NON-MAYRIAN MODELS OF BACTERIAL SPECIATION

Consider first the Species-less model, where species are formed and extinguished at very high rates, with each nascent species living only a short time before going extinct (17, 84). Each such nascent species is ecologically homogeneous, and so these ecotypes could incur periodic selection events that would purge within-ecotype diversity, as in the cohesive Stable Ecotype model. However, because of their short existence, these species have very limited opportunity for adaptations leading to periodic selection (16). Thus, within-species diversity in the Species-less model would be limited not by recurrent diversity-purging forces, but instead by the limited amount of time from the founding of an ecotype by a single mutant (or recombinant) cell until the ecotype's extinction (Fig. 2). If some species manage to spawn a new species before they go extinct, a large clade of such species could be dominated by cladogenesis events with very few anagenesis events.

The Species-less model may apply when the habitats to which a new species adapts are ephemeral, so that when a habitat disappears, the species specialized to the habitat also disappears. Martin Polz and colleagues have suggested that short-lived ecotypes specialized to ephemeral habitats may constitute a major part of marine bacterial diversity (61). They term these ecotypes "opportunitrophs," where each has colonized and specialized to a particular particle of marine snow. If there is limited opportunity for a specialist of one form of snow to colonize another such particle, there will be parallel niche invasions by many opportunitroph ecotypes, each extremely short-lived.

Mark Achtman has suggested that pathogens may frequently follow the Species-less model when they evolve new immune-escape adaptations (1). An established serotype might have only limited success because most individuals of a host population have developed immunity, but the evolution of a novel serotype not recognized by the hosts' immune system will have a flare of success before it, in turn, becomes widely recognized. Provided that each

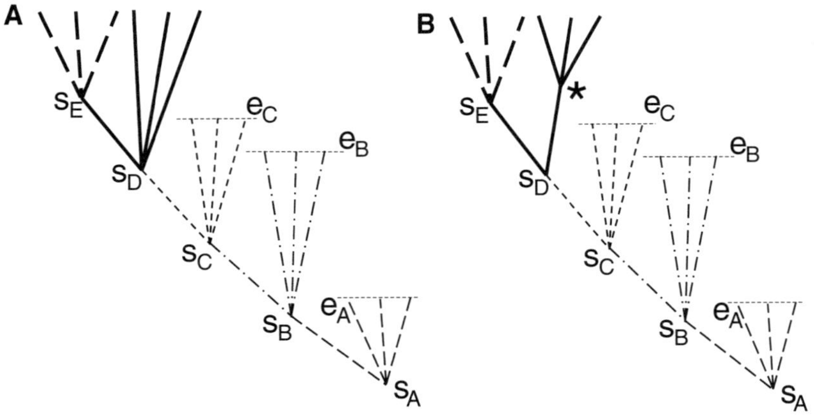


**Figure 2.** The Species-less model of bacterial diversification. In the Species-less model, the diversity within an ecotype is not limited by periodic selection but instead by the short time from the ecotype's invention as a single mutant until its extinction. Each ecotype is represented in the figure by a unique line style; the origination and extinction of each ecotype *i* are indicated by $s_i$ and $e_i$, respectively. (A) In the absence of periodic selection, each extant ecotype that has given rise to another ecotype is a paraphyletic group, and each recent ecotype that has not yet given rise to another ecotype is monophyletic. If two closely related ecotypes represent a monophyletic-paraphyletic pair (as in the case of ecotypes D and E, in bold), then we may conclude that a periodic selection event has not occurred in the parental ecotype since the origin of the daughter ecotype. (B) If instead a periodic selection event has occurred in the parental ecotype since the founding of the daughter ecotype, then the parent and daughter ecotypes will be sister monophyletic groups. Observing that pairs of most-closely-related ecotypes usually form monophyletic-paraphyletic pairs would indicate that the origin of new ecotypes is more frequent than periodic selection events in established ecotypes.

immunity-defined ecotype has a short lifetime and can coexist with its parental ecotype, this will constitute a Species-less dynamic.

Consider next how we might test the Species-less model (Fig. 2), where diversity within an ecotype is not recurrently purged by periodic selection. If one lineage within such a species should give rise to a new species, the parental species will appear in a phylogeny as a paraphyletic lineage (i.e., the ecotypes will be nested; Fig. 2A). That is, the most recent common ancestor of the parental ecotype will include among its descendants the more basal parental ecotype as well as the nascent, daughter ecotype. Such a result would indicate that the parental ecotype incurred no periodic selection events after the origination of the daughter ecotype. If it had, both parental and daughter ecotypes would appear as sister, monophyletic clades (Fig. 2B). This analysis requires that we have the molecular resolution to identify the most recent products of speciation, possibly through analyzing the thousands of genes shared across full-genome sequences.

Another non-Mayrian model of diversification is the Nano-Niche model (17). Here ecotypes are formed rapidly and invade extremely similar niches, such that the ecotypes use entirely the same set of resources but in different proportions. While each ecotype may incur its own periodic selection events, the sharing of resources allows for the possibility that there will eventually be a periodic selection event emanating from one ecotype that can completely extinguish its closest relatives (Fig. 1C). The Nano-Niche model invokes cohesion, but it is not the single, ecologically homogeneous species at which cohesion applies; the cohesion applies across multiple, young ecotypes.

Finally, the Recurrent Niche Invasion model takes into account the role of mobile genetic elements, such as plasmids or phage, in determining bacterial niches (17). For example, in the case of *Rhizobium*, a bacterial lineage may acquire a symbiosis plasmid, which will adapt it as an endosymbiotic mutualist to a particular set of legume hosts; then the lineage may lose that plasmid and acquire another symbiosis plasmid, thereby adapting it to another set of legumes (65). In general, a cell is converted from one ecotype to another by acquiring and/or losing a niche-specifying plasmid or phage, and so a lineage moves back and forth between memberships in different, previously existing ecotypes.

If there is a high rate of conversion of lineages from one ecotype to another through plasmid or phage transfer, the molecular approaches to ecotype demarcation we have discussed will fail to identify ecotypes. This is because each time a cell changes from one ecotype to another (by acquisition and loss of a plasmid or phage), this event implicitly transfers the alleles at every chromosomal locus from membership in the cell's current ecotype to membership in its new ecotype, all without any recombination beyond the transfer of the plasmid (Fig. 3). In this way, two ecotypes could coexist for eons and yet never be distinguished as separate sequence clusters for niche-neutral, chromosomal genes. This role of plasmids and phage in instantly and recurrently pulling bacterial lineages from one ecotype to another may be unique to the prokaryotes.

In the Recurrent Niche Invasion model, cohesion may stretch across all the bacteria that can potentially utilize the same set of niche-specifying plasmids (or phage). At any one time, this will include bacteria that are in different plasmid-specified ecotypes. The cohesion will transcend the various plasmid-based ecotypes if each chromosome-based adaptation within each such ecotype is adaptive in the context of all the plasmids utilized by these bacteria.

Alternatively, cohesion could be limited to each plasmid-defined ecotype. This would occur if each ecotype accumulates its own chromosomally based improvements, which are useful only in the context of its own plasmid and are deleterious in the context of other ecotypes' plasmids. In this case, the ecotypes will no longer share plasmids, as each transfer of a plasmid from one ecotype to another will result in a recipient that is not well adapted to bearing the transferred plasmid. Each ecotype will thus incur intrapopulation cohesion and will diverge from other plasmid-based ecotypes irreversibly; that their niche-specifying adaptations were originally all on plasmids will over time become irrelevant. This model predicts that chromosome-based clades will be signifi-

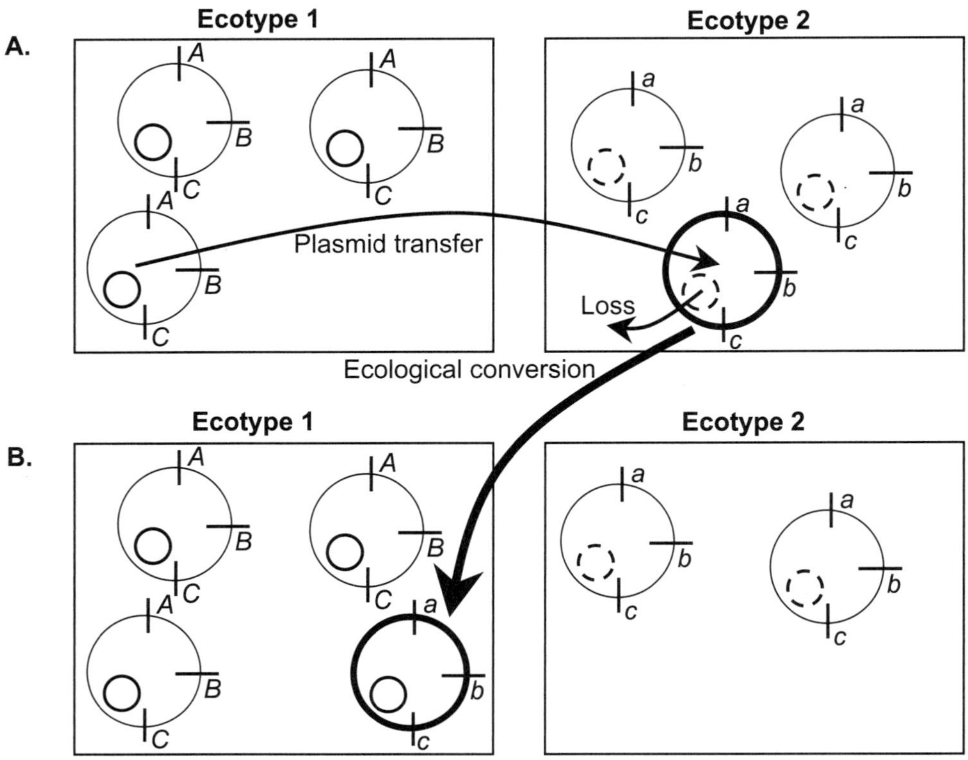


**Figure 3.** Ecological conversion from one plasmid-defined ecotype to another. Each cell is represented by a large circle, and its chromosomal genotype is represented as *ABC* or *abc*. Ecotype 1 is determined by a plasmid represented by a small solid-line circle, and ecotype 2 is determined by a plasmid represented by a small dashed circle. (A) A transfer of an ecotype 1-determining plasmid into a recipient member of ecotype 2 (indicated by the bold circle), along with subsequent loss of the ecotype 2-determining plasmid. (B) The recipient is converted to become a member of ecotype 1, so all of its chromosomal genes are effectively transferred into ecotype 1.

cantly different in their associations with different niche-specifying plasmids, as seen in nonagricultural *Rhizobium* (88).

Whether plasmid- or phage-based ecotypes are in nature cohesive as single ecotypes, or are instead cohesive as pools of related ecotypes, has not been adequately explored. There are known cases where a chromosome-defined clade of strains can hold a diversity of niche-specifying plasmids, e.g., as in the case of the insect-killing plasmids of *Bacillus thuringiensis* (76) or the virulence-conferring phage in *Listeria monocytogenes* (58). This is consistent with cohesion occurring at the level of multiple plasmid-based ecotypes. Another interpretation is that different chromosome-based subclades are adapted preferentially to different niche-specifying plasmids, but they are very recently divergent and are somewhat generalist in their plasmid preferences.

## COHESION THROUGH SHARING OF NICHE-TRANSCENDING GENES

Beyond the harmful effects of recombination on niche-specifying genes, zoologists and botanists have long understood a significant adaptive effect of recombination—this is the recombination of "niche-transcending genes" (16). Occasionally, a mutation that is adaptive in the context of all an animal species' populations may appear. This niche-transcending gene may adaptively spread across the species range while populations maintain their adaptive divergence in niche-specifying genes. The tension between niche-specifying and niche-transcending adaptations may be illustrated by human populations in the early agricultural era. The rates of recombination between distant human populations in this time would have been low enough to allow populations to maintain their adaptive divergence, in skin color for example, while allowing the sharing of niche-transcending genes that provided widely useful adaptations to agricultural living (34). In many cases, recombination within and between closely related animal and plant species has allowed sharing of niche-transcending genes while allowing each population or species to maintain its niche-specifying adaptations (20, 35, 63).

Likewise, niche-transcending genes appear to have recombined and spread within the bacterial world. The dynamics of their spread is predicted by the Adapt Globally Act Locally model (46). Here a niche-transcending gene initiates a periodic selection event within its original ecotype. The niche-transcending gene then recombines on a small segment into another ecotype, where the recipient will outcompete other members of the recipient's ecotype and cause a purging of diversity there. The Adapt Globally Act Locally model predicts that ecotypes can maintain their ecological distinctness and their sequence divergence across the genome, while becoming homogenized for the segment of DNA carrying the niche-transcending gene.

Recently, comparative genomics has produced evidence for the spread of niche-transcending genes among ecologically distinct bacteria populations. This is seen, for example, in two *Synechococcus* ecotypes (A and B′) that are adapted to different temperatures and concentrations of mineral nutrients in a Yellowstone hot spring (2, 7), and are consequently found in different microhabitats (85). That they have coexisted for a long time is evidenced by their 78% average nucleotide identity (7). However, the ecotypes share a 35-kb segment containing the nitrogen fixation pathway, which is nearly identical in the two ecotypes. We may conclude that the *nif* pathway has recently recombined across these ecotypes, either directly or by way of other organisms. The two ecotypes, adapted to different thermal and nutrient zones, have thus each become better adapted to their respective stations by sharing these genes.

One could alternatively interpret this as a periodic selection event within a single population (32, 60). For example, one might hypothesize that (i) the A and B′ *Synechococcus* groups are really members of a single, ecologically homogeneous population; (ii) one individual within this population acquired the *nif* region by horizontal genetic transfer; (iii) this organism and its descendants outcompeted other individuals within the population; and (iv) because of recombination, the only segment to become homogenized was the *nif* region. However, there are several problems with this hypothesis (46). First, the A and B′ clades are adapted to 65° and 60°, respectively (2), and accordingly live in different microhabitats (62, 83), so it is difficult to see how an improvement in an A organism would cause that organism to outcompete members of the B′ group, or vice versa. Also, if A and B′ were a single population, it is difficult to understand why they would appear to us as two long-coexisting lineages with an average nucleotide identity of only 78%. Finally, recombination in bacteria has been shown to be infrequent enough such that selection for an adaptation will purge diversity genome-wide, not just in the selected chromosomal region, as occurs in the highly sexual animals and plants (10; Hanage, personal communication). Nevertheless, this single-population periodic selection explanation has been invoked, mistakenly I believe, to explain a chromosomal region of homogeneity within the ecologically diverse *E. coli* (32) and a similar observation in ecologically distinct groups of *Halorobrum* (60).

Horizontal genetic transfer of niche-transcending genes is understood to be an important part of adaptive evolution in bacteria, as shown by the great number of genomic islands shared across both close and distant relatives (30, 89). The sharing of niche-transcending genes between ecotypes does not appear to reverse their adaptive divergence. Rather, it allows the recipient to build on its unique, preexisting set of adaptations either to invade a new niche or to improve its performance in its current niche (13). For example, Tom Whittam and his colleagues showed that enterotoxigenic *E. coli*, which attack the epithelial cells of the small intestine, have shared the class 5 fimbriae by horizontal genetic transfer with *Burkolderia cepacia* (3). The within-human niche of *B. cepacia* is quite distinct from that of enterotoxigenic *E. coli*, as *B. cepacia* resides in the lungs and attacks the respiratory epithelium. Acquiring the niche-transcending class 5 fimbriae genes has not caused donor and recipient to converge ecologically, and this is likely true for other transfers of niche-transcending genes.

It would be interesting to quantify the relative importance of niche-specifying versus niche-transcending adaptations. In some cases, the set of niche-specifying adaptations separating a set of closely related populations may be severely limited. This is observed, for example, in pitcher plant mosquitoes, where niche-specifying adaptations simply adjust the season of a population's activity, such that all the various populations across a latitudinal range are active in nearly the same range of temperature conditions (but at different times of year) (8). In this case, all adaptations to this shared set of conditions are niche transcending. Perhaps this applies also in bacteria for host specificity within a closely related group of pathogens, if adaptation to a particular host membrane molecule is the extent of niche specificity, and all other adaptations are niche transcending. At the other extreme, nearly every gene in the genome may be niche specifying. This is likely the case for hot spring bacteria populations living on a sharp temperature gradient, where most genes are adjusted to optimize protein activity within an organism's temperature range (56).

## WHITHER COHESION?

Let's now revisit the central Mayrian notion of species and speciation—that divergence within a species is constrained by cohesive forces but divergence between species is not. We will consider each of the three levels of cohesion in turn.

Consider first to what extent individual, ecologically homogeneous populations are subject to intrapopulation cohesion. In the world of animals and plants, each such population is subject to cohesion genome-wide primarily by drift. In the bacterial world, some pathogenic and endosymbiont populations will also be subject to strong drift. However, periodic selection is the most likely force of cohesion for free-living populations, at least those that survive long enough for adaptive mutations to occur. In the Species-less model, with short-lived ecotypes, there may be no opportunities for adaptive improvement and consequent purging of diversity; so these short-lived ecotypes would not be cohesive. It is not clear whether cohesive, long-lived ecotypes are common at all in the bacterial world, or whether the bacterial world is primarily populated by short-lived ecotypes that never have a chance at cohesion.

Second, recombination does not seem likely to be a cohesive force that quashes speciation in either the macroorganisms or bacteria. Most clearly in bacteria, recombination is not sufficient to prevent adaptive divergence in niche-specifying genes, and sexual isolation is not required for bacterial speciation; two nascent ecotypes do not even need the physical separation of residence in different microhabitats. Likewise, animals and plants are able to diverge into ecologically distinct populations and maintain the integrity of their niche specificity, in spite of modest recombination between them. Owing to the obligate nature of recombination in animals and plants, the rate of recombination between populations must be reduced from that within populations, but apparently residence in different microhabitats is sufficient sexual isolation in many cases. Studies of the adaptive divergence between adjacent, interbreeding animal and plant populations indicate that these populations have the dynamic properties of species: each is ecologically distinct and appears to be irreversibly separate from other such populations, at least to the extent that we can predict the future of divergence from past coexistence. Provided that these populations can continue to live in spatially separated and contrasting habitats, it is reasonable to hypothesize their irreversible separateness. Like the case for bacteria, macroorganismal populations appear not to be subject to species-quashing cohesion at the hand of recombination.

There is, however, one circumstance where ecological divergence among long-coexisting, interbreeding populations of animals and plants can cause them to lose their niche specificity through a speciation-quashing event. Futuyma has hypothesized that during bouts of catastrophic global climate change, divergent populations tracking their optimal environments will inevitably find themselves sympatric (28). If in sympatry there is no reproductive isolation between populations, the accumulated genetic differences between them may be reset to zero. In this case, the potential to exchange genes across ecologically divergent populations may act in the Mayrian sense of speciation-quashing cohesion. I will note, however, that this model does not force cohesion for populations that differ in their optimal microenvironments and are able to find their own particular optimal environments as populations track climate change.

There are also some circumstances in the bacterial world where ecologically distinct populations are prevented from diverging. In the Recurrent Niche Invasion model, where ecotypes are determined only by highly mobile genetic elements such as plasmids or phage, the chromosomal genomes of different ecotypes may never diverge. Also, in the Nano-Niche model, closely related ecotypes, which use the same set of resources but in different proportions, may coexist and experience their separate periodic selection events. However, a speciation-quashing adaptive mutation may allow a highly adaptive mutant to outcompete and extinguish not just other members of its own ecotype but also closely related ecotypes that use the same resources.

Third, cohesion by sharing of niche-transcending genes may be a red herring for species studies. The tradition of species studies in zoology and botany has focused on the origins of a total blockage of interbreeding and recombination among populations. As I have argued, the principal consequence of interbreeding is that it enables ecologically distinct populations to share niche-transcending genes, while interbreeding appears unable to prevent irreversible divergence between populations (in both macroorganisms and bacteria). By focusing on the stoppage of gene flow, the speciology of animals and plants has implicitly emphasized the end of sharing of niche-transcending genes, and has deemphasized the quotidian origins of ecological distinctness among interbreeding populations. In contrast, studies of speciation in bacteria have focused on the origins of niche-specifying adaptations that distinguish newly divergent species, by investigating the ecological dimensions of speciation (18, 36, 69, 83) and the roles of horizontal genetic transfer (7, 57) and homologous recombination (27, 66) in bacterial speciation.

This emphasis on the origins of ecological divergence was forced on bacteriology because bacteria can acquire niche-transcending genes potentially from any organism; so it would be futile to study the end of sharing niche-transcending adaptations in bacteria. It appears that, fortuitously, bacteriology has produced a paradigm of value for species studies in macroorganisms as well as bacteria—that our focus should be on the origins of ecological diversity and not on barriers to recombination.

**Acknowledgments.** This work was supported by NSF FIBR grant EF-0328698 and research funds from Wesleyan University.

## REFERENCES

1. **Achtman, M., and M. Wagner.** 2008. Microbial diversity and the genetic nature of microbial species. *Nat. Rev. Microbiol.* **6:**431–440.
2. **Allewalt, J. P., M. M. Bateson, N. P. Revsbech, K. Slack, and D. M. Ward.** 2006. Effect of temperature and light on growth of and photosynthesis by *Synechococcus* isolates typical of those predominating in the Octopus Spring microbial mat community of Yellowstone National Park. *Appl. Environ. Microbiol.* **72:**544–550.

3. **Anantha, R. P., A. L. McVeigh, L. H. Lee, M. K. Agnew, F. J. Cassels, D. A. Scott, T. S. Whittam, and S. J. Savarino.** 2004. Evolutionary and functional relationships of colonization factor antigen I and other class 5 adhesive fimbriae of enterotoxigenic *Escherichia coli. Infect. Immun.* **72:**7190–7201.
4. **Atwood, K. C., L. K. Schneider, and F. J. Ryan.** 1951. Periodic selection in *Escherichia coli. Proc. Natl. Acad. Sci. USA* **37:**146–155.
5. **Barraclough, T. G., M. Hughes, N. Ashford-Hodges, and T. Fujisawa.** 2009. Inferring evolutionarily significant units of bacterial diversity from broad environmental surveys of single-locus data. *Biol. Lett.* **5:**425–428.
6. **Barrett, E. L., R. E. Solanes, J. S. Tang, and N. J. Palleroni.** 1986. *Pseudomonas fluorescens* biovar V: its resolution into distinct component groups and the relationship of these groups to other *P. fluorescens* biovars, to *P. putida*, and to psychrotrophic pseudomonads associated with food spoilage. *J. Gen. Microbiol.* **132:**2709–2721.
7. **Bhaya, D., A. R. Grossman, A. S. Steunou, N. Khuri, F. M. Cohan, N. Hamamura, M. Melendrez, M. M. Bateson, D. M. Ward, and J. F. Heidelberg.** 2007. Population level functional diversity in a microbial community revealed by comparative genomic and metagenomic analyses. *ISME J.* **1:**703–713.
8. **Bradshaw, W. E., and C. M. Holzapfel.** 2008. Genetic response to rapid climate change: it's seasonal timing that matters. *Mol. Ecol.* **17:**157–166.
9. **Cohan, F. M.** 1994. The effects of rare but promiscuous genetic exchange on evolutionary divergence in prokaryotes. *Am. Naturalist* **143:**965–986.
10. **Cohan, F. M.** 2005. Periodic selection and ecological diversity in bacteria, p. 78–93. *In* D. Nurminsky (ed.), *Selective Sweep.* Landes Bioscience, Georgetown, TX.
11. **Cohan, F. M.** 2002. Population structure and clonality of bacteria, p. 161–163. *In* M. Pagel (ed.), *Encyclopedia of Evolution*, vol. 1. Oxford University Press, New York, NY.
12. **Cohan, F. M.** 1996. The role of genetic exchange in bacterial evolution. *ASM News* **62:** 631–636.
13. **Cohan, F. M.** 2002. Sexual isolation and speciation in bacteria. *Genetica* **116:**359–370.
14. **Cohan, F. M.** 2002. What are bacterial species? *Annu. Rev. Microbiol.* **56:**457–487.
15. **Cohan, F. M., A. Koeppel, and D. Krizanc.** 2006. Sequence-based discovery of ecological diversity within *Legionella*, p. 367–376. *In* N. P. Cianciotto, Y. Abu Kwaik, P. H. Edelstein, B. S. Fields, D. F. Geary, T. G. Harrison, C. Joseph, R. M. Ratcliff, J. E. Stout, and M. S. Swanson (ed.), Legionella*: State of the Art 30 Years after Its Recognition.* ASM Press, Washington, DC.
16. **Cohan, F. M., and A. F. Koeppel.** 2008. The origins of ecological diversity in prokaryotes. *Curr. Biol.* **18:**R1024–R1034.
17. **Cohan, F. M., and E. B. Perry.** 2007. A systematics for discovering the fundamental units of bacterial diversity. *Curr. Biol.* **17:**R373–R386.
18. **Connor, N., J. Sikorski, A. P. Rooney, S. Kopac, A. F. Koeppel, A. Burger, S. G. Cole, E. B. Perry, D. Krizanc, N. C. Field, M. Slaton, and F. M. Cohan.** 2010. The ecology of speciation in *Bacillus*. *Appl. Environ. Microbiol.* **76:**1349–1358.
19. **Corander, J., P. Marttinen, J. Siren, and J. Tang.** 2008. Enhanced Bayesian modelling in BAPS software for learning genetic structures of populations. *BMC Bioinformatics* **9:** 539.
20. **Coyne, J. A.** 2009. *Why Evolution Is True*, p. 189. Oxford University Press, Oxford, United Kingdom.
21. **Coyne, J. A., and H. A. Orr.** 2004. *Speciation.* Sinauer Associates, Sunderland, MA.
22. **Darwin, C.** 1859. *On the Origin of Species by Means of Natural Selection, or the Preservation of Favoured Races in the Struggle for Life.* John Murray, London. United Kingdom.
23. **de Queiroz, K.** 2005. Ernst Mayr and the modern concept of species. *Proc. Natl. Acad. Sci. USA* **102**(Suppl. 1):6600–6607.
24. **Dieckmann, U., M. Doebeli, J. A. J. Metz, and D. Tautz (ed.).** 2004. *Adaptive Speciation.* Cambridge University Press, Cambridge, United Kingdom.

25. **Doolittle, W. F.** 2009. Eradicating typological thinking in prokaryotic systematics and evolution. *Cold Spring Harb. Symp. Quant. Biol.* **74:**197–204.
26. **Dykhuizen, D. E., D. Brisson, S. Sandigursky, G. P. Wormser, J. Nowakowski, R. B. Nadelman, and I. Schwartz.** 2008. The propensity of different *Borrelia burgdorferi* sensu stricto genotypes to cause disseminated infections in humans. *Am. J. Trop. Med. Hyg.* **78:**806–810.
27. **Fraser, C., W. P. Hanage, and B. G. Spratt.** 2007. Recombination and the nature of bacterial speciation. *Science* **315:**476–480.
28. **Futuyma, D. J.** 1987. On the role of species in anagenesis. *Am. Naturalist* **130:**465–473.
29. **Gevers, D., F. M. Cohan, J. G. Lawrence, B. G. Spratt, T. Coenye, E. J. Feil, E. Stackebrandt, Y. Van de Peer, P. Vandamme, F. L. Thompson, and J. Swings.** 2005. Re-evaluating prokaryotic species. *Nat. Rev. Microbiol.* **3:**733–739.
30. **Gogarten, J. P., W. F. Doolittle, and J. G. Lawrence.** 2002. Prokaryotic evolution in light of gene transfer. *Mol. Biol. Evol.* **19:**2226–2238.
31. **Goris, J., K. T. Konstantinidis, J. A. Klappenbach, T. Coenye, P. Vandamme, and J. M. Tiedje.** 2007. DNA-DNA hybridization values and their relationship to whole-genome sequence similarities. *Int. J. Syst. Evol. Microbiol.* **57:**81–91.
32. **Guttman, D. S., and D. E. Dykhuizen.** 1994. Detecting selective sweeps in naturally occurring *Escherichia coli*. *Genetics* **138:**993–1003.
33. **Haldane, J. B. S.** 1932. *The Causes of Evolution.* Longmans, Green, and Co., London, United Kingdom.
34. **Hawks, J., E. T. Wang, G. M. Cochran, H. C. Harpending, and R. K. Moyzis.** 2007. Recent acceleration of human adaptive evolution. *Proc. Natl. Acad. Sci. USA* **104:** 20753–20758.
35. **Hey, J.** 2001. *Genes, Categories, and Species: the Evolutionary and Cognitive Cause of the Species Problem.* Oxford University Press, Oxford, United Kingdom.
36. **Hunt, D. E., L. A. David, D. Gevers, S. P. Preheim, E. J. Alm, and M. F. Polz.** 2008. Resource partitioning and sympatric differentiation among closely related bacterioplankton. *Science* **320:**1081–1085.
37. **Kämpfer, P., R. M. Kroppenstedt, and W. Dott.** 1991. A numerical classification of the genera *Streptomyces* and *Streptoverticillium* using miniaturized physiological tests. *J. Gen. Microbiol.* **137:**1831–1891.
38. **Kanhere, A., and M. Vingron.** 2009. Horizontal gene transfers in prokaryotes show differential preferences for metabolic and translational genes. *BMC Evol. Biol.* **9:**9.
39. **Kassen, R., and T. Bataillon.** 2006. Distribution of fitness effects among beneficial mutations before selection in experimental populations of bacteria. *Nat. Genet.* **38:**484–488.
40. **Koch, A. L.** 1974. The pertinence of the periodic selection phenomenon to prokaryote evolution. *Genetics* **77:**127–142.
41. **Koeppel, A., E. B. Perry, J. Sikorski, D. Krizanc, W. A. Warner, D. M. Ward, A. P. Rooney, E. Brambilla, N. Connor, R. M. Ratcliff, E. Nevo, and F. M. Cohan.** 2008. Identifying the fundamental units of bacterial diversity: a paradigm shift to incorporate ecology into bacterial systematics. *Proc. Natl. Acad. Sci. USA* **105:**2504–2509.
42. **Kuo, C. H., N. A. Moran, and H. Ochman.** 2009. The consequences of genetic drift for bacterial genome complexity. *Genome Res.* **19:**1450–1454.
43. **Levin, B. R.** 1981. Periodic selection, infectious gene exchange and the genetic structure of *E. coli* populations. *Genetics* **99:**1–23.
44. **Maiden, M. C. J., J. A. Bygraves, E. Feil, G. Morelli, J. E. Russell, R. Urwin, Q. Zhang, J. Zhou, K. Zurth, D. A. Caugant, I. M. Feavers, M. Achtman, and B. G. Spratt.** 1998. Multilocus sequence typing: a portable approach to the identification of clones within populations of pathogenic microorganisms. *Proc. Natl. Acad. Sci. USA* **95:**3140–3145.
45. **Majewski, J.** 2001. Sexual isolation in bacteria. *FEMS Microbiol. Lett.* **199:**161–169.

46. **Majewski, J., and F. M. Cohan.** 1999. Adapt globally, act locally: the effect of selective sweeps on bacterial sequence diversity. *Genetics* **152:**1459–1474.
47. **Majewski, J., P. Zawadzki, P. Pickerill, F. M. Cohan, and C. G. Dowson.** 2000. Barriers to genetic exchange between bacterial species: *Streptococcus pneumoniae* transformation. *J. Bacteriol.* **182:**1016–1023.
48. **Mallet, J.** 2008. Hybridization, ecological races and the nature of species: empirical evidence for the ease of speciation. *Philos. Trans. R. Soc. Lond. B Biol. Sci.* **363:**2971–2986.
49. **Mallet, J.** 2008. Mayr's view of Darwin: was Darwin wrong about speciation? *Biol. J. Linn. Soc. Lond.* **95:**3–16.
50. **Manning, S. D., A. S. Motiwala, A. C. Springman, W. Qi, D. W. Lacher, L. M. Ouellette, J. M. Mladonicky, P. Somsel, J. T. Rudrik, S. E. Dietrich, W. Zhang, B. Swaminathan, D. Alland, and T. S. Whittam.** 2008. Variation in virulence among clades of *Escherichia coli* O157:H7 associated with disease outbreaks. *Proc. Natl. Acad. Sci. USA* **105:**4868–4873.
51. **Mayr, E.** 1963. *Animal Species and Evolution.* Belknap Press of Harvard University Press, Cambridge, MA.
52. **Mayr, E.** 1982. *The Growth of Biological Thought: Diversity, Evolution, and Inheritance.* Harvard University Press, Cambridge, MA.
53. **Mayr, E.** 1944. *Systematics and the Origin of Species from the Viewpoint of a Zoologist.* Columbia University Press, New York, NY.
54. **McVean, G., P. Awadalla, and P. Fearnhead.** 2002. A coalescent-based method for detecting and estimating recombination from gene sequences. *Genetics* **160:**1231–1241.
55. **Meglitsch, P.** 1954. On the nature of species. *Syst. Zool.* **3:**491–503.
56. **Miller, S. R., and R. W. Castenholz.** 2000. Evolution of thermotolerance in hot spring cyanobacteria of the genus *Synechococcus. Appl. Environ. Microbiol.* **66:**4222–4229.
57. **Ochman, H., and L. M. Davalos.** 2006. The nature and dynamics of bacterial genomes. *Science* **311:**1730–1733.
58. **Orsi, R. H., M. L. Borowsky, P. Lauer, S. K. Young, C. Nusbaum, J. E. Galagan, B. W. Birren, R. A. Ivy, Q. Sun, L. M. Graves, B. Swaminathan, and M. Wiedmann.** 2008. Short-term genome evolution of *Listeria monocytogenes* in a non-controlled environment. *BMC Genomics* **9:**539.
59. **Palenik, B., Q. Ren, V. Tai, and I. T. Paulsen.** 2009. Coastal *Synechococcus* metagenome reveals major roles for horizontal gene transfer and plasmids in population diversity. *Environ. Microbiol.* **11:**349–359.
60. **Papke, R. T., O. Zhaxybayeva, E. J. Feil, K. Sommerfeld, D. Muise, and W. F. Doolittle.** 2007. Searching for species in haloarchaea. *Proc. Natl. Acad. Sci. USA* **104:**14092–14097.
61. **Polz, M. F., D. E. Hunt, S. P. Preheim, and D. M. Weinreich.** 2006. Patterns and mechanisms of genetic and phenotypic differentiation in marine microbes. *Philos. Trans. R. Soc. Lond. B Biol. Sci.* **361:**2009–2021.
62. **Ramsing, N. B., M. J. Ferris, and D. M. Ward.** 2000. Highly ordered vertical structure of *Synechococcus* populations within the one-millimeter-thick photic zone of a hot spring cyanobacterial mat. *Appl. Environ. Microbiol.* **66:**1038–1049.
63. **Rieseberg, L. H., O. Raymond, D. M. Rosenthal, Z. Lai, K. Livingstone, T. Nakazato, J. L. Durphy, A. E. Schwarzbach, L. A. Donovan, and C. Lexer.** 2003. Major ecological transitions in wild sunflowers facilitated by hybridization. *Science* **301:**1211–1216.
64. **Schemske, D. W.** 2000. Understanding the origin of species. *Evolution* **54:**1069–1073.
65. **Segovia, L., D. Piñero, R. Palacios, and E. Martínez-Romero.** 1991. Genetic structure of a soil population of nonsymbiotic *Rhizobium leguminosarum. Appl. Environ. Microbiol.* **57:**426–433.
66. **Sheppard, S. K., N. D. McCarthy, D. Falush, and M. C. Maiden.** 2008. Convergence of *Campylobacter* species: implications for bacterial evolution. *Science* **320:**237–239.

67. **Sikorski, J.** 2008. Populations under microevolutionary scrutiny: what will we gain? *Arch. Microbiol.* **189:**1–5.
68. **Sikorski, J., and E. Nevo.** 2007. Patterns of thermal adaptation of *Bacillus simplex* to the microclimatically contrasting slopes of 'Evolution Canyons' I and II, Israel. *Environ. Microbiol.* **9:**716–726.
69. **Smith, N. H., S. V. Gordon, R. de la Rua-Domenech, R. S. Clifton-Hadley, and R. G. Hewinson.** 2006. Bottlenecks and broomsticks: the molecular evolution of *Mycobacterium bovis*. *Nat. Rev. Microbiol.* **4:**670–681.
70. **Soyer, Y., R. H. Orsi, L. D. Rodriguez-Rivera, Q. Sun, and M. Wiedmann.** 2009. Genome wide evolutionary analyses reveal serotype specific patterns of positive selection in selected *Salmonella* serotypes. *BMC Evol. Biol.* **9:**264.
71. **Stackebrandt, E.** 2002. From species definition to species concept: population genetics is going to influence the systematics of Prokaryotes. *WFCC Newsletter* **35:**1–4.
72. **Stackebrandt, E., and J. Ebers.** 2006. Taxonomic parameters revisited: tarnished gold standards. *Microbiol. Today* **33:**152–155.
73. **Stackebrandt, E., W. Frederiksen, G. M. Garrity, P. A. Grimont, P. Kämpfer, M. C. Maiden, X. Nesme, R. Rossello-Mora, J. Swings, H. G. Truper, L. Vauterin, A. C. Ward, and W. B. Whitman.** 2002. Report of the ad hoc committee for the re-evaluation of the species definition in bacteriology. *Int. J. Syst. Evol. Microbiol.* **52:**1043–1047.
74. **Staley, J. T.** 2006. The bacterial species dilemma and the genomic-phylogenetic species concept. *Philos. Trans. R. Soc. Lond. B Biol. Sci.* **361:**1899–1909.
75. **Stamos, D. N.** 2007. *Darwin and the Nature of Species.* State University of New York Press, Albany, NY.
76. **Swiecicka, I., and J. Mahillon.** 2005. The clonal structure of *Bacillus thuringiensis* isolates from north-east Poland does not correlate with their *cry* gene diversity. *Environ. Microbiol.* **7:**34–39.
77. **Templeton, A.** 1989. The meaning of species and speciation: a genetic perspective, p. 3–27. *In* D. Otte and J. Endler (ed.), *Speciation and Its Consequences.* Sinauer Associates, Sunderland, MA.
78. **Vandamme, P., B. Pot, M. Gillis, P. de Vos, K. Kersters, and J. Swings.** 1996. Polyphasic taxonomy, a consensus approach to bacterial systematics. *Microbiol. Rev.* **60:** 407–438.
79. **Vos, M., and X. Didelot.** 2009. A comparison of homologous recombination rates in bacteria and archaea. *ISME J.* **3:**199–208.
80. **Walk, S. T., E. W. Alm, L. M. Calhoun, J. M. Mladonicky, and T. S. Whittam.** 2007. Genetic diversity and population structure of *Escherichia coli* isolated from freshwater beaches. *Environ. Microbiol.* **9:**2274–2288.
81. **Walk, S. T., E. W. Alm, D. M. Gordon, J. L. Ram, G. A. Toranzos, J. M. Tiedje, and T. S. Whittam.** 2009. Cryptic lineages of the genus *Escherichia. Appl. Environ. Microbiol.* **75:**6534–6544.
82. **Ward, D. M.** 1998. A natural species concept for prokaryotes. *Curr. Opin. Microbiol.* **1:**271–277.
83. **Ward, D. M., M. M. Bateson, M. J. Ferris, M. Kühl, A. Wieland, A. Koeppel, and F. M. Cohan.** 2006. Cyanobacterial ecotypes in the microbial mat community of Mushroom Spring (Yellowstone National Park, Wyoming) as species-like units linking microbial community composition, structure and function. *Philos. Trans. R. Soc. Lond. B Biol Sci.* **361:**1997–2008.
84. **Ward, D. M., and F. M. Cohan.** 2005. Microbial diversity in hot spring cyanobacterial mats: pattern and prediction, p. 185–202. *In* W. P. Inskeep and T. McDermott (ed.), *Geothermal Biology and Geochemistry in Yellowstone National Park.* Thermal Biology Institute, Bozeman, MT.
85. **Ward, D. M., F. M. Cohan, D. Bhaya, J. F. Heidelberg, M. Kühl, and A. Grossman.** 2008. Genomics, environmental genomics and the issue of microbial species. *Heredity* **100:**207–219.

86. **Wayne, L. G., D. J. Brenner, R. R. Colwell, P. A. D. Grimont, O. Kandler, M. I. Krichevsky, W. E. C. Moore, R. G. E. Murray, E. Stackebrandt, M. P. Starr, and H. G. Trüper.** 1987. Report of the ad hoc committee on reconciliation of approaches to bacterial systematics. *Int. J. Syst. Bacteriol.* **37:**463–464.
87. **Welch, R. A., V. Burland, G. Plunkett III, P. Redford, P. Roesch, D. Rasko, E. L. Buckles, S.-R. Liou, A. Boutin, J. Hackett, D. Stroud, G. F. Mayhew, D. J. Rose, S. Zhou, D. C. Schwartz, N. T. Perna, H. L. T. Mobley, M. S. Donnenberg, and F. R. Blattner.** 2002. Extensive mosaic structure revealed by the complete genome sequence of uropathogenic *Escherichia coli. Proc. Natl. Acad. Sci. USA* **99:**17020–17024.
88. **Wernegreen, J. J., E. E. Harding, and M. A. Riley.** 1997. *Rhizobium* gone native: unexpected plasmid stability of indigenous *Rhizobium leguminosarum. Proc. Natl. Acad. Sci. USA* **94:**5483–5488.
89. **Whittam, T. S., and A. C. Bumbaugh.** 2002. Inferences from whole-genome sequences of bacterial pathogens. *Curr. Opin. Genet. Dev.* **12:**719–725.
90. **Williamson, S. H., M. J. Hubisz, A. G. Clark, B. A. Payseur, C. D. Bustamante, and R. Nielsen.** 2007. Localizing recent adaptive evolution in the human genome. *PLoS Genet.* **3:**e90.
91. **Wu, D., P. Hugenholtz, K. Mavromatis, R. Pukall, E. Dalin, N. N. Ivanova, V. Kunin, L. Goodwin, M. Wu, B. J. Tindall, S. D. Hooper, A. Pati, A. Lykidis, S. Spring, I. J. Anderson, P. D'haeseleer, A. Zemla, M. Singer, A. Lapidus, M. Nolan, A. Copeland, C. Han, F. Chen, J. F. Cheng, S. Lucas, C. Kerfeld, E. Lang, S. Gronow, P. Chain, D. Bruce, E. M. Rubin, N. C. Kyrpides, H. P. Klenk, and J. A. Eisen.** 2009. A phylogeny-driven genomic encyclopaedia of Bacteria and Archaea. *Nature* **462:**1056–1060.
92. **Yoon, C. K.** 2009. *Naming Nature: the Clash between Instinct and Science.* Norton, New York, NY.

*Section II*

# *ESCHERICHIA COLI* AND MOLECULAR EPIDEMIOLOGY OF PATHOGENIC STRAINS

*Population Genetics of Bacteria: a Tribute to Thomas S. Whittam*
Edited by S. T. Walk and P. C. H. Feng


*Chapter 6*

# The Niche of *Escherichia coli*

***Elizabeth W. Alm, Seth T. Walk, and David M. Gordon***

## INTRODUCTION

In the summer of 1884, German pediatrician Theodor Escherich began investigating the bacterial community of the gastrointestinal (GI) tract of human neonates and infants (24, 72). Escherich was interested in defining what microorganisms were present in healthy individuals, what physiologic roles these microbes played, and if some of them influenced the onset and progression of pediatric diseases. Using cultivation techniques developed in Robert Koch's laboratory, Escherich demonstrated that voided meconium, a dark greenish mass of desquamated cells, mucus, and bile that accumulates in the fetal intestine during development, is sterile. Using temporal culture experiments and Christian Gram's new staining technique, Escherich further showed that the human GI tract becomes colonized by microorganisms as early as 4 to 7 hours and as late as 12 to 18 hours after birth. In a sketch of a microscopic preparation of a 27-hour-old infant's meconium, Escherich portrayed a diverse assemblage of microorganisms and named one member of this community *Bacterium coli commune* (24). This bacterium, later renamed *Escherichia coli* in honor of Dr. Escherich, has become the most studied organism to date.

Since its discovery, *E. coli* has been referred to as a "normal inhabitant" of the GI tract of humans and animals. While large amounts of epidemiologic data support this finding, *E. coli* does not require a host to reproduce, and individual strains, identified by their genotypes, are gained and lost from a host's GI tract over time. Therefore, *E. coli* as a species is not simply a GI tract specialist, but rather a hardy generalist that is only sometimes associated with the complex GI tract microbiota of humans and animals. In this chapter, we present an overview of the habitats where this organism can be isolated and

***Elizabeth W. Alm*** • Department of Biology, Central Michigan University, 157 Brooks Hall, Mount Pleasant, MI 48859. ***Seth T. Walk*** • Department of Internal Medicine, University of Michigan Health System, 4618 Medical Science Building II, Ann Arbor, MI 48109. ***David M. Gordon*** • Research School of Biology, Australian National University, Building 116, Gould Wing, Canberra, ACT 0200, Australia.

review some of the data that characterize the boundaries of its broad ecological niche.

## A DEFINITION OF *E. COLI*

*E. coli* is the type species of the genus *Escherichia*, which in turn is the type genus of the family *Enterobacteriaceae* (25). At least six different species have been classified into the genus *Escherichia* (*E. albertii, E. blattae, E. coli, E. fergusonii, E. hermannii*, and *E. vulneris*), based primarily on phenotypic data and the use of DNA-DNA hybridization experiments (8, 9, 25, 26). It is necessary to begin this chapter with a discussion of "what is *E. coli*?" because, conceptually speaking, *E. coli* ecology is characterized by the habitats where this organism can be sampled, and so distinguishing the species from close relatives is critical. With this in mind, we will first define the breadth of the *E. coli* species based on genetic polymorphisms and phenotypic variation.

Available nucleotide sequence data for representative isolates of the six *Escherichia* species suggest that the classification of *E. blattae, E. hermannii*, or *E. vulneris* ought to be revised. For example, phylogenetically informative polymorphisms (genetic variation) in the 16S rRNA-encoding gene suggest that *E. coli* is more closely related to *Salmonella* spp. than *E. coli* is to either *E. blattae* or *E. hermannii* (16, 63, 65), and these data have recently been used to suggest that *E. blattae* be switched to the genus *Shimwellia* (65). Also, sequences of a number of housekeeping loci (multilocus sequence typing, or MLST) support the misclassification of *E. hermannii* (Table 1). Similar data for *E. vulneris* are not entirely congruent. Data from the 16S rRNA-encoding gene support the classification of *E. vulneris* in the genus *Escherichia* (16, 65), but data at 16S-23S gene spacer loci and a number of MLST loci do not support this classification (Table 1). In general, *E. coli* appears to be more closely related to *Salmonella* than to *E. vulneris*. Unlike *E. blattae, E. hermannii*, and *E. vulneris*, the remaining three named species (*E. coli, E. albertii*, and *E. fergusonii*) consistently cluster monophyletically when data from the 16S rRNA-encoding gene and housekeeping loci are analyzed (46, 84). In summary, the genus *Escherichia* is not as genetically diverse as once thought and contains only three named species: *E. albertii, E. fergusonii*, and *E. coli*.

Other phylogenetic lineages of *Escherichia*, referred to as cryptic clades, have recently been identified and may represent novel *Escherichia* species (84).

**Table 1.** Phylogenetically incongruent housekeeping loci of *E. vulneris* and *E. hermannii*

| Gene designation | Protein | Reference |
|---|---|---|
| *tuf* | Elongation factor Tu | 62 |
| *atpD* | F-ATPase $\beta$-subunit | 62 |
| *gap* | Glyceraldehyde-3-phosphate dehydrogenase | 53 |
| *ompA* | Outer membrane protein 3A | 53 |
| *dnaJ* | Chaperone, heat shock protein 40 | 63 |

A total of five cryptic clades (named *Escherichia* clades I to V) were discovered by MLST, and although all are phylogenetically tractable, there is evidence for gene flow (recombination) among them and between cryptic clades and *E. coli, E. fergusonii*, and *E. albertii* (hereafter we will refer to these three as simply the named *Escherichia* species). Interestingly, the number of recombination events identified to date does not correlate well with the phylogenetic distance between isolates. At least eight examples were found where cryptic clades and named *Escherichia* species share an allele of a housekeeping locus, suggesting that these core genome genes have undergone recent recombination. In addition, the presence of *E. coli* pan-genome loci was not well correlated with genetic distance, suggesting that isolates of the cryptic clades and named *Escherichia* species are equally likely to gain similar genetic material. The frequency of gene flow among these lineages and the phylogenetic scale at which it is occurring suggest that boundaries to gene flow among *Escherichia* species are minimal and also that clades may not occupy entirely different ecological niches.

That being said, the cryptic clades were originally discovered by MLST of bacterial isolates that fit the biochemical profile of *E. coli*. Therefore, the presently available data may be heavily biased toward shared ecological characteristics and may not accurately account for the selective pressures operating at the population, species, or genus level. Evidence for this possibility comes in the form of sampling frequencies from different habitats and growth media. For example, six isolates representing three *Escherichia* clades (clades III, IV, and V) were found in a single environmental *E. coli* collection based on their ability to grow on mTEC agar (lactose fermentation in the presence of sodium lauryl sulfate and sodium deoxycholate). These isolates were identified in a collection of just 205 randomly selected isolates from freshwater beaches along Lake Huron, Michigan. This frequency of isolation (3%) is in stark contrast to the frequency of cryptic clades found by Wirth and colleagues, where only two isolates representing one clade (clade V) were identified in a global *E. coli* MLST analysis (90). An appropriate denominator for the frequency found by Wirth and colleagues is difficult to estimate, but when one considers that the authors hand-selected the most diverse genotypes from numerous collections (the *E. coli* Reference [ECOR] collection [61a], 462 strains from healthy and diseased hosts, 15 isolates representing the genetic diversity of *E. coli* from wild animals in Australia, and at least 259 pathogenic isolates), a conservative estimate would be 2 in 3,336 or 0.06% (the ECOR collection alone represents the genetic diversity of a collection of >2,600 isolates). This observation suggests that cryptic clades may be abundant in habitats like the environment, but that previous attempts to detect this diversity did not utilize the appropriate growth conditions/media. It stands to reason that additional MLST analyses of isolates collected from a diversity of selective growth media and from undersampled habitats like the environment may result in the discovery of even more *Escherichia* diversity.

If constrained by the traditional criterion that large gaps in phenotypic characters must exist between species (see chapter 5, this volume), then *E. coli* and the five cryptic clades represent the same species, because they are not easily differentiated with commonly used biochemical markers (84). Perhaps other, yet-to-be-assayed markers will differentiate these bacteria in the future, but the possibility currently remains that *E. coli* and the cryptic clades have no measureable phenotypic "uniqueness." Therefore, it is important to recognize that all collections of isolates that were assembled based solely on a phenotypic definition may include cryptic clade representatives. It stands to reason that the current perception of the "*E. coli* niche" may be too broad and that a more descriptive understanding would be obtained if the species was phylogenetically defined (or perhaps "ecotype" defined). The data to evaluate this hypothesis are not currently available, and so we will discuss the niche of *E. coli* in this chapter with respect to the broader view that may include the niches of the cryptic clades.

## GENETIC STRUCTURE OF *E. COLI*

Early multilocus enzyme electrophoresis (MLEE) studies demonstrated the existence of "subspecific" structure in *E. coli* (88). More extensive MLEE analysis, together with other techniques of genetic analysis, confirmed the existence of this subspecific structure (19, 37, 43, 55). Traditionally, there are four recognized subspecies or phylogroups of *E. coli* that have been designated A, B1, B2, and D. However, as the diversity of strains that have been characterized increases, the status and monophyly of these phylogroups are increasingly being called into question (37, 77). Strains belonging to phylogroups A and B1 are clearly sister groups, but whether they should be considered as two distinct phylogroups is unclear. Group B2 strains are clearly monophyletic, while strains traditionally considered as group D strains are not monophyletic and represent at least two distinct clades. There is an additional phylogroup known as group E, of which the strain O157:H7 is the most famous member. Gordon et al. (37) and Walk et al. (84) report on what appears to be a sixth, as yet undescribed, phylogroup. The diversity and phylogroup structure of *E. coli* are illustrated in Color Plate 1.

Medical microbiologists have long appreciated the nature of clonal groups, recognizing that all strains denoted as O157:H7 are not identical but that they share a common set of characteristics and are related by descent. However, until recently it has been difficult to recognize clonal groups among commensal isolates of *E. coli*. The MLST method has become a well-established means of characterizing a variety of bacterial species (57). A number of housekeeping genes (members of the core genome) are selected, usually seven, and the nucleotide sequence of a 300- to 700-bp region of each gene is determined. For each gene, unique sequences are assigned a different allele number and each unique combination of alleles at the seven loci corresponds to a sequence type (ST).

The ST diversity in *E. coli* is extensive. For example, in one of the MLST databases (http://mlst.ucc.ie/) there are at present (July 2009) data for 2,533 isolates, and these represent 1,251 STs. Most (77%) of the STs are represented by a single isolate. However, some STs are very common. For example, isolates of ST10 or ST95 each account for about 9% of the total isolates in the database.

Using allele profile data, it is possible to identify strains that are related by descent. The relationships are determined by discovering which STs differ from other STs at a single locus. The software eBURST implements the method most often used to discover these relationships (28). The groups of STs found to be related by descent are usually denoted as clonal complexes (CCs), and a founder ST can be inferred as the ST with the greatest number of single-locus variants associated with it. The largest CC in *E. coli* (based on the MLST scheme http://mlst.ucc.ie/) is CC10, all of which are members of phylogroup A (Fig. 1A). ST10 is the founder of CC10 and has directly given rise to at least 42 variants that differ from it by a single locus. Some of these single-locus variants have given rise to their own suite of single-locus variants. Isolates belonging to CC95, of which ST95 is the founder, have also given rise to a number of single-locus variants, but not as many as ST10, despite being as frequently represented in the database (Fig. 1B). CC95 isolates are members of phylogroup B2.

Members of a CC appear to share similar ecological and life-history traits. Most isolates of CC10 have been recovered from the feces of asymptomatic hosts, although some have been implicated as the cause of intestinal and extraintestinal infection. Members of CC95 can be isolated from feces, but they are frequently found to be the cause of neonatal meningitis, urinary tract infection, and sepsis. Members of the ST29 complex belong to phylogroup B1 and many, if not all, are capable of causing diarrhea.

Jaureguy et al. (48) have suggested that CCs may be more relevant units of study for the clinically significant bacteria than the higher category of phylogroups. An increasing number of studies focus on particular STs and CCs, for example, CC95 and CC69. However, it is also important to remember that many STs cannot be assigned to a CC. For example, in the database (http://mlst.ucc.ie/) about 30% of STs cannot be assigned to a CC.

## WITHIN-HOST DIVERSITY

How many *E. coli* genotypes typically inhabit a host at a single point in time? Studies attempting to answer this question usually isolate *E. coli* from feces. Classic approaches to isolation involve the dilution streaking of fecal material from a sample swab onto media such as MacConkey agar. Less commonly, the fecal material will be subsampled and the feces suspended in a saline solution, serially diluted, and plated onto an appropriate medium. Some number of colonies with the appropriate characteristics is then chosen at random and presumptive confirmation of their identity as *E. coli* is made. For example, isolates that are lactose positive and citrate negative and produce indole are considered to be *E. coli*.

Colonies presumptively identified as *E. coli* are further characterized using a variety of methods. These include phenotypic methods such as antibiotic resistance profiling, biotyping, serotyping, or MLEE. DNA-based fingerprinting

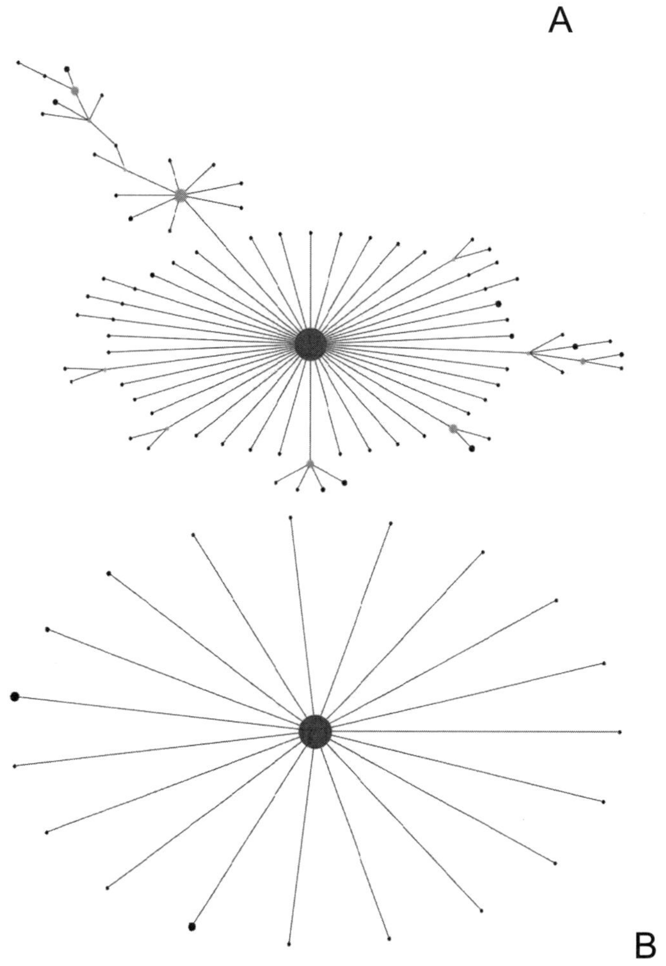


**Figure 1.** Graphical representation of CCs of *E. coli* as determined by eBURST. Central circle denotes the ancestor of the CC: ST10 (A) and ST95 (B). Each dot denotes an ST, and the size of the dot indicates the number of representatives of the ST in the database (http://mlst.ucc.ie).

approaches include random amplified polymorphic DNA, pulsed-field gel electrophoresis, and repetitive extragenic palindromic PCR methods. The number of colonies selected may range from 5 to over 80. The number of colonies selected for characterization is an important determinant of the number of distinct genotypes that can potentially be detected. For example, there is a 90% chance that at least one example of a genotype that occurs at a frequency of 40% in a sample will be detected when 5 colonies are selected, while if 45 colonies are selected, there is a 90% chance of detecting a genotype that represents 5% of the *E. coli* cells present.

These methods typically detect two or three genotypes per sample, but the discovery of nine or more different genotypes is not uncommon (Fig. 2). Usually one genotype is much more abundant than all other genotypes, although the relative abundance of the more common genotypes can be similar (Fig. 3). Most studies that have examined the extent of within-host diversity are only capable of detecting genotypes that represent about 10% of the *E. coli* cells in a sample. Therefore, *E. coli* genotypes may exist in a host at much lower frequencies. However, until the application of high-throughput sequencing technologies to *E. coli* diversity studies, selective plating is the only method of identifying these low-frequency genotypes. Selective plating may involve the use of antibiotics, carbon sources rarely used by *E. coli* (e.g., adonitol), high salt concentration, or extreme pH conditions. Selective plating reveals that there can be genotypes present in a sample that occur at frequencies of <0.01% (36).

Thus, a human host typically harbors two or three genotypes, and together these genotypes represent the majority (>90%) of the *E. coli* cells to be found in a sample. The available data also suggest that there may be quite a large number of strains present at low frequencies. First, nine or more genotypes may be detected in a sample at a single point in time. Second, selective plating reveals the presence of genotypes that exist at very low frequencies (<0.01%). Finally, the number of samples that contain rare genotypes (i.e., <5%) also indicates that there may be a rather large pool of genotypes that are present at frequencies that make them unlikely to be detected most of the time.

Some results suggest that the diversity (number of genotypes) varies between humans and animals and with geographic locality (77). Studies concern-

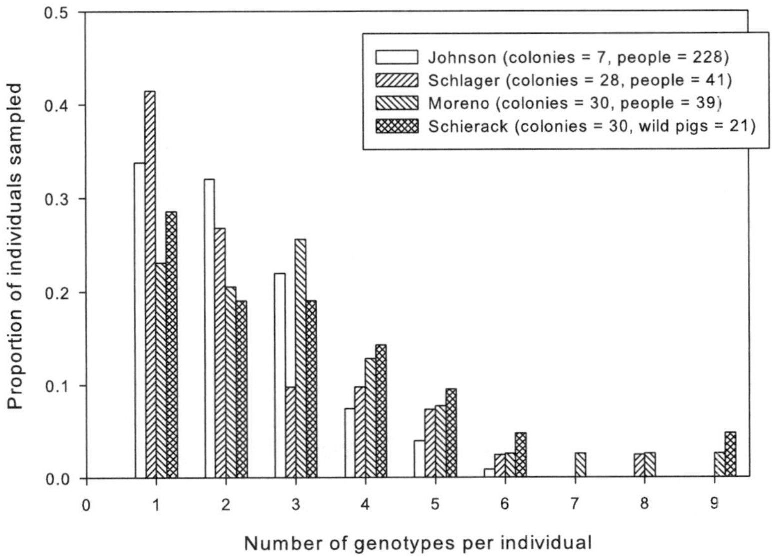


**Figure 2.** The frequency distribution of the number of genotypes recovered per individual host. Data are for isolates from humans (50, 60, 71) and wild boars (70).

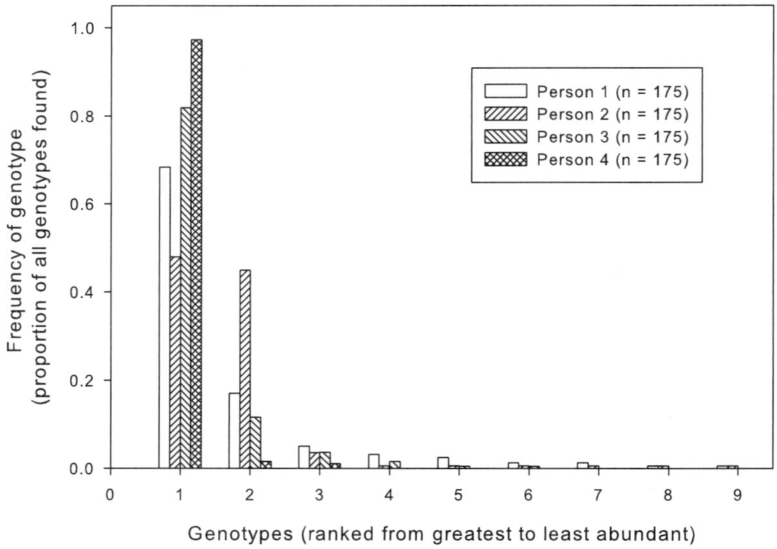


**Figure 3.** The relative abundance of the genotypes recovered from four human hosts (36). The total number of isolated *E. coli* colonies analyzed for each person (*n*) was 175.

ing the clonal diversity of nonhuman vertebrates are uncommon. However, what studies have been done suggest that the same basic patterns observed in humans occur in most species (66).

The biotic and abiotic conditions found in the GI tract vary greatly from mouth to anus. Although there have been relatively few studies, the available data indicate that different *E. coli* genotypes may be isolated from the upper and lower GI tract (21, 70). In a study of *E. coli* isolated from wild boar (*Sus scrofa*), it was found that 51% of the genotypes detected in the jejunum were also observed in the ileum, while 36% could also be detected in the colon. Similarly, 50% of the genotypes seen in the ileum were also recovered from the colon. All genotypes found in the small intestine should, in theory, also be isolated from the colon and feces. However, given that *E. coli* cell densities in the small intestine are typically 100- to 1,000-fold lower than densities in the colon, cells from the small intestine are far less likely to be detected among the cells residing in the colon.

As well as axial variation, there is good evidence to suggest radial variation in the distribution of genotypes. *E. coli* strains differ markedly in their ability to adhere to intestinal cells, and as a consequence, there will be strains intimately associated with intestinal cells or the mucous layer (68). The extent to which there are strains associated with only the luminal contents of the intestine is unknown. Nor do we completely understand the degree to which cells of normally adherent *E. coli* are represented in fecal material. It is well known that in cattle harboring O157:H7, the pathogen is more likely to be detected

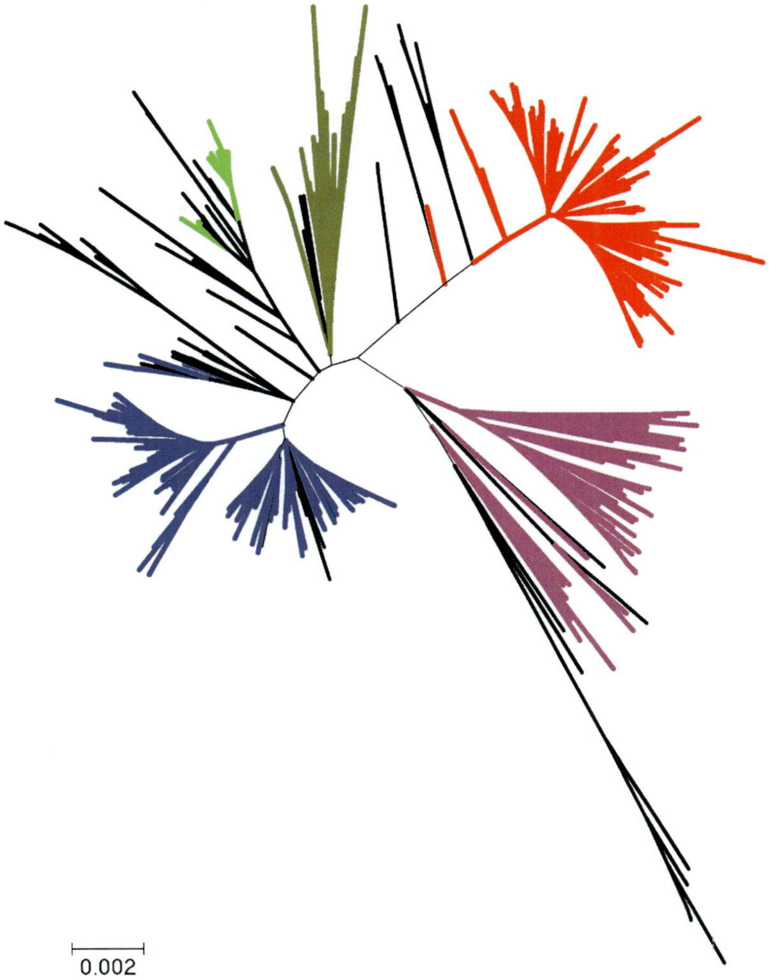


**Color Plate 1 (chapter 6).** Genetic diversity of *E. coli*. Dendrogram based on nucleotide sequence data for seven genes (37). Phylogroup A strains are denoted in green, B1 strains in blue, E strains in olive, B2 strains in red, and D strains in pink. Strains not assigned to a phylogroup are indicated by a black line.

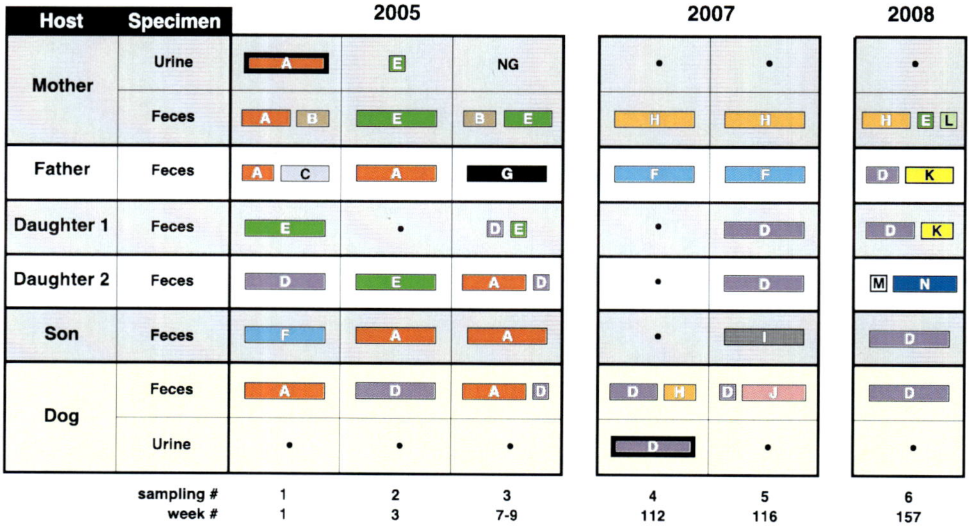


**Color Plate 2 (chapter 7).** Distribution of *E. coli* clones by host over time among five human household members and their pet dog. Each color represents a different unique clone (14 total clones, arbitrarily designated A to N). Width of each rectangle is proportional to the number of colonies accounted for by the particular clone among the total number of colonies analyzed from the particular sample (usually five colonies per sample). Sequential sampling # and week #, counting from the mother's initial UTI episode, is shown below the chart. Rectangles with heavy black borders indicate symptomatic UTI episodes (mother, sampling #1; dog, sampling #4). Gray and white background = humans; buff background = dog. Adapted with permission from reference 17.

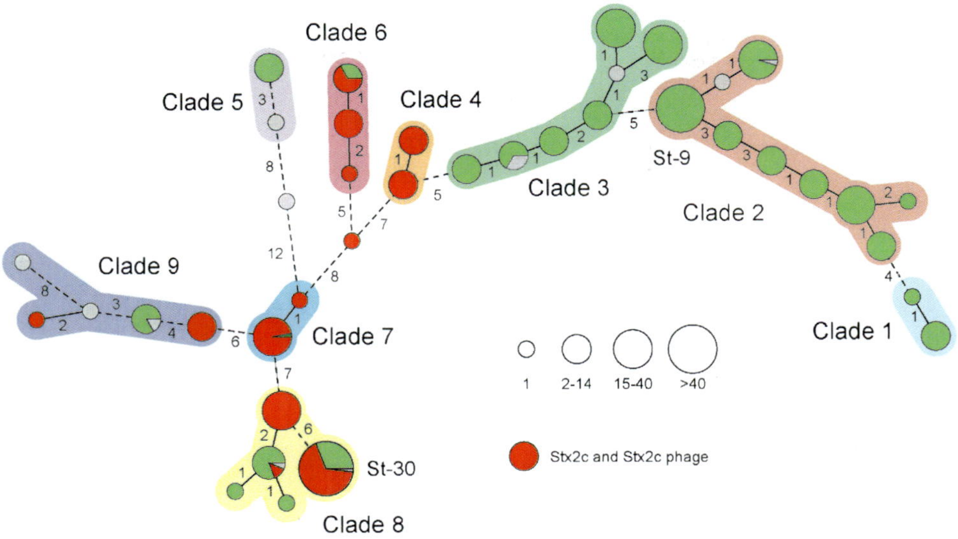


**Color Plate 3 (chapter 17).** Minimum spanning tree of 39 genotypes of *E. coli* O157:H7 based on SNPs detected at 96 loci. The circle sizes indicate the frequencies estimated for 538 strains from clinical cases. The pie charts show the percentage of O157 strains of each SNP genotype positive for the common ($stx_2$) Shiga toxin gene (green) or for the variant ($stx_{2c}$) Shiga toxin and its phage genes (red). Gray circles mark strains with no detectable *stx* genes or missing data ($n$ = 14). The numbers between each SNP genotype represent the number of SNP differences; those with more than four differences are indicated by a dashed line. Clade 9 represents strains with ancestral traits (SOR positive, GUD positive); all other clades have strains with derived traits (SOR negative, GUD negative). Illustration created by Thomas S. Whittam, 2007.

when swabs of the rectal-anal junction are examined as compared to an examination of fecal material from the same animal (18).

Few studies have investigated the extent to which the *E. coli* clonal composition of a host varies with time. The available evidence indicates that particular strains can be detected at regular intervals in a host for 1 to 2 years, and in one case the same strain was detected in a host over 6 years (17). Other strains are detected on multiple occasions but at irregular intervals, while some are detected once and never observed again (15). As well as having no good understanding of the clonal diversity that is typically found in a single host individual, we have very little idea of the rate at which individual strains are gained and lost from a host, and hence the rate at which the *E. coli* community of a host turns over. The available evidence also suggests that even clones that reside in a host for extended periods are eventually replaced, but we have little idea of the factors that lead to one clone being replaced by another. Are these replacement events a consequence of one strain outcompeting another in a constant environment or do changes in the gut environment favor one strain over another? Indeed, for those strains that are detected once and never again, why does a strain succeed in achieving densities sufficient to be detected, only to apparently become extinct?

As already stated, there are scant data with which to determine the factors that govern the clonal composition and stability in a host. It seems that the *E. coli* themselves can influence clonal community composition. It appears that in hosts where a phylogroup B2 or D strain is present at high frequencies, fewer strains will be detected compared to hosts where the most abundant strains are phylogroup A or B1 strains (60). It also appears that at least some phylogroup B2 strains have the ability to persist in a host for longer periods than strains of the other phylogroups (61).

## AMONG-HOST DIVERSITY

At the host population level, the degree of strain diversity observed in *E. coli* is high. Indeed, at first approximation, it is not unreasonable to state that every host individual harbors a different strain of *E. coli*. There is little understanding of the factors that affect *E. coli* population diversity at the population level. Strain sharing has been shown to be more common among members of the same household or among sexual partners (50). Particular STs are also quite prevalent among humans (35). For example, for humans living in Canberra, Australia, strains belonging to ST69 were observed in 30% of the 57 people found to harbor a phylogroup D strain. Similarly, ST95 is a phylogroup B2 strain, and strains belonging to this ST were found in 16% of 120 humans examined. At the genotype level, spatial structure does not appear to explain a significant amount of the observed variation among *E. coli* isolated from hosts living in geographically distinct populations (34, 39, 88). However, when individuals are sampled for *E. coli* from a host population over time, temporal effects can account for a significant amount of the observed variation (87).

Unfortunately, no study has examined the same population over time, and consequently it is not possible to determine if the temporal effects are a consequence of turnover in the host population or systematic changes in the relative abundance of particular genotypes.

The relative abundance of strains belonging to the four phylogroups of *E. coli* varies substantially among humans living in different cities (77) (Fig. 4). The reasons for these outcomes are unknown but may relate to climatic and socioeconomic factors, and the potential for sampling bias cannot be excluded. The relative abundance of strains of the four phylogroups also differs with respect to their source of isolation. For example, strains belonging to phylogroups B2 and D are less frequently isolated from the environment (33, 83) or from fish, frogs, and reptiles than are strains from phylogroups A or B1 (38). In mammals, B2 strains are more frequently recovered from hosts possessing hindgut modifications for microbial fermentation than are strains of the other phylogroups (38).

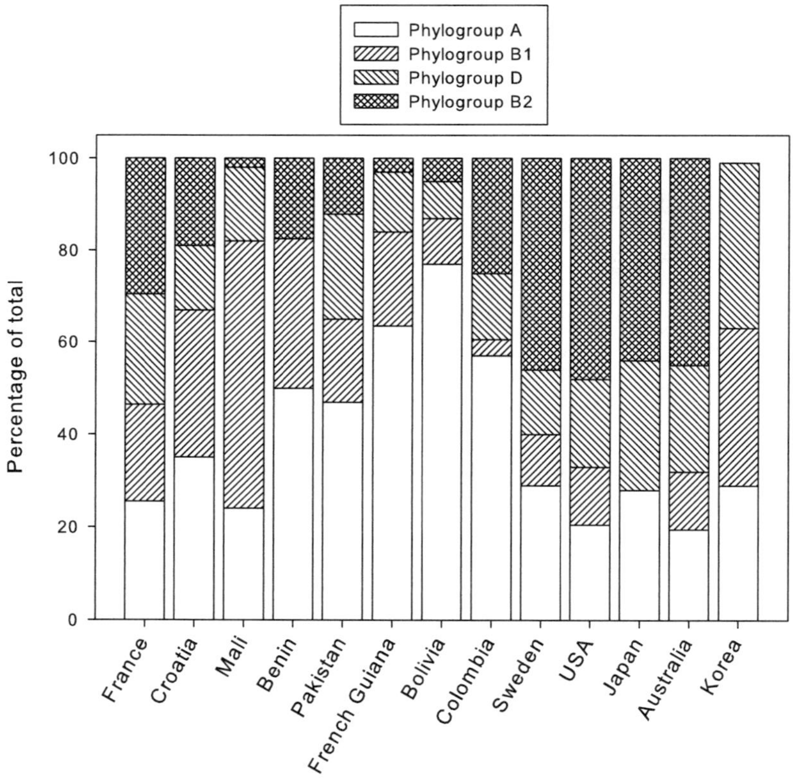


**Figure 4.** The relative frequency of strains of the four main *E. coli* phylogroups recovered from people living in different cities. Data are from references 77 and 80.

## *E. COLI* IN MAMMALS AND OTHER ANIMALS

Gordon and Cowling (38) reported on the results of a survey of more than 2,300 nondomesticated vertebrates living in Australia. These authors defined the prevalence of *E. coli* as the fraction of hosts in which *E. coli* could be detected in their feces. The method of isolation was such that *E. coli* would be detected provided it represented, on average, at least 1% of the bacterial cell population capable of growth on a MacConkey agar plate.

*E. coli* was found in 56% of the mammalian hosts examined. Mammal species live in different climates and differ in their body mass ($\log_{10}$) and diet. Each of these factors was found to be a statistically significant predictor of the presence of *E. coli* in a host individual. In general, *E. coli* is less likely to be isolated from carnivores than omnivores or herbivores, and the likelihood of isolation from a mammal increases with the body mass ($\log_{10}$) of the host. *E. coli* is less likely to be isolated from hosts living in tropical or desert climates compared to hosts living in grassland or temperate climates. This pattern of variation was consistent across host orders and does not appear to be a host effect, as the prevalence of *E. coli* in species of the same genus differed depending on climatic zone. Maximum summer temperatures in the desert regions of central Australia are typically in excess of 35°C, the relative humidity is less than 25%, and median rainfall is less than 20 mm a month. It seems likely that these conditions will adversely affect the survival of *E. coli* in the external environment and may, in turn, limit its transmission among host individuals. However, why *E. coli* is less prevalent in mammals living in the tropics is unknown.

*E. coli* was detected in 23% of the birds examined. Many of the birds were collected from localities where they lived in close association with humans, while others were collected from localities away from significant levels of human habitation. Diet varies substantially among host species, and although not as extensive as in mammals, among-species variation in body mass is considerable. Human association, diet, and $\log_{10}$ body mass were found to be significant predictors of the prevalence of *E. coli* in birds. Birds living in close association with humans were about two times more likely to have *E. coli* than birds living away from human habitation. The probability of a bird harboring *E. coli* increased with its body mass. The prevalence of *E. coli* was lowest in exclusively seed- or fruit-eating species and higher in those species that include nectar, insects, or vertebrates in their diet.

In general, *E. coli* was unlikely to be detected in ectothermic vertebrates. *E. coli* was isolated from 10% of the fish examined and from 12% of the frogs. In reptiles, *E. coli* was isolated from 33% of the crocodiles examined, 4% of the turtles, 2% of the snakes, and 10% of the lizards. As was the case for birds, association with humans also appeared to explain some of the variation in the prevalence of *E. coli* among frog and lizard species. *E. coli* was unusually common in individuals of two species of lizards living in inner Sydney, New South Wales, and two species of tree frog living in the suburbs of Cairns, Queensland.

It is unlikely that the extent to which a host is exposed to *E. coli* is the sole factor responsible for the variation in prevalence observed among different species. The variation may also be a consequence of host characteristics that affect the ability of *E. coli* to establish a population in the GI tract. Among Australian insectivorous bats and carnivorous marsupials (*Dasyuridae*), the digestive tract is dominated by the small intestine, the colon is short and often poorly differentiated, and a cecum is usually lacking or is poorly developed (45). Short intestines, coupled with nutrient-rich food, result in rapid gut transit times. For example, in the *Dasyuridae* gut, transit times increase with body size and vary from about 1 hour for the 18-g *Sminthopsis crassicaudata* to 13 hours for the 1,000-g *Dasyurus* (45).

Microbial fermentation enables mammals consuming plant tissue to increase the nutritional value of the material they ingest, but it requires the presence of a fermentation chamber. In kangaroos and wallabies (*Macropodidae*), as well as the rat kangaroos (*Potoroidae*), the foregut provides the primary site of microbial fermentation, although these species also have a cecum (45). The hindgut is the site of microbial fermentation in the majority of Australian herbivores (45). As in animals with longer intestines, gut transit times are significantly longer in herbivores compared to carnivores and, as in carnivores, increase with body mass. For example, food transit times are about 35 hours in the 1-kg common ringtail possum (*Pseudocheirus peregrinus*) and about 100 hours in the 10-kg koala (*Phascolarctos cinereus*).

Several mathematical models have been developed with the purpose of exploring the dynamics of bacterial populations in the GI tract. These models have investigated the process of bacterial establishment using chemostat models (6) and plug flow reactor models (75). The chemostat models describe a physical system most closely resembling that of a cecum, while the plug flow reactor models describe a typical carnivore gut. These models have shown that establishment is less likely when transit times are short.

The evidence indicates that the presence of *E. coli* in a host depends on its GI morphology and dynamics as well as the host's diet. However, disentangling these relationships will be difficult as these factors are usually confounded. For example, the smallest Australian carnivores are exclusively insectivorous, while the diet of the larger dasyurids is more catholic. Addressing the impact of gut morphology and dynamics will require carefully designed in vivo experiments.

## *E. COLI* IN THE ENVIRONMENT

The total burden of waterborne disease on a global scale is quite high. The most common diseases associated with water are fecal-oral-transmitted enteric infections (diarrheal illness), and the World Health Organization estimates that 1.8 million people die from these illnesses each year. Eighty-eight percent of diarrheal illness is attributed to unsafe water supply and inadequate sanitation and hygiene. Public health agencies charged with ensuring a safe water supply depend on a means to determine whether or not water has been contaminated with feces that may contain human enteric pathogens. Worldwide, the most commonly used indication of fecal contamination of water is the presence of *E. coli* (79, 91).

In the anaerobic environment of the lower intestine, facultative *E. coli* is in low abundance in comparison to anaerobes (23), but *E. coli* is abundant in fecal matter excreted from the host, at densities of approximately $10^7$ cells/gram (42). Since *E. coli* is a characteristic component of feces, Schardinger proposed in 1892 that its presence in water could indicate the presence of fecal contamination and therefore the potential presence of enteric pathogens (54). In 1966, Bonde proposed that an ideal fecal indicator organism would have specific characteristics related to its relationship to enteric pathogens and to its fate in the environment, including a clear connection to enteric pathogens and an inability to proliferate to any significant degree in the environment (7). The belief that *E. coli* fulfilled these indicator criteria led to its development as an indicator organism for water quality monitoring (22).

Although the indicator concept and criteria have undergone a number of modifications over the years, *E. coli* remains the preferred indicator of fecal contamination for water quality monitoring (78, 91). When there is a known source of fecal contamination, such as sewage effluent, *E. coli* abundance in water does predict GI illness (12, 13). A growing body of literature suggests, however, that *E. coli* may be present in the environment without a fecal source, and that in many extraintestinal environments *E. coli* populations are self-sustaining and are a part of the natural microbiota. Therefore, the assumption that the presence of *E. coli* in the environment represents a risk of enteric illness is not always true.

Michael Savageau proposed that the *E. coli* species is maintained in two habitats: the lower intestine of warm-blooded animals, which he calls the primary habitat; and a secondary habitat of water, sediment, and soil (69). As *E. coli* organisms cycle between these two habitats, selective pressures shape the genetic and phenotypic structure of the population. However, it is not clear whether primary, secondary, or a combination of these two habitats is the greater influence on the species. It is tempting to hypothesize that different lineages experience selective stimuli from different habitats, but this has yet to be sufficiently addressed.

## *E. COLI* IN THE "SECONDARY" HABITAT

The primary habitat of the animal intestine and the secondary, extraintestinal habitat may be strikingly different. For example, the human colon is a highly regulated environment: temperatures range between 34.4 and 37.8°C (76), pH varies little from 6.8 (59), redox potentials range from –260 to –200 mV (5); and the colon is constantly moist and dark. In addition to abiotic stimuli, a plethora of bacterial species comprising the GI tract microbiota have coevolved with their hosts, forming symbiotic relationships, like commensalism (44). The density of microorganisms in the colon is among the greatest known and ranges from $10^8$ cells/gram of cecal contents to $10^{11}$ to $10^{12}$ cells/gram of feces. Species composition along the length of the large intestine is dynamic, as facultative anaerobes decrease and obligate anaerobes increase. As many as 15,000 microbial species are estimated to be present in the human intestine, yet they represent only nine bacterial and archaeal phyla (summarized by Dethlefsen et al. [20]). Therefore, *E. coli* competes with a highly diverse assemblage of microorganisms inside the host, and outside the host may experience a highly dynamic environment with wide variations in temperature, pH, oxygen potential, moisture, and exposure to sunlight. In addition, it is likely that the microbial communities of extraintestinal environments are highly different from the intestinal community, and potentially quite different from one another.

Due to the considerable difference in basic parameters in the primary versus the secondary habitat, it has long been presumed that *E. coli* merely persists, and actually exhibits a net negative growth rate, in the secondary habitat (69). In vitro and microcosm studies have supported die-off of *E. coli* under secondary habitat conditions; for example, *E. coli* has been reported to persist for about 1 day in water (27), 0.5 to 2 days in sediment (81), and 1 to 5 days in soil (82). Although there has been general consensus that *E. coli* exhibits die-off in the secondary environment (69, 89), the results of other studies suggest that *E. coli* is capable of prolonged survival or even growth in extraintestinal environments. Variables such as moderately acidic pH, higher oxygen tension, the presence of organic matter (1), cooler temperatures (58), association with sediments (52), protection from sunlight (30, 31), and the absence of predators (4) have been shown to extend the survival of *E. coli* in water.

The underlying assumption in the use of *E. coli* as an indicator of fecal contamination is that *E. coli* detected in the environment originated relatively recently from fecal sources. Beginning in the 1980s, studies from the tropics demonstrated the presence of *E. coli* in systems that were likely not contaminated by human or animal feces. *E. coli* densities above the maximum contaminant levels for potable water were detected in the Mameyes River watershed, which originated in a protected cloud rainforest in Puerto Rico. Diffusion chamber studies indicated that the *E. coli* organisms were still metabolically active and respiring for as long as 5 days (14, 49, 56). A study of water collected in the axilae of bromeliads growing in the same rainforest also revealed abundant *E. coli*, in the absence of any human or animal source (67). In Hawaii, studies suggested that fecal coliforms were abundant in remote mountain streams and likely originated from the surrounding soils and rhizosphere during rain events (32, 41). In light of these studies, two alternative origins for *E. coli* were suggested: *E. coli* may have originated from fecal sources at some time in the past, but over time adapted and became part of the natural microbiota of these tropical systems; or *E. coli* was always part of the natural microbiota in the tropics.

As tropical environments are characterized by warmer temperatures and stable concentrations of nutrients, which would be more similar to an intestinal environment, consensus evolved to include the concept that perhaps *E. coli* could persist year-round in tropical environments. Natural populations of *E. coli* were then reported from the subtropics. In an investigation of high *E. coli* levels in a river near Fort Lauderdale, Florida, Solo-Gabriele and coworkers

(73) concluded that populations of *E. coli* were multiplying within the soil on the banks of the river and were then washed into the river during rain events. Also, a study of coliform blooms (defined as coliforms in excess of $10^4$ CFU/100 ml) that occurred between 2002 and 2004 on Lake Burragorang and Lake Burley Griffin in New South Wales, Australia, concluded that the *E. coli* strains responsible for the blooms were unlikely to have originated from fecal contamination; rather, they represented strains of *E. coli* that had evolved a free-living lifestyle (64).

Results from the tropics and subtropics were rapidly followed by reports of natural populations of *E. coli* in temperate regions. The initial reports were of high densities of *E. coli* in shoreline sand of the Laurentian Great Lakes (3, 29, 40, 85, 86). The high densities of *E. coli* in sand led many to speculate that, as in the tropics and subtropics, *E. coli* is an environmental organism even in some of the coldest temperate regions. Ishii and colleagues (47) examined populations of *E. coli* in the soils of Minnesota in the United States. They were able to isolate *E. coli* strains with similarity values of ≥92% repeatedly from the same site over a 1-year period. These isolates made up a significant fraction of all *E. coli* strains isolated from the soil, and the authors found that these "naturalized" *E. coli* strains from different soils were distinct from each other and from water isolates, and could be readily distinguished from *E. coli* from animal feces. In addition to soils, unique populations of naturalized *E. coli* have been found associated with the macroalga *Cladophora* (10, 11) in Lake Michigan and with epilithic periphyton communities in Lake Superior (51). When *E. coli* isolates from Lake Huron beaches were incubated in autoclaved beach sand, either in microcosms in the laboratory at 19.5°C or in diffusion chambers buried along the beach, they were able to increase in number from low initial densities to $10^4$ to $10^5$ CFU/gram of sand. Densities reached maximum by day 6 in the diffusion chambers and persisted until the experiment was concluded on day 48 (2).

The Great Lakes watershed is large (201,460 miles$^2$) and is home to approximately 10 and 31% of the U.S. and Canadian populations, respectively, according to the U.S. Environmental Protection Agency (www.epa.gov/glnpo/factsheet.html). There are a large number of *E. coli* point sources into this large, temperate freshwater ecosystem (urban, industrial, and agricultural runoff), yet some *E. coli* genotypes appear to be overrepresented in this environment. Population genetic characterization of a collection of 190 *E. coli* isolates from six beaches along Lake Huron identified 27 common (sampled more than once) genotypes (83). One of these genotypes was independently sampled seven times over the entire 35 months of the study. Representative isolates of this genotype were collected from five of the six beaches and collected from the water column and from sand cores along the beach. Recovery of a single genotype from such a large ecosystem with a diversity of inputs is highly unlikely by chance and suggests that certain strains of *E. coli* are adapted to life outside of the GI tract.

In addition to the nonrandom distribution of *E. coli* genotypes along Lake Huron, analysis of nucleotide polymorphisms at MLST loci also suggested that certain genotypes may be adapted, or under the influence of selection (83). For example, 30 strains were found to be identical by MLEE of 18 enzyme loci. The majority of these isolates ($n = 23$) as well as other closely related isolates ($n = 21$) resolved by MLST were named the ET 1 clade (the ET designation is used to distinguish electrophoretic types or individual MLEE genotypes). In total, these 44 isolates represented 27 STs that clustered monophyletically with high bootstrap support, indicating that they were distinct from other phylogenetic lineages. When ET 1 clade isolates were analyzed for the abundance and distribution of synonymous (silent) and nonsynonymous polymorphisms, it was discovered that all 44 isolates shared a single allele at the *lysP* locus. The *lysP* gene encodes a lysine-specific permease that is upregulated in growth in acidic medium, anaerobiosis, and high concentrations of lysine (74). It is tempting to hypothesize that the *lysP* allele carried by all isolates of the ET 1 clade is advantageous in the Lake Huron habitat. It is equally plausible that *lysP* is tightly linked to another locus that is under the influence of selection, and that the ET 1 *lysP* allele has hitchhiked to high frequency in the population. More data are required to test these hypotheses, but the lack of genetic diversity at the *lysP* locus among the 44 isolates of the ET 1 clade is remarkable and is consistent with the action of purifying selection on *E. coli* strains that are common outside of the host GI tract.

## SUMMARY

*E. coli* is widespread among warm-blooded animals and humans, and the dynamics of its colonization of the GI tract is dependent on a number of host and microbial determinants. *E. coli* is also a member of microbial communities outside of the host GI tract, like tropical and temperate ecosystems. The biology of *E. coli* and most enteric microbes has been inadequately studied, especially in the context of these ex vivo environments. However, adaptation to stimuli in transit between hosts is an essential life-history trait of these organisms. With this in mind, future work that details the influence of secondary habitats on the population genetic structure of *E. coli* is poised to significantly influence our understanding of microbial ecology and evolution.

## REFERENCES

1. **Allen, L. A., S. M. Pasley, and M. A. Pierce.** 1952. Some factors affecting the viability of faecal bacteria. *J. Gen. Microbiol.* **7:**36–43.
2. **Alm, E. W., J. Burke, and E. Hagan.** 2006. Persistence and potential replication of the fecal indicator bacteria, *Escherichia coli*, in the shoreline sand at Lake Huron. *J. Great Lakes Res.* **32:**401–405.
3. **Alm, E. W., J. Burke, and A. Spain.** 2003. Fecal indicator bacteria are abundant in wet sand at freshwater beaches. *Water Res.* **37:**3978–3982.
4. **Anderson, I. C., M. W. Rhodes, and H. I. Kator.** 1983. Seasonal variation in survival of *Escherichia coli* exposed in situ in membrane diffusion chambers containing filtered and nonfiltered estuarine water. *Appl. Environ. Microbiol.* **45:**1877–1883.
5. **Aw, T. Y.** 2003. Cellular redox: a modulator of intestinal epithelial cell proliferation. *News Physiol. Sci.* **18:**201–204.

6. **Ballyk, M. M., D. A. Jones, and H. L. Smith.** 2001. Microbial competition in reactors with wall attachment. *Microb. Ecol.* **41:**210–221.
7. **Bonde, G. J.** 1966. Bacteriological methods for estimation of water pollution. *Health Lab. Sci.* **3:**124–128.
8. **Brenner, D. J., B. R. Davis, A. G. Steigerwalt, C. F. Riddle, A. C. McWhorter, S. D. Allen, J. J. Farmer III, Y. Saitoh, and G. R. Fanning.** 1982. Atypical biogroups of *Escherichia coli* found in clinical specimens and description of *Escherichia hermannii* sp. nov. *J. Clin. Microbiol.* **15:**703–713.
9. **Brenner, D. J., A. C. McWhorter, J. K. Knutson, and A. G. Steigerwalt.** 1982. *Escherichia vulneris*: a new species of *Enterobacteriaceae* associated with human wounds. *J. Clin. Microbiol.* **15:**1133–1140.
10. **Byappanahalli, M. N., D. A. Shively, M. B. Nevers, M. J. Sadowsky, and R. L. Whitman.** 2003. Growth and survival of *Escherichia coli* and enterococci populations in the macro-alga *Cladophora* (Chlorophyta). *FEMS Microbiol. Ecol.* **46:**203–211.
11. **Byappanahalli, M. N., R. L. Whitman, D. A. Shively, J. Ferguson, S. Ishii, and M. J. Sadowsky.** 2007. Population structure of *Cladophora*-borne *Escherichia coli* in nearshore water of Lake Michigan. *Water Res.* **41:**3649–3654.
12. **Cabelli, V. J., A. P. Dufour, M. A. Levin, L. J. McCabe, and P. W. Haberman.** 1979. Relationship of microbial indicators to health effects at marine bathing beaches. *Am. J. Public Health* **69:**690–696.
13. **Cabelli, V. J., A. P. Dufour, L. J. McCabe, and M. A. Levin.** 1982. Swimming-associated gastroenteritis and water quality. *Am. J. Epidemiol.* **115:**606–616.
14. **Carrillo, M., E. Estrada, and T. C. Hazen.** 1985. Survival and enumeration of the fecal indicators *Bifidobacterium adolescentis* and *Escherichia coli* in a tropical rain forest watershed. *Appl. Environ. Microbiol.* **50:**468–476.
15. **Caugant, D. A., B. R. Levin, and R. K. Selander.** 1981. Genetic diversity and temporal variation in the *E. coli* population of a human host. *Genetics* **98:**467–490.
16. **Cilia, V., B. Lafay, and R. Christen.** 1996. Sequence heterogeneities among 16S ribosomal RNA sequences, and their effect on phylogenetic analyses at the species level. *Mol. Biol. Evol.* **13:**451–461.
17. **Clermont, O., M. Lescat, C. L. O'Brien, D. M. Gordon, O. Tenaillon, and E. Denamur.** 2008. Evidence for a human-specific *Escherichia coli* clone. *Environ. Microbiol.* **10:**1000–1006.
18. **Cobbold, R. N., D. D. Hancock, D. H. Rice, J. Berg, R. Stilborn, C. J. Hovde, and T. E. Besser.** 2007. Rectoanal junction colonization of feedlot cattle by *Escherichia coli* O157:H7 and its association with supershedders and excretion dynamics. *Appl. Environ. Microbiol.* **73:**1563–1568.
19. **Desjardins, P., B. Picard, B. Kaltenbock, J. Elion, and E. Denamur.** 1995. Sex in *Escherichia coli* does not disrupt the clonal structure of the population: evidence from random amplified polymorphic DNA and restriction-fragment-length polymorphism. *J. Mol. Evol.* **41:**440–448.
20. **Dethlefsen, L., P. B. Eckburg, E. M. Bik, and D. A. Relman.** 2006. Assembly of the human intestinal microbiota. *Trends Ecol. Evol.* **21:**517–523.
21. **Dixit, S. M., D. M. Gordon, X. Y. Wu, T. Chapman, K. Kailasapathy, and J. J. Chin.** 2004. Diversity analysis of commensal porcine *Escherichia coli*—associations between genotypes and habitat in the porcine gastrointestinal tract. *Microbiology* **150:**1735–1740.
22. **Dufour, A. P. (ed.).** 1977. Escherichia coli*: the Fecal Coliform*, p. 48–58. *In* A. W. Hoadley and B. J. Dutka (ed.), *Bacterial Indicators/Health Hazards Associated with Water.* American Society for Testing and Materials, Philadelphia, PA.
23. **Eckburg, P. B., E. M. Bik, C. N. Bernstein, E. Purdom, L. Dethlefsen, M. Sargent, S. R. Gill, K. E. Nelson, and D. A. Relman.** 2005. Diversity of the human intestinal microbial flora. *Science* **308:**1635–1638.
24. **Escherich, T.** 1988. The intestinal bacteria of the neonate and breast-fed infant. 1884. *Rev. Infect. Dis.* **10:**1220–1225.

25. **Farmer, J. J., III, B. R. Davis, F. W. Hickman-Brenner, A. McWhorter, G. P. Huntley-Carter, M. A. Asbury, C. Riddle, H. G. Wathen-Grady, C. Elias, G. R. Fanning, A. G. Steigerwalt, C. M. O'Hara, G. K. Morris, P. B. Smith, and D. G. Brenner.** 1985. Biochemical identification of new species and biogroups of *Enterobacteriaceae* isolated from clinical specimens. *J. Clin. Microbiol.* **21:**46–76.
26. **Farmer, J. J., III, G. R. Fanning, B. R. Davis, C. M. O'Hara, C. Riddle, F. W. Hickman-Brenner, M. A. Asbury, V. A. Lowery III, and D. J. Brenner.** 1985. *Escherichia fergusonii* and *Enterobacter taylorae*, two new species of *Enterobacteriaceae* isolated from clinical specimens. *J. Clin. Microbiol.* **21:**77–81.
27. **Faust, M. A., A. E. Aotaky, and M. T. Hargadon.** 1975. Effect of physical parameters on the in situ survival of *Escherichia coli* MC-6 in an estuarine environment. *Appl. Microbiol.* **30:**800–806.
28. **Feil, E. J., B. C. Li, D. M. Aanensen, W. P. Hanage, and B. G. Spratt.** 2004. eBURST: inferring patterns of evolutionary descent among clusters of related bacterial genotypes from multilocus sequence typing data. *J. Bacteriol.* **186:**1518–1530.
29. **Francy, D. S., and A. M. Gifford.** 2002. *Escherichia coli* in the swash zone at four Ohio bathing beaches. U.S. Geological Survey fact sheet FS-134-02. U.S. Geological Survey, Columbus, OH.
30. **Fujioka, R. S., H. H. Hashimoto, E. B. Siwak, and R. H. Young.** 1981. Effect of sunlight on survival of indicator bacteria in seawater. *Appl. Environ. Microbiol.* **41:**690–696.
31. **Fujioka, R. S., and O. T. Narikawa.** 1982. Effect of sunlight on enumeration of indicator bacteria under field conditions. *Appl. Environ. Microbiol.* **44:**395–401.
32. **Fujioka, R. S., K. Tenno, and S. Kansako.** 1988. Naturally occurring fecal coliforms and fecal streptococci in Hawaii's freshwater streams. *Toxicity Assessment* **3:**613–630.
33. **Garcia-Aljaro, C., E. Moreno, A. Andreu, G. Prats, and A. R. Blanch.** 2009. Phylogroups, virulence determinants and antimicrobial resistance in $stx_2$ gene-carrying *Escherichia coli* isolated from aquatic environments. *Res. Microbiol.* **160:**585–591.
34. **Gordon, D. M.** 1997. The genetic structure of *Escherichia coli* populations in feral house mice. *Microbiology* **143**(Pt. 6)**:**2039–2046.
35. **Gordon, D. M.** 2010. Strain typing and the ecological structure of *Escherichia coli*. *J. AOAC Int.* **93:**974–984.
36. **Gordon, D. M., S. Bauer, and J. R. Johnson.** 2002. The genetic structure of *Escherichia coli* populations in primary and secondary habitats. *Microbiology* **148:**1513–1522.
37. **Gordon, D. M., O. Clermont, H. Tolley, and E. Denamur.** 2008. Assigning *Escherichia coli* strains to phylogenetic groups: multi-locus sequence typing versus the PCR triplex method. *Environ. Microbiol.* **10:**2484–2496.
38. **Gordon, D. M., and A. Cowling.** 2003. The distribution and genetic structure of *Escherichia coli* in Australian vertebrates: host and geographic effects. *Microbiology* **149:**3575–3586.
39. **Gordon, D. M., and J. Lee.** 1999. The genetic structure of enteric bacteria from Australian mammals. *Microbiology* **145**(Pt. 10)**:**2673–2682.
40. **Haack, S. K., L. R. Fogarty, and C. Wright.** 2003. *Escherichia coli* and enterococci at beaches in the Grand Traverse Bay, Lake Michigan: sources, characteristics, and environmental pathways. *Environ. Sci. Technol.* **37:**3275–3282.
41. **Hardina, C. M., and R. S. Fujioka.** 1991. Soil: the environmental source of *Escherichia coli* and enterococci in Hawaii's streams. *Environ. Toxicol. Water Qual.* **6:**185–195.
42. **Hartl, D. L., and D. E. Dykhuizen.** 1984. The population genetics of *Escherichia coli*. *Annu. Rev. Genet.* **18:**31–68.
43. **Herzer, P. J., S. Inouye, M. Inouye, and T. S. Whittam.** 1990. Phylogenetic distribution of branched RNA-linked multicopy single-stranded DNA among natural isolates of *Escherichia coli*. *J. Bacteriol.* **172:**6175–6181.
44. **Hooper, L. V., and J. I. Gordon.** 2001. Commensal host-bacterial relationships in the gut. *Science* **292:**1115–1118.

45. **Hume, I. D.** 1999. *Marsupial Nutrition.* Cambridge University Press, Cambridge, United Kingdom.
46. **Hyma, K. E., D. W. Lacher, A. M. Nelson, A. C. Bumbaugh, J. M. Janda, N. A. Strockbine, V. B. Young, and T. S. Whittam.** 2005. Evolutionary genetics of a new pathogenic *Escherichia* species: *Escherichia albertii* and related *Shigella boydii* strains. *J. Bacteriol.* **187:**619–628.
47. **Ishii, S., W. B. Ksoll, R. E. Hicks, and M. J. Sadowsky.** 2006. Presence and growth of naturalized *Escherichia coli* in temperate soils from Lake Superior watersheds. *Appl. Environ. Microbiol.* **72:**612–621.
48. **Jaureguy, F., L. Landraud, V. Passet, L. Diancourt, E. Frapy, G. Guigon, E. Carbonnelle, O. Lortholary, O. Clermont, E. Denamur, B. Picard, X. Nassif, and S. Brisse.** 2008. Phylogenetic and genomic diversity of human bacteremic *Escherichia coli* strains. *BMC Genomics* **9:**560.
49. **Jimenez, L., I. Muniz, G. A. Toranzos, and T. C. Hazen.** 1989. Survival and activity of *Salmonella typhimurium* and *Escherichia coli* in tropical freshwater. *J. Appl. Bacteriol.* **67:**61–69.
50. **Johnson, J. R., K. Owens, A. Gajewski, and C. Clabots.** 2008. *Escherichia coli* colonization patterns among human household members and pets, with attention to acute urinary tract infection. *J. Infect. Dis.* **197:**218–224.
51. **Ksoll, W. B., S. Ishii, M. J. Sadowsky, and R. E. Hicks.** 2007. Presence and sources of fecal coliform bacteria in epilithic periphyton communities of Lake Superior. *Appl. Environ. Microbiol.* **73:**3771–3778.
52. **LaLiberte, P., and D. J. Grimes.** 1982. Survival of *Escherichia coli* in lake bottom sediment. *Appl. Environ. Microbiol.* **43:**623–628.
53. **Lawrence, J. G., H. Ochman, and D. L. Hartl.** 1991. Molecular and evolutionary relationships among enteric bacteria. *J. Gen. Microbiol.* **137:**1911–1921.
54. **Leclerc, H., D. A. Mossel, S. C. Edberg, and C. B. Struijk.** 2001. Advances in the bacteriology of the coliform group: their suitability as markers of microbial water safety. *Annu. Rev. Microbiol.* **55:**201–234.
55. **Lecointre, G., L. Rachdi, P. Darlu, and E. Denamur.** 1998. *Escherichia coli* molecular phylogeny using the incongruence length difference test. *Mol. Biol. Evol.* **15:**1685–1695.
56. **Lopez-Torres, A. J., T. C. Hazen, and G. A. Toranzos.** 1987. Distribution and *in situ* survival and activity of *Klebsiella pneumoniae* and *Escherichia coli* in a tropical rain forest watershed. *Curr. Microbiol.* **15:**213–218.
57. **Maiden, M. C., J. A. Bygraves, E. Feil, G. Morelli, J. E. Russell, R. Urwin, Q. Zhang, J. Zhou, K. Zurth, D. A. Caugant, I. M. Feavers, M. Achtman, and B. G. Spratt.** 1998. Multilocus sequence typing: a portable approach to the identification of clones within populations of pathogenic microorganisms. *Proc. Natl. Acad. Sci. USA* **95:**3140–3145.
58. **McFeters, G. A., and D. G. Stuart.** 1972. Survival of coliform bacteria in natural waters: field and laboratory studies with membrane-filter chambers. *Appl. Microbiol.* **24:**805–811.
59. **McNeil, N. I., K. L. Ling, and J. Wager.** 1987. Mucosal surface pH of the large intestine of the rat and of normal and inflamed large intestine in man. *Gut* **28:**707–713.
60. **Moreno, E., J. R. Johnson, T. Perez, G. Prats, M. A. Kuskowski, and A. Andreu.** 2009. Structure and urovirulence characteristics of the fecal *Escherichia coli* population among healthy women. *Microbes Infect.* **11:**274–280.
61. **Nowrouzian, F. L., I. Adlerberth, and A. E. Wold.** 2006. Enhanced persistence in the colonic microbiota of *Escherichia coli* strains belonging to phylogenetic group B2: role of virulence factors and adherence to colonic cells. *Microbes Infect.* **8:**834–840.
61a. **Ochman, H., and R. K. Selander.** 1984. Standard reference strains of *Escherichia coli* from natural populations. *J. Bacteriol.* **157:**690–693.
62. **Paradis, S., M. Boissinot, N. Paquette, S. D. Belanger, E. A. Martel, D. K. Boudreau, F. J. Picard, M. Ouellette, P. H. Roy, and M. G. Bergeron.** 2005. Phylogeny of the

*Enterobacteriaceae* based on genes encoding elongation factor Tu and F-ATPase β-subunit. *Int. J. Syst. Evol. Microbiol.* **55:**2013–2025.
63. **Pham, H. N., K. Ohkusu, N. Mishima, M. Noda, M. Monir Shah, X. Sun, M. Hayashi, and T. Ezaki.** 2007. Phylogeny and species identification of the family *Enterobacteriaceae* based on *dnaJ* sequences. *Diagn. Microbiol. Infect. Dis.* **58:**153–161.
64. **Power, M. L., J. Littlefield-Wyer, D. M. Gordon, D. A. Veal, and M. B. Slade.** 2005. Phenotypic and genotypic characterization of encapsulated *Escherichia coli* isolated from blooms in two Australian lakes. *Environ. Microbiol.* **7:**631–640.
65. **Priest, F. G., and M. Barker.** 2010. Gram-negative bacteria associated with brewery yeasts: reclassification of *Obesumbacterium proteus* biogroup 2 as *Shimwellia pseudoproteus* gen. nov., sp. nov., and transfer of *Escherichia blattae* to *Shimwellia blattae* comb. nov. *Int. J. Syst. Evol. Microbiol.* **60:**828–833.
66. **Pupo, G. M., R. Lan, and P. R. Reeves.** 2000. Multiple independent origins of *Shigella* clones of *Escherichia coli* and convergent evolution of many of their characteristics. *Proc. Natl. Acad. Sci. USA* **97:**10567–10572.
67. **Rivera, S. C., T. C. Hazen, and G. A. Toranzos.** 1988. Isolation of fecal coliforms from pristine sites in a tropical rain forest. *Appl. Environ. Microbiol.* **54:**513–517.
68. **Rolhion, N., and A. Darfeuille-Michaud.** 2007. Adherent-invasive *Escherichia coli* in inflammatory bowel disease. *Inflamm. Bowel Dis.* **13:**1277–1283.
69. **Savageau, M. A.** 1983. *Escherichia coli* habitats, cell types, and molecular mechanisms of gene control. *Am. Nat.* **122:**732–744.
70. **Schierack, P., A. Romer, J. Jores, H. Kaspar, S. Guenther, M. Filter, J. Eichberg, and L. H. Wieler.** 2009. Isolation and characterization of intestinal *Escherichia coli* clones from wild boars in Germany. *Appl. Environ. Microbiol.* **75:**695–702.
71. **Schlager, T. A., J. O. Hendley, A. L. Bell, and T. S. Whittam.** 2002. Clonal diversity of *Escherichia coli* colonizing stools and urinary tracts of young girls. *Infect. Immun.* **70:**1225–1229.
72. **Shulman, S. T., H. C. Friedmann, and R. H. Sims.** 2007. Theodor Escherich: the first pediatric infectious diseases physician? *Clin. Infect. Dis.* **45:**1025–1029.
73. **Solo-Gabriele, H. M., M. A. Wolfert, T. R. Desmarais, and C. J. Palmer.** 2000. Sources of *Escherichia coli* in a coastal subtropical environment. *Appl. Environ. Microbiol.* **66:**230–237.
74. **Steffes, C., J. Ellis, J. Wu, and B. P. Rosen.** 1992. The *lysP* gene encodes the lysine-specific permease. *J. Bacteriol.* **174:**3242–3249.
75. **Stemmons, E. D., and H. L. Smith.** 2000. Competition in a chemostat with wall attachment. *SIAM J. Appl. Math.* **61:**567–595.
76. **Sund-Levander, M., C. Forsberg, and L. K. Wahren.** 2002. Normal oral, rectal, tympanic and axillary body temperature in adult men and women: a systematic literature review. *Scand. J Caring Sci.* **16:**122–128.
77. **Tenaillon, O., D. Skurnik, B. Picard, and E. Denamur.** 2010. The population genetics of commensal *Escherichia coli. Nat. Rev. Microbiol.* **8:**207–217.
78. **U.S. Environmental Protection Agency.** 1986. Ambient water quality criteria for bacteria—1986. Criteria and Standards Division, Office of Water Regulations and Standards, U.S. Environmental Protection Agency, Washington, DC.
79. **U.S. Environmental Protection Agency.** 1989. Drinking water: national primary water regulations; total coliforms (including fecal coliforms and *E. coli*). Final Rule. *Fed. Reg.* **54:**27544–27568.
80. **Unno, T., D. Han, J. Jang, S. N. Lee, G. Ko, H. Y. Choi, J. H. Kim, M. J. Sadowsky, and H. G. Hur.** 2009. Absence of *Escherichia coli* phylogenetic group B2 strains in humans and domesticated animals from Jeonnam Province, Republic of Korea. *Appl. Environ. Microbiol.* **75:**5659–5666.
81. **Van Donsel, D. J., and E. E. Geldreich.** 1971. Relationships of salmonellae to fecal coliforms in bottom sediments. *Water Res.* **5:**1079–1087.
82. **Van Donsel, D. J., E. E. Geldreich, and N. A. Clarke.** 1967. Seasonal variations in survival of indicator bacteria in soil and their contribution to storm-water pollution. *Appl. Microbiol.* **15:**1362–1370.

83. **Walk, S. T., E. W. Alm, L. M. Calhoun, J. M. Mladonicky, and T. S. Whittam.** 2007. Genetic diversity and population structure of *Escherichia coli* isolated from freshwater beaches. *Environ. Microbiol.* **9:**2274–2288.
84. **Walk, S. T., E. W. Alm, D. M. Gordon, J. L. Ram, G. A. Toranzos, J. M. Tiedje, and T. S. Whittam.** 2009. Cryptic lineages of the genus *Escherichia*. *Appl. Environ. Microbiol.* **75:**6534–6544.
85. **Whitman, R. L., and M. B. Nevers.** 2003. Foreshore sand as a source of *Escherichia coli* in nearshore water of a Lake Michigan beach. *Appl. Environ. Microbiol.* **69:**5555–5562.
86. **Whitman, R. L., D. A. Shively, H. Pawlik, M. B. Nevers, and M. N. Byappanahalli.** 2003. Occurrence of *Escherichia coli* and enterococci in *Cladophora* (Chlorophyta) in nearshore water and beach sand of Lake Michigan. *Appl. Environ. Microbiol.* **69:**4714–4719.
87. **Whittam, T. S.** 1989. Clonal dynamics of *Escherichia coli* in its natural habitat. *Antonie van Leeuwenhoek* **55:**23–32.
88. **Whittam, T. S., H. Ochman, and R. K. Selander.** 1983. Geographic components of linkage disequilibrium in natural populations of *Escherichia coli*. *Mol. Biol. Evol.* **1:**67–83.
89. **Winfield, M. D., and E. A. Groisman.** 2003. Role of nonhost environments in the lifestyles of *Salmonella* and *Escherichia coli*. *Appl. Environ. Microbiol.* **69:**3687–3694.
90. **Wirth, T., D. Falush, R. Lan, F. Colles, P. Mensa, L. H. Wieler, H. Karch, P. R. Reeves, M. C. Maiden, H. Ochman, and M. Achtman.** 2006. Sex and virulence in *Escherichia coli*: an evolutionary perspective. *Mol. Microbiol.* **60:**1136–1151.
91. **World Health Organization.** 2008. *Guidelines for Drinking-Water Quality*, 3rd ed., vol. 1. World Health Organization, Geneva, Switzerland.

*Population Genetics of Bacteria: a Tribute to Thomas S. Whittam*
Edited by S. T. Walk and P. C. H. Feng


*Chapter 7*

# Molecular Epidemiology and Population Genetics of Extraintestinal Pathogenic *Escherichia coli*

*James R. Johnson*

## *ESCHERICHIA COLI* AS AN EXTRAINTESTINAL PATHOGEN

*E. coli* is an important extraintestinal pathogen for humans, companion animals, and production animals, causing considerable morbidity and mortality, increased health care costs, and economic losses (41, 54). *E. coli* can infect diverse extraintestinal organs and tissues, thereby producing a wide range of clinical syndromes.

In humans, the main site of extraintestinal *E. coli* disease is the urinary tract. *E. coli* is the primary agent of urinary tract infections (UTIs), which include cystitis (bladder infection), pyelonephritis (kidney infection), prostatitis, and febrile UTI. *E. coli* also is a leading cause of neonatal meningitis and intra-abdominal infections, and occasionally infects the lungs, skin, muscles, skeletal system, and other body sites (54). From any of these primary sites of infection, or from intravascular catheters or via translocation across the intestinal epithelium from the lumen into the lymphatics, *E. coli* can enter the bloodstream and cause bacteremia, which often results in the sepsis syndrome (54).

In cats and dogs, *E. coli* causes certain extraintestinal infections similar to those observed in humans, including UTI, metritis (uterine infection), pneumonia, and bacteremia (14, 41). In poultry, *E. coli* causes skin infection (cellulitis) plus a range of often fatal invasive syndromes that typically arise from a primary respiratory tract focus (air sacculitis, swollen head syndrome), with or without extension into the bloodstream (colibacillosis, colisepticemia), pericardium (pericarditis), and the tissues surrounding the liver (perihepatitis) (15, 70, 71). Swine and cattle can develop *E. coli* sepsis and metritis; cows also experience *E. coli* mastitis (udder infection) (11, 14).

***James R. Johnson*** • Veterans Administration Medical Center and University of Minnesota, Minneapolis, MN 55417.

## SPECIALIZATION OF ExPEC

The strains of *E. coli* that cause such extraintestinal infection, in whatever host species, have long been recognized to differ, collectively, from those that colonize the intestinal tract or cause diarrhea in members of the same species. This has led to the "special pathogenicity" hypothesis, according to which those strains that cause localized or systemic extraintestinal infections possess a heightened ability to do so, compared with other *E. coli* strains (2, 53, 70). Such strains have been the subject of considerable study in an effort to better understand their distinguishing features, virulence mechanisms, host and syndrome range, evolutionary origins, reservoirs, and transmission pathways. This has led to the creation of various taxonomic categories, including extraintestinal pathogenic *E. coli* (ExPEC) for the group as a whole, with subdivisions such as uropathogenic *E. coli* (UPEC), neonatal meningitis *E. coli* (NEMEC), sepsis-associated *E. coli* (SEPEC), and avian pathogenic *E. coli* (APEC) (4, 15, 53). In contrast, strains that are not commonly associated with disease, that exhibit low or no pathogenicity in animal models of extraintestinal infection, and that are largely confined to the intestinal tract as colonizers are often referred to as commensals.

These labels, although helpful in some respects, are inconsistently or loosely defined and potentially misleading. First, they may suggest strict segregation of pathogenic strains according to preferred site of infection or host species. On the contrary, many pathogenic *E. coli* strains can cause infection in multiple host species and at diverse anatomical sites and can produce a wide range of clinical syndromes (19–21, 36, 46, 71, 74, 75). Second, they imply a categorical distinction between pathogenic and nonpathogenic strains. In actuality, extraintestinal virulence often occurs along a continuum, depending on the particular host species and site of infection or model system (31). Third, for clinical isolates they implicitly equate immediate source of isolation (e.g., urine or bloodstream versus feces) with intrinsic virulence capability. This is problematical because some strains that cause extraintestinal infection, including even the most invasive clinical syndromes, do so not because of their intrinsic virulence potential but because of host defense defects (22, 29), whereas the fecal *E. coli* population in uninfected hosts often includes potential pathogens with the capability of causing serious, even fatal, disease if they should gain access to relevant anatomical sites (17, 55).

Finally, such labels imply that the special ability of these strains to cause extraintestinal disease is their raison d'être. However, even the most virulent ExPEC clones spend a much greater proportion of their existence (averaged over all their member cells) as intestinal commensals or periurethral or bladder colonizers than as pathogens (17, 56–59, 61). Indeed, it is conceivable that extraintestinal virulence is largely an accidental by-product of adaptive changes that were evolutionarily selected primarily because they confer increased fitness within the commensal niche (40, 60).

If source of isolation is insufficient by itself to define a given *E. coli* strain as an extraintestinal pathogen (i.e., ExPEC) versus a commensal, or as having a special ability to cause infection in a particular host or at a particular body site, how then can these distinctions be made? A common approach is the epidemiological one, where large, well-defined populations of isolates from distinct sources (e.g., different clinical syndromes or host groups, or fecal versus clinical isolates) are compared according to a broad range of traits, thereby identifying traits that are statistically associated with each source (34, 35). The strain in question is then tested for such source-distinguishing traits to infer its pathogenicity or host or syndrome specificity based on its trait profile. The effectiveness of this approach depends on the discriminatory power of the assessed traits, which in turn depends on the strength of the traits' associations with particular source groups and the degree of difference between the populations that were compared to establish the traits as predictors of source or pathogenicity. Notwithstanding the limitations of the epidemiological approach, it offers the important advantage of incorporating naturally occurring infections in the authentic host population.

A more direct, albeit more cumbersome, method for assessing a strain's virulence potential involves animal infection experiments. Here, a nonhuman host (i.e., mammal, bird, fish, insect, or worm) is challenged with a defined inoculum of the strain in question and outcomes such as the extent, severity, and time course of infection, or survival versus death, are assessed (3, 18, 69). Unlike the epidemiological approach, this experimental approach provides a direct readout regarding a strain's virulence. However, it also involves artificial conditions that are unrepresentative of naturally occurring infections, including use of a proxy host species. Therefore, inferences regarding the applicability of the experimental findings to naturally occurring infections in the authentic host require extrapolations that often are of uncertain validity.

## VIRULENCE FACTORS

Of the diverse *E. coli* traits that have been assessed in such epidemiological and experimental studies, those that are most strongly associated with extraintestinal virulence are the so-called virulence factors of ExPEC (5, 16) (Fig. 1). These virulence factors, which are nonessential for growth in vitro, give the organism a selective advantage when growing in close association with the host, particularly within the pathogenic niche, but in some instances also within the intestinal (commensal) niche (12). It is primarily such virulence factors that are thought to allow what otherwise would be a strictly commensal strain to become virulent. The wide assortment of ExPEC virulence factors confer the ability to attach to or invade epithelial and endothelial cells, injure or disrupt the function of host immune cells, resist the effects of complement and other host defense systems, synthesize and/or scavenge iron and other critical nutrients, swim to new locations, and modulate the host inflammatory response.

There are many recognized or suspected virulence factors of ExPEC, including diverse adhesins, toxins, siderophore systems, nutrient synthesis enzymes, protectins (such as polysaccharide capsules and lipopolysaccharide), motility-conferring flagellae, and multiple other secreted or surface proteins

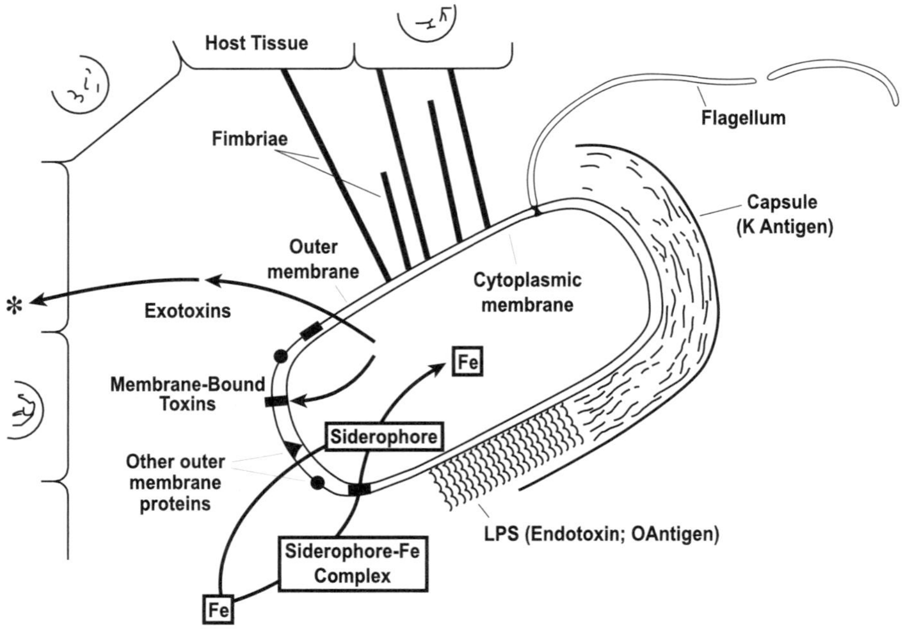


**Figure 1.** Diagram of a representative ExPEC cell and its virulence factors. Small solid circles, triangles, and rectangles represent membrane-associated proteins that contribute to complement resistance (serum survival), host cell invasion, and other virulence-promoting functions. LPS, lipopolysaccharide. Adapted with permission from reference 5.

(Fig. 1). For some such traits, a function that plausibly could promote virulence is well defined, and a significant contribution to virulence has been established in an experimental model, including through fulfillment of Koch's molecular postulates (35). These require virulence attenuation upon inactivation of the trait and restoration of virulence upon reintroduction of the trait (7). However, the actual roles that such traits play during the natural pathogenesis process or within the commensal niche are largely undefined.

Several broad generalizations can be made regarding the distribution of virulence factors among individual ExPEC strains. First, ExPEC strains typically express representatives of multiple different functional categories of virulence factors. As a result, inactivation of a given virulence factor may diminish, but rarely abolishes, virulence under experimental conditions. Similarly, expression of the virulence factor in a nonpathogenic strain may confer some increase in virulence, but rarely to the level observed in the original ExPEC strain. This suggests that virulence in ExPEC is multiply determined by the combinatorial effect of multiple virulence factors (i.e., is a "team effort").

Second, ExPEC strains commonly display multiple representatives of a particular functional class of virulence factors. For example, it is not unusual for an ExPEC strain to exhibit multiple adhesins (e.g., P, S, and type 1 fimbriae), siderophore systems (e.g., the yersiniabactin and aerobactin systems), and toxins (e.g., $\alpha$-hemolysin, cytotoxic necrotizing factor, and secreted autotransporter toxin) (28, 37). Whether the various co-occurring virulence factors from a given functional class act additively or cooperatively, act at different anatomical sites or stages of the infectious process, and/or extend the organism's host range or niche capabilities is poorly understood.

Third, different ExPEC strains exhibit radically different combinations of virulence factors, evidence that multiple *E. coli* genotypes can cause extraintestinal virulence (28, 37). A strain's specific repertoire of virulence factors also provides some information regarding the strain's relationship to other ExPEC strains, since strains with similar profiles are often closely related, whereas those with dissimilar profiles are often more distantly related. However, this association is imprecise due to the horizontal mobility and genetic instability of virulence genes (as discussed below), which allows for both convergent evolution to similar virulence profiles in distant lineages and diversity of profiles within a given lineage.

Fourth, certain virulence factors, and combinations thereof, are associated (to a variable degree) with certain clinical syndromes, anatomical sites of infection, and/or host species. These associations support the hypothesis of niche adaptation, including (as discussed above) the existence of specific pathotypes of ExPEC such as UPEC, NEMEC, APEC, etc. However, even the strongest of these associations, such as that of the K1 capsule with (human) neonatal meningitis, are incomplete (e.g., a subset of NEMEC isolates lack K1 capsule, making them similar to isolates from other pathogenic syndromes and hosts, and even to nonpathogenic fecal isolates [30]).

Fifth, regardless of the specific extraintestinal syndrome or host, ExPEC-associated virulence factors are, with few exceptions, quite distinct from those associated with the several varieties of diarrhea-causing *E. coli*. This is understandable from a physiological perspective, since the pathogenic processes that produce diarrheal disease, and the corresponding host defense elements, are vastly different from those involved in establishment of extraintestinal infection.

Although the presence/absence of specific virulence genes explains much of a strain's extraintestinal virulence, other aspects of virulence factors must be taken into account. One is virulence gene expression, which may be absent or differentially regulated under certain environmental conditions and/or growth phases (62). For example, several types of fimbrial adhesins are subject to phase variation, which can occur in a coordinated fashion. Differential expression of type 1 fimbriae and P fimbriae may importantly influence a strain's tropism for different anatomical sites, like the urinary bladder versus kidney, and may vary both among strains and at different stages of the infectious process for a given strain (9). Additionally, coordinate regulation occurs between certain "antagonistic" virulence factors from different functional categories, like flagellae (motility) and type 1 fimbriae (attachment/stasis) (63). This suggests that ExPEC strains can shift between different pathogenic (and perhaps commensal) lifestyles by utilizing different components of their virulence factor repertoire at different times.

Another important aspect of virulence genes, aside from presence/absence and expression, is point mutations. These may significantly shift the function of the gene product in a virulence-promoting manner, as best documented for *fimH*, which encodes the mannose-binding adhesin molecule of type 1 fimbriae. Certain mutant FimH variants, containing single amino acid substitutions in critical regions of the peptide, bind much more avidly to the bladder epithelium than does the ancestral protein and confer enhanced urovirulence in a mouse model of UTI (68). Such mutations have been termed "pathoadaptive," since they increase the organism's fitness within the pathogenic niche.

## PHYLOGENETIC ISSUES

Aside from virulence factors, a fundamental aspect of ExPEC is their distinctive localization within the *E. coli* phylogenetic tree (Fig. 2). Phylogenetically, *E. coli* is a moderately clonal species (6, 76). The accumulation over time of independent point mutations has produced a continuously branching, tree-like phylogenetic structure, within which horizontal gene transfer has created multiple web-like interconnections among the various branches. The main trunks that diverge from the (inferred) common ancestor of all *E. coli* are often referred to as major phylogenetic groups A, B1, B2, and D (13). However, this taxonomy is somewhat oversimplified. For example, group D is not monophyletic; instead, it exhibits two well-separated subgroups (32). Also, extensive recombination among groups A, B1, and D has blurred their boundaries, particularly for certain highly recombined strains (76). Additionally, other phylogenetic groups exist, like the lineage comprising diarrheagenic *E. coli* of serotypes O157:H7 and O55:H7, which is as distant from the major recognized groups as these groups are from one another.

Group B2 is the most phylogenetically distinct and homogeneous *E. coli* phylogenetic group. Most ExPEC strains belong to group B2, and most group B2 strains are ExPEC. The two group D lineages account for the next-largest share of ExPEC strains. In contrast, groups A and B1 comprise primarily commensal and diarrheagenic variants, including *Shigella* (76). The minority of ExPEC strains that are from groups A and B1 appear to have arisen through the accumulation of multiple virulence genes via horizontal transfer within otherwise commensal-appearing lineages. A notable exception is a well-known serogroup O78-associated lineage within group B1 that is a prominent extraintestinal pathogen of poultry and sometimes causes serious infections in humans (32).

## PHYLOGENY VERSUS VIRULENCE GENES

Projection of virulence profiles onto the *E. coli* phylogenetic tree shows that many of the recognized ExPEC virulence genes are concentrated within a specific major phylogenetic group or certain subdivisions thereof, in patterns

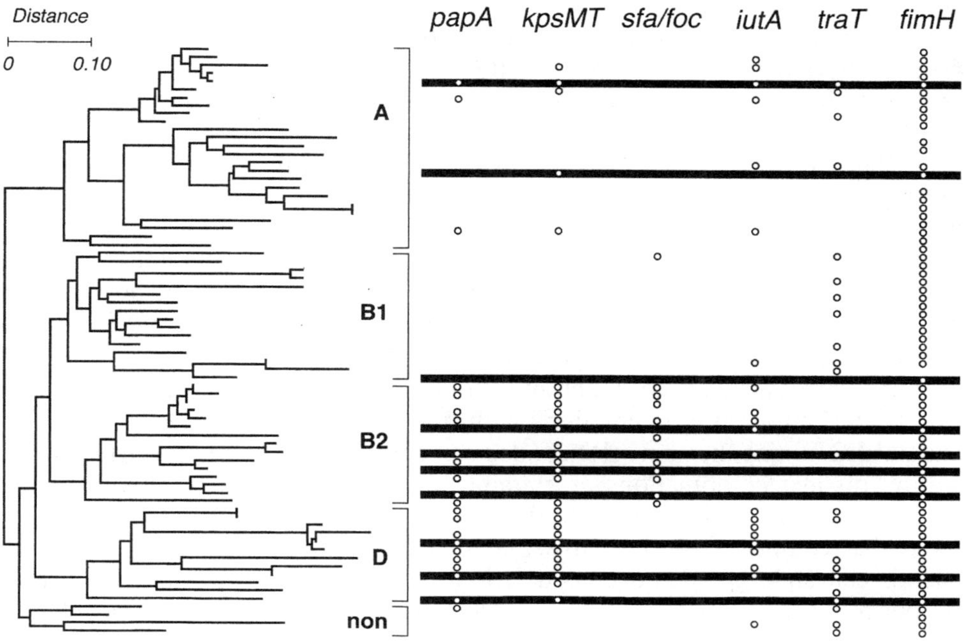


**Figure 2.** Phylogenetic distribution of *E. coli* extraintestinal virulence genes in relation to clinical source. Dendrogram at left depicts phylogenetic relationships for the 72 members of the *E. coli* Reference (ECOR) collection, as inferred based on MLEE using 38 metabolic enzymes (13). The four major phylogenetic groups (A, B1, B2, and D) of *E. coli* and the nonaligned strains ("non") are bracketed and labeled. Bullets at right indicate presence of putative virulence genes (*papA*, P fimbriae; *kpsMT*, group II capsule synthesis; *sfa/foc*, S and F1C fimbriae; *iutA*, aerobactin receptor; *traT*, serum resistance; and *fimH*, type 1 fimbriae). Horizontal bars at right indicate the 10 ECOR strains that are symptomatic infection isolates, all from humans with UTI. The remaining strains, except for one asymptomatic bacteriuria isolate, are fecal isolates from healthy human or animal hosts. Note the concentration of (chromosomal) virulence genes *papA*, *kpsMT*, and *sfa/foc* within phylogenetic groups B2 and D, but their occasional joint appearance also in distant lineages, consistent with coordinate horizontal transfer. The more scattered phylogenetic distribution of *traT* is consistent with this gene's plasmid location, whereas *fimH* is nearly universally prevalent, consistent with its presence in other species of *Enterobacteriaceae*, presumably reflecting an origin in a shared enterobacterial ancestor. Note the concentration of UTI isolates within phylogenetic groups B2 and D, and the association of virulence genes with the UTI isolates. Neighbor joining tree dendrogram adapted with permission from reference 13. Entirety of figure and legend reprinted with permission from reference 33.

suggesting predominantly vertical transmission (34, 43) (Fig. 2). Others are more broadly and sporadically distributed, in patterns suggesting extensive horizontal transmission plus instability. Still others exhibit a combination of these patterns, with their abundance within widely separated lineages suggesting infrequent horizontal transfer events, followed by vertical transmission within the recipient lineages. Many virulence genes also commonly occur together regardless of lineage, suggesting a genetic linkage.

The genetic basis for these patterns has been clarified by the finding that virulence genes (in ExPEC and many other pathogens) commonly occur both within specialized chromosomal regions, often termed fitness islands or pathogenicity-associated islands, and on plasmids (10, 64). Fitness islands, which typically contain multiple genes or open reading frames of both known and unknown function, are characterized by a GC content different from that of the rest of the genome (suggesting a possible non-*E. coli* origin), presence of inverted repeats and other mobility-promoting elements, and occurrence near tRNA sites, suggesting interactions with transducing bacteriophages. In different ExPEC strains the same tRNA sites can contain quite dissimilar islands that exhibit distinct combinations of extraintestinal virulence genes. (Interestingly, in diarrheagenic strains the same tRNA sites can contain entirely different islands that comprise strictly diarrhea-associated virulence genes [8].) Genomic islands can undergo spontaneous deletion and may be shuttled between strains by phages. While residing in situ, islands also can undergo extensive internal remodeling via insertions, deletions, and rearrangements.

Virulence plasmids, which are often conjugally transferable, sometimes contain antimicrobial resistance genes and microcin systems in addition to virulence genes, thus conferring multiple forms of fitness advantage to the host strain (38). Depending on their immediate genetic environment (e.g., occurring within a transposon), certain virulence genes can migrate between plasmids and the chromosome or may become fixed in one location or the other.

The considerable diversity among ExPEC strains with respect to their complement of fitness islands and virulence plasmids, plus the extreme plasticity of these mobile genetic platforms with respect to specific virulence gene content, likely accounts for the observed wide range of ExPEC virulence factor profiles. These same phenomena likely also facilitate the ongoing evolution of new ExPEC variants, which if ecologically successful can expand and emerge as prominent new pathogens. Molecular detection by probe arrays or whole-genome sequencing of a broad range of virulence genes, plasmid elements, and backbone genome regions can facilitate the epidemiological and evolutionary analysis of *E. coli* infections and ExPEC (44, 50, 73).

## SEROTYPING AND OTHER TYPING METHODS

The initial recognition of clonality within *E. coli* followed the development of serotyping methods that could detect diverse somatic (O), capsular (K), and flagellar (H) antigens, certain combinations of which were noted to occur with excess frequency among clinical isolates (49). Early surveys documented associations of certain O:K:H serotypes with specific clinical syndromes and virulence-associated bacterial traits (e.g., hemolysin production and erythrocyte agglutination), leading to the concept of "virulent clones." The subsequent introduction of more genetically based typing methods such as multilocus enzyme electrophoresis (MLEE) and multilocus sequence typing (MLST) showed that the traditionally recognized virulent clones of ExPEC, as identified by their distinctive O:K:H serotypes and virulence profiles, correspond largely with specific evolutionary lineages within *E. coli* phylogenetic groups B2 and D (28, 32). These virulence-associated lineages, now frequently referred to as clonal groups or clonal clusters, are significantly overrepresented within clinical collections, compared with fecal isolates.

Notably, MLEE and MLST have shown that in some instances the serotype can misrepresent clonal relationships due to the horizontal mobility of the O, K, and H antigen-encoding genes (39, 52, 72, 74). As with mobile virulence genes and virulence profiles, this process can lead to strains in distant lineages having the same serotype and to those within a given lineage having different serotypes.

Repetitive element PCR profiling and random amplified polymorphic DNA analysis are genetic typing methods that resolve at approximately the same level of similarity as do MLEE, MLST, and O:K:H serotyping (i.e., the clonal group), but are simpler and cheaper (27, 65, 67). However, they are much less reproducible, are not portable across laboratories, and although capable of resolving different individual lineages, do not reliably quantify the degree of phylogenetic similarity among lineages.

In contrast, pulsed-field gel electrophoresis (PFGE) analysis, which is relatively insensitive to broad phylogenetic relationships, provides resolution down to the level of individual clones within the larger clonal groups (20). This can be useful in epidemiological studies, since the genetic diversity of *E. coli* is sufficient that each clonal group typically comprises multiple individual clones, as defined based on their members' indistinguishable or highly similar PFGE profiles. Inferences of host-to-host transmission or broad-host-range capability are better supported when isolates from different sources can be shown to be similar according to PFGE profile, not just by virulence gene profile and clonal group membership (17). However, like PCR-based genomic profiling methods and serotyping, PFGE analysis is not well suited for characterizing the degree of similarity among different lineages, except possibly within a clonal group.

## FECAL RESERVOIR

Abundant evidence indicates that the host's own intestinal *E. coli* population is the immediate reservoir from which ExPEC strains emerge to cause most extraintestinal infections, particularly UTI (34). In adults, children, and dogs with acute UTI, the causative *E. coli* strain frequently can be detected in the host's fecal microbiota at the time of the UTI episode, and often as the predominant strain (Color Plate 2). In other instances it may be absent from the feces but present in other areas of the body, like the vagina or periurethral area, suggesting possible antecedent (undetected) fecal colonization or introduction from an external source, such as a sex partner.

ExPEC strains, especially those from group B2, may have an enhanced ability to achieve dominance within the human (and possibly canine) intestinal tract and to persist there, compared with non-ExPEC strains, which in contrast tend to be transient colonizers and to occur as minor components within a polyclonal population (45). The greater abundance that ExPEC clones often achieve within the fecal reservoir may combine with these strains' greater intrinsic virulence potential to increase their likelihood of causing extraintestinal disease. Among different healthy human populations, the intestinal prevalence of group B2 and ExPEC strains varies with the country's latitude and/or level of industrialization for undefined reasons that, if better understood, perhaps could be exploited for developing preventive interventions.

## TRANSMISSION

If the host's fecal microbiota is the main immediate reservoir for ExPEC infections, and if this population exhibits continuous turnover, then external sources and transmission pathways may lead to the host's repeated acquisition of intestinal ExPEC strains and, thus, predispose to infections. Abundant evidence indicates that cohabiting hosts (for example, family members and their pet dogs and cats) share *E. coli* clones extensively, in patterns highly suggestive of host-to-host transmission (17, 25) (Color Plate 2). Significantly, clones that are from group B2 and have extensive virulence factor profiles also tend to be the most extensively shared and to persist longest within households.

ExPEC also may be disseminated through the human population via the food supply, primarily meat products (23). Meat-associated *E. coli* strains appear to originate primarily from the corresponding food animals and, consequently, uncommonly exhibit human disease-associated virulence factors or qualify as ExPEC. Nonetheless, some such strains represent classic ExPEC clonal groups known for causing extraintestinal infections in humans and pets, and thus presumably represent a food-borne disease threat (23, 24). Moreover, many meat-associated strains are extensively resistant to antimicrobials, including first-line therapeutic agents. The extent to which this is due to agricultural antimicrobial use and leads to antimicrobial-resistant infections in humans is highly controversial, but is extremely important to define because of the profound public health, political, and economic implications (66).

## EPIDEMIC CLONES

Extraintestinal *E. coli* infections are usually regarded as sporadic events, unrelated to one another. However, over the past approximately 2 decades appreciation has grown regarding the existence of epidemic clonal groups of ExPEC. Some of these clones are multiply antimicrobial resistant; this sometimes has been responsible for bringing their members to attention and stimulating the specialized testing that showed their clonal commonality. Several such drug-resistant epidemic clonal groups appear to have emerged recently and to have disseminated rapidly and broadly among the human and companion animal populations, causing community-wide or globally distributed infections that often are appreciated as such only well after the fact, based on retrospective strain typing in research laboratories. The timing and geographic scale of the rise and fall of epidemic clones vary considerably, from localized brief surges, perhaps coinciding with multiple introductions of different contaminated products or other point-source phenomena, to more sustained (i.e., over years) and extensive expansions.

In recent years, each successively identified epidemic ExPEC clonal group has been more extensively antimicrobial resistant and/or more extensively distributed than its predecessors. Perhaps the first such clonal group to be recognized was *E. coli* serotype O15:K52:H1, from phylogenetic group D. This clonal group's distinctive resistance profile (to trimethoprim-sulfamethoxazole and four other antimicrobials) called attention to a community-wide outbreak in the vicinity of London, England, in 1986–1987 (48). Subsequent surveys showed this clonal group to be globally distributed, although no further obvious outbreaks have been observed (48).

Approximately a decade later, in the late 1990s, a new epidemic clonal group, *E. coli* "clonal group A" (CGA), was first recognized (42). CGA, which somewhat resembles *E. coli* O15:K52:H1 with respect to phylogenetic background (group D), resistance profile, and virulence profile, was identified as a prominent cause of antimicrobial-resistant UTIs at multiple university student health clinics, and of acute pyelonephritis in women across the United States. Clustered infections within a California community due to CGA isolates with the same PFGE profile suggested possible point-source spread, perhaps through a contaminated product. The subsequent identification of CGA strains in retail foods and food animals supported possible food-borne transmission (23, 51). CGA later was found to be globally distributed, with its highest prevalence in the United States (26).

In the past several years, multiple investigators have reported yet another recently emerged clonal group, i.e., sequence type ST131 (O25:H4), as a prominent cause of antimicrobial-resistant extraintestinal infections across Europe, North America, and Asia (47). In contrast to *E. coli* O15:K52:H1 and CGA, ST131 is from phylogenetic group B2. Its members nearly always exhibit fluoroquinolone resistance, often combined with resistance to extended-spectrum cephalosporins, which usually is mediated by the CTX-M-15 extended-spectrum $\beta$-lactamase. Compared with the earlier-recognized epidemic clonal groups, this newcomer's high prevalence around the globe, resistance to first-line antimicrobial agents, and possibly greater pathogenicity (as implied by its group B2 background and extensive virulence profile) make it an even more concerning threat. Especially in view of the current near absence of new antimicrobial development against gram-negative pathogens, studies are urgently needed to clarify the basis for the emergence of such multidrug-resistant ExPEC clonal groups and to define their reservoirs and modes of dissemination so that appropriate interventions can be introduced.

## HUMAN EFFECTS ON ExPEC ECOLOGY

Humans are one of the main host species for ExPEC and have altered the planet in ways that probably have profoundly affected the global *E. coli* population. Although these effects remain poorly understood, they may have significantly increased both the risk of *E. coli* extraintestinal infections and the likelihood that such infections will involve antimicrobial-resistant pathogens. Host-to-host transmission of ExPEC and resistant strains probably increases with greater mixing of human and animal populations, both within and among locales. The modern "global village" phenomenon, with enormous rates of rapid human and animal migration across national boundaries, means that an organism (whether pathogen or commensal) anywhere in the world is now only a few plane rides away from anywhere else. This plus the globally distributed nature of the food supply have contributed to the dissemination of epidemic ExPEC clonal groups and other virulent and antimicrobial-resistant *E. coli*.

As developing countries become more affluent, dietary shifts toward increased meat consumption may expose consumers to more meat-associated *E. coli*, thereby increasing the likelihood of ExPEC acquisition. Additionally, possible shifts in the intestinal milieu that favor group B2 strains may result from the decreased dietary fiber content and other metabolic characteristics of a Westernized diet.

Moreover, for the past approximately 60 years of the "antimicrobial era," antimicrobial use in food animals (whether for growth promotion or treatment or prevention of infection) and humans (whether by prescription or over the counter, and whether appropriate or inappropriate) has provided unprecedented selection pressure for resistant strains in the intestinal tract of the antimicrobial-consuming hosts. Such strains, some of which are ExPEC, and/or their resistance elements can be transmitted to others via direct contact, the environment, or the food supply. Concentrated food animal production and processing operations, which are increasingly common in industrialized countries and often involve extensive antimicrobial use, provide ideal settings for selection of resistant *E. coli*, transmission among hosts, and contamination of meat products with organisms from large numbers of animals (1). The predictable combined effect of these processes is to amplify, admix, and disseminate antimicrobial-resistant *E. coli* and ExPEC clones on an unprecedented scale, with as yet undefined consequences for human health.

## SUMMARY

The distinctive *E. coli* strains that cause most extraintestinal infections in humans and food and companion animals, which often are collectively referred to as ExPEC, differ from typical commensal fecal *E. coli* and diarrheagenic strains according to their phylogenetic background (being predominantly from groups B2 and D) and specialized virulence gene content. Better understandings of ExPEC strains' pathogenetic mechanisms, reservoirs, transmission pathways, antimicrobial resistance development, and evolutionary origins are needed to guide the development of effective preventive measures against these important opportunistic pathogens.

## REFERENCES


1. **Barza, M., and S. L. Gorbach.** 2002. The need to improve antimicrobial use in agriculture. Ecological and human health consequences. A report of the Facts about Antibiotics in Animals and the Impact on Resistance (FAAIR) project. *Clin. Infect. Dis.* **34**(Suppl. 3)**:**S71–S143.
2. **Caugant, D. A., B. R. Levin, G. Lidin-Janson, T. S. Whittam, C. Svanborg Edén, and R. K. Selander.** 1983. Genetic diversity and relationships among strains of *Escherichia coli* in the intestine and those causing urinary tract infections. *Prog. Allergy* **33:**203–227.
3. **Diard, M., S. Baeriswyl, O. Clermont, S. Gouriou, B. Picard, F. Taddel, E. Denamur, and I. Matic.** 2007. *Caenorhabditis elegans* as a simple model to study phenotypic and genetic virulence determinants of extraintestinal pathogenic *Escherichia coli. Microbes Infect.* **9:**214–223.
4. **Dobrindt, U., G. Blum-Oehler, T. Hartsch, G. Gottschalk, E. Z. Ron, R. Fünfstück, and J. Hacker.** 2001. S-Fimbria-encoding determinant $sfa_I$ is located on pathogenicity island $III_{536}$ of uropathogenic *Escherichia coli* strain 536. *Infect. Immun.* **69:**4248–4256.
5. **Eisenstein, B. I., and G. W. Jones.** 1988. The spectrum of infections and pathogenic mechanisms of *Escherichia coli. Adv. Intern. Med.* **33:**231–252.
6. **Elena, S. F., T. S. Whittam, C. L. Winkworth, M. A. Riley, and R. E. Lenski.** 2005. Genomic divergence of *Escherichia coli* strains: evidence for horizontal transfer and variation in mutation rates. *Int. Microbiol.* **8:**271–278.
7. **Falkow, S.** 1988. Molecular Koch's postulates applied to microbial pathogenicity. *Rev. Infect. Dis.* **10:**S274–S276.
8. **Groisman, E. A., and H. Ochman.** 1996. Pathogenicity islands: bacterial evolution in quantum leaps. *Cell* **87:**791–794.
9. **Gunther, N. W., IV, V. Lockatell, D. E. Johnson, and H. L. Mobley.** 2001. In vivo dynamics of type 1 fimbria regulation in uropathogenic *Escherichia coli* during experimental urinary tract infection. *Infect. Immun.* **69:**2838–2846.
10. **Hacker, J., and E. Carniel.** 2001. Ecological fitness, genomic islands and bacterial pathogenicity. A Darwinian view of the evolution of microbes. *EMBO Rep.* **2:**376–381.
11. **Harel, J., C. Forget, J. Saint-Amand, F. Daigle, D. Dubreil, M. Jacques, and J. M. Fairbrother.** 1992. Molecular cloning of a determinant coding for fimbrial antigen F165, a Prs-like fimbrial antigen for porcine septicaemic *Escherichia coli. J. Gen. Microbiol.* **138:**1495–1502.
12. **Herías, M. V., T. Midtvedt, L. Å. Hanson, and A. E. Wold.** 1995. Role of *Escherichia coli* P fimbriae in intestinal colonization in gnotobiotic rats. *Infect. Immun.* **63:**4781–4789.
13. **Herzer, P. J., S. Inouye, M. Inouye, and T. S. Whittam.** 1990. Phylogenetic distribution of branched RNS-linked multicopy single-stranded DNA among natural isolates of *Escherichia coli. J. Bacteriol.* **172:**6175–6181.
14. **Highland, M. A., B. A. Byrne, C. DebRoy, E. M. Samitz, T. S. Peterson, and K. L. Oslund.** 2009. Extraintestinal pathogenic *Escherichia coli*-induced pneumonia in three kittens and fecal prevalence in a clinically healthy cohort population. *J. Vet. Diagn. Invest.* **21:**609–615.
15. **Janben, T., C. Schwarz, P. Preikschat, M. Voss, H. C. Philipp, and L. H. Wieler.** 2001. Virulence-associated genes in avian pathogenic *Escherichia coli* (APEC) isolated from internal organs of poultry having died from colibacillosis. *Int. J. Med. Microbiol.* **291:**371–378.
16. **Johnson, J. R.** 1991. Virulence factors in *Escherichia coli* urinary tract infection. *Clin. Microbiol. Rev.* **4:**80–128.

17. **Johnson, J. R., C. Clabots, and M. A. Kuskowski.** 2008. Multiple-host sharing, long-term persistence, and virulence of *Escherichia coli* clones from human and animal household members. *J. Clin. Microbiol.* **46:**4078–4082.
18. **Johnson, J. R., O. Clermont, M. Menard, M. A. Kuskowski, B. Picard, and E. Denamur.** 2006. Experimental mouse lethality of *Escherichia coli* isolates, in relation to accessory traits, phylogenetic group, and ecological source. *J. Infect. Dis.* **194:**1141–1150.
19. **Johnson, J. R., P. Delavari, and T. T. O'Bryan.** 2001. *Escherichia coli* O18:K1:H7 isolates from acute cystitis and neonatal meningitis exhibit common phylogenetic origins and virulence factor profiles. *J. Infect. Dis.* **183:**425–434.
20. **Johnson, J. R., P. Delavari, A. L. Stell, T. S. Whittam, U. Carlino, and T. A. Russo.** 2001. Molecular comparison of extraintestinal *Escherichia coli* isolates from the same electrophoretic lineages from humans and domestic animals. *J. Infect. Dis.* **183:**154–159.
21. **Johnson, J. R., A. Gajewski, A. J. Lesse, and T. A. Russo.** 2003. Extraintestinal pathogenic *Escherichia coli* as a cause of invasive non-urinary infections. *J. Clin. Microbiol.* **41:**5798–5802.
22. **Johnson, J. R., M. A. Kuskowski, T. T. O'Bryan, and J. N. Maslow.** 2002. Epidemiological correlates of virulence genotype and phylogenetic background among *Escherichia coli* blood isolates from adults with diverse-source bacteremia. *J. Infect. Dis.* **185:**1439–1447.
23. **Johnson, J. R., M. A. Kuskowski, K. Smith, T. T. O'Bryan, and S. Tatini.** 2005. Antimicrobial-resistant and extraintestinal pathogenic *Escherichia coli* in retail foods. *J. Infect. Dis.* **191:**1040–1049.
24. **Johnson, J. R., J. S. McCabe, D. G. White, B. Johnston, M. A. Kuskowski, and P. McDermott.** 2009. Molecular analysis of *Escherichia coli* from retail meats (2002–2004) from the United States National Antimicrobial Resistance Monitoring System. *Clin. Infect. Dis.* **49:**195–201.
25. **Johnson, J. R., S. Miller, B. Johnston, C. Clabots, and C. DebRoy.** 2009. Sharing of *Escherichia coli* sequence type ST131 and other multidrug-resistant and urovirulent *E. coli* strains among dogs and cats within a household. *J. Clin. Microbiol.* **47:**3721–3725.
26. **Johnson, J. R., A. C. Murray, M. A. Kuskowski, S. Schubert, M.-F. Prère, B. Picard, R. Colodner, R. Raz, and the Trans-Global Initiative for Antimicrobial Resistance Analysis (TIARA) Investigators.** 2005. Distribution and characteristics of *Escherichia coli* clonal group A. *Emerg. Infect. Dis.* **11:**141–145.
27. **Johnson, J. R., and T. T. O'Bryan.** 2000. Improved repetitive-element PCR fingerprinting for resolving pathogenic and nonpathogenic phylogenetic groups within *Escherichia coli. Clin. Diagn. Lab. Immunol.* **7:**265–273.
28. **Johnson, J. R., T. T. O'Bryan, M. A. Kuskowski, and J. N. Maslow.** 2001. Ongoing horizontal and vertical transmission of virulence genes and *papA* alleles among *Escherichia coli* blood isolates from patients with diverse-source bacteremia. *Infect. Immun.* **69:**5363–5374.
29. **Johnson, J. R., I. Orskov, F. Orskov, P. Goullet, B. Picard, S. L. Moseley, P. L. Roberts, and W. E. Stamm.** 1994. O, K, and H antigens predict virulence factors, carboxylesterase B pattern, antimicrobial resistance, and host compromise among *Escherichia coli* strains causing urosepsis. *J. Infect. Dis.* **169:**119–126.
30. **Johnson, J. R., E. Oswald, T. T. O'Bryan, M. A. Kuskowski, and L. Spanjaard.** 2002. Phylogenetic distribution of virulence-associated genes among *Escherichia coli* isolates associated with neonatal meningitis in The Netherlands. *J. Infect. Dis.* **185:**774–784.
31. **Johnson, J. R., K. Owens, A. Gajewski, and M. A. Kuskowski.** 2005. Bacterial characteristics in relation to clinical source among *Escherichia coli* isolates from women with acute cystitis or pyelonephritis and uninfected women. *J. Clin. Microbiol.* **43:**6064–6072.

32. **Johnson, J. R., K. L. Owens, C. R. Clabots, S. J. Weissman, and S. B. Cannon.** 2006. Phylogenetic relationships among clonal groups of extraintestinal pathogenic *Escherichia coli* as assessed by multi-locus sequence analysis. *Microbes Infect.* **8:**1702–1713.
33. **Johnson, J. R., and T. A. Russo.** 2002. Extraintestinal pathogenic *Escherichia coli* (ExPEC): the "other bad *E. coli*." *J. Lab. Clin. Med.* **139:**155–162.
34. **Johnson, J. R., and T. A. Russo.** 2005. Molecular epidemiology of extraintestinal pathogenic (uropathogenic) *Escherichia coli*. *Int. J. Med. Microbiol.* **295:**383–404.
35. **Johnson, J. R., and T. A. Russo.** 2004. Chapter 8.6.1.4. Molecular epidemiology of extraintestinal pathogenic *Escherichia coli*. *In* A. Böck, R. Curtiss III, J. B. Kaper, P. D. Karp, F. C. Neidhardt, T. Nyström, J. M. Slauch, C. L. Squires, and D. Ussery (ed.), EcoSal—Escherichia coli *and* Salmonella: *Cellular and Molecular Biology*. http://www.ecosal.org. ASM Press, Washington, DC.
36. **Johnson, J. R., and T. A. Russo.** 2002. Uropathogenic *Escherichia coli* as agents of diverse non-urinary tract extraintestinal infections. *J. Infect. Dis.* **186:**859–864.
37. **Johnson, J. R., and A. L. Stell.** 2000. Extended virulence genotypes of *Escherichia coli* strains from patients with urosepsis in relation to phylogeny and host compromise. *J. Infect. Dis.* **181:**261–272.
38. **Johnson, T. J., K. E. Siek, S. J. Jonson, and L. K. Nolan.** 2006. DNA sequence of a ColV plasmid and prevalence of selected plasmid-encoded virulence genes among avian *Escherichia coli* strains. *J. Bacteriol.* **188:**745–758.
39. **Kapur, V., D. G. White, R. A. Wilson, and T. S. Whittam.** 1992. Outer membrane protein patterns mark clones of *Escherichia coli* O2 and O78 strains that cause avian septicemia. *Infect. Immun.* **60:**1687–1691.
40. **Le Gall, T., O. Clermont, S. Gouriou, B. Picard, X. Nassif, E. Denamur, and O. Tenaillon.** 2007. Extraintestinal virulence is a coincidental by-product of commensalism in B2 phylogenetic group *Escherichia coli* strains. *Mol. Biol. Evol.* **24:**2373–2384.
41. **Ling, G. V.** 1995. *Urinary Tract Infections.* Mosby, St. Louis, MO.
42. **Manges, A. R., J. R. Johnson, B. Foxman, T. T. O'Bryan, K. E. Fullerton, and L. W. Riley.** 2001. Widespread distribution of urinary tract infections caused by a multidrug-resistant *Escherichia coli* clonal group. *N. Engl. J. Med.* **345:**1007–1013.
43. **Maslow, J. N., T. S. Whittam, C. F. Gilks, R. A. Wilson, M. E. Mulligan, K. S. Adams, and R. D. Arbeit.** 1995. Clonal relationships among bloodstream isolates of *Escherichia coli*. *Infect. Immun.* **63:**2409–2417.
44. **McNamara, S. E., U. Srinivasan, L. Zhang, T. S. Whittam, C. F. Marrs, and B. Foxman.** 2009. Comparison of probe hybridization array typing to multilocus sequence typing for pathogenic *Escherichia coli*. *J. Clin. Microbiol.* **47:**596–602.
45. **Moreno, E., J. R. Johnson, T. Perez, G. Prats, M. A. Kuskowski, and A. Andreu.** 2009. Structure and urovirulence characteristics of the fecal *Escherichia coli* population among healthy women. *Microbes Infect.* **11:**274–280.
46. **Ngeleka, M., J. K. P. Kwaga, D. G. White, T. S. Whittam, C. Riddell, R. Goodhope, A. A. Potter, and B. Allen.** 1996. *Escherichia coli* cellulitis in broiler chickens: clonal relationships among strains and analysis of virulence-associated factors of isolates from diseased birds. *Infect. Immun.* **64:**3118–3126.
47. **Nicolas-Chanoine, M.-H., J. Blanco, V. Leflon-Guibout, R. Demarty, M. P. Alonso, M. M. Caniça, Y.-J. Park, J.-P. Lavigne, J. Pitout, and J. R. Johnson.** 2008. Intercontinental emergence of *Escherichia coli* clone O25:H4-ST131 producing CTX-M-15. *J. Antimicrob. Chemother.* **61:**273–281.
48. **Olesen, B., F. Scheutz, M. Menard, M. N. Skov, H. J. Kolmos, M. A. Kuskowski, and J. R. Johnson.** 2009. Three-decade epidemiological analysis of *Escherichia coli* O15:K52:H1. *J. Clin. Microbiol.* **47:**1857–1862.
49. **Orskov, I., F. Orskov, A. Birch-Andersen, M. Kanamori, and C. Svanborg-Edén.** 1982. O, K, H and fimbrial antigens in *Escherichia coli* serotypes associated with pyelonephritis and cystitis. *Scand. J. Infect. Dis. Suppl.* **33:**18–25.

50. **Palaniappan, R. U., Y. Zhang, D. Chiu, A. Torres, C. DebRoy, T. S. Whittam, and Y. F. Chang.** 2006. Differentiation of *Escherichia coli* pathotypes by oligonucleotide spotted array. *J. Clin. Microbiol.* **44:**1495–1501.
51. **Ramanchandani, M., A. R. Manges, C. DebRoy, S. P. Smith, J. R. Johnson, and L. W. Riley.** 2005. Possible animal origin of human-associated, multidrug-resistant, uropathogenic *Escherichia coli. Clin. Infect. Dis.* **40:**251–257.
52. **Reid, S. D., R. K. Selander, and T. S. Whittam.** 1999. Sequence diversity of flagellin (*fliC*) alleles in pathogenic *Escherichia coli. J. Bacteriol.* **181:**153–160.
53. **Russo, T. A., and J. R. Johnson.** 2000. A proposal for an inclusive designation for extraintestinal pathogenic *Escherichia coli*: ExPEC. *J. Infect. Dis.* **181:**1753–1754.
54. **Russo, T. A., and J. R. Johnson.** 2003. Medical and economic impact of extraintestinal infections due to *Escherichia coli*: an overlooked epidemic. *Microbes Infect.* **5:**449–456.
55. **Sannes, M. R., M. A. Kuskowski, K. Owens, A. Gajewski, and J. R. Johnson.** 2004. Virulence factor profiles and phylogenetic background of *Escherichia coli* isolates from veterans with bacteremia versus uninfected control patients. *J. Infect. Dis.* **190:**2121–2128.
56. **Schlager, T. A., J. O. Hendley, A. L. Bell, and T. S. Whittam.** 2002. Clonal diversity of *Escherichia coli* colonizing stools and urinary tracts of young girls. *Infect. Immun.* **70:**1225–1229.
57. **Schlager, T. A., J. O. Hendley, R. A. Wilson, V. Simon, and T. S. Whittam.** 1999. Correlation of periurethral bacterial flora with bacteriuria and urinary tract infection in children with neurogenic bladder receiving intermittent catheterization. *Clin. Infect. Dis.* **28:**346–350.
58. **Schlager, T. A., J. O. Jendley, J. A. Lohr, and T. S. Whittam.** 1993. Effect of periurethral colonization on the risk of urinary tract infection in healthy girls after their first urinary tract infections. *Pediatr. Infect. Dis. J.* **12:**988–993.
59. **Schlager, T. A., J. R. Johnson, L. M. Ouellette, and T. S. Whittam.** 2008. *Escherichia coli* colonizing the neurogenic bladder are similar to widespread clones causing disease in patients with normal bladder function. *Spinal Cord* **46:**633–638.
60. **Schlager, T. A., T. S. Whittam, J. L. Bhang, C. L. Wobbe, and A. Stapleton.** 2003. Variation in frequency of the virulence-factor gene in *Escherichia coli* clones colonizing the stools and urinary tracts of healthy prepubertal girls. *J. Infect. Dis.* **188:**1059–1064.
61. **Schlager, T. A., T. S. Whittam, J. O. Hendley, J. Bhang, R. Grady, and A. Stapleton.** 2000. Expression of virulence factors among *Escherichia coli* isolated from the periurethra and urine of children with neurogenic bladder on intermittent catheterization. *Pediatr. Infect. Dis. J.* **19:**37–41.
62. **Schlager, T. A., T. S. Whittam, J. O. Hendley, R. J. Hollis, M. A. Pfaller, R. A. Wilson, and A. Stapleton.** 1995. Comparison of expression of virulence factors by *Escherichia coli* causing cystitis and *E. coli* colonizing the periurethra of healthy girls. *J. Infect. Dis.* **172:**772–777.
63. **Sims, A., and H. Mobley.** 2008. Multiple genes repress motility in uropathogenic *Escherichia coli* constitutively expressing type 1 fimbriae. *J. Bacteriol.* **190:**3747–3756.
64. **Skyberg, J. A., T. J. Johnson, J. R. Johnson, C. Clabots, C. M. Logue, and L. K. Nolan.** 2006. Acquisition of avian pathogenic *Escherichia coli* plasmids by a commensal *E. coli* isolate enhances its abilities to kill chicken embryos, grow in human urine, and colonize the murine kidney. *Infect. Immun.* **74:**6287–6292.
65. **Versalovic, J., V. Kapur, T. Koeuth, G. H. Mazurek, T. S. Whittam, J. M. Musser, and J. R. Lupski.** 1995. DNA fingerprinting of pathogenic bacteria by fluorophore-enhanced repetitive sequence-based polymerase chain reaction. *Arch. Pathol. Lab. Med.* **119:**23–29.
66. **Walk, S. T., J. M. Mladnicky, J. A. Middleton, A. J. Heidt, J. R. Cunningham, P. C. Bartlett, K. Sato, and T. S. Whittam.** 2007. The influence of antibiotic selection on the population genetic composition of *Escherichia coli* from conventional and organic dairy farms. *Appl. Environ. Microbiol.* **73:**5982–5989.

67. **Wang, G., T. S. Whittam, C. M. Berg, and D. E. Berg.** 1993. RAPD (arbitrary primer) PCR is more sensitive than multilocus enzyme electrophoresis for distinguishing related bacterial strains. *Nucleic Acids Res.* **21:**5930–5933.
68. **Weissman, S. J., V. Beskhlebnaya, V. Chesnokova, S. Chattopadhyay, W. E. Stamm, T. M. Hooton, and E. V. Sokurenko.** 2007. Differential stability and trade-off effects of pathoadaptive mutations in the *Escherichia coli* FimH adhesin. *Infect. Immun.* **75:** 3548–3555.
69. **White, D., M. Dho-Moulin, R. Wilson, and T. S. Whittam.** 1993. Clonal relationships and variation in virulence among *Escherichia coli* strains of avian origin. *Microb. Pathog.* **14:**399–409.
70. **White, D. G., R. A. Wilson, D. A. Emery, K. V. Nagaraja, and T. S. Whittam.** 1993. Clonal diversity among strains of *Escherichia coli* incriminated in turkey colisepticemia. *Vet. Microbiol.* **34:**19–34.
71. **White, D. G., R. A. Wilson, A. S. Gabriel, M. Saco, and T. S. Whittam.** 1990. Genetic relationships among strains of avian *Escherichia coli* associated with swollen-head syndrome. *Infect. Immun.* **58:**3613–3620.
72. **Whittam, T. A., and R. A. Wilson.** 1988. Genetic relationships among pathogenic *Escherichia coli* of serogroup O157. *Infect. Immun.* **56:**2467–2473.
73. **Whittam, T. S., and A. C. Bumbaugh.** 2002. Inferences form whole-genome sequences of bacterial pathogens. *Curr. Opin. Genet. Dev.* **12:**719–725.
74. **Whittam, T. S., and R. A. Wilson.** 1988. Genetic relationships among pathogenic strains of avian *Escherichia coli*. *Infect. Immun.* **56:**2458–2466.
75. **Whittam, T. S., M. L. Wolfe, and R. A. Wilson.** 1989. Genetic relationships among *Escherichia coli* isolates causing urinary tract infections in humans and animals. *Epidemiol. Infect.* **102:**37–46.
76. **Wirth, T., D. Falush, R. Lan, F. Colles, P. Mensa, L. H. Wieler, H. Karch, P. R. Reeves, M. C. J. Maiden, H. Ochman, and M. Achtman.** 2006. Sex and virulence in *Escherichia coli*: an evolutionary perspective. *Mol. Microbiol.* **60:**1136–1151.

*Population Genetics of Bacteria: a Tribute to Thomas S. Whittam*
Edited by S. T. Walk and P. C. H. Feng


*Chapter 8*

# Epidemiology of Argentinean Shiga Toxin-Producing *Escherichia coli*

***Marta Rivas, Isabel Chinen, Elizabeth Miliwebsky, Lucía Galli, Horacio A. Repetto, and Marcelo Masana***

Shiga toxin-producing *Escherichia coli* (STEC) is an important food-borne pathogen that can cause nonbloody diarrhea, hemorrhagic colitis, and hemolytic uremic syndrome (HUS) (14). HUS, a life-threatening complication that occurs in 5 to 10% of patients, is characterized by hemolytic anemia, thrombocytopenia, and renal failure (11). No specific treatment exists for postenteric HUS, and the mortality rate among children with this type of HUS is 3 to 5% (14). The dominant STEC serotype is O157:H7, which was identified as a human pathogen in 1982, and since then has been responsible for numerous outbreaks and sporadic cases in different parts of the world (42). However, there have been increasing reports of non-O157 strains associated with gastrointestinal infections (2). These serotypes differ in their frequency and severity of human disease, suggesting differences in their virulence characteristics (16).

The production of Shiga toxin 1, Shiga toxin 2, and/or their variants (Stx1c, Stx2c, Stx2-O118, Stx2-OX3a, $Stx2d_{activatable}$, and Stx2f) is the primary virulence trait responsible for human disease. Another virulence-associated factor of most STEC isolates is a 94-kDa outer membrane protein, called intimin, which is encoded by the *eae* gene on a ca. 34-kb chromosomal pathogenicity island termed the locus of enterocyte effacement (LEE). This locus is associated with intimate adherence to epithelial cells, initiation of host signal transduction pathways, and formation of attaching-and-effacing intestinal lesions (33). However, the presence of this island is not essential for pathogenesis, as a wide

***Marta Rivas, Isabel Chinen, Elizabeth Miliwebsky, and Lucía Galli*** • Servicio Fisiopatogenia, Instituto Nacional de Enfermedades Infecciosas–ANLIS "Dr. C. G. Malbrán," Ciudad Autónoma de Buenos Aires, Argentina. ***Horacio A. Repetto*** • Servicio de Pediatría, Hospital Nacional "Prof. Alejandro Posadas," Departamento de Pediatría, Facultad de Medicina, Universidad de Buenos Aires, Pcia. de Buenos Aires, Argentina. ***Marcelo Masana*** • Instituto Tecnología de Alimentos, Centro de Investigación de Agroindustria, Instituto Nacional de Tecnología Agropecuaria, INTA, Pcia. de Buenos Aires, Argentina.

number of LEE-negative STEC strains are capable of causing human disease (2). The products of several genes have been suggested to play a role in virulence. These products include adhesins, toxins, proteases, iron acquisition systems, lipopolysaccharide, and flagellin. In addition, most STEC strains produce an enterohemolysin (EHEC-Hly), encoded by a large plasmid-borne (90-kb) gene known as *ehxA*, that has been associated with severe clinical disease in humans (33).

Cattle and other ruminants have been implicated as the main reservoir of O157:H7 and non-O157 strains (3). STEC infections are transmitted to humans through contaminated food and water and direct contact with infected persons or animals (47).

## CLINICAL FEATURES OF HUS

HUS was given its name by Gasser et al. (9) in 1955. They reported five patients who presented with microangiopathic hemolytic anemia, thrombocytopenia, and acute renal failure. None of them had associated gastroenteritis.

However, the first report of a group of children who presented with diarrhea and developed HUS was published by Gianantonio et al. in 1962 (10) as part of a cohort of children with acute renal failure treated by peritoneal dialysis. Two years later, they reported a group of more than 50 children with postdiarrheal HUS (11). At that time, it became evident that this clinical entity was prevalent in Argentina, and from 1996 its incidence has been systematically registered, ranging between 450 and 500 new cases per year. The major contribution of Gianantonio and his team was to establish peritoneal dialysis as a treatment in the acute period, which produced a dramatic decrease in mortality from 27 to 5%.

HUS is a systemic thrombotic microangiopathy caused by different etiologies and mechanisms that results in endothelial lesions leading to thrombosis of the microcirculation in most organs. After the first description regarding the association between *E. coli*, Stx, and HUS by Karmali et al. in 1983 (15), the occurrence of the typical or epidemic form (D+) has been documented in our country (6, 27), representing at present more than 95% of the cases (34).

In Argentina, HUS is the leading cause of acute renal failure among children. The mainstay of treatment in this phase is supportive therapy that generally includes control of fluid and electrolyte imbalance, use of dialysis if required, control of hypertension, and blood transfusion as required (31). The mortality in the acute stage has been lower than 5% since 1978. Children usually die because of severe involvement of the central nervous system, intestine, or myocardium and its complications, or from intercurrent infections. Of the 95% who survive, approximately one-third are at risk for having chronic sequelae (35).

HUS is also the second-leading cause of chronic renal failure (35, 43). The more frequent renal lesion is characterized by the hyperfunction of nephrons remaining after the acute necrotizing lesion. Three courses of progression to end-stage renal failure have been described. (i) Children with more severe forms do not recover from acute renal failure and enter directly into a dialysis and transplantation program. (ii) A second group recovers renal function partially, with persistent proteinuria and frequently hypertension; progression to end-stage renal failure occurs in 2 to 5 years. (iii) The third group may recover normal serum creatinine and creatinine clearance, with persistent proteinuria. They are at risk of progressing to chronic renal failure and end-stage disease more than 5 years, and sometimes as long as 20 years, after the acute disease (35).

In Argentina, HUS is the cause of 20% of all kidney transplants among children and adolescents; no recurrence has been observed in children with the epidemic form and the clinical course is similar to that of children transplanted for urological causes (7, 35).

## SURVEILLANCE AND DISEASE TRENDS

Important differences exist in the incidence of STEC infections and HUS throughout South America. A regional network for surveillance purposes does not yet exist, and data are restricted to only a few countries. If we compare the HUS situation in the region, the annual incidence in Argentina is five times higher than in Chile and three times higher than in Uruguay, but the magnitude of the problem is still unknown in most countries.

In Argentina, data on human STEC infections are gathered through different strategies: (i) the National Health Surveillance System, which collects data on HUS cases (since 2000 the report has been mandatory, and must be immediate and individualized); (ii) the Sentinel Surveillance System through 25 HUS sentinel units; (iii) the Laboratory-Based Surveillance System through the National Diarrheal and Foodborne Pathogens Network; and (iv) molecular surveillance through PulseNet Latin America and Caribbean.

In Argentina, postdiarrheal HUS is endemic. Over the last 10 years, approximately 500 HUS cases were reported annually. The incidence has ranged between 7.8 and 17 cases per 100,000 children less than 5 years of age and the lethality between 2 and 5% (Fig. 1).

In 2008, 543 cases were reported in the country. The incidence rate was 17 cases per 100,000 children under 5 years old, with higher incidences in the central and southern regions (Fig. 2). Regarding the age distribution, 54% of subjects were under 2 years old, with a median age of 36 months, and 54% were female. The lethality was 2%. The disease had a seasonal pattern: 70% of the cases occurred during the warmest months, from October to March. The National Reference Laboratory (NRL) received samples from approximately 40% of the HUS cases reported.

Since 2003, the molecular surveillance of STEC using the pulsed-field gel electrophoresis (PFGE) methodology has been implemented through PulseNet Latin America and Caribbean. Argentinean databases for O157 and non-O157 STEC strains were created by the NRL. In the period from 1988 to 2008, a total of 565 XbaI-PFGE patterns corresponding to 1,346 STEC O157 strains,

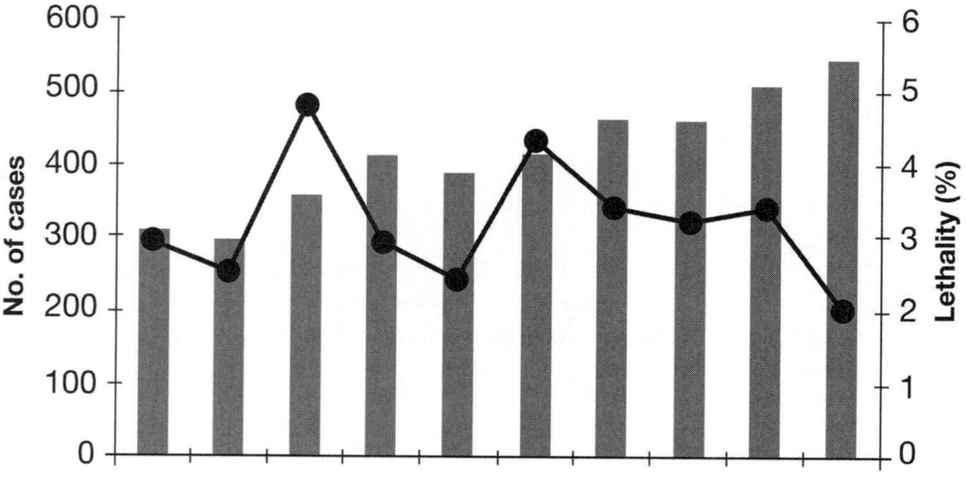


| Year | 1999 | 2000 | 2001 | 2002 | 2003 | 2004 | 2005 | 2006 | 2007 | 2008 |
|---|---|---|---|---|---|---|---|---|---|---|
| Incidence | 7.8 | 8.4 | 10.4 | 12.2 | 11.5 | 12.5 | 13.9 | 13.9 | 15.0 | 17.0 |

**Figure 1.** Number of HUS cases, incidence rates, and percentages of lethality in Argentina, 1999 to 2008.

isolated from human infections ($n$ = 1,103), food ($n$ = 110), animals ($n$ = 112), and environmental samples ($n$ = 21), were included. The most frequent XbaI-PFGE patterns associated with human strains are shown in Fig. 3.

Among the non-O157 STEC strains, 322 XbaI-PFGE patterns corresponding to 463 isolates were established. To date, the patterns are classified and codified according to the following O groups: O8, O22, O26, O91, O103, O111, O113, O121, O130, O145, O174, and O178.

In our country, the *E. coli* O157 strains showed a high degree of diversity; however, strains with similar patterns have been isolated in different regions throughout the years. Strains that yielded identical patterns were grouped in clusters when they were linked by time or place. This molecular surveillance allowed the identification of clusters of sporadic cases as part of diffuse outbreaks. The identity of the strains, grouped in clusters or outbreaks by XbaI-PFGE, was further confirmed by PFGE using BlnI as second enzyme.

Two XbaI-PFGE patterns are prevalent, AREXHX01.0011 ($n$ = 140) and AREXHX01.0022 ($n$ = 91), representing 10.4 and 6.8% of the database, respectively. Some XbaI-PFGE patterns, including 0011 and 0022, are common among strains isolated from samples of different origin (Table 1), showing that they are widespread in the human population, food, reservoirs, and the environment. Furthermore, pattern 0011 and other closely related patterns are part of the hypervirulent clone described in other countries, as Smi-H in Sweden and EXHX01.0047 in the United States. Some XbaI-PFGE patterns are re-

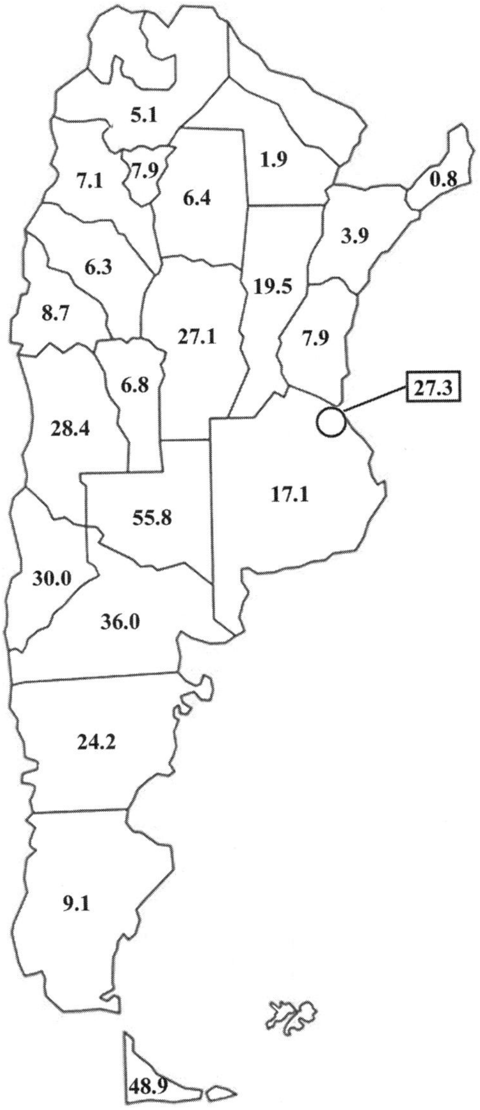


**Figure 2.** Incidence rates (cases per 100,000 children under 5 years old) of HUS in different Argentinean provinces, 2008.

stricted to strains of particular origin (human, food, or animal), suggesting that those isolates could belong to different lineages of O157 strains, as was described in several studies (17, 18).

As part of PulseNet International, the NRL receives consultations from the Centers for Disease Control and Prevention (CDC). As an example, the PFGE pattern (EXHX01.0124) established during the spinach outbreak in the United States has been previously recognized in Argentina (AREXHX01.0057), associated with strains of different origin isolated in different regions of the country throughout the years.

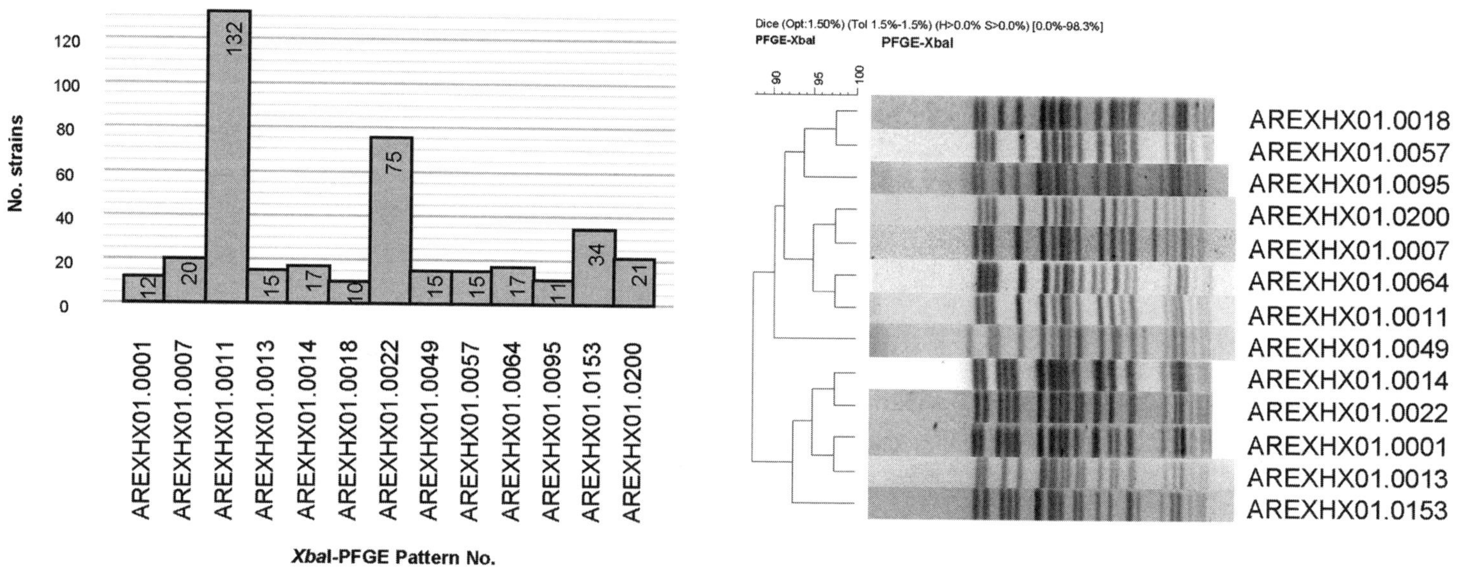


**Figure 3.** XbaI-PFGE patterns associated with human STEC O157 strains.

**Table 1.** PFGE patterns yielded by STEC O157 strains of human, food, animal, and environmental origin

| XbaI-PFGE pattern | No. of strains |
|---|---|
| AREXHX01.0011 | 132 human, 9 food, 5 animal |
| AREXHX01.0022 | 75 human, 12 food, 1 animal, 2 environmental |
| AREXHX01.0012 | 12 human, 3 environmental |
| AREXHX01.0018 | 10 human, 8 animal |
| AREXHX01.0139 | 6 human, 7 food |
| AREXHX01.0076 | 5 human, 6 animal |
| AREXHX01.0043 | 4 human, 3 food, 1 animal, 1 environmental |
| AREXHX01.0267 | 2 human, 1 food |
| AREXHX01.0080 | 1 human, 2 food, 1 environmental |
| AREXHX01.0092 | 2 human, 1 animal |

## DIAGNOSIS

At the NRL, three diagnostic criteria are used to establish the STEC infection: (i) isolation and characterization of STEC strains, (ii) detection of free fecal Stx (StxMF), and (iii) serological tests to detect Stx antibodies. However, isolation of the STEC strain is considered the definitive diagnostic procedure.

Fecal samples are collected and routinely cultured for intestinal pathogens at Local Clinical Laboratories. The clinical specimens and isolates are sent to the NRL for further characterization. Fecal samples are plated onto sorbitol MacConkey agar directly, and after enrichment at 37°C for 4 h in trypticase soy broth supplemented with cefixime (50 ng/ml) and potassium tellurite (25 mg/ml). The confluent growth zone and colonies are screened for $stx_1$, $stx_2$, and $rfb_{O157}$ genes by a multiplex PCR (19). If the PCR is negative, immunomagnetic separation is performed. Isolates with $stx_1$ and/or $stx_2$ genes are identified by standard biochemical tests, serotyped, and characterized by phenotypic and genotypic techniques. For STEC typing, $stx_1$ and $stx_2$ genotypes and *eae* types are determined by PCR-restriction fragment length polymorphism (PCR-RFLP). Phage typing is performed by the method described by Ahmed et al. (1) in 1987 and extended by Khakhria et al. in 1990 (16a), and PFGE is carried out using the 24-h CDC protocol, with minor modifications (3a). The phenotypic and genotypic characterization and the typing of STEC strains are performed by the methods presented in Fig. 4. The StxMF detection is done by cytotoxicity assays on Vero cells using Stx1- and Stx2-specific monoclonal antibodies and serological tests to detect Stx antibodies by seroneutralization of Stx cytotoxic effect on Vero cells (36).

## STEC IN HUMAN DISEASES

In 1988, Novillo et al. (27) showed evidence of STEC infection in 49 patients with a diagnosis of idiopathic HUS. StxMF was detected in 15 of 49 (31%) patients and Stx-neutralizing antibodies in 11 out of 18 (61%).

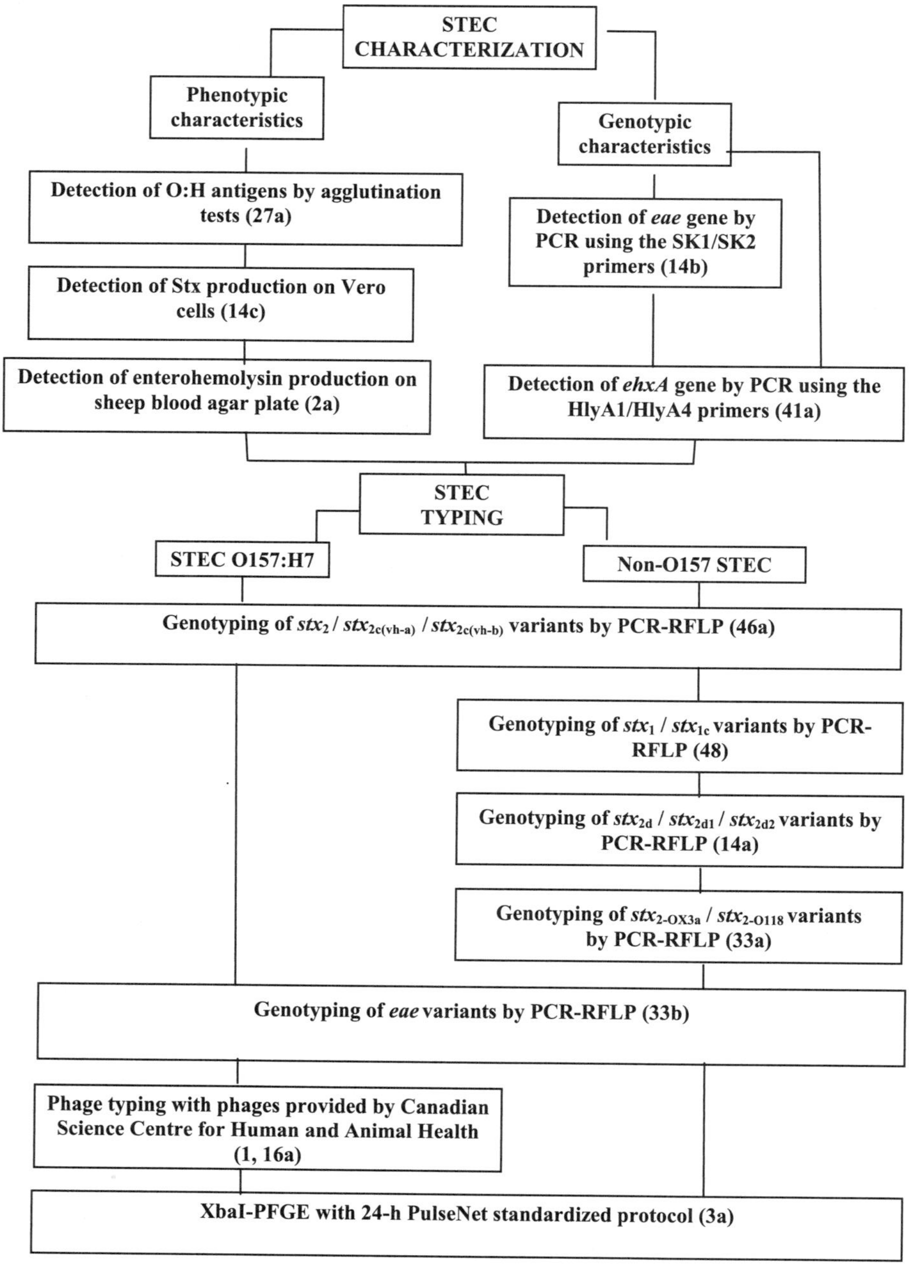


**Figure 4.** Flow chart for the phenotypic and genotypic characterization of STEC strains.

In 1989, López et al. (21), studying children with HUS, found cumulative evidence of STEC infection in 85% of the cases with a low incidence of *E. coli* O157:H7 (2%). However, in studies in the 1990s, evidence of STEC infection was found in 59% of Argentinean HUS cases, and *E. coli* O157 was the predominant serogroup (25). *Shigella dysenteriae* type 1 was never isolated from HUS cases.

Our group has conducted a prospective study in 95 household contacts of 34 children with HUS to determine the frequency of STEC infection (36). The control group consisted of 34 children with acute gastroenteritis who did not develop HUS. Cumulative evidence of STEC infection was found in 38.2% of HUS patients, 31.6% of family members, and 29.4% of control children. Stx2 was prevalent in STEC strains and StxMF. Both parents had the same infection rate, higher than those detected in siblings and grandparents. Of 16 patients without evidence of infection, 3 had household contacts with StxMF and 13 with Stx antibodies. These results showed the wide STEC dissemination in the Argentinean population and indicated that person-to-person transmission may play an important role in the high incidence of HUS in our country.

During a prospective study, the phenotypic and genotypic features of 103 STEC strains isolated from 99 children with HUS and bloody and nonbloody diarrhea, and the clonal relatedness of *E. coli* O157:H7 strains using typing techniques were established (38). The strains belonged to 18 different serotypes, and 59% were of serotype O157:H7. The $stx_2$ gene was identified in 90.3% and the $stx_1$ gene in 9.7% of them. Among the 61 *E. coli* O157 strains, 93.4% harbored the $stx_2$ and $stx_{2c(vh\text{-}a)}$ genes; PT4 (39.3%) and PT2 (29.5%) were the predominant phage types. By XbaI-PFGE, a total of 41 different patterns with 80% similarity were identified and seven clusters with identical profiles were established. Isolates with indistinguishable PFGE patterns were also identical by phage typing and *stx* genotyping. However, no epidemiological linkage could be demonstrated between the cases.

In 2007, during a study carried out in Neuquén City (13a), 908 fecal samples of adults and pediatric patients with bloody and nonbloody diarrhea were screened by multiplex PCR. A total of 11 (1.2%) positive samples were found. The rate of STEC detection in diarrhea was low in comparison with the high number of HUS cases reported in the province of Neuquén in the same period. These results suggest that the strains of this region may have a higher pathogenic potential, which allows a rapid evolution to HUS, hampering STEC detection at the first stage of diarrhea.

In the period from 2004 to 2008, a total of 846 STEC infections were confirmed in Argentina, mainly associated with STEC O157. However, non-O157 infections were also detected (Fig. 5). A total of 848 STEC strains were isolated and confirmed at the NRL from 426 HUS cases, 223 bloody and 102 nonbloody diarrhea cases, and 97 cases of other pathology. In 2 cases, a coinfection with two different STEC serotypes was detected. The most prevalent serotypes identified were O157:H7 (635; 74.9%) and O145:NM (113; 13.3%). Other O groups, such as O121 (19; 2.2%), O26 (15; 1.8%), O174 (9; 1.1%), O111 (5; 0.6%), and O8 (4; 0.5%), among others, were also detected. Among

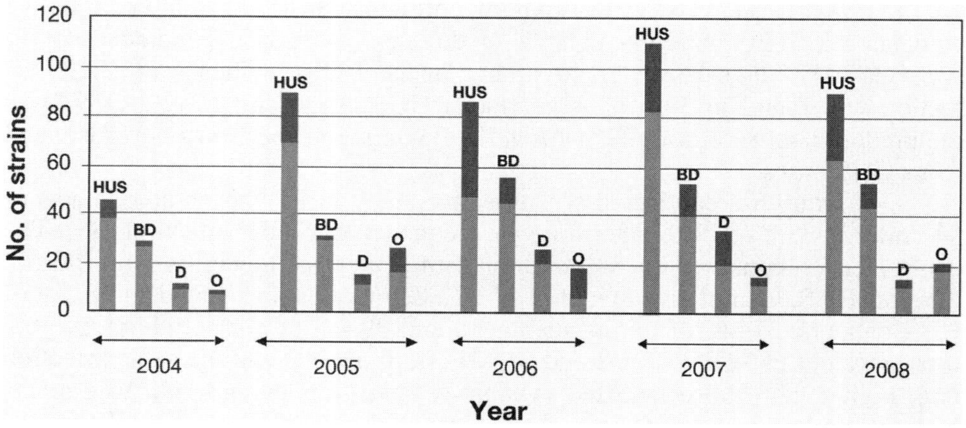


**Figure 5.** Distribution of STEC O157 (light gray, lower bars) and non-O157 (dark gray, upper bars) strains in human infections in Argentina, 2004 to 2008.

STEC O157:H7 strains, the genetic profile *stx*$_2$ and *stx*$_{2c(vh\text{-}a)}$, *eae, ehxA* (517; 81.4%) prevailed, followed by *stx*$_2$, *eae, ehxA* (89; 14%). PT4 (40%), PT49 (25%), and PT2 (20%) were the phage types found most frequently. The strains showed a high antimicrobial susceptibility (97.6%). Among the non-O157 STEC strains, genetic profiles were more diverse, but *stx*$_2$, *eae, ehxA* (141; 66.2%) and *stx*$_1$, *eae, ehxA* (24; 11.3%) were the most prevalent. The antimicrobial susceptibility was 81.2%. The intimin subtypes were specifically associated with some O:H serotypes. Similar to other studies, Int-$\gamma$ was associated with serotypes O157:H7/NM and O145:NM; Int-$\beta$ with O26:H11/NM and O145:H25; Int-$\varepsilon$ with O103:H2 and O121:H19; and Int-$\theta$ with O111:NM.

## OUTBREAKS OF STEC

In Argentina, outbreaks are identified through the surveillance system of HUS and STEC-associated diseases. The definition of an outbreak used for this analysis is two or more linked cases. PFGE and phage typing are used to establish the clonal relatedness of the isolates. In the period from 2002 to 2009, a total of 12 outbreaks of bloody diarrhea and HUS cases associated with O157 and non-O157 STEC strains occurred at kindergartens, in families, and in the community (Table 2).

In 2008, the sentinel unit of La Pampa Province, working together with the Epidemiology Branch and the Food Laboratory, was able to establish a clonal relatedness among STEC O157 strains isolated from an HUS case, an asymptomatic household contact, and foods. The isolated strains presented the same genetic profile and belonged to PT49; and PFGE patterns (XbaI-PFGE AREXHX01.0344 and XbaI-PFGE AREXHX01.0153) with a high degree of similarity were established, with only one band difference, and they were considered part of the same cluster (08LPEXH-1).

**Table 2.** Outbreaks in Argentina associated with O157 and non-O157 STEC strains

| Year | Outbreak site | Location | No. of patients with bloody diarrhea/HUS | No. of asymptomatic patients | Serotype | Genetic profile | XbaI-PFGE pattern | Source |
|---|---|---|---|---|---|---|---|---|
| 2002 | Kindergarten | Buenos Aires | 1/0 | 1 | O157:H7 | *$stx_2$* and *$stx_{2c(vh\text{-}a)}$*, *eae*, *ehxA* | AREXHX01.0226 | Unknown |
| 2003 | Kindergarten | Mar del Plata | 13/1 | 0 | O26:H11 | *$stx_1$*, *eae*, *ehxA* | AREVCX01.0037 | Person-to-person |
| | | | | | O103:H2 | *$stx_1$*, *eae*, *ehxA* | AREXWX01.0016 | |
| 2004 | Kindergarten | Entre Ríos | 3/1 | 2 | O157:H7 | *$stx_2$* and *$stx_{2c(vh\text{-}a)}$*, *eae*, *ehxA* | AREXHX01.0243 | Swimming pool |
| 2005 | Kindergarten | Rosario | 27/4 | 1 | O145:NM | *$stx_2$*, *eae*, *ehxA* | ARENMX01.0006 | Swimming pool |
| | | | | | ONT:HNT | *$stx_{2d2(vh\text{-}b)}$* | | |
| 2006 | Kindergarten | Buenos Aires | 1/0 | 5 | O174:H21 | *$stx_{2d2(vh\text{-}b)}$* | AREZDX01.0015 | Unknown |
| 2006 | Family | Neuquén | 0/1 | 4 | O145:NM | *$stx_2$*, *eae*, *ehxA* | ARENMX01.0027 | Person-to-person |
| 2007 | Community | Neuquén | 3/1 | 0 | O157:H7 | *$stx_2$*, *eae*, *ehxA* | AREXHX01.0200 | Unknown |
| 2008 | Kindergarten | Buenos Aires | 1/2 | 0 | O157:H7 | *$stx_2$* and *$stx_{2c(vh\text{-}a)}$*, *eae*, *ehxA* | AREXHX01.0489 | Unknown |
| 2008 | Family | La Pampa | 0/1 | 1 | O157:H7 | *$stx_2$* and *$stx_{2c(vh\text{-}a)}$*, *eae*, *ehxA* | AREXHX01.0153<br>AREXHX01.0344 | Bovine meat, fresh sausage |
| 2008 | Family | Neuquén | 0/1 | 3 | O157:H7 | *$stx_2$* and *$stx_{2c(vh\text{-}a)}$*, *eae*, *ehxA* | AREXHX01.0331 | Person-to-person |
| 2008 | Kindergarten | Río Negro | 0/1 | 1 | O157:H7 | *$stx_2$* and *$stx_{2c(vh\text{-}a)}$*, *eae*, *ehxA* | AREXHX01.0011 | Unknown |
| 2009 | Kindergarten | Córdoba | 0/2 | 6 | O157:H7 | *$stx_2$* and *$stx_{2c(vh\text{-}a)}$*, *eae*, *ehxA* | AREXHX01.0427 | Unknown |

Miliwebsky et al. (26) described the detection and duration of fecal shedding of O157 and non-O157 STEC strains in symptomatic and asymptomatic cases during outbreaks in day care centers. In different events, it was observed that STEC O157:H7 strains were shed by approximately 30 days, and O26 and O145 STEC strains were shed during 37 and 19 days, respectively. This study contributed to defining control strategies to allow professional health workers, teachers, and parents to be educated on reducing risky behavior, to decrease the risk of STEC infection.

## CASE-CONTROL STUDY

In 2001 to 2002, the first prospective case-control study was conducted to evaluate risk factors for sporadic STEC infection among Argentinean children in two cities, Mendoza and Buenos Aires, and their surroundings (39). A total of 150 cases and 299 controls were enrolled. The median age of cases was 1.8 years, and 58% were girls. Serotype O157:H7 was the most commonly isolated STEC. Analysis of single variable associations identified dietary habits and animal exposures linked to illness. Many beef-related exposures were significantly associated with STEC infection. Eating beef outside the home and eating undercooked beef anyplace was associated with illness. Eleven percent of case-children but only 5% of controls consumed *jugo de carne* (liquid squeezed from a tender, usually lightly cooked piece of beef and spoon-fed). Living in or visiting a place with farm animals, contact with farm animals (including horses, pigs, poultry, and cattle), and contact with cattle manure were associated with illness. Risky exposures suggesting person-to-person transmission from young children included contact with a child <5 years old, attending day care or kindergarten, and contact with a child <5 years old with diarrhea. Wearing diapers was also linked to illness. No significant differences between cases and controls were found in the distribution of most variables that relate to socio-economic status (e.g., number of bedrooms, water supply, garbage disposal, educational level of parents). The protective factors identified were eating more than the median number of fruits and vegetables, male sex, and the respondent always washing hands after handling raw beef. On multivariable logistic regression analysis, significant risk factors for STEC infection were eating undercooked beef outside the home (odds ratio [OR], 17.63; 95% confidence interval [CI] 1.6 to 197.4; $P = 0.02$), living in or visiting a place with farm animals (OR, 6.61; 95% CI, 1.5 to 28.8; $P = 0.01$), contact with a child <5 years old with diarrhea (OR, 3.29; 95% CI, 1.0 to 10.4; $P = 0.04$), and having nonparental household income (OR, 2.21; 95% CI, 1.2 to 4.0; $P = 0.01$). Eating ground beef at home (e.g., meatballs, empanadas, or meat pie) remained protective (OR, 0.36; 95% CI, 0.1 to 0.9; $P = 0.03$). With this model, the fixed adjustment factors had significant estimated protective associations as follows: eating more than the median number of fruits and vegetables (OR, 0.31; 95% CI, 0.1 to 0.6; $P = 0.0007$), male sex (OR, 0.34; 95% CI, 0.2 to 0.7; $P =$ 0.001), having a nonparent respondent (OR, 0.34; 95% CI, 0.2 to 0.7; $P =$

0.001), and the respondent always washing hands after handling raw beef (OR, 0.23; 95% CI, 0.1 to 0.6; $P$ = 0.001).

As part of the case-control study, in 2002 we established the first relationship between a sporadic HUS case and the consumption of home-prepared hamburger 48 h before diarrhea onset (37). The isolates recovered from the child and from the food were characterized as STEC O157:H7 $stx_2$ and $stx_{2c(vh\text{-}a)}$, *eae, ehxA* of PT4 with identical restriction patterns (XbaI-PFGE AREXHX01.0011 and BlnI-AREXHA26.0040).

## STEC IN FOOD

In 2004, the Argentinean Food Codex established the use of the USDA/FSIS methodology (Revision 3, effective 10/25/02, 2002) for sampling, isolation, and characterization of *E. coli* O157:H7/NM in meat products. For STEC non-O157, the NRL developed and validated a PCR methodology for detection and isolation in meat samples that includes an internal amplification control.

STEC strains isolated from food are sent to the NRL by different official food control laboratories for characterization and typing. Since 1996, 261 *E. coli* strains, isolated mainly from meat and meat products, have been submitted. A total of 126 strains were identified and characterized as STEC O157:H7 and 135 as non-O157. Nineteen O groups with different virulence profiles were detected (Table 3).

In different surveys, O157 and non-O157 STEC strains were isolated from meat and meat products using different methodologies.

In 2000, Parma et al. (32) described the genotypes and serotypes of STEC strains isolated from cattle and foods. Stx2, *eae*-negative strains were prevalent in foods. All 21 STEC strains were distributed among 11 serotypes, some of which (O20:H19, O91:H21, O113:H21, O116:H21, O117:H7, O171:H2, and

**Table 3.** Prevalent STEC O groups of strains detected in food

| O group | No. (%) of strains |
|---|---|
| O157 | 126 (48.3) |
| O174 | 21 (8.0) |
| O8 | 15 (5.7) |
| O20 | 13 (5.0) |
| O113 | 7 (2.7) |
| O91 | 5 (1.9) |
| O22 | 3 (1.1) |
| O25 | 3 (1.1) |
| O171 | 3 (1.1) |
| O26 | 2 (0.8) |
| O136 | 2 (0.8) |
| ONT | 52 (19.9) |
| Others | 9 (3.4) |
| **Total** | **261 (100)** |

OX3:H21) were shared between bovine and food strains. These O groups have been isolated from cases of hemorrhagic colitis and HUS worldwide.

At retail outlets, *E. coli* O157:H7 was detected in 3.9% of 279 meat products (4), corresponding to 3.8% of ground beef samples, 4.8% of fresh sausages, and 3.3% of dry sausages. All 11 isolates harbored both the *eae* and *ehxA* genes and were genotyped as $stx_2$ and $stx_{2c(vh\text{-}a)}$ ($n$ = 4), $stx_1$ and $stx_{2c(vh\text{-}a)}$ ($n$ = 5), $stx_1$ ($n$ = 1), and $stx_{2c(vh\text{-}a)}$ ($n$ = 1). Gómez et al. (13) described the detection of non-O157 STEC in 8.4% of frozen hamburgers and in 0.9% of soft cheese, and Oteiza et al. (29) established a 3% prevalence of STEC in morcilla, a typical Argentinean sausage, sampled at local retail markets. Two (2%) strains were characterized as STEC O157:H7 harboring the $stx_2$ and $stx_{2c(vh\text{-}a)}$, *eae, ehxA* genes, and one (1%) strain as STEC O26:H11 carrying the $stx_1$, *eae, ehxA* genes. Roldán et al. (40) sampled 250 ground beef and hamburger samples at retail outlets and 150 milk samples from bulk tank milk from dairy barns, which were analyzed by selective enrichment and immunomagnetic separation. STEC O157:H7 $stx_2$, *eae, ehxA*-positive strains were isolated from 1.2% of the beef samples. The milk samples were STEC negative.

Chinen et al. (5) described the isolation and characterization of STEC O157:H7 from cooked and uncooked beef and chicken burgers and from chicken carcasses. Samples were collected during sampling procedures triggered by episodes of infection reported in 2001 and 2002 in Buenos Aires City. Twenty STEC O157:H7 strains were isolated from 19 (6.8%) out of 279 samples of beef and chicken burgers, and 4 strains from 4 (10.3%) out of 39 chicken carcasses. The prevalent *stx* genotype was $stx_2$ and $stx_{2c(vh\text{-}a)}$ (14 strains; 58%). All strains were characterized as *eae* and *ehxA* positive. By XbaI-PFGE, the strains yielded 10 different patterns. Eighteen out of 24 strains were grouped in four clusters: #1 (4 strains; AREXHX01-0043), #2 (4 strains; AREXHX01-0022), #3 (8 strains; AREXHX01-0139), and #4 (2 strains; AREXHX01-0200). Identical strains by phage typing, *stx* genotyping, and PFGE were detected in uncooked and cooked beef and chicken burgers in different restaurants, which had been collected on the same or different sampling dates. These findings help to underline the importance of STEC O157 detection in meat products, to improve active surveillance and to define control strategies in order to prevent new cases of STEC infection.

## STEC IN RESERVOIRS

The first isolation of STEC O157 from cattle occurred in Argentina in 1977 (28), when three *E. coli* O157 strains were recovered from feces of 13 calves aged 1 to 3 weeks with colibacillosis in Buenos Aires Province. Since that first report, cattle have been established as the main reservoir for STEC worldwide.

A first survey of the prevalence of STEC in cattle was performed by Sanz et al. (41). In that study, 126 calves and 118 cows were investigated. STEC strains were recovered from calves with diarrhea (23%), healthy calves (29%), and cows at farms (22%) and at abattoirs (44%). Stx genotyping showed that

*$stx_1$* was predominant in calves, and *$stx_2$* strains in cows. Strains isolated from healthy calves also carried the *eae* gene in a larger proportion (75%) than strains recovered from cows at slaughter (2%). According to the authors, the lower prevalence of the *eae*-positive strains in older animals would decrease the health risk for persons consuming meat contaminated with STEC. In a later study, Parma et al. (32) characterized 86 STEC strains from cattle into 31 serotypes. Strains of serotypes linked to hemorrhagic colitis or HUS cases were isolated from diarrheic calves (O26:H11, O103:H2, O103:NM, and O111:NM), while O103:NM was identified in strains isolated from cows.

Accurate data on the prevalence of STEC O157 in cattle were possible only after immunomagnetic separation was implemented in different surveys. A prevalence of STEC O157:H7 of 3.8% was established in 288 fecal samples from a beef cattle farm in Gualeguaychú, Argentina (44). The most prevalent *stx* genotype identified was *$stx_1$* and *$stx_{2c(vh\text{-}a)}$* in 7 of 11 isolates (64%) from cattle. In the same study, *E. coli* O157 strains from surface water troughs were also recovered (5.1%), but no genetic correlation between bovine and water strains was found.

Feedlot facilities are not as common in Argentina as in other meat-producing countries. Therefore, few studies on STEC carriage in feedlots have been conducted. Padola et al. (30) reported the recovery of O2:H25, O15:H21, O25:H19, O145:NM, O146:NM, O146:H21, O157:H7, and O175:H8 serotypes in grain-fed animals, which had not been identified in graze-fed bovines in previous surveys. In this study, four STEC O157 strains harbored the *$stx_2$*, *eae, ehxA* genes. During a more recent study in a feedlot facility, O. López (personal communication) pointed out a higher prevalence of STEC O157 (38.4%), mainly characterized as *$stx_2$*, *eae, ehxA*. In this study the prevalence of non-O157 STEC strains was 45.2%.

Contamination with STEC at abattoirs has been less investigated in Argentina. A first study by Gioffré et al. (12) analyzed 160 fecal samples from steers at slaughter in order to evaluate two PCR procedures for STEC detection. Only eight (5%) non-O157 STEC strains were recovered. Serotype O8:H19 was predominant (four strains), and one isolate was characterized as O26:H11, often associated with human disease. Strains of serotype O26:H11 were also isolated from 76 diarrheic calves on local farms (24). In this study, non-O157 STEC strains were isolated from 15.8% of calves with diarrhea, with a majority of strains (60%) carrying the *$stx_1$* gene.

Between July 1999 and December 2000, Meichtri et al. (23) conducted another study to establish STEC prevalence in cattle, analyzing stools and rectal swabs from 200 young steers. Eighty-six STEC strains were isolated from 39% of PCR-positive animals. The prevalent serotype was O8:H19 (12.8%), while 51.2% of the isolates belonged to serotypes previously associated with hemorrhagic colitis or HUS worldwide (Table 4). Remarkably, one STEC O157:H7 strain (0.5%) was isolated and characterized, though no specific detection methodology for STEC O157 was employed. The virulence profile of the STEC strains revealed that the *$stx_2$* genotype was most prevalent (79.1%), followed by *$stx_1$* and *$stx_2$* (14%) and then *$stx_1$* (7%). Four (4.7%) strains of serotypes,

**Table 4.** STEC serotypes isolated from fecal samples from young steers at the abattoir

| Associated with human disease worldwide | Associated with human disease in Argentina | Associated with food outbreaks worldwide, including Argentina |
|---|---|---|
| O8:H21 | O8:H21 | O103:H2 |
| O25:H2 | O25:H2 | O145:NM |
| O91:H21 | O91:H21 | O157:H7 |
| O103:H2 | O103:H2 | |
| O113:H21 | O121:H19 | |
| O121:H19 | O113:H21 | |
| O145:NM | O145:NM | |
| O145:H21 | O157:H7 | |
| O157:H7 | | |
| ONT:H7 | | |
| ONT:H16 | | |
| ONT:H19 | | |
| ONT:H21 | | |
| ONT:H28 | | |
| ONT:NM | | |
| ONT:HNT | | |

O2:NM, O112:H2, O145:NM, and O157:H7, carried the $stx_2$, *eae, ehxA* genes associated with higher virulence.

Recently, a more extensive investigation of STEC O157 and non-O157 STEC in nine selected beef-exporting abattoirs in Argentina has been conducted. In this study, a prevalence of 4.1% (95% CI, 2.9 to 5.6) for STEC O157 was found in the fecal content of 811 bovines at slaughter (22). This study also established the contamination of STEC O157 in carcasses as 2.6% (95% CI, 1.6 to 3.9). Among 54 STEC O157:H7 strains isolated, the predominant *stx* genotype was $stx_2$ and $stx_{2c(vh\text{-}a)}$ (56%), which also is the most prevalent *stx* genotype (>80%) in STEC O157 postenteric HUS cases in Argentina.

Likewise, the prevalence of non-O157 STEC in fecal and carcass samples was estimated as 22 and 9%, respectively (22a). Six serotypes (O8:H19, O179:H19, ONT:H2, O130:H11, O113:H21, and ONT:H7) accounted for half of all 307 non-O157 STEC isolates, whereas 4.2% of strains belonged to O groups with the highest pathogenic potential (O103 and O145). With respect to the routes of contamination of the carcasses with non-O157 STEC serotypes, this study showed that identical serotypes were recovered in carcasses and feces of the same animal in 4% (3/73) of the cases. The most frequent genotype was $stx_1$ and $stx_2$, *ehxA, saa* (Table 5), while the $stx_2$ and $stx_{2c(vh\text{-}a)}$, *eae, ehxA* genotype prevalent in HUS cases in Argentina was identified in two isolates from fecal samples.

In another study, our group (5a) compared the subtypes of STEC O157 strains isolated between November 2006 and April 2008 from HUS ($n$ = 122), bloody diarrhea ($n$ = 69), and nonbloody diarrhea ($n$ = 30) patients, asymptomatic carriers ($n$ = 5), and bovines at abattoirs ($n$ = 54) and established the relationship of the strains. Among strains of both origins, $stx_2$ and $stx_{2c(vh\text{-}a)}$ genotype and PT4, PT2, and PT49 phage types were prevalent. Identical profile

**Table 5.** Comparison of virulence profiles identified in non-O157 STEC isolated from fecal and carcass samples from cattle at the abattoir

| Virulence profile | No. (%) of strains isolated in: | |
|---|---|---|
| | Fecal samples ($n$ = 197) | Carcass samples ($n$ = 64) |
| $stx_1$, $stx_2$, *ehxA*, *saa* | 35 (17.7) | 17 (26.6) |
| $stx_{2c(vh\text{-}b)}$ | 35 (17.7) | 9 (14.0) |
| $stx_2$, *ehxA*, *saa* | 33 (16.8) | 11 (17.2) |
| $stx_2$, *ehxA* | 17 (8.6) | 10 (15.6) |
| $stx_{2c(vh\text{-}a)}$ | 19 (9.6) | 3 (4.7) |
| $stx_1$, $stx_{2c(vh\text{-}b)}$, *ehxA*, *saa* | 11 (5.6) | 3 (4.7) |
| $stx_2$, $stx_{2c(vh\text{-}b)}$, *ehxA*, *saa* | 9 (4.6) | 4 (6.25) |
| $stx_1$, *eae*, *ehxA* | 11 (5.6) | 1 (1.56) |
| $stx_1$, $stx_2$, $stx_{2c(vh\text{-}b)}$, *ehxA*, *saa* | 9 (4.6) | 3 (4.7) |
| $stx_1$ | 9 (4.6) | 2 (3.1) |
| $stx_2$ | 9 (4.6) | 1 (1.56) |

combinations in order of *stx* genotype, phage type, and restriction pattern were found in some human and bovine strains, and this finding enhanced the importance of cattle as a reservoir of STEC.

Other animals besides cattle can play a role as STEC reservoirs and in the transmission of the infection, either at home or in fairs and zoos. In 2004, STEC was recovered from wild animals kept in captivity in Argentina (20). In this study, 50.8% of samples collected by rectal swabs at the Zoo and Botanical Garden of La Plata City were STEC positive. Ten species of the order Cetartiodactyla, including alpaca, Patagonian cavy, red deer, and antelope, among others, and one species of the Rodentia order were recognized as new carriers of STEC. Among the seven serotypes identified, the most frequently detected was O146:H28 (24%), previously associated with human infections. Interestingly, toxin type $Stx_{1c}$ was first identified in Argentina, often in combination with $stx_{2\text{-}O118}$ or $stx_{2\text{-}OX3a}$ genes. Thus, zoo mammals can act as reservoirs for strains carrying these particular markers.

Domestic animals, such as cats and dogs, have been also investigated for their role as STEC reservoirs. The analysis of fecal samples from 450 dogs and 149 cats allowed the identification of some STEC serotypes that can cause infection in humans, such as O22:H8, O22:NM, O91:H21, and O157:NM (1a). Another study of STEC in asymptomatic humans and their pets reported a 3.4 and 3.6% prevalence in dogs and cats, respectively (7a). *E. coli* O157:H7 strains (2/6; 33.3%) harboring the $stx_2$ and $stx_{2c(vh\text{-}a)}$, *eae, ehxA* genotype and belonging to PT2 were identified. The PFGE patterns established had been previously included in the *E. coli* O157 database associated with HUS and bloody diarrhea cases. These findings add evidence that pets can be carriers of infection for the susceptible population.

## PUTATIVE VIRULENCE FACTORS

Great progress has been made in recent years in the identification of virulence factors of STEC strains, mainly in O157. However, in view of the increasing number of reports of LEE-negative STEC infections, there is now a need for comprehensive data on the molecular basis for their virulence; hence, prevalence of adhesins encoded outside the LEE region and other virulence markers in these types of strains should be determined.

Several proteins were proposed to be novel adhesive factors; these include ToxB, Saa (an autoagglutinating adhesin identified in LEE-negative strains), Sfp (sorbitol-fermenting enterohemorrhagic *E. coli* O157 fimbriae), Iha (an adherence-conferring protein similar to *Vibrio cholerae* IrgA), Efa1 (EHEC factor for adherence), and LPF (long polar fimbriae, closely related to LPF of *Salmonella enterica* serovar Typhimurium). Moreover, additional toxins that may play a role in pathogenesis have also been described, including subtilase cytotoxin, CDT-V (cytolethal distending toxin), and East1 (enteroaggregative *E. coli* heat-stable enterotoxin 1).

In a study conducted by Toma et al. (45), the distribution of the abovementioned putative adhesins was investigated in 139 STEC strains of different serotypes and of human, animal, and food origins. The most prevalent adhesin was that encoded by the *iha* gene (127/139; 91%), which was distributed in all seropathotypes. The genes *toxB* and *efa1* were present mainly in strains of seropathotypes A and B (LEE positive). The *saa* gene was present only in strains of seropathotypes C, D, and E (LEE negative). The fimbrial gene $lpfA_{O113}$ was present in all seropathotypes except for seropathotype A, while *sfpA* was not present in any of the strains studied. The distribution of STEC adhesins depended mainly on serotypes and not on the source of isolation.

Adherence mechanisms of LEE-negative STEC strains are poorly understood. Saa is widely used as a marker of adhesion in the absence of LEE; meanwhile, Toma et al. (46) demonstrated an Saa-independent mechanism of adherence.

In a recent study conducted in our laboratory, a total of 200 LEE-negative STEC strains isolated from humans ($n = 47$) and cattle ($n = 153$) were investigated for the presence of five virulence markers that included fimbrial and nonfimbrial adhesins. Moreover, three toxins—subtilase cytotoxin, CDT-V, and East1—were studied in addition to the Shiga toxins (8). The most prevalent virulence marker found was that encoded by the $lpfA_{O113}$ gene (199/200; 99%). Distinct virulence profiles were found in the LEE-negative STEC strains analyzed, and differences were observed between cattle and human strains (Table 6). Eleven different virulence profiles were identified among the 47 human strains studied. Out of the 153 bovine strains, 11 distinct virulence profiles were also detected. However, the $lpfA_{O113}$, *iha, fimA* profile, coincidentally, was the most prevalent in strains of both origins.

Out of 11 profiles identified in human strains, 9 were associated with HUS. The profile $lpfA_{O113}$, *iha, fimA* was identified in strains of different serotypes isolated from HUS, diarrhea cases, and asymptomatic contacts. Of those HUS cases, two patients died during the acute phase of the disease and the strains belonged to the O143:NM and ONT:NM serotypes. Moreover, two other HUS patients died and the isolated strains harbored the $lpfA_{O113}$, *iha, saa, ehxA,*

**Table 6.** Comparison of virulence profiles identified in LEE-negative STEC strains isolated from cattle and humans

| Virulence profile | No. (%) of strains isolated from: Cattle ($n$ = 153) | Humans ($n$ = 47) | $P$ value |
|---|---|---|---|
| *lpfA*$_{O113}$, *iha, fimA* | 49 (32) | 19 (40.4) | NS[a] |
| *lpfA*$_{O113}$, *iha, saa, ehxA, subA, fimA* | 32 (21) | 6 (12.8) | NS |
| *lpfA*$_{O113}$, *iha, saa, ehxA, fimA* | 29 (19) | 3 (6.4) | <0.05 |
| *lpfA*$_{O113}$, *iha, saa, ehxA, subA, cdt-V, fimA* | 17 (11) | 8 (17.2) | NS |
| *lpfA*$_{O113}$, *astA* | 11 (7.2) | 0 (0) | <0.05 |
| *lpfA*$_{O113}$, *ehxA, fimA* | 5 (3.5) | 1 (2.1) | NS |
| *lpfA*$_{O113}$, *fimA* | 4 (2.6) | 2 (4.3) | NS |
| *lpfA*$_{O113}$, *iha, astA, fimA* | 2 (1.3) | 4 (8.5) | <0.05 |
| *lpfA*$_{O113}$, *saa, ehxA, fimA* | 2 (1.3) | 1 (2.1) | NS |
| *lpfA*$_{O113}$, *iha, saa, ehxA, subA, cdt-V, astA, fimA* | 1 (0.6) | 0 (0) | NS |
| *lpfA*$_{O113}$, *ehxA, astA, fimA* | 1 (0.6) | 0 (0) | NS |
| *lpfA*$_{O113}$, *iha, astA* | 0 (0) | 1 (2.1) | NS |
| *lpfA*$_{O113}$, *saa, ehxA, subA, fimA* | 0 (0) | 1 (2.1) | NS |
| *iha* | 0 (0) | 1 (2.1) | NS |

[a]NS, not significant ($P > 0.05$).

*subAB, cdt-V, fimA* genes and belonged to the O113:H21 and O174:H28 serotypes.

These results show that bovine LEE-negative STEC strains possess genes encoding virulence factors present in human LEE-negative STEC strains associated with disease.

## CONCLUSIONS

STEC strains are widespread in Argentina, and infect humans and animals and contaminate food products. Postenteric HUS is endemic, with more than 500 cases reported each year. Different serotypes and genotypes are detected, and the severity of clinical symptoms caused by STEC has been associated with its virulence profile, especially the *stx* genotype. *E. coli* O157:H7 is the main serotype isolated, but other serotypes are detected. The $stx_2$ and $stx_{2c(vh\text{-}a)}$, *eae, ehxA* genotype is prevalent among O157 strains and $stx_2$, *eae, ehxA* among non-O157 strains. PT4 and PT49 are the predominant phage types. The STEC strains showed a high diversity because different PFGE restriction patterns are recognized. However, identical patterns are identified in different regions throughout the years, suggesting that they could be part of diffuse outbreaks.

Various studies performed in Argentina, at both farms and abattoirs, have confirmed cattle as an important reservoir for STEC, and probably the main route for the infection of the population. Prevalence figures reported in cattle from Argentina are within the limits of various reports from other countries with similar rearing practices. However, a relevant fact from the more recent studies is that the prevalent STEC O157 genotype in cattle is also prevalent in postenteric HUS cases in the country.

Many STEC infections among children in Argentina could be prevented by avoiding consumption of undercooked beef, by limiting exposure to farm animals and their environment, by avoiding exposure to children with diarrhea, and by washing hands after handling raw beef. Implementation of preventive measures by industry, government, and consumers could result in a decrease in the incidence of STEC infections and thereby a decrease in the incidence of HUS, with its associated human and economic cost.

**Acknowledgments.** The data reported here are a summary of the efforts of many individuals and working groups who have collaborated to develop a clearer picture of STEC infections in Argentina.

## REFERENCES

1. **Ahmed, R., C. Bopp, A. Borczyk, and S. Kasatiya.** 1987. Phage-typing scheme for *Escherichia coli* O157:H7. *J. Infect. Dis.* **155:**806–809.

1a. **Bentancor, A., M. V. Rumi, M. V. Gentilini, C. Sardoy, K. Irino, A. Agostini, and A. Cataldi.** 2007. Shiga toxin-producing and attaching and effacing *Escherichia coli* in cats and dogs in a high hemolytic uremic syndrome incidence region in Argentina. *FEMS Microbiol. Lett.* **267:**251–256.

2. **Bettelheim, K. A.** 2007. The non-O157 Shiga-toxigenic (Verocytotoxigenic) *Escherichia coli*; under-rated pathogens. *Crit. Rev. Microbiol.* **33:**67–87.

2a. **Beutin, L., M. A. Montenegro, I. Ørskov, F. Ørskov, J. Prada, S. Zimmermann, and R. Stephan.** 1989. Close association of verotoxin (Shiga-like toxin) production with enterohemolysin production in strains of *Escherichia coli*. *J. Clin. Microbiol.* **27:**2559–2564.

3. **Caprioli, A., S. Morabito, H. Brugreb, and E. Oswald.** 2005. Enterohaemorrhagic *Escherichia coli*: emerging issues on virulence and modes of transmission. *Vet. Res.* **36:** 289–311.

3a. **Centers for Disease Control and Prevention.** 2007. One-day (24–28 h) standardized laboratory protocol for molecular subtyping of *Escherichia coli* O157:H7 by pulsed field gel electrophoresis. CDC, Atlanta, GA.

4. **Chinen, I., J. D. Tanaro, E. Miliwebsky, L. H. Lound, G. Chillemi, S. Ledri, A. Baschkier, M. Scarpin, E. Manfredi, and M. Rivas.** 2001. Detection of Shiga toxin-producing *Escherichia coli* O157:H7 in retail meat in Argentina. *J. Food Prot.* **64:**1346–1351.

5. **Chinen, I., S. Epszteyn, C. L. Melamed, L. Aguerre, E. Martinez Espinosa, M. M. Motter, A. Baschkier, E. Manfredi, E. Miliwebsky, and M. Rivas.** 2009. Shiga toxin-producing *Escherichia coli* O157 in beef and chicken burgers, and chicken carcasses in Buenos Aires, Argentina. *Int. J. Food Microbiol.* **132:**167–171.

5a. **D'Astek, B., et al.** 2009. *In Abstracts of the 7th International Symposium on Shiga Toxin (Verocytotoxin)-Producing* Escherichia coli *Infections.*

6. **De Cristofano, M. G., A. Fayad, J. Ferraris, C. Cortinez, J. Ramirez, A. Ascione, and C. A. Gianantonio.** 1986. Sindrome urémico hemolítico de la infancia. Su relación con la presencia de verotoxina libre fecal. *Arch. Argent. Pediatr.* **84:**339–342.

7. **Ferraris, J. R., J. A. Ramirez, S. Ruiz, M. G. Caletti, G. Vallejo, J. J. Piantanida, J. L. Araujo, and E. T. Sojo.** 2002. Shiga toxin-associated hemolytic uremic syndrome: absence of recurrence after renal transplantation. *Pediatr. Nephrol.* **17:**809–814.

7a. **Gallego, V., N. Deza, C. Carbonari, J. Gugliada, J. Stambullian, and H. Repetto.** 2006. Detección de *Escherichia coli* productor de toxina Shiga (STEC) en adultos asintomáticos que conviven con mascotas. *In Abstracts of the 6th International Symposium on Shiga Toxin (Verocytotoxin)-Producing* Escherichia coli *Infections.*

8. **Galli, L., E. Miliwebsky, K. Irino, G. A. Leotta, and M. Rivas.** 2010. Virulence profile comparison between LEE-negative Shiga toxin-producing *Escherichia coli* (STEC) strains isolated from cattle and humans. *Vet. Microbiol.* **143:**307–313.
9. **Gasser, V. C., E. Gautier, A. Steck, E. E. Seibenmann, and R. Oeschlin.** 1955. Hämolytische-urämische syndrome: bilaterale nierenindennekrosen bei akuten erworbenen hämolytischen anemia. *Schweiz. Med. Wochensch.* **85:**905–909.
10. **Gianantonio, C. A., M. Vitacco, J. Mendilaharzu, F. Mendilaharzu, and A. Rutty.** 1962. Acute renal failure in infancy and childhood. Clinical course and treatment of 41 patients. *J. Pediatr.* **61:**660–678.
11. **Gianantonio, C., M. Vitacco, F. Mendilaharzu, A. Rutty, and J. Mendilaharzu.** 1964. The hemolytic uremic syndrome. *J. Pediatr.* **64:**478–491.
12. **Gioffré, A., L. Meichtri, E. Miliwebsky, A. Baschkier, G. Chillemi, M. I. Romano, S. Sosa Estani, A. Cataldi, R. Rodríguez, and M. Rivas.** 2002. Detection of Shiga toxin-producing *Escherichia coli* by PCR in cattle in Argentina. Evaluation of two procedures. *Vet. Microbiol.* **87:**301–313.
13. **Gómez, D., E. Miliwebsky, C. Fernandez Pascua, A. Baschkier, E. Manfredi, M. Zotta, F. Nario, A. Piquín, M. Sanz, A. Etcheverría, N. Padola, A. Parma, and M. Rivas.** 2002. Aislamiento y characterization de *Escherichia coli* productor de toxina Shiga en hamburguesas supercongeladas y quesos de pasta blanda. *Rev. Argent. Microbiol.* **34:**66–71.

13a. **Gonzalez, G., L. Pianciola, V. Di Russo, E. Miliwebsky, N. Deza, C. Carbonari, I. Chinen, M. Rivas.** 2009. Screening of Shiga toxin-producing *Escherichia coli* by PCR in patients with diarrhea in Neuquén City, Argentina. Abstr. P04.1.6, p. 60. *In Abstracts of the 7th International Symposium on Shiga Toxin (Verocytotoxin)-Producing* Escherichia coli *Infections.*

14. **Griffin, P. M., P. Mead, and S. Sivapalasingam.** 2002. *Escherichia coli* O157:H7 and other enterohemorrhagic *Escherichia coli*, p. 627–642. *In* M. J. Blaser, P. D. Smith, J. I. Ravdin, H. B. Greenberg, and R. L. Guerrant (ed.), *Infections of the Gastrointestinal Tract*, 2nd ed. Lippincott Williams & Wilkins, Philadelphia, PA.

14a. **Jelacic, J. K., T. Damrow, G. S. Chen, S. Jelacic, M. Bielaszewska, M. Ciol, H. M. Caravalho, A. R. Melton-Celsa, A. D. O'Brien, and P. I. Tarr.** 2003. Shiga toxin-producing *Escherichia coli* in Montana: bacterial genotypes and clinical profiles. *J. Infect. Dis.* **188:**719–729.

14b. **Karch, H., H. Bohm, H. Shmidt, F. Gunzer, S. Aleksic, and J. Heesemann.** 1993. Clonal structure and pathogenicity of Shiga-like toxin-producing fermenting *Escherichia coli* O157:H-. *J. Clin. Microbiol.* **31:**1200–1205.

14c. **Karmali, M. A., M. Petric, C. Lim, R. Cheung, and G. S. Arbus.** 1985. Sensitive method for detecting low numbers of verotoxin-producing *Escherichia coli* in mixed cultures by use of colony sweeps and polymyxin extraction of verotoxin. *J. Clin. Microbiol.* **22:**614–619.

15. **Karmali, M. A., M. Petric, B. T. Steele, and C. Lim.** 1983. Sporadic cases of haemolytic-uraemic syndrome associated with faecal cytotoxin and cytotoxin-producing *Escherichia coli* in stools. *Lancet* **321:**619–620.
16. **Karmali, M. A., M. Mascarenhas, S. Shen, K. Ziebell, S. Johnson, R. Reid-Smith, J. Isaac-Renton, C. Clark, K. Rahn, and J. B. Kaper.** 2003. Association of genomic O island 122 of *Escherichia coli* EDL 933 with Verocytotoxin-producing *Escherichia coli* seropathotypes that are linked to epidemic and/or serious disease. *J. Clin. Microbiol.* **41:**4930–4940.

16a. **Khakhria, R., D. Duck, and H. Lior.** 1990. Extended phage-typing scheme for *Escherichia coli* O157:H7. *Epidemiol. Infect.* **105:**511–520.

17. **Kim, L., J. Nietfeldt, and A. K. Benson.** 1999. Octamer-based genome scanning distinguishes a unique subpopulation of *Escherichia coli* O157:H7 strains in cattle. *Proc. Natl. Acad. Sci. USA* **96:**13288–13293.

18. **Laing, C. R., C. Buchanan, E. N. Taboada, Y. Zhang, M. A. Karmali, J. E. Thomas, and V. P. Gannon.** 2009. In silico genomic analyses reveal three distinct lineages of *Escherichia coli* O157:H7, one of which is associated with hyper-virulence. *BMC Genomics* **10:**287–298.
19. **Leotta, G. A., I. Chinen, S. Epszteyn, E. Miliwebsky, I. C. Melamed, M. Motter, M. Ferrer, E. Marey, and M. Rivas.** 2005. Validación de una técnica de PCR múltiple para la detección de *Escherichia coli* productor de toxina Shiga. *Rev. Argent. Microbiol.* **37:**1–10.
20. **Leotta, G. A., N. Deza, J. Origlia, C. Toma, I. Chinen, E. Miliwebsky, S. Iyoda, S. Sosa-Estani, and M. Rivas.** 2006. Detection and characterization of Shiga toxin-producing *Escherichia coli* in captive non-domestic mammals. *Vet. Microbiol.* **118:**151–157.
21. **López, E. L., M. Diaz, S. Grinstein, S. Devoto, F. Mendilaharzu, B. E. Murray, S. Ashkenazi, E. Rubeglio, M. Woloj, M. Vasquez, M. Turco, L. K. Pickering, and T. G. Cleary.** 1989. Hemolytic uremic syndrome and diarrhea in Argentine children: the role of Shiga-like toxins. *J. Infect. Dis.* **160:**469–475.
22. **Masana, M. O., G. Leotta, L. Del Castillo, B. D'Astek, P. M. Palladino, L. Galli, E. Vilacoba, C. Carbonari, H. R. Rodríguez, and M. Rivas.** 2010. Prevalence, characterization, and genotypic analysis of *Escherichia coli* O157:H7/NM from selected beef exporting abattoirs of Argentina. *J. Food Prot.* **73:**649–656.
22a. **Masana, M. O., G. A. Leotta, O. M. Palladino, L. L. Del Castillo, C. C. Carbonari, B. A. D'Astek, E. Miliwebsky, E. Vilacoba, L. Galli, K. Irino, and M. Rivas.** 2009. Origen de la contaminación con distintos serotipos de STEC no-O157 en plantas frigoríficas de Argentina. *In Abstracts of the the XII Congreso Argentino de Ciencia y Tecnología de Alimentos.*
23. **Meichtri, L., E. Miliwebsky, A. Gioffré, I. Chinen, A. Baschkier, G. Chillemi, B. E. Guth, M. O. Masana, A. Cataldi, H. R. Rodríguez, and M. Rivas.** 2004. Shiga toxin-producing *Escherichia coli* in healthy young beef steers from Argentina: prevalence and virulence properties. *Int. J. Food Microbiol.* **96:**189–198.
24. **Mercado, E. C., A. Gioffré, S. M. Rodríguez, A. Cataldi, K. Irino, A. M. Elizondo, A. L. Cipolla, M. I. Romano, R. Malena, and M. A. Méndez.** 2004. Non-O157 Shiga toxin-producing *Escherichia coli* isolated from diarrhoeic calves in Argentina. *J. Vet. Med. B Infect. Dis. Vet. Public Health* **51:**82–88.
25. **Miliwebsky, E., L. Balbi, D. Gómez, R. Wainsztein, M. Cueto Rua, C. Roldán, M. G. Caletti, N. A. Leardini, A. Baschkier, G. Chillemi, and M. Rivas.** 1999. Síndrome urémico hemolítico en niños de Argentina: su asociación con la infección por *Escherichia coli* productor de toxina Shiga. *Bioquím. Patol. Clín.* **63:**113–121.
26. **Miliwebsky, E., N. Deza, I. Chinen, E. Martinez Espinosa, D. Gomez, E. Pedroni, L. Caprile, A. Bashckier, E. Manfredi, G. Leotta, and M. Rivas.** 2007. Prolonged fecal shedding of Shiga toxin-producing *Escherichia coli* among children attending day-care centers in Argentina. *Rev. Argent. Microbiol.* **39:**90–92.
27. **Novillo, A. A., L. E. Voyer, R. Cravioto, M. C. Freire, G. Castaño, R. Wainstein, and N. Binztein.** 1988. Haemolytic uraemic syndrome associated with faecal cytotoxin and verotoxin neutralizing antibodies. *Pediatr. Nephrol.* **2:**288–290.
27a. **Ørskov, F., and I. Ørskov.** 1984. Serotyping of *Escherichia coli*, p. 43–112. *In* T. Bergan (ed.), *Methods in Microbiology*, vol. 14. Academic Press, London, United Kindom.
28. **Ørskov, F., I. Ørskov, and J. A. Villar.** 1987. Cattle as reservoirs of verotoxin-producing *Escherichia coli* O157:H7. *Lancet* **2:**276.
29. **Oteiza, J. M., I. Chinen, E. Miliwebsky, and M. Rivas.** 2006. Isolation and characterization of Shiga toxin-producing *Escherichia coli* from precooked sausages (morcillas). *Food Microbiol.* **23:**283–288.
30. **Padola, N. L, M. E. Sanz, J. E. Blanco, M. Blanco, J. Blanco, A. I. Etcheverría, G. H. Arroyo, M. A. Usera, and A. E. Parma.** 2004. Serotypes and virulence genes

of bovine Shigatoxigenic *Escherichia coli* (STEC) isolated from a feedlot in Argentina. *Vet. Microbiol.* **100:**3–9.
31. **Palermo, M. S., R. A. Exeni, and G. C. Fernández.** 2009. Hemolytic uremic syndrome: pathogenesis and update of interventions. *Expert Rev. Anti Infect. Ther.* **7:**697–707.
32. **Parma, A. E., M. E. Sanz, J. E. Blanco, J. Blanco, M. R. Viñas, M. Blanco, N. L. Padola, and A. I. Etcheverría.** 2000. Virulence genotypes and serotypes of verotoxigenic *Escherichia coli* isolated from cattle and foods in Argentina. *Eur. J. Epidemiol.* **16:**757–762.
33. **Paton, A.W., and J. C. Paton.** 1998. Detection and characterization of Shiga toxigenic *Escherichia coli* by using multiplex PCR assays for $stx_1$, $stx_2$, *eaeA*, enterohemorrhagic *E. coli hlyA*, $rfb_{O111}$, and $rfb_{O157}$. *J. Clin. Microbiol.* **36:**598–602.
33a. **Pièrard, D., G. Muyldermans, L. Moriau, D. Stevens, and S. Lauwers.** 1998. Identification of new verocytotoxin type 2 variant B-subunit genes in human and animal *Escherichia coli* isolates. *J. Clin. Microbiol.* **36:**3317–3322.
33b. **Ramachandran, V., K. Brett, M. A. Hornitzky, M. Dowton, K. A. Bettelheim, M. J. Walker, and S. P. Djordjevic.** 2003. Distribution of intimin subtypes among *Escherichia coli* isolates from ruminant and human sources. *J. Clin. Microbiol.* **41:**5022–5032.
34. **Repetto, H. A.** 1997. Epidemic hemolytic-uremic syndrome in children. *Kidney Int.* **52:** 1708–1719.
35. **Repetto, H. A.** 2005. Long-term course and mechanisms of progression of renal disease in hemolytic uremic syndrome. *Kidney Int.* **68**(Suppl. 97)**:**S1–S5.
36. **Rivas, M., L. E. Voyer, M. Tous, M. F. de Mena, N. Leardini, R. Wainsztein, R. Callejo, B. Quadri, S. Corti, and V. Prado.** 1996. Verocytotoxin-producing *Escherichia coli* infection in family members of children with hemolytic uremic syndrome. *Medicina* (B. Aires) **56:**119–125.
37. **Rivas, M., M. G. Caletti, I. Chinen, S. M. Refi, C. D. Roldán, G. Chillemi, G. Fiorilli, A. Bertolotti, L. Aguerre, and S. Sosa-Estani.** 2003. Home-prepared hamburger as the source for a sporadic case of hemolytic uremic syndrome, Argentina. *Emerg. Infect. Dis.* **9:**1184–1186.
38. **Rivas, M., E. Miliwebsky, I. Chinen, C. D. Roldán, L. Balbi, B. García, G. Fiorilli, S. Sosa-Estani, J. Kincaid, J. Rangel, P. M. Griffin, and the Case-Control Study Group.** 2006. Characterization and epidemiologic subtyping of Shiga toxin-producing *Escherichia coli* strains isolated from hemolytic uremic syndrome and diarrhea cases in Argentina. *Foodborne Pathog. Dis.* **3:**88–96.
39. **Rivas, M., S. Sosa-Estani, J. Rangel, M. G. Caletti, P. Vallés, C. D. Roldán, L. Balbi, M. C. Marsano de Mollar, D. Amoedo, E. Miliwebsky, I. Chinen, R. M. Hoekstra, P. Mead, and P. M. Griffin.** 2008. Risk factors for sporadic Shiga toxin-producing *Escherichia coli* infections in Argentine children. *Emerg. Infect. Dis.* **14:**763–771.
40. **Roldán, M. L., I. Chinen, J. L. Otero, E. Miliwebsky, N. Alfaro, P. Burns, and M. Rivas.** 2007. Aislamiento, caracterización y subtipificación de cepas de *Escherichia coli* O157:H7 a partir de productos cárnicos y leche. Argentina. *Rev. Argent. Microbiol.* **39:** 113–119.
41. **Sanz, M. E., M. R. Viñas, and A. E. Parma.** 1998. Prevalence of bovine verotoxin-producing *Escherichia coli* in Argentina. *Eur. J. Epidemiol.* **14:**399–403.
41a. **Schmidt, H., L. Beutin, and H. Karch.** 1995. Molecular analysis of the plasmid-encoded hemolysin of *Escherichia coli* O157:H7 strain EDL933. *Infect. Immun.* **63:** 1055–1061.
42. **Snedeker, K. G., D. J. Shaw, M. E. Locking, and R. J. Prescott.** 2009. Primary and secondary cases in *Escherichia coli* O157 outbreaks: a statistical analysis. *BMC Infect. Dis.* **9:**144.
43. **Spizzirri, F. D., R. A. Rahman, N. Bibiloni, J. D. Ruscasso, and O. R. Amore.** 1997. Childhood hemolytic uremic syndrome in Argentina: long-term follow-up and prognostic features. *Pediatr. Nephrol.* **11:**156–160.
44. **Tanaro, J. D., G. A. Leotta, L. H. Lound, L. Galli, M. C. Piaggio, C. C. Carbonari, S. Araujo, and M. Rivas.** 2010. *Escherichia coli* O157 in bovine feces and surface water streams in a beef cattle farm of Argentina. *Foodborne Pathog. Dis.* **7:**475–477.

45. **Toma, C., E. Martínez Espinosa, T. Song, E. Miliwebsky, I. Chinen, S. Iyoda, M. Iwanaga, and M. Rivas.** 2004. Distribution of putative adhesins in Shiga toxin-producing *Escherichia coli* of different seropathotypes. *J. Clin. Microbiol.* **42:**4937–4946.
46. **Toma, C., N. Nakasone, E. Miliwebsky, N. Higa, M. Rivas, and T. Suzuki.** 2008. Differential adherence of Shiga toxin-producing *Escherichia coli* harboring *saa* to epithelial cells. *Int. J. Med. Microbiol.* **298:**571–578.
46a. **Tyler, S. D., W. M. Johnson, H. Lior, G. Wang, and K. R. Rozee.** 1991. Identification of verotoxin type 2 variant B subunit genes in *Escherichia coli* by the polymerase chain reaction and restriction fragment length polymorphism analysis. *J. Clin. Microbiol.* **29:** 1339–1343.
47. **World Health Organization.** 1997. Prevention and control of enterohaemorrhagic *Escherichia coli* (EHEC) infections. Report of a WHO consultation. World Health Organization, Geneva, Switzerland.
48. **Zhang, W. L., M. Bielaszewska, T. Kuezius, and H. Karch.** 2002. Identification, characterization, and distribution of Shiga toxin 1 gene variant (*stx*1c) in *Escherichia coli* isolated from humans. *J. Clin. Microbiol.* **40:**1441–1446.

*Section III*

# COMMON THEMES AMONG *ESCHERICHIA COLI* PATHOGENS

*Population Genetics of Bacteria: a Tribute to Thomas S. Whittam*
Edited by S. T. Walk and P. C. H. Feng


*Chapter 9*

# Gene Acquisition and Loss in the Phylogenetic Lineages of the Invasive *Escherichia coli*

***Alyssa C. Bumbaugh and David W. Lacher***

*Shigella* and enteroinvasive *Escherichia coli* (EIEC) are a class of enteric pathogens that have evolved the ability to invade epithelial cells and cause severe intestinal illness. For historical reasons, these bacteria are referred to as belonging to two genera, *Shigella* and *Escherichia*; however, this classification is artificial based upon molecular evidence, and they are more appropriately referred to collectively as invasive *E. coli*. All known *Shigella* species are invasive and cause bacillary dysentery, whereas only certain serotypes of *E. coli*, the EIEC strains, have the same ability.

There are four species of *Shigella* (*S. boydii, S. dysenteriae, S. flexneri*, and *S. sonnei*) that have been recognized historically because of the severity of disease and their clinical importance. Worldwide, shigellosis is responsible for more than 1.1 million deaths each year (38), and in the United States, the incidence rate is estimated to be approximately 1 in 25,000 (8), with the highest number of infections occurring in children less than 12 years of age. The four *Shigella* species designations are based on differences in the expression of specific somatic antigens (O types), clinical presentation, and biochemical properties that allow for the identification and distinction between isolates. The number of O types ranges from a single type of *S. sonnei* to 18 O types for *S. boydii*. Disease caused by specific *Shigella* strains varies geographically and has changed historically. At the present time, *S. flexneri* type 2a is the most common cause of shigellosis in developing countries, whereas *S. sonnei* continues to account for most infections in industrialized countries.

Like other types of *E. coli, Shigella* is transmitted by the fecal-oral route, usually as a result of direct person-to-person transfer or through contact with or ingestion of contaminated food and water. The infectious dose is very low,

***Alyssa C. Bumbaugh*** • Department of Biology, Shippensburg University, Shippensburg, PA 17257. ***David W. Lacher*** • Division of Molecular Biology, Office of Applied Research and Safety Assessment, Center for Food Safety and Applied Nutrition, U.S. Food and Drug Administration, Laurel, MD 20708.

with ingestion of less than 10 (reported for *S. dysenteriae*) to 1,000 (reported for *S. flexneri* type 2a) bacteria causing symptomatic infections in human volunteer studies (77). Strains of *Shigella* have been found to colonize only humans and nonhuman primates; no alternative species of animal reservoirs have been identified. However, houseflies (*Musca domestica*) have been implicated as mechanical vectors of transmission (44).

EIEC strains are similar to *Shigella* and were first identified in Italy in the 1940s (18). Isolates within the EIEC pathotype have been found to harbor the same virulence plasmid as *Shigella* (32). Also like *Shigella*, EIEC have been involved in several large outbreaks of acute gastroenteritis in the United States (26, 32, 71, 81) and have also been implicated in traveler's diarrhea (84). In the developing world, EIEC infections contribute to endemic rates of diarrheal disease; EIEC strains are typically isolated in 1 to 5% of the cases of acute diarrhea in children (14, 21, 43, 64, 75, 78), although incidence rates vary with season (64) and socioeconomic conditions (78).

## INVASIVE PHENOTYPE

Invasive *E. coli* strains have a characteristic form of pathogenesis involving invasion of the mucosal epithelial cells of the large intestine. The molecular and cellular events underlying epithelial cell invasion by the invasive *E. coli* have been intensively studied and reviewed (23, 31, 56, 67, 68). Briefly, invasion occurs via bacterium-directed phagocytosis with the major events as follows: contact of bacteria with the surface of the epithelial cell induces rearrangements of the cytoskeleton of the host cell; local membrane ruffling; and uptake of the bacteria (11). Inside the cell, the bacteria escape from the endosomal vacuole by lysing the membrane, after which they enter the cytoplasm and multiply. The intracellular bacteria move through the cytoplasm by polymerizing actin filaments. This movement results in protrusions from an infected cell's membrane that contain bacterial cells at the tip. These protrusions can then be engulfed by adjacent cells. In this way, the invasive bacteria can multiply and spread from cell to cell without being exposed to the extracellular environment.

## INVASIVE GENOTYPE

The molecular components underlying the invasive phenotype are encoded on the ~200-kb pINV plasmid (Fig. 1). The pINV plasmids vary in size and composition, but in general they include an entry region containing 35 genes organized into at least four transcriptional units (56). These genes are homologous to the genes of the SPI-1 island of *Salmonella* (25) and include a type III secretion system (T3SS), secreted proteins, molecular chaperones, and regulators encoded by *virB-ipgD, icsB-mxiE, mxiM-spa13*, and *spa47-spa40*. The pINV plasmid also carries genes for the intracellular actin-based motility of the

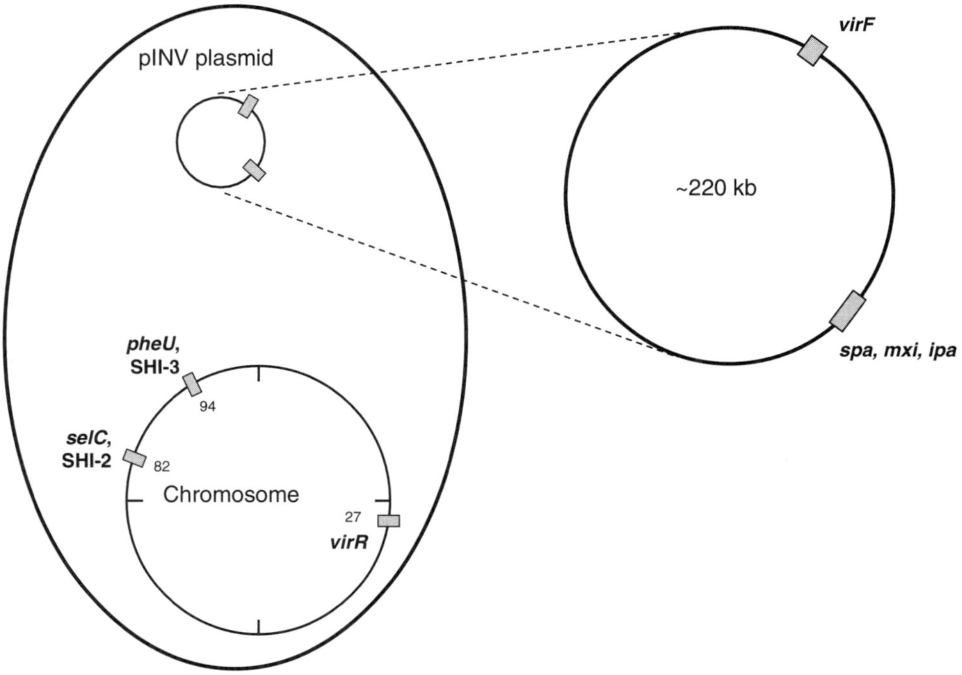


**Figure 1.** General genetic composition of an invasive *E. coli* isolate. The pINV plasmid encodes genes involved in host cell entry and it is found in all *Shigella* and EIEC. The chromosome contains the *virR* locus, which regulates the acquired pINV plasmid. Example PAIs (SHI-2 and SHI-3) are incorporated in or near tRNA sites. The distribution of PAIs and other virulence loci is variable among the invasive *E. coli*.

invasive *E. coli*, a variety of plasmid antigens, and other suspected virulence-related proteins. Although most of the molecular reports describing pINV have been conducted using *S. flexneri*, it is clear that many of the genes on pINV are critical to cell invasion and are required for full virulence of all invasive *E. coli*.

Extensive work has been done to identify the regulatory pathway that ultimately results in the expression of the pINV T3SS. The regulation of pINV loci is temperature dependent (19), with a chromosomal locus, *virR* (48), binding upstream of the pINV *virF* locus, and causing a conformational change in the topology of the DNA. VirF is then responsible for the transcriptional regulation of *virB* (2), also located on pINV. VirB acts as a transcriptional activator for the regulation of the genes encoding the T3SS as well as additional effector molecules. In some instances, the plasmid can become integrated into the chromosome, resulting in a reduction of *virB* transcription and, ultimately, a non-invasive phenotype that could provide a selective advantage for these pathogens under certain circumstances (9).

## PHYLOGENETIC RELATIONSHIPS AMONG *E. COLI* AND *SHIGELLA* STRAINS

In addition to their ability to invade epithelial cells, invasive *E. coli* strains often share other phenotypic properties; they usually do not decarboxylate lysine and, with a few exceptions, they are nonmotile (5, 70). Together these observations encouraged the notion that invasive strains are evolutionarily related and represent a specialized natural group of bacteria. Ochman et al. (55) used multilocus enzyme electrophoresis to assess genetic variation at enzyme-encoding loci in a diverse, global collection of *E. coli* and *Shigella*. The results of this study showed that in terms of genetic distance, there is a very close relationship between *Shigella* and *E. coli*, and that the assignment of *Shigella* to separate species, or even a distinct genus, is unwarranted from an evolutionary standpoint. In 1997, Pupo et al. (60) supported and extended these findings by examining 32 strains including representatives from the enteropathogenic *E. coli* (EPEC), enterohemorrhagic *E. coli* (EHEC), and enterotoxigenic *E. coli* (ETEC) pathotypes as well as 12 *Shigella* and 5 EIEC strains. The bacteria were characterized by multilocus enzyme electrophoresis for 10 enzyme-encoding genes and the nucleotide sequence for part of the *mdh* gene. Independently, Ochman and Pupo found that *Shigella* strains are within the genetic diversity of *E. coli* and form at least two distinct clusters corresponding to *S. boydii* and *S. flexneri* serotypes.

A later study by Pupo et al. (61) represents the most extensive evolutionary analysis of *Shigella* at the DNA sequence level. In this work, they determined the nucleotide sequence of 7,160 bp representing eight housekeeping loci from four regions of the genome. Phylogenies constructed separately for each region were very similar in topology, with all but five of the *Shigella* strains falling into one of three main clusters. There was only a small amount of nucleotide diversity within clusters, and most of the divergence occurred between clusters. Because the same genetic relationships were seen for the genes in each genomic region, and the clusters were all supported by high bootstrap confidence limits, Pupo et al. concluded that these clusters are robust and mark distinct phylogenetic groups (61). The serotypes identified for each group are summarized in Table 1. A phylogenetic analysis by Escobar-Paramo et al. (17) based on four chromosomal and three virulence plasmid loci provided additional support for the reported phylogenetic groups.

Lan et al. (40) identified four additional invasive clusters (Groups 4 to 7) (Table 1) when the phylogenetic analysis was extended to include the EIEC pathotype. This study generated nucleotide sequence data for four housekeeping and two plasmid loci from 32 isolates representing 12 EIEC serotypes. Out of these four groups, three had more than one O type (Table 1), with both O124 and O164 found in Groups 4 and 5. Since mucate formation and acetate utilization can differentiate *Shigella* and EIEC, these biochemical markers were examined in the context of the phylogeny. Group 4 isolates are mucate negative and Group 6 isolates are positive for both mucate formation and acetate utilization (40). It was also noted that fewer serotypes exist within the EIEC clusters

**Table 1.** Classification of *Shigella* and EIEC serotypes into phylogenetic groups

| Phylogenetic group | Serotypes[a] |
|---|---|
| Group 1 | B1, B2, B3, B4, B6, B8, B10, B14, B18 |
| | D3, D4, D5, D6, D7, D9, D11, D12, D13 |
| | F6, F6A |
| Group 2 | B5, B7, B9, B11, B15, B16, B17 |
| | D2 |
| Group 3 | B12 |
| | F1A, F1B, F2A, F2B, F3A, F3B, F3C, F4A, F4B, F5, FX, FY |
| Group 4 | O28, O29, O124, O136, O164 |
| Group 5 | O124, O135, O152, O164 |
| Group 6 | O143, O167 |
| Group 7 | O144 |
| Others | B13 |
| | D1, D8, D10 |
| | SS |

[a]B, *S. boydii*; D, *S. dysenteriae*; F, *S. flexneri*; SS, *S. sonnei*.

(Groups 4 to 7) than the *Shigella* clusters (Groups 1 to 3). This low level of divergence suggests a more recent origin for the EIEC within the *Escherichia* genus. Lan et al. (40) also suggested that EIEC is intermediate between *E. coli* and *Shigella*.

Building upon the earlier reports, Yang et al. (88) sequenced 23 housekeeping loci in *Shigella* and EIEC. This report suggested the division of Group 1 serotypes into three subclusters. Subcluster 1 contains *S. dysenteriae* types 3, 4, 6, 9, 11, 12, and 13. Subcluster 2 comprises *S. boydii* types 1, 3, 6, 8, 10, and 18 and *S. dysenteriae* type 5. Subcluster 3 consists of *S. boydii* types 2, 4, and 14 and *S. flexneri* type 6. This analysis also indicated that *S. dysenteriae* type 1 and EHEC O157:H7 isolates share an immediate ancestor (88).

Additional studies have also provided support for the existence of outlier serotypes. Hyma et al. (35) performed an extensive study with *S. boydii* type 13. Results from this analysis confirmed that some *S. boydii* type 13 isolates are actually *Escherichia albertii*, a distinct species separate from *E. coli*, while other *S. boydii* type 13 isolates are members of the *E. coli*. A multilocus variable-number tandem repeat analysis (MLVA) study (27) of 68 *Shigella* isolates of various serotypes indicated that *S. sonnei* isolates were the least variable, supporting the single clonal origin hypothesis for *S. sonnei*. While the MLVA study examined numbers of tandem repeats as opposed to nucleotide sequence, the phylogenetic clusters generated with these data are in agreement with those reported by Pupo et al. (61) and Lan et al. (40) with the exception of a *S. dysenteriae* type 8 isolate being found in Group 3. MLVA also supported *S. boydii* type 13 as an outlier of the *E. coli* (27).

A comparative genomic analysis by Touchon et al. (80) identified a collection of just less than 2,000 stable, core genes for *E. coli, Shigella*, and *Escherichia fergusonii*. These core genes were used to generate a phylogeny that clearly identified invasive *E. coli* Groups 1 and 3 and the clonal *S. sonnei* lineage. A Group 2 genome is currently unavailable and, therefore, not represented in the analysis. This study provided evidence that the invasive *E. coli* have emerged recently with respect to other *E. coli* lineages (80).

Data supporting the phylogenetic lineages of the invasive *E. coli* were also generated in the Whittam lab. We used a multilocus sequence typing (MLST) scheme based on 15 housekeeping loci (35) in 27 *Shigella* and 15 EIEC isolates and identified the 7 reported evolutionary lineages of invasive *E. coli* (T. Whittam, unpublished data) (Fig. 2). The MLST loci used and isolates examined were independent from the earlier publications. The MLST groupings of serotypes in our study were concordant with Pupo et al. (61) and Lan et al. (40), suggesting that it is a reliable approach to determine the relationships of the invasive *E. coli* and that the serotype of the isolate is often, but not always, an indicator of the phylogenetic lineage.

Taken together, the collective molecular evolutionary analyses support three main conclusions. First, bacteria that belong to four traditional species of *Shigella* (*S. boydii, S. dysenteriae, S. flexneri*, and *S. sonnei*) cluster within the genetic diversity of *E. coli* and form a single pathotype of *E. coli*. Second, *Shigella* strains do not form a single lineage within *E. coli*; that is, they are not a monophyletic group but instead have multiple origins within *E. coli*. Finally, the recognized species of *Shigella* (classification based on O type and biochem-

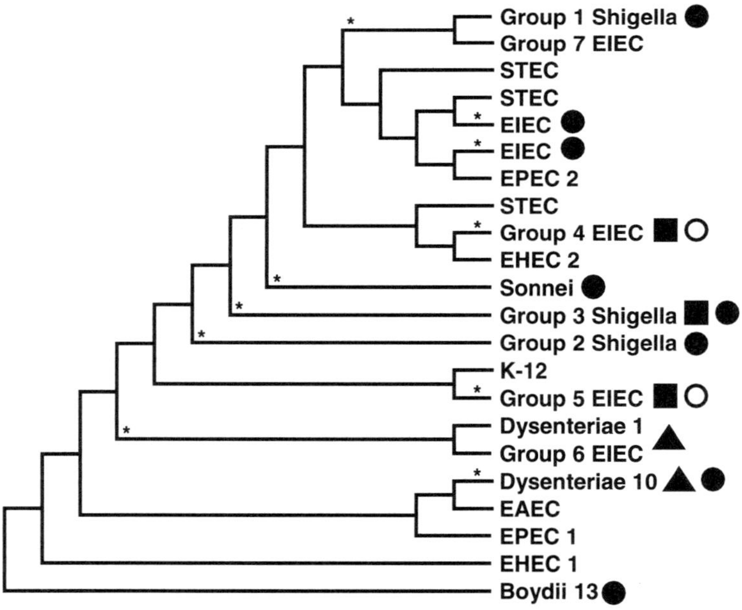


**Figure 2.** General phylogenetic relationships between the invasive *E. coli* and other *E. coli* pathotypes. Patterns of gene acquisition are indicated as follows: * = pINV; square = SHI-1; circle = aerobactin operon; triangle = *shuA*. The unfilled symbols indicate variable presence of the virulence genes within the group. EAEC, enteroaggregative *E. coli*; STEC, Shiga toxin-producing *E. coli*.

ical properties) do not represent natural subgroups within *E. coli*, but instead the species are dispersed in genetically distinct groups (Table 1) that are not in accordance with the phenotypic and antigenic properties that define the species classification.

## GENOMIC EVOLUTION: GENE ACQUISITION

The acquisition of new genes by horizontal transfer has played a major role in the adaptation and ecological specialization of bacterial lineages (41). It has been estimated, for example, that ~18% of the current genome of *E. coli* K-12 represents foreign DNA acquired by horizontal transfers since the divergence of *E. coli* and *Salmonella enterica* (42). Gene acquisitions have also contributed to the variation in virulence among strains and closely related bacterial species (30, 65). In *E. coli* and *S. enterica*, pathogenicity islands (PAIs) have been acquired at different times, thus generating a variety of pathogens with distinct virulence genes and mechanisms of pathogenesis (29, 53, 54).

In the evolution of the invasive *E. coli*, gene acquisition has been important in two ways: first, with the spread of the pINV plasmid that encodes invasive ability; and second, with the presumed acquisition of a variety of mobile PAIs (Fig. 1 and 2). Based on the sequence analysis of three genes, Lan and Reeves (39) found that the *Shigella* invasion plasmid can be classified into two homogeneous sequence types, called pINVA and pINVB. The plasmid sequence types have an interesting alignment with the *Shigella* groups (Table 1): pINVA occurs in Group 1 strains, whereas pINVB occurs in Group 3 strains. Both types are found in Group 2 strains, several EIEC strains, as well as in *S. sonnei* and *S. dysenteriae* types 1 and 10. This pattern supports the hypothesis that there have been several lateral transfers of the pINV plasmids to create new invasive lineages.

A study by Escobar-Paramo et al. (17) suggested that the pINV plasmid was acquired only once in the evolutionary history of the invasive *E. coli*, with subsequent exchange of plasmid genes occurring between the groups. This model was based upon sequence data from four chromosomal and three virulence plasmid loci derived from a collection of 51 *Shigella* and EIEC strains. To test an alternative prediction, Yang et al. (88) sequenced five plasmid genes in an independent collection of 38 *Shigella* and EIEC strains and determined that Groups 1 and 2 have a different ancestral plasmid than Group 3. This result supports the earlier report of two plasmid forms with multiple origins of pINV in the evolution of the invasive *E. coli*. Sequence data indicate that the plasmids are diverse and were once transferable since the *tra* operon is present in EIEC but not *Shigella*. This indicates that transfer of pINV between different lineages could be possible.

Comparative genomic analyses have also examined the pINV in detail. In the genome analyses of strain Sb227 (*S. boydii* type 4), Sd197 (*S. dysenteriae* type 1), and Ss046 (*S. sonnei*), Yang et al. (87) reported that all of the entry genes harbored on the plasmids of each isolate were flanked by insertion sequences (ISs). The Nie et al. (51) comparison of the pINV plasmids from *S.*

*flexneri* types 2a and 5b (strains Sf301 and Sf8401, respectively) indicated that these Group 3 isolates have similar plasmids. The development of the Blast2Network (B2N) by Brilli et al. (7) allowed for determination of plasmid homology in all sequenced *E. coli, Salmonella,* and *Shigella* plasmids. This phylogenetic profiling resulted in a distinct clade of *Shigella* plasmids with the single exception of the ColJS plasmid from *S. sonnei*. The profiling also identified gene clusters specific to the pINV plasmids (T3SS, *mxi, spa, ipa, ipg,* and *osp*) as well as others that were shared among the three genera (antibiotic resistance and sex pilus-related genes) (7).

In addition to pINV, five PAIs have been identified and characterized among invasive strains. One encodes a *Shigella* enterotoxin, three carry operons involved in iron scavenging, and the fifth specifies O-antigen modification (Table 2). It has been suggested by Escobar-Paramo et al. (16) that a specific chromosomal background is necessary not only for the acquisition but also for the expression of virulence genes in *E. coli* and *Shigella*. This report found that the strains causing severe, acute diarrheal diseases, such as ETEC, EHEC, and EIEC, have a more derived genetic background as compared to the ancestral background found in the pathovars causing mild or chronic forms of diarrheal disease.

**Table 2.** PAIs identified in *Shigella* strains

| Island | Serotype (strain) | tRNA (min) | Size in kb (no. of ORFs) | Associated virulence factors | Reference(s) |
|---|---|---|---|---|---|
| SHI-1 | *S. flexneri* type 2a (YSH6000) | *pheV* (67) | 46.6 (31) | ShET1 enterotoxin (*set1*), autotransporter protease (*sigA*), mucinase (*pic*) | 3, 63 |
| SHI-2 | *S. flexneri* type 2a (SA100)<br>*S. flexneri* type 5a (M90T) | *selC* (82)<br>*selC* (82) | ~30.0 (35)<br>23.8 (24) | Aerobactin operon (*iucA-D*), aerobactin receptor (*iutA*), ColV immunity (*shiD*) | 83<br>50 |
| SHI-3 | *S. boydii* type 5 (0-1392) | *pheU* (94) | 20.9 (16) | Aerobactin operon (*iucA-D*), aerobactin receptor (*iutA*) | 62 |
| SHI-4 (SRL) | *S. flexneri* type 2a (YSH6000) | *serX* (24) | 66.3 (59) | Multiple antibiotic resistance genes, ferric dicitrate transport (*fec* operon) | 46 |
| SHI-O | *S. flexneri* type 1a (Y53) | *thrW* (6) | 10.6 (5) | Serotype F1a O-antigen modification | 1 |

SHI-1 (*Shigella* island 1; formerly known as *she*) is a 46.6-kb PAI located at the 3′ end of the *pheV* tRNA gene in *S. flexneri* (Table 2). The island encodes several virulence factors including ShET1 (*Shigella* enterotoxin 1), whose activity was originally isolated in culture filtrates of a plasmid-cured strain that caused significant fluid accumulation in rabbit ileal loops (20). ShET1 is encoded by *set1A* and *set1B* and is associated predominantly with *S. flexneri* type 2a strains (3, 4, 52, 63). SHI-1 also encodes SigA, a cytopathic protease that contributes to intestinal fluid accumulation, and Pic, a protease with mucinase and hemagglutinin activities. Interestingly, *set1* and *pic* have overlapping reading frames, with *set1* being transcribed from the complementary strand. The island contains many intact and truncated mobile genetic elements, plasmid-related sequences, and several open reading frames (ORFs) with high sequence similarity to those found on O islands in the *E. coli* O157:H7 genome.

SHI-2 is located on the chromosome near the *selC* tRNA locus, the site of insertion of PAIs in several other enteric pathogens. SHI-2 has been characterized in strains of *S. flexneri* types 2a and 5a (50, 83). The two variants of the island that have been studied differ in length, but both encode an aerobactin system for iron acquisition, immunity to colicin V, and several other proteins (Table 2). It has been hypothesized that proteins produced by SHI-2 enhance fitness by facilitating bacterial survival in competitive situations with other bacteria.

SHI-3 is a 21-kb PAI that also carries genes for the synthesis and transport of aerobactin, as well as a P4 prophage-like integrase gene and numerous IS elements (62). This island was found in an *S. boydii* type 5 strain (0-1392) and is located at the *pheU* tRNA locus in some *S. boydii* isolates but not in others. Although the aerobactin operon is thought to be advantageous in certain environments, an *S. boydii* aerobactin synthesis mutant (0-1392 *iucB*) did not differ from wild type in tissue culture assays of invasion and intercellular spread (62).

SHI-4 (or SRL, for *Shigella* resistance locus) is a 66-kb PAI composed of antibiotic resistance genes (*tet, cat, oxa-1*, and *aadA1*) closely linked to the *fec* operon, a ferric dicitrate iron transport system. The PAI is embedded in a larger (~99-kb) multiple resistance deletable element flanked by IS*91*-like sequences that is capable of precise excision (46). SHI-4 appears to be widely disseminated among *Shigella*, although its distribution in light of the phylogenetic classification of the invasive *E. coli* is unknown. The *fec* system is one of several iron uptake systems whose primary role in virulence may be in the uptake of iron from the intestinal lumen, where exogenous citrate is available for chelation (46).

An additional island described by Adhikari et al. (1) consists of serotype conversion genes in an *S. flexneri* type 1a strain. These genes occur on a unique segment of the chromosome, have many of the characteristics of a PAI, and have thus been referred to as the SHI-O island (36). Sequence analysis suggests that the present transposon-like structure of SHI-O was originally part of a bacteriophage that integrated near the *thrW-proA* attachment site (~6 min in the K-12 genome). Interestingly, the opposite end of the element shows homology to the *dsdC* gene in *E. coli*, which maps to minute 53, suggesting that the chromosome of the *S. flexneri* type 1a strain examined has undergone additional genomic rearrangements.

## PHYLOGENETIC DISTRIBUTION OF ACQUIRED VIRULENCE FACTORS

Virulence determinants of the invasive *E. coli* are most often examined in isolates of clinical importance, but it is also informative to investigate the occurrence and distribution of virulence genes from an evolutionary perspective. A study by Purdy et al. (62) used PCR assays to examine the distribution of SHI-2 and SHI-3 along with known integration sites and reported the results using the traditional species classification (comparison of species groupings as opposed to phylogenetic groupings). By examining the results using a phylogenetic perspective, it is suspected that SHI-2 occurs in Group 3 in the *selC* site, SHI-3 occurs in Group 2 and some Group 3 strains in the *pheU* site, and neither island occurs in Group 1 strains. SHI-2 is also found in *S. dysenteriae* type 1, and both SHI-2 and SHI-3 appear to be found in *S. sonnei*. The presence of an aerobactin operon is indicative of SHI-2 and SHI-3; however, a survey of the distribution of the *iucC* locus shows that it occurs in the invasive and other *E. coli* pathotypes (16). Genomic sequences indicate the presence of SHI-2 in *S. sonnei* (but not in the *selC* site) and SHI-3 in the *pheU* site of the *S. boydii* type 4 genome strain (87). There are inversions that have occurred near the SHI-2 and SHI-O islands of the *S. flexneri* type 5b genome as compared to the two *S. flexneri* type 2a genomes. The SHI-2 PAI spans the chromosomal origin in strains Sf301 (*S. flexneri* type 2a) and Sf8401 (*S. flexneri* type 5b). The distance between the *iut/iuc* operon and the origin differs in these two strains due to a subsequent inversion within an inversion in the Sf8401 genome (51).

Two additional studies examined the distribution of the SHI-1 and SHI-4 PAIs in *Shigella* and EIEC strains. Al-Hasani et al. (3) used PCR to detect *sigA* and *pic* in enteropathogens. When placed in a phylogenetic perspective, the results of this study found *sigA* to have a wider distribution among different *Shigella* types, whereas *pic* appears to be localized to the Group 3 strains. The genome sequences of Sd197, Sb227, Ss046, and Sf8401 indicate that SHI-1 is absent in *S. dysenteriae* type 1 and *S. flexneri* type 5b (instead, a general secretion pathway island is present at the *pheV* site) and the *pic/setAB* genes are not found in *S. boydii* type 4 or *S. sonnei* (51, 87). The molecular epidemiology of the SHI-4 PAI was also investigated by Turner and colleagues (82). PCR amplification of three marker loci was used to screen for SHI-4, which appeared to be widespread among the clonal groups. Since the isolates used in this study were selected on the basis of resistance to multiple antibiotics (82), it is possible that the actual distribution of these loci among *Shigella* and EIEC may be different in a randomly sampled collection of strains. Sampling bias may have occurred, as only a portion of the population was surveyed.

Other reports of gene acquisition stem from comparisons of available genome sequences. The *Shigella* genomes (Table 3) (37, 51, 85, 87, 89) carry

**Table 3.** Invasive *E. coli* genome sequences publicly available through the NCBI Genomes database

| Species | Serotype (strain) | Size (Mbp) | Phylogenetic group | Accession no. | Reference[b] |
|---|---|---|---|---|---|
| *S. boydii* | 18 (3083-94) | 4.6 | 1 | NC_010658 | Rasko et al. [a] |
| | 4 (Sb227) | 4.5 | 1 | NC_007613 | 87 |
| *S. dysenteriae* | 1 (Sd197) | 4.4 | Other | NC_007606 | 87 |
| | 4 (1012)[a] | 5.2 | 1 | NZ_AAMJ00000000 | Rasko et al. [b] |
| *S. flexneri* | 2a (2457T) | 4.6 | 3 | NC_004741 | 85 |
| | 2a (Sf301) | 4.6 | 3 | NC_004337 | 37 |
| | 5b (Sf8401) | 4.6 | 3 | NC_008258 | 51 |
| | X var (2002017) | 4.6 | 3 | CP001383 | 89 |
| *S. sonnei* | - (Ss046) | 4.8 | Other | NC_007348 | 87 |
| *E. coli* | O144 (53638)[a] | 5.1 | 7 | NZ_AAKB00000000 | Rasko et al. [c] |

[a] In progress.
[b] Unpublished data from [a] D. A. Rasko, M. Rosovitz, A. T. Maurelli, G. Myers, R. Seshardri, R. Cer, L. Jiang, J. Ravel, and Y. Sebastian; [b] D. A. Rasko, M. Rosovitz, G. Myers, R. Seshardri, R. Cer, L. Jiang, J. Ravel, W. F. Fricke, and Y. Sebastian; [c] D. A. Rasko, M. Rosovitz, C. Brinkley, G. S. A. Myers, R. Seshardri, R. Z. Cer, L. Jiang, Y. Sebastian, and J. Ravel.

multiple copies of *ipaH* (chromosomal and plasmid borne), which have a conserved carboxyl terminus but are variable at the amino terminus, suggesting horizontal transfer. A phylogenetic study found *ipa* to be widespread in the invasive *E. coli* (16). Yang et al. (87) described regions similar to the O islands identified in the genomic sequences of *E. coli* O157 (33, 57). These regions consisted of genes encoding RTX-like exoproteins and transporters, iron compound ABC transporters, and adhesins. A type II secretion system was also identified in *S. dysenteriae* type 1 strain Sd197, providing evidence for active secretion of the Shiga toxin (87). An osmotic tolerance region (encoded by *bet*) was identified near SHI-O in *S. flexneri* type 5b strain Sf8401, perhaps providing for enhanced survival (51).

Two iron acquisition systems have been reported in *S. dysenteriae* type 1 and later confirmed with the publication of the genome sequence of strain Sd197. Runyen-Janecky et al. (66) identified an iron acquisition locus, *sit*, and found the distribution to be widespread among the invasive *E. coli*. It has been suggested that this locus may be on a previously unidentified PAI (66). In addition to the PAIs, other virulence-associated loci have been reported in *Shigella*. ShuA, a heme-binding protein, was characterized in *S. dysenteriae* serotype 1 (49) and also reported in two EIEC serogroups (O136 and O143) (86). In independent reports, *S. dysenteriae* types 1 (16) and 10 were found to be positive for *shuA* (88). Results from a study in the Whittam laboratory identified the presence of *shuA* in *S. dysenteriae* types 1 and 10, both Group 6 serotypes, and two *S. boydii* type 13 strains (3556-77 and 3054-94) (Whittam, unpublished). The report by Hyma et al. (35) showed a relatively close relationship between the *S. dysenteriae* type 1/EIEC cluster and a cluster containing some atypical *S. boydii* type 13 isolates.

Other virulence-associated proteins like the serine protease autotransporter of *Enterobacteriaceae* (SPATEs) are important for the pathogenesis of the invasive *E. coli*. A report by Boisen et al. (6) examined the distribution of both class I and class II SPATEs in *S. flexneri* and *S. sonnei*. All of the *Shigella* isolates were PCR positive for at least one SPATE. The Group 3 serotypes were positive for *sat* and variably positive for the class II SPATEs *sepA* and *pic*, indicating that these genes are not always linked (6). *S. flexneri* type 6 and *S. sonnei* were positive for *sigA* but were negative for *sepA* and *pic*. This difference in SPATE acquisition is most likely due to the independent evolutionary origins of the *Shigella* groups.

A report by Talukder and colleagues (74) provided evidence for the initial acquisition of pINV followed by later gain of additional virulence loci. In the recently emerged *S. flexneri* type 1c, Talukder et al. (74) found that all of these isolates were positive for *Shigella* enterotoxin 2 (ShET2 or *senA*), encoded by pINV, while none were positive for ShET1 (*set1*), which is encoded by the SHI-1 PAI. The distribution of the *set1A* locus suggests lateral transfer could be responsible for the inconsistent dispersal within Group 3. Genetic characterization of the *set1A* locus in the Whittam laboratory showed that the gene was present in six isolates of MLST Groups 3, 4, and 5. Nucleotide sequence analysis showed there was little variation at the nucleotide level, with only two variable sites out of 436 base pairs (Whittam, unpublished).

There are few reports on the distribution of virulence elements in the EIEC groups. However, it appears that the EIEC stably maintain the large virulence plasmid conferring the invasive phenotype. Lan et al. (40) found that the gene repertoire of pINV is variable in EIEC, since *sepA* and *senA* are not always present. SPATEs also have a variable presence in the EIEC. In studying 22 EIEC isolates, Boisen et al. (6) reported sporadic distribution of the class I *sigA* locus and absence of the class II *pic* locus. The acquisition of chromosomally encoded (such as *shuA*) and PAI-associated virulence loci (such as aerobactin operons) is much more variable within the EIEC isolates. Because of this, it is possible that these loci have been acquired recently in the evolutionary history of this pathotype.

In both *Shigella* and EIEC, the acquisition of pINV seems to be the important "first step" in the divergence from the other pathogenic *E. coli* lineages. Possessing the pINV plasmid confers some virulence attributes; however, obtaining PAIs (Table 2), toxins, and antibiotic resistance genes through horizontal gene transfer events has proven to provide an even greater selective advantage for certain members of the invasive *E. coli*. The acquisition of pINV has occurred independently a minimum of 11 times in the evolutionary history of *E. coli* as the invasive lineages arose (Fig. 2), and there is heterogeneity within the identified groups as demonstrated by the differences in the pINV plasmids and variable distribution of the five characterized PAIs and other virulence loci. While gene acquisition is a major contributor to the pathogenesis caused by the invasive *E. coli*, this is only one side of the evolutionary story.

## GENOMIC EVOLUTION: GENE LOSS

In contrast to gene acquisition, there is growing evidence that gene loss has been important in the adaptive radiation of *Shigella* and EIEC. There are several metabolic and structural traits in *Shigella* that appear to have been lost in parallel at multiple times in the divergence of invasive clones. For example, *S. dysenteriae* isolates are distinguished biochemically from the other *Shigella* by their inability to ferment mannitol (15). It is reported that some of these loss-of-function phenotypes will result from major gene deletions such as those described with the lysine decarboxylase regions of *S. flexneri*, while others will be caused by IS inactivation or point mutations such as frameshifts and nonsense substitutions resulting in pseudogenes. The large, universal deletions that enhance virulence are referred to as "black holes" (47), the loss-of-function counterpart to PAIs, and are examples of pathogenicity-adaptive, or patho adaptive, mutations (72). Pathoadaptive evolution enhances bacterial virulence without horizontal transfer of specific virulence factors; instead, the increase in bacterial virulence results from random mutations in genes or pathways that are incompatible with pathogenesis. The incompatible genes themselves have been designated as antivirulence loci (AVL) (59).

Maurelli et al. (47) presented evidence that the universal loss of lysine decarboxylase activity (encoded by *cadA*) enhanced the virulence of *Shigella* species because cadaverine, a product of the reaction catalyzed by lysine decarboxylase, inhibits the activity of the *Shigella* enterotoxin. The *cadA* locus is absent in strains Sf301 and Sb227 (*S. flexneri* type 2a and *S. boydii* type 4, respectively) and function is eliminated in strains Sd197 and Ss046 (*S. dysenteriae* type 1 and *S. sonnei*, respectively). A frameshift mutation is responsible for the loss of function in strain Sd197 and an IS insertion eliminates the gene function in strain Ss046 (87). Another AVL example in *Shigella* is the loss or inactivation of the *nadA* and *nadB* genes, which are responsible for the biosynthesis of quinolinate (59). Quinolinate, an intermediate of the *nadAB* pathway, blocks invasion of the invasive *E. coli* by impacting T3SS-mediated activities of invasion and cell-to-cell spread (59). The inactivation of *nadAB* in a collection of 14 *Shigella* and EIEC strains was demonstrated to occur independently through IS elements, deletions, and point mutations (58). Studies of both examples of AVL (*cadA* and *nadAB*) provide evidence for the independent, multiple origins of the invasive lineages and convergent evolution of those lineages.

Based upon the established invasive *E. coli* phylogenetic groups and reported biochemical characteristics, the loss of gene function encoded by the mannitol operon has occurred at least four times in the evolutionary history of *Shigella*, since the *S. dysenteriae* types appear in four different phylogenetic lineages (Table 1). Because this loss occurs within an operon of housekeeping function, it seems likely that there is a selective advantage for inactivating the metabolic pathway. The MLST study conducted in the Whittam laboratory found that the mannitol-1-phosphate dehydrogenase locus (*mtlD*) was either absent or inactivated by an IS2-like element for the *S. dysenteriae* types and *S. flexneri* type 6 strains. The *mtlA* locus of the operon was absent in all *S. dys-*

*enteriae* types and *S. flexneri* type 6 strains that were examined, and the *mtlR* locus was present only in *S. dysenteriae* types 2 and 3 (Whittam, unpublished). These observations suggest that natural selection has favored inactivation of the mannitol operon. Five serotypes (*S. dysenteriae* types 3, 4, 6, 9, and 11) have an IS*2*-like element inserted near the 5′ end of the *mtlD* gene between bases 208 and 209. An early pathogenesis study showed that the incorporation of the *E. coli* chromosomal region bounded by *xyl* and *rha* (which includes *mtl*) into an *S. flexneri* type 2a background led to a loss of fatal infection in the starved guinea pig model (22). Genes within this region encode the aerobactin-binding protein and receptor (28) in *S. flexneri* type 2a and the structural genes for Shiga toxin in *S. dysenteriae* type 1 (69). Taken together, this suggests the possibility of AVL in the mannitol operon. Talukder et al. (73) identified a subgroup of atypical *S. flexneri* type 4 isolates that are also mannitol negative, further suggesting the selective, independent loss of this metabolic pathway. *Shigella* are also biochemically inert for the utilization of other carbohydrates such as sorbitol. The genome sequences indicate inactivation of this metabolic pathway by the creation of pseudogenes (*srlE* in Sf8401 and *srlA* in Sf301) (51). The genome sequences of Sd197, Sb227, and Ss064 all predict biochemical diversity in carbohydrate and amino acid metabolic pathways (87), and the loss of metabolic functions among the *Shigella* could be due to convergent evolution.

Other candidate AVL that might be undergoing independent selective loss are the phosphoprotein phosphatases encoded by *prpA* and *prpB*. These phosphatases are part of the signaling pathway that degrades misfolded proteins and are predicted to respond to environmental stress. A study of *prpA* and *prpB* in 58 *Shigella* isolates found that gene inactivation occurred through the accumulation of IS elements, deletions in the gene, deletions in the genomic region spanning the genes, or nonsense mutations (45). While it is not clear if it would be advantageous to lose phosphatases, it does seem that there are parallel, independent inactivation events that have occurred as the lineages have evolved. Li et al. (45) predicted that a single deletion event occurred in *S. dysenteriae* type 1 strains for both genes. Out of the 58 isolates, 86% were defective in one gene and 71% were defective for both genes, with over half being inactivated by IS elements (45). A comparative genomic analysis posits that there was a loss of the *prpB* locus in the ancestor to the EHEC-*S. dysenteriae* serotype 1 lineage (34).

A study in the Whittam laboratory identified a loss in the genomic region of an *S. dysenteriae* type 1 strain, 3823-69, that included the *hca* operon (Whittam, unpublished). This operon consists of five catabolic genes arranged as a putative operon (*hcaA1A2CBD*) and two additional genes transcribed in the opposite direction that encode a potential permease (*hcaT*) and a regulator (*hcaR*) (12). The products of these genes are involved in the dioxygenolytic pathway for initial catabolism of 3-phenylpropionic acid in *E. coli* K-12, and bacterial-mediated degradation of this aromatic compound occurs in the intestinal tract (13). A study measuring acid resistance and acid sensitivity found that propionic acid was more inhibitory to *S. flexneri* than other acids (76). The genome sequences of strains Sf301 (37), Sd197, Sb227, and Ss046 (87) identified the *hcaD* locus as a pseudogene, and a search of the unfinished genome of *S. dysenteriae* M131649 identified a similar deletion of ORFs in this metabolic pathway (Whittam, unpublished). Genomic analysis found that the catabolic pathway of phenylpropionate and its derivatives has been lost in all the extraintestinal pathogenic *E. coli* found in the *E. coli* Reference Collection (ECOR) group B2 (80). It is possible that there is a selective advantage for the loss of this pathway not only in the *Shigella*, but in other *E. coli* pathovars as well.

Other functional losses occur in structural components like fimbriae and flagella. In the reported *Shigella* genome sequences (strains Sf301, Sd197, Sb227, and Ss064), none of the 14 fimbriae loci identified in the genomic sequence of *E. coli* O157:H7 strain Sakai were present (87) and the genes encoding the type 1 fimbriae (*fimBEAICFGH*) were variably present with at least some loss of the gene cluster in EIEC and *Shigella* (24). Studies by Tominaga et al. (79) and others (Whittam, unpublished) have provided insight into the modes of flagellar inactivation in *Shigella*. In a study of the *flhDC* operon in 46 *Shigella* isolates, it was found that the operon was completely deleted in 11 of these isolates (79). For the 35 isolates harboring the operon, the mode of inactivation was examined within the phylogenetic framework, with Group 1 having inactivation occurring via IS*911*, a 5-bp deletion in *flhD*, and a 2-bp substitution in *flhC*, which results in a nonsense mutation. Group 2 isolates were inactivated with IS*Sfl4* in the operon promoter region and IS*600* in the *otsA* and *flhD* genes. Group 3 isolates had an IS*1*F element in *flhD* or an entire deletion (*S. flexneri* types 3a and 3c) (79). Taken together, these results showed that defects in the *flhDC* genes and/or the promoter region contribute to the loss of motility in *Shigella*. Tominaga's study (79) found that an *S. dysenteriae* type 7 isolate had an intact operon but with two amino acid changes in *flhD* and one in *flhC*. An independent restriction fragment length polymorphism pattern and sequence analysis of an *S. dysenteriae* type 7 isolate detected a 7-bp duplication in *fliC*, a nonsense mutation in *S. boydii* type 14, and loss of *fliC* in *S. flexneri* type 6 (Whittam, unpublished). Additionally, *flhDC* was intact for *S. boydii* type 13, *S. dysenteriae* type 8, and the *S. sonnei* clone; however, the nonmotile phenotype could be attributable to an IS in *S. sonnei* (Whittam, unpublished). Because there are cryptic *fliC* genes in *Shigella*, it is probable that there is a selective loss of function in either or both of the *fhlDC* and *fli* operons, supporting the hypothesis of multiple gene loss/inactivation events occurring in parallel in the invasive *E. coli*.

The identification of additional AVL and losses of function may lie in comparative genomics as more sequences are added to the databases. These comparative analyses and reports described above provide insight into the modes of gene inactivation within the invasive *E. coli*. A comparative genomic hybridization microarray was used to compare pathogenic *E. coli* (including EIEC, *S. boydii, S. flexneri*, and *S. sonnei*) and K-12 (24). The genomes of the invasive strains were highly diversified because of indels with large alterations of chromosome structure caused by horizontal gene transfers and deletions. The reported genome sequence of the *S. flexneri* type 2a strain 2457T indicates the presence of 37 islands, 372 pseudogenes, and 6.7% of the genome consisting of ISs (85). As with the genome sequences of *E. coli* O157:H7 Sakai and EDL933 isolates (33, 57), there are distinct differences that exist between the sequenced *S. flexneri* type 2a genomes (37, 85). With the addition and comparative analysis of Sd197, Sb227, and Ss046 with the other reported genomes, there is striking evidence of potential gene expression differences, as these *Shigella* strains have over 200 pseudogenes, 300 to 700 copies of IS elements, indels, translocations, and inversions (87). These events are not limited to the chromosome, as pINV also has IS-associated deletions and rearrangements.

The rate of gene loss in the invasive *E. coli* is faster than that of other pathogenic lineages (34). This might be due to ineffective purifying selection acting on a reduced population since *Shigella* have a limited host range and small infective dose. In comparing the protein-coding genes of *E. coli* K-12 to six *E. coli* and six *Shigella* genomes, 255 genes were reported to be absent in *S. sonnei* and 543 genes were absent in *S. dysenteriae* (34). A comparative analysis of the *S. flexneri* type 2a and *E. coli* K-12 genomes was part of a larger investigation of gene death in obligate pathogens. This analysis indicated that there was a 15% gene loss in *S. flexneri* type 2a as compared to K-12, with 78% of the loss due to pseudogenes (10). Many of these lost functions were in metabolic pathways and cellular processes such as secretion. The findings of Dagan et al. (10) predicted that slow, gradual gene nonfunctionalization occurs in which pathways are initially damaged followed by a subsequent rapid domino effect of mutation accumulation in other pathway genes.

## SUMMARY AND FUTURE CONSIDERATIONS

Gene acquisition, loss, and rearrangements are major contributing factors in the evolution of virulence. Based on these findings, one can conclude that the genome sequence for a given isolate is not necessarily representative of all isolates within the species or even within a particular serotype. Lan and Reeves (39) propose that a true "species" genome can only be obtained by having the sequence of all the DNA that is important for a species. For example, to have a "species" genome for *E. coli*, one would have to know the nucleotide sequence for at least one representative of each distinct pathovar. In the time span of about 10 years this proposed idea of a "species" genome keeps growing larger and larger. With the increased number of genome sequencing projects and subsequent comparisons, it is now suggested that there is not likely to be a model strain for the species due to the vast diversity in genome content allowing for adaptation of each pathovar to its own particular niche (80).

It is very clear that gene acquisition and loss have allowed for establishment and maintenance of invasive lineages of *E. coli*. Hershberg et al. (34) suggested that the invasive lineages first acquired the pINV plasmid, followed by a loss or inactivation of the *tra* genes, leading to an obligate intracellular lifestyle. Because of the multiple acquisitions of the plasmid, convergent evolution occurred in parallel as *Shigella* and EIEC began to acquire and lose genes. Even within the well-established groups, there is evidence of independent evolution, as there is variable distribution of virulence loci and different modes of gene inactivation of the same genes or operons.

Nie et al.'s (51) report of the *S. flexneri* type 5b genome sequence indicated that there are differences in the genome structure and content within the same phylogenetic group. As suggested, these differences may contribute to differences in pathogenicity and epidemiology. For example, there are differences in SHI-O, SHI-1 is not present in *S. flexneri* type 5b, and the shift of the SHI-2 PAI to a different chromosomal tRNA integration site could lead to a difference in gene expression (51). This brings about an important point: members of a phylogenetic group might not cause the same severity of disease. While we strive to learn more about the presence of virulence factors and identify AVL, we must not forget about the epidemiology of the invasive *E. coli*. In developing nations the Group 3 isolates are more prevalent, with particular serotypes causing the majority of disease. *S. sonnei*, common in industrialized nations, causes a milder form of shigellosis. Could the difference in virulence be due to the absence of the SHI-1 PAI? Do the epidemiological differences correlate to groups being better adapted to their particular host populations? What intrinsic genetic differences in the host populations drive the evolution of the pathogen? While we tend to focus on virulence and pathogenesis, we should also consider survival outside of the host and adaptations for efficient transmission, as these might provide additional targets for control and prevention of shigellosis. It is also important to consider that in a single case of disease, multiple genetic backgrounds could be harbored by the bacterial population (even if they are of the same serotype or different serotypes) and could collectively be responsible for the clinical response. This could mean that molecular analysis of a single isolate is providing only incomplete information about the virulence factors associated with a case of shigellosis.

Ultimately, our quest is to identify the genetic attributes that allow for increased fitness of the invasive *E. coli*. Due to the multiple origins of these pathogens within the *E. coli*, we might not be able to determine a "one size fits all" model; however, examples of convergent evolution abound. We know that both gene acquisition and loss have played a major role in the adaptation of these radiations; selection has favored both the establishment of certain virulence factors in particular lineages and the independent deletion of antivirulence functions.

## REFERENCES

1. **Adhikari, P., G. Allison, B. Whittle, and N. K. Verma.** 1999. Serotype 1a O-antigen modification: molecular characterization of the genes involved and their novel organization in the *Shigella flexneri* chromosome. *J. Bacteriol.* **181:**4711–4718.
2. **Adler, B., C. Sasakawa, T. Tobe, S. Makino, K. Komatsu, and M. Yoshikawa.** 1989. A dual transcriptional activation system for the 230 kb plasmid genes coding for virulence-associated antigens of *Shigella flexneri*. *Mol. Microbiol.* **3:**627–635.
3. **Al-Hasani, K., B. Adler, K. Rajakumar, and H. Sakellaris.** 2001. Distribution and structural variation of the *she* pathogenicity island in enteric bacterial pathogens. *J. Med. Microbiol.* **50:**780–786.

4. **Al-Hasani, K., K. Rajakumar, D. Bulach, R. Robins-Browne, B. Adler, and H. Sakellaris.** 2001. Genetic organization of the *she* pathogenicity island in *Shigella flexneri* 2a. *Microb. Pathog.* **30:**1–8.
5. **Andrade, A., J. A. Giron, J. M. Amhaz, L. R. Trabulsi, and M. B. Martinez.** 2002. Expression and characterization of flagella in nonmotile enteroinvasive *Escherichia coli* isolated from diarrhea cases. *Infect. Immun.* **70:**5882–5886.
6. **Boisen, N., F. Ruiz-Perez, F. Scheutz, K. A. Krogfelt, and J. P. Nataro.** 2009. High prevalence of serine protease autotransporter cytotoxins among strains of enteroaggregative *Escherichia coli*. *Am. J. Trop. Med. Hyg.* **80:**294–301.
7. **Brilli, M., A. Mengoni, M. Fondi, M. Bazzicalupo, P. Lio, and R. Fani.** 2008. Analysis of plasmid genes by phylogenetic profiling and visualization of homology relationships using Blast2Network. *BMC Bioinformatics* **9:**551.
8. **Centers for Disease Control and Prevention.** 2010. Preliminary FoodNet data on the incidence and infection with pathogens transmitted commonly through food—10 states, 2009. *MMWR Morb. Mortal. Wkly. Rep.* **59:**418–422.
9. **Colonna, B., M. Casalino, P. A. Fradiani, C. Zagaglia, S. Naitza, L. Leoni, G. Prosseda, A. Coppo, P. Ghelardini, and M. Nicoletti.** 1995. H-NS regulation of virulence gene expression in enteroinvasive *Escherichia coli* harboring the virulence plasmid integrated into the host chromosome. *J. Bacteriol.* **177:**4703–4712.
10. **Dagan, T., R. Blekhman, and D. Graur.** 2006. The "domino theory" of gene death: gradual and mass gene extinction events in three lineages of obligate symbiotic bacterial pathogens. *Mol. Biol. Evol.* **23:**310–316.
11. **Dehio, C., M. C. Prévost, and P. J. Sansonetti.** 1995. Invasion of epithelial cells by *Shigella flexneri* induces tyrosine phosphorylation of cortactin by a pp60$^{c\text{-}src}$-mediated signalling pathway. *EMBO J.* **14:**2471–2482.
12. **Diaz, E., A. Ferrandez, and J. L. Garcia.** 1998. Characterization of the *hca* cluster encoding the dioxygenolytic pathway for initial catabolism of 3-phenylpropionic acid in *Escherichia coli* K-12. *J. Bacteriol.* **180:**2915–2923.
13. **Diaz, E., A. Ferrandez, M. A. Prieto, and J. L. Garcia.** 2001. Biodegradation of aromatic compounds by *Escherichia coli*. *Microbiol. Mol. Biol. Rev.* **65:**523–569.
14. **Echeverria, P., O. Sethabutr, O. Serichantalergs, U. Lexomboon, and K. Tamura.** 1992. *Shigella* and enteroinvasive *Escherichia coli* infections in households of children with dysentery in Bangkok. *J. Infect. Dis.* **165:**144–147.
15. **Edwards, P. R., and W. H. Ewing.** 1986. *Edwards and Ewing's Identification of Enterobacteriaceae*, 4th ed. Elsevier, New York, NY.
16. **Escobar-Paramo, P., O. Clermont, A. B. Blanc-Potart, H. Bui, C. Le Bouguenec, and E. Denamur.** 2004. A specific genetic background is required for acquisition and expression of virulence factors in *Escherichia coli*. *Mol. Biol. Evol.* **21:**1085–1094.
17. **Escobar-Paramo, P., C. Giudicelli, C. Parsot, and E. Denamur.** 2003. The evolutionary history of *Shigella* and enteroinvasive *Escherichia coli* revised. *J. Mol. Evol.* **57:** 140–148.
18. **Ewing, W. H., and J. L. Gravatti.** 1947. *Shigella* types encountered in the Mediterranean area. *J. Bacteriol.* **53:**191–195.
19. **Falconi, M., B. Colonna, G. Prosseda, G. Micheli, and C. O. Gualerzi.** 1998. Thermoregulation of *Shigella* and *Escherichia coli* EIEC pathogenicity. A temperature-dependent structural transition of DNA modulates accessibility of *virF* promoter to transcriptional repressor H-NS. *EMBO J.* **17:**7033–7043.
20. **Fasano, A., F. R. Noriega, D. R. Maneval, Jr., S. Chanasongcram, R. Russell, S. Guandalini, and M. M. Levine.** 1995. *Shigella* enterotoxin 1: an enterotoxin of *Shigella flexneri* 2a active in rabbit small intestine *in vivo* and *in vitro*. *J. Clin. Invest.* **95:**2853–2861.
21. **Faundez, G., G. Figueroa, M. Troncoso, and F. C. Cabello.** 1988. Characterization of enteroinvasive *Escherichia coli* strains isolated from children with diarrhea in Chile. *J. Clin. Microbiol.* **26:**928–932.

22. **Formal, S. B., E. H. LaBrec, T. H. Kent, and S. Falkow.** 1965. Abortive intestinal infection with an *Escherichia coli-Shigella flexneri* hybrid strain. *J. Bacteriol.* **89:**1374–1382.
23. **Frischknecht, F., and M. Way.** 2001. Surfing pathogens and the lessons learned for actin polymerization. *Trends Cell Biol.* **11:**30–38.
24. **Fukiya, S., H. Mizoguchi, T. Tobe, and H. Mori.** 2004. Extensive genomic diversity in pathogenic *Escherichia coli* and *Shigella* strains revealed by comparative genomic hybridization microarray. *J. Bacteriol.* **186:**3911–3921.
25. **Galan, J. E.** 1996. Molecular genetic bases of *Salmonella* entry into host cells. *Mol. Microbiol.* **20:**263–271.
26. **Gordillo, M. E., G. R. Reeve, J. Pappas, J. J. Mathewson, H. L. DuPont, and B. E. Murray.** 1992. Molecular characterization of strains of enteroinvasive *Escherichia coli* O143, including isolates from a large outbreak in Houston, Texas. *J. Clin. Microbiol.* **30:**889–893.
27. **Gorge, O., S. Lopez, V. Hilaire, O. Lisanti, V. Ramisse, and G. Vergnaud.** 2008. Selection and validation of a multilocus variable-number tandem-repeat analysis panel for typing *Shigella* spp. *J. Clin. Microbiol.* **46:**1026–1035.
28. **Griffiths, E., P. Stevenson, T. L. Hale, and S. B. Formal.** 1985. Synthesis of aerobactin and a 76,000-dalton iron-regulated outer membrane protein by *Escherichia coli* K-12-*Shigella flexneri* hybrids and by enteroinvasive strains of *Escherichia coli*. *Infect. Immun.* **49:**67–71.
29. **Groisman, E. A., and H. Ochman.** 1993. Cognate gene clusters govern invasion of host epithelial cells by *Salmonella typhimurium* and *Shigella flexneri*. *EMBO J.* **12:**3779–3787.
30. **Hacker, J., and J. B. Kaper.** 1999. The concept of pathogenicity islands, p. 1–11. *In* J. B. Kaper and J. Hacker (ed.), *Pathogenicity Islands and Other Mobile Virulence Elements.* ASM Press, Washington, DC.
31. **Hale, T. L.** 1991. Genetic basis of virulence in *Shigella* species. *Microbiol. Rev.* **55:** 206–224.
32. **Harris, J. R., J. Mariano, J. G. Wells, B. J. Payne, H. D. Donnell, and M. L. Cohen.** 1985. Person-to-person transmission in an outbreak of enteroinvasive *Escherichia coli*. *Am. J. Epidemiol.* **122:**245–252.
33. **Hayashi, T., K. Makino, M. Ohnishi, K. Kurokawa, K. Ishii, K. Yokoyama, C. G. Han, E. Ohtsubo, K. Nakayama, T. Murata, M. Tanaka, T. Tobe, T. Iida, H. Takami, T. Honda, C. Sasakawa, N. Ogasawara, T. Yasunaga, S. Kuhara, T. Shiba, M. Hattori, and H. Shinagawa.** 2001. Complete genome sequence of enterohemorrhagic *Escherichia coli* O157:H7 and genomic comparison with a laboratory strain K-12. *DNA Res.* **8:**11–22.
34. **Hershberg, R., H. Tang, and D. A. Petrov.** 2007. Reduced selection leads to accelerated gene loss in *Shigella*. *Genome Biol.* **8:**R164.
35. **Hyma, K. E., D. W. Lacher, A. M. Nelson, A. C. Bumbaugh, J. M. Janda, N. A. Strockbine, V. B. Young, and T. S. Whittam.** 2005. Evolutionary genetics of a new pathogenic *Escherichia* species: *Escherichia albertii* and related *Shigella boydii* strains. *J. Bacteriol.* **187:**619–628.
36. **Ingersoll, M., E. A. Groisman, and A. Zychlinsky.** 2002. Pathogenicity islands of *Shigella*. *Curr. Top. Microbiol. Immunol.* **264:**49–65.
37. **Jin, Q., Z. Yuan, J. Xu, Y. Wang, Y. Shen, W. Lu, J. Wang, H. Liu, J. Yang, F. Yang, X. Zhang, J. Zhang, G. Yang, H. Wu, D. Qu, J. Dong, L. Sun, Y. Xue, A. Zhao, Y. Gao, J. Zhu, B. Kan, K. Ding, S. Chen, H. Cheng, Z. Yao, B. He, R. Chen, D. Ma, B. Qiang, Y. Wen, Y. Hou, and J. Yu.** 2002. Genome sequence of *Shigella flexneri* 2a: insights into pathogenicity through comparison with genomes of *Escherichia coli* K12 and O157. *Nucleic Acids Res.* **30:**4432–4441.
38. **Kotloff, K. L., J. P. Winickoff, B. Ivanoff, J. D. Clemens, D. L. Swerdlow, P. J. Sansonetti, G. K. Adak, and M. M. Levine.** 1999. Global burden of *Shigella* infections: implications for vaccine development and implementation of control strategies. *Bull. W. H. O.* **77:**651–666.

39. **Lan, R., and P. R. Reeves.** 2000. Intraspecies variation in bacterial genomes: the need for a species genome concept. *Trends Microbiol.* **8:**396–401.
40. **Lan, R., M. C. Alles, K. Donohoe, M. B. Martinez, and P. R. Reeves.** 2004. Molecular evolutionary relationships of enteroinvasive *Escherichia coli* and *Shigella* spp. *Infect. Immun.* **72:**5080–5088.
41. **Lawrence, J. G.** 1999. Gene transfer, speciation, and the evolution of bacterial genomes. *Curr. Opin. Microbiol.* **2:**519–523.
42. **Lawrence, J. G., and H. Ochman.** 1998. Molecular archaeology of the *Escherichia coli* genome. *Proc. Natl. Acad. Sci. USA* **95:**9413–9417.
43. **Levine, M. M., C. Ferreccio, V. Prado, M. Cayazzo, P. Abrego, J. Martinez, L. Maggi, M. M. Baldini, W. Martin, D. Maneval, B. Kay, L. Guers, H. Lior, S. S. Wasserman, and J. P. Nataro.** 1993. Epidemiologic studies of *Escherichia coli* diarrheal infections in a low socioeconomic level peri-urban community in Santiago, Chile. *Am. J. Epidemiol.* **138:**849–869.
44. **Levine, O. S., and M. M. Levine.** 1991. Houseflies (*Musca domestica*) as mechanical vectors of shigellosis. *Rev. Infect. Dis.* **13:**688–696.
45. **Li, B., E. W. Brown, C. D'Agostino, J. E. LeClerc, and T. A. Cebula.** 2005. Structure and distribution of the phosphoprotein phosphatase genes, *prpA* and *prpB*, among *Shigella* subgroups. *Microbiology* **151:**2671–2683.
46. **Luck, S. N., S. A. Turner, K. Rajakumar, H. Sakellaris, and B. Adler.** 2001. Ferric dicitrate transport system (Fec) of *Shigella flexneri* 2a YSH6000 is encoded on a novel pathogenicity island carrying multiple antibiotic resistance genes. *Infect. Immun.* **69:** 6012–6021.
47. **Maurelli, A. T., R. E. Fernandez, C. A. Bloch, C. K. Rode, and A. Fasano.** 1998. "Black holes" and bacterial pathogenicity: a large genomic deletion that enhances the virulence of *Shigella* spp. and enteroinvasive *Escherichia coli*. *Proc. Natl. Acad. Sci. USA* **95:**3943–3948.
48. **Maurelli, A. T., and P. J. Sansonetti.** 1988. Identification of a chromosomal gene controlling temperature-regulated expression of *Shigella* virulence. *Proc. Natl. Acad. Sci. USA* **85:**2820–2824.
49. **Mills, M., and S. M. Payne.** 1995. Genetics and regulation of heme iron transport in *Shigella dysenteriae* and detection of an analogous system in *Escherichia coli* O157:H7. *J. Bacteriol.* **177:**3004–3009.
50. **Moss, J. E., T. J. Cardozo, A. Zychlinsky, and E. A. Groisman.** 1999. The *selC*-associated SHI-2 pathogenicity island of *Shigella flexneri*. *Mol. Microbiol.* **33:**74–83.
51. **Nie, H., F. Yang, X. Zhang, J. Yang, L. Chen, J. Wang, Z. Xiong, J. Peng, L. Sun, J. Dong, Y. Xue, X. Xu, S. Chen, Z. Yao, Y. Shen, and Q. Jin. 2006.** Complete genome sequences of *Shigella flexneri* 5b and comparison with *Shigella flexneri* 2a. *BMC Genomics* **7:**173.
52. **Noriega, F. R., F. M. Liao, S. B. Formal, A. Fasano, and M. M. Levine.** 1995. Prevalence of *Shigella* enterotoxin 1 among *Shigella* clinical isolates of diverse serotypes. *J. Infect. Dis.* **172:**1408–1410.
53. **Ochman, H., and E. A. Groisman.** 1996. Distribution of pathogenicity islands in *Salmonella* spp. *Infect. Immun.* **64:**5410–5412.
54. **Ochman, H., and E. A. Groisman.** 1994. The origin and evolution of species differences in *Escherichia coli* and *Salmonella typhimurium*. *EXS* **69:**479–493.
55. **Ochman, H., T. S. Whittam, D. A. Caugant, and R. K. Selander.** 1983. Enzyme polymorphism and genetic population structure in *Escherichia coli* and *Shigella*. *J. Gen. Microbiol.* **129:**2715–2726.
56. **Parsot, C., and P. J. Sansonetti.** 1999. The virulence plasmid of shigellae: an archipelago of pathogenicity islands?, p. 151–165. *In* J. B. Kaper and J. Hacker (ed.), *Pathogenicity Islands and Other Mobile Virulence Elements.* ASM Press, Washington, DC.
57. **Perna, N. T., G. Plunkett III, V. Burland, B. Mau, J. D. Glasner, D. J. Rose, G. F. Mayhew, P. S. Evans, J. Gregor, H. A. Kirkpatrick, G. Posfai, J. Hackett, S. Klink, A. Boutin, Y. Shao, L. Miller, E. J. Grotbeck, N. W. Davis, A. Lim, E. T. Dimalanta,**

**K. D. Potamousis, J. Apodaca, T. S. Anantharaman, J. Lin, G. Yen, D. C. Schwartz, R. A. Welch, and F. R. Blattner.** 2001. Genome sequence of enterohaemorrhagic *Escherichia coli* O157:H7. *Nature* **409:**529–533.

58. **Prunier, A.-L., R. Schuch, R. E. Fernandez, and A. T. Maurelli.** 2007. Genetic structure of the *nadA* and *nadB* antivirulence loci in *Shigella* spp. *J. Bacteriol.* **189:**6482–6486.
59. **Prunier, A.-L., R. Schuch, R. E. Fernandez, K. L. Mumy, H. Koehler, B. A. McCormick, and A. T. Maurelli.** 2007. *nadA* and *nadB* of *Shigella flexneri* 5a are antivirulence loci responsible for the synthesis of quinolate, a small molecule inhibitor of *Shigella* pathogenicity. *Microbiology* **153:**2363–2372.
60. **Pupo, G. M., D. K. Karaolis, R. Lan, and P. R. Reeves.** 1997. Evolutionary relationships among pathogenic and nonpathogenic *Escherichia coli* strains inferred from multilocus enzyme electrophoresis and *mdh* sequence studies. *Infect. Immun.* **65:**2685–2692.
61. **Pupo, G. M., R. Lan, and P. R. Reeves.** 2000. Multiple independent origins of *Shigella* clones of *Escherichia coli* and convergent evolution of many of their characteristics. *Proc. Natl. Acad. Sci. USA* **97:**10567–10572.
62. **Purdy, G. E., and S. M. Payne.** 2001. The SHI-3 iron transport island of *Shigella boydii* 0-1392 carries the genes for aerobactin synthesis and transport. *J. Bacteriol.* **183:**4176–4182.
63. **Rajakumar, K., C. Sasakawa, and B. Adler.** 1997. Use of a novel approach, termed island probing, identifies the *Shigella flexneri she* pathogenicity island which encodes a homolog of the immunoglobulin A protease-like family of proteins. *Infect. Immun.* **65:**4606–4614.
64. **Ram, S., S. Khurana, S. B. Khurana, S. Sharma, and D. V. Vadehra.** 1990. Seasonal fluctuations in the occurrence of enteroinvasive *Escherichia coli* diarrhoea. *Indian J. Med. Res.* **91:**258–262.
65. **Rode, C. K., L. J. Melkerson-Watson, A. T. Johnson, and C. A. Bloch.** 1999. Type-specific contributions to chromosome size differences in *Escherichia coli. Infect. Immun.* **67:**230–236.
66. **Runyen-Janecky, L. J., S. A. Reeves, E. G. Gonzales, and S. M. Payne.** 2003. Contribution of the *Shigella flexneri* Sit, Iuc, and Feo iron acquisition systems to iron acquisition in vitro and in cultured cells. *Infect. Immun.* **71:**1919–1928.
67. **Sansonetti, P. J., H. d'Hauteville, C. Ecobichon, and C. Pourcel.** 1983. Molecular comparison of virulence plasmids in *Shigella* and enteroinvasive *Escherichia coli. Ann. Microbiol.* (Paris) **134A:**295–318.
68. **Sansonetti, P. J., and C. Egile.** 1998. Molecular bases of epithelial cell invasion by *Shigella flexneri. Antonie van Leeuwenhoek* **74:**191–197.
69. **Sekizaki, T., S. Harayama, G. M. Brazil, and K. N. Timmis.** 1987. Localization of *stx*, a determinant essential for high-level production of Shiga toxin by *Shigella dysenteriae* serotype 1, near *pyrF* and generation of *stx* transposon mutants. *Infect. Immun.* **55:**2208–2214.
70. **Silva, R. M., M. R. Toledo, and L. R. Trabulsi.** 1980. Biochemical and cultural characteristics of invasive *Escherichia coli. J. Clin. Microbiol.* **11:**441–444.
71. **Snyder, J. D., J. G. Wells, J. Yashuk, N. Puhr, and P. A. Blake.** 1984. Outbreak of invasive *Escherichia coli* gastroenteritis on a cruise ship. *Am. J. Trop. Med. Hyg.* **33:**281–284.
72. **Sokurenko, E. V., D. L. Hasty, and D. E. Dykhuizen.** 1999. Pathoadaptive mutations: gene loss and variation in bacterial pathogens. *Trends Microbiol.* **7:**191–195.
73. **Talukder, K. A., M. A. Islam, D. K. Dutta, F. Hassan, A. Safa, G. B. Nair, and D. A. Sack.** 2002. Phenotypic and genotypic characterization of serologically atypical strains of *Shigella flexneri* type 4 isolated in Dhaka, Bangladesh. *J. Clin. Microbiol.* **40:**2490–2497.
74. **Talukder, K. A., Z. Islam, M. A. Islam, D. K. Dutta, A. Safa, M. Ansaruzzaman, A. S. Faruque, S. N. Shahed, G. B. Nair, and D. A. Sack.** 2003. Phenotypic and

genotypic characterization of provisional serotype *Shigella flexneri* 1c and clonal relationships with 1a and 1b strains isolated in Bangladesh. *J. Clin. Microbiol.* **41:**110–117.
75. **Taylor, D. N., P. Echeverria, O. Sethabutr, C. Pitarangsi, U. Leksomboon, N. R. Blacklow, B. Rowe, R. Gross, and J. Cross.** 1988. Clinical and microbiologic features of *Shigella* and enteroinvasive *Escherichia coli* infections detected by DNA hybridization. *J. Clin. Microbiol.* **26:**1362–1366.
76. **Tetteh, G. L., and L. R. Beuchat.** 2001. Sensitivity of acid-adapted and acid-shocked *Shigella flexneri* to reduced pH achieved with acetic, lactic, and propionic acids. *J. Food Prot.* **64:**975–981.
77. **Todd, E. C., J. D. Greig, C. A. Bartleson, and B. S. Michaels.** 2008. Outbreaks where food workers have been implicated in the spread of foodborne disease, part 4. Infective doses and pathogen carriage. *J. Food Prot.* **71:**2339–2373.
78. **Toledo, M. R. F., and L. R. Trabulsi.** 1990. Frequency of enteroinvasive *Escherichia coli* in children with diarrhea and healthy controls, in Sao Paulo, SP, Brazil. *Rev. Microbiol. Sao Paulo* **21:**1–4.
79. **Tominaga, A., R. Lan, and P. R. Reeves.** 2005. Evolutionary changes in the *flhDC* flagellar master operon in *Shigella* strains. *J. Bacteriol.* **187:**4295–4302.
80. **Touchon, M., C. Hoede, O. Tenaillon, V. Barbe, S. Baeriswyl, P. Bidet, E. Bingen, S. Bonacorsi, C. Bouchier, O. Bouvet, A. Calteau, H. Chiapello, O. Clermont, S. Cruveiller, A. Danchin, M. Diard, C. Dossat, M. El Karoui, E. Frapy, L. Garry, J. M. Ghigo, A. M. Gilles, J. Johnson, C. Le Bouguenec, M. Lescat, S. Mangenot, V. Martinez-Jehanne, I. Matic, X. Nassif, S. Oztas, M. A. Petit, C. Pichon, Z. Rouy, C. Saint Ruf, D. Schneider, J. Tourret, B. Vacherie, D. Vallenet, C. Medigue, E. P. C. Rocha, and E. Danamur.** 2009. Organized genome dynamics in the *Escherichia coli* species results in highly diverse adaptive paths. *PLoS Genet.* **5:**e1000344.
81. **Tulloch, E. F., Jr., K. J. Ryan, S. B. Formal, and F. A. Franklin.** 1973. Invasive enteropathic *Escherichia coli* dysentery. An outbreak in 28 adults. *Ann. Intern. Med.* **79:** 13–17.
82. **Turner, S. A., S. N. Luck, H. Sakellaris, K. Rajakumar, and B. Adler.** 2003. Molecular epidemiology of the SRL pathogenicity island. *Antimicrob. Agents Chemother.* **47:** 727–734.
83. **Vokes, S. A., S. A. Reeves, A. G. Torres, and S. M. Payne.** 1999. The aerobactin iron transport system genes in *Shigella flexneri* are present within a pathogenicity island. *Mol. Microbiol.* **33:**63–73.
84. **Wanger, A. R., B. E. Murray, P. Echeverria, J. J. Mathewson, and H. L. DuPont.** 1988. Enteroinvasive *Escherichia coli* in travelers with diarrhea. *J. Infect. Dis.* **158:**640–642.
85. **Wei, J., M. B. Goldberg, V. Burland, M. M. Venkatesan, W. Deng, G. Fournier, G. F. Mayhew, G. Plunkett III, D. J. Rose, A. Darling, B. Mau, N. T. Perna, S. M. Payne, L. J. Runyen-Janecky, S. Zhou, D. C. Schwartz, and F. R. Blattner.** 2003. Complete genome sequence and comparative genomics of *Shigella flexneri* serotype 2a strain 2457T. *Infect. Immun.* **7:**2775–2786.
86. **Wyckoff, E. E., D. Duncan, A. G. Torres, M. Mills, K. Maase, and S. M. Payne.** 1998. Structure of the *Shigella dysenteriae* haem transport locus and its phylogenetic distribution in enteric bacteria. *Mol. Microbiol.* **28:**1139–1152.
87. **Yang, F., J. Yang, X. Zhang, L. Chen, Y. Jiang, Y. Yan, X. Tang, J. Wang, Z. Xiong, J. Dong, Y. Xue, Y. Zhu, X. Xu, L. Sun, S. Chen, H. Nie, J. Peng, J. Xu, Y. Want, Z. Yuan, Y. Wen, Z. Yao, Y. Shen, B. Qiang, Y. Hou, J. Yu, and Q. Jin.** 2005. Genome dynamics and diversity of *Shigella* species, the etiologic agents of bacillary dysentery. *Nucleic Acids Res.* **33:**6445–6458.
88. **Yang, J., H. Nie, L. Chen, X. Zhang, F. Yang, X. Xu, Y. Zhu, J. Yu, and Q. Jin.** 2007. Revisiting the molecular evolutionary history of *Shigella* spp. *J. Mol. Evol.* **64:** 71–79.
89. **Ye, C., R. Lan, S. Xia, J. Zhang, Q. Sun, S. Zhang, H. Jing, L. Wang, Z. Li, Z. Zhou, A. Zhao, Z. Cui, J. Cao, D. Jin, L. Huang, Y. Wang, X. Lou, X. Bai, Y. Wang, Q. Xu, and J. Xu.** 2010. Emergence of a new multidrug-resistant serotype X variant in an epidemic clone of *Shigella flexneri. J. Clin. Microbiol.* **48:**419–426.

*Population Genetics of Bacteria: a Tribute to Thomas S. Whittam*
Edited by S. T. Walk and P. C. H. Feng


*Chapter 10*

# Fitness Islands in Uropathogenic *Escherichia coli*

***Amanda L. Lloyd and Harry L. T. Mobley***

The rapidly increasing number of sequenced bacterial genomes has opened the door for novel and comparative genomic approaches between pathogenic and nonpathogenic strains, and is providing insight into the exchange of genetic information between and within species. Genome sequencing efforts to date have been primarily focused on bacterial pathogens of medical or veterinary importance (76). However, with the ongoing reductions in large-scale DNA sequencing costs, genome sequences of multiple strains within the same species, of both pathogenic and nonpathogenic origins, are now becoming available for comparative genomic studies.

Comparison of the genome sequences of multiple strains is essential because the makeup of strains within a given pathotype can be heterogeneous. When one considers the development of a vaccine, for example, one must know whether the antigen selected is truly representative of all such strains; that is, is the antigen present in every strain causing a syndrome, or, if always present, is the sequence highly conserved? The more strains surveyed by sequencing or by comparative genomic hybridization, the more confident one may become regarding the virulence gene profile of a given bacterial pathotype.

*Escherichia coli* is an extremely versatile organism, ranging from a harmless, and perhaps beneficial, inhabitant of the human intestinal tract to a pathogen capable of causing at least six well-defined clinical presentations (46). Six distinct pathotypes of *E. coli* cause diarrheal diseases in humans: enterohemorrhagic *E. coli* (EHEC), enteropathogenic *E. coli* (EPEC), enterotoxigenic *E. coli* (ETEC), enteroaggregative *E. coli* (EAEC), diffusely adherent *E. coli* (DAEC), and enteroinvasive *E. coli* (EIEC). Extraintestinal pathogenic *E. coli* (ExPEC), a classification first proposed in 2000 to describe *E. coli* isolates capable of causing disease outside the intestinal tract (85), includes uropathogenic *E. coli* (UPEC), neonatal meningitis *E. coli* (NMEC), and avian patho-

***Amanda L. Lloyd and Harry L. T. Mobley*** • Department of Microbiology and Immunology, University of Michigan Medical School, Ann Arbor, MI 48109.

genic *E. coli* (APEC). Other clinical presentations of ExPEC infection include intra-abdominal infections, pneumonia (often nosocomial in origin), intravascular-device infection, osteomyelitis, and soft tissue infection (85). Bacteremia can result from infection at any of these sites (24).

## GENOME SIZE VARIABILITY OF *E. COLI*

The size of sequenced *E. coli* genomes varies from 4.6 Mb for the fecal/commensal *E. coli* K-12 strain MG1655 to 5.5 Mb for EHEC strain O157:H7 EDL933. The majority of these differences are attributed to the insertion of large regions of genomic DNA (80). The genome of pathogenic *E. coli* consists of a conserved core genome, which encodes the basic metabolic and cellular processes necessary for survival, and a flexible gene pool (i.e., pan-genome), which consists of strain- and pathotype-specific genes that afford additional properties permitting colonization of specialized niches (18). The majority of the differences in genome size between *E. coli* strains are attributed to this flexible gene pool.

Rasko and colleagues recently examined 17 sequenced *E. coli* genomes, including 4 UPEC genomes (CFT073, F11, 536, UTI89), 2 EHEC isolates (EDL933, Sakai), 3 EPEC strains (E22, E110019, B171), 2 ETEC genomes (E24377A, B7A), 2 EAEC strains (042, 101-1), 1 APEC isolate (APECO1), the laboratory-adapted K-12 strains MG1655 and W3110, and the commensal strain HS. This study revealed that the core genome of *E. coli*—that is, those genes conserved across all 17 sequenced *E. coli* genomes—is approximately 2,200 genes (76). The estimated size of the pan-genome of *E. coli*, defined as all genes present in any *E. coli* strain, is approximately 13,000 genes.

## HORIZONTAL GENE TRANSFER AND GENOMIC ISLANDS

Bacterial genomes are in a constant state of flux (56). Horizontal gene transfer (HGT), the process by which genetic material is transferred between bacterial species independent of cell division, is the primary means of generating bacterial diversity within and between closely related species (54). Evolutionarily, point mutations generate diversity slowly, whereas the acquisition or deletion of large regions of DNA results in evolution in "quantum leaps" (21, 25). The genome of the fecal/commensal *E. coli* K-12 strain MG1655 has been significantly shaped by horizontally acquired genetic material: 547 kb of DNA has been introduced in at least 234 lateral transfer events since *E. coli* diverged from *Salmonella*. In total, approximately 18% of the chromosome of *E. coli* K-12 strain MG1655 appears to have been horizontally acquired (55).

Natural selection plays a critical role following HGT, as genetic material that contributes traits beneficial to the recipient strain will be retained and stably maintained in the genome. Consequently, HGT is coupled with genome reduction, as bacterial genomes cannot sustain uncontrolled increase in size (89), and genes of a lower selective value are lost (67). These large regions of DNA mobilized between strains during HGT are referred to as "genomic islands"

(GIs). Pathogenicity islands (PAIs) are a subset of GIs known to encode one or more virulence genes. PAIs were first described in 1994 in UPEC strain 536, when Blum and colleagues identified two regions of DNA encoding virulence factors, including hemolysin and P fimbriae, which they termed PAIs (8). Spontaneous deletion of these PAIs from the chromosome of 536 resulted in a substantial reduction in virulence during in vivo studies.

Much of the early work on PAIs was conducted by Hacker and colleagues, who defined PAIs as GIs that meet most, if not all, of the following criteria: (i) they encode one or more virulence genes; (ii) they are present in pathogenic strains but absent or rare in nonpathogenic strains; (iii) they have a different G+C content relative to the backbone of the genome; (iv) they consist of large, often unstable chromosomal regions, often >30 kb; (v) they are associated with tRNA genes and/or insertion elements at their boundaries; (vi) they are often flanked by direct repeats; and (vii) they frequently contain (often cryptic) mobility genes, such as insertion elements, integrases, transposases, or origins of plasmid replication (29). HGT results in an unusually high degree of similarity between the donor and recipient strains (55, 67) that differ from their core and pan-genomes. Genomic insertions of <30 kb are referred to as "genomic islets"; those that encode genes involved in virulence have been termed "pathogenicity islets" (31). This chapter will examine genes present on genomic islands and islets of UPEC, with a primary focus on genes contributing to the overall fitness of the organism rather than traditional virulence genes. Although some of these regions may be considered PAIs, for the sake of this review, the more general term "genomic island" will be used to refer to any horizontally acquired chromosomal insertion of >30 kb in the *E. coli* K-12 backbone. Particular reference will be made to the sequenced UPEC strains CFT073 (106), 536 (9), and UTI89 (15) and the sequenced but incompletely assembled and annotated UPEC strain F11.

## UPEC AND URINARY TRACT INFECTIONS

Uncomplicated urinary tract infections (UTIs) are the most common bacterial infection in otherwise healthy young women, and UPEC are the most common etiological agents. Greater than 80% of community-acquired UTIs and approximately half of nosocomial UTIs are caused by UPEC (81, 98). The clinical spectrum of UTIs is broad and may include asymptomatic bacteriuria (ABU), cystitis (bladder infection), or pyelonephritis (kidney infection). Uncomplicated UTI occurs by an ascending route, following bacterial colonization of the periurethral area with intestinal microbiota and ascension of the urethra to the bladder. Cystitis, the most common clinical presentation of UTI, is defined clinically as $>10^5$ CFU/ml of midstream urine, although women may experience symptomatic UTI with lower levels of bacteriuria. In some patients, UPEC continue their ascension up the ureters to the kidneys, resulting in pyelonephritis. A potentially severe complication of UTI is bacteremia, in which the bacteria gain access to the bloodstream from the renal tubules of the kidneys. Due to anatomical differences (shorter urethra), women are significantly more likely than men to suffer from UTIs (23).

Fecal/commensal *E. coli* strains are unable to colonize the human urinary tract (61), despite inhabiting the intestines alongside UPEC strains and having an equal likelihood of causing periurethral contamination. To successfully transition from the intestinal tract to the urinary tract, UPEC require additional factors beyond those found in the fecal/commensal *E. coli* strain K-12. Indeed, UPEC have acquired these factors horizontally, by the acquisition of large segments of DNA from related species, in the form of genomic islands and islets. We have previously shown that GIs comprise 672 kb (12.8%) of the UPEC strain CFT073 genome (58); this number is an underestimation of the effect of HGT on the evolution of this strain, as this value does not include the numerous genomic insertions of <30 kb. Similarly, UPEC strain CFT073 contains 13 GIs of >30 kb (58), as illustrated in Fig. 1. Deletion of GIs in UPEC strains CFT073 (57) and 536 (9) results in attenuation in vivo. The acquisition of multiple GIs by UPEC appears to contribute not only to virulence, but to the overall fitness of the organism.

## BACTERIAL FITNESS

The Darwinian view holds that evolution is driven by the need to increase the overall fitness of the organism (37). Bacterial fitness is considered to be a set of characteristics or factors that enhance the survival, spread, and/or transmission of an organism within a specific ecological niche (75). Consequently, the acquisition of GIs that increase the fitness, adaptability, or functionality of a bacterium provides a competitive advantage over strains without these extra genes, and these particular GIs can be referred to as "fitness islands" (30). Over evolutionary timescales, acquisition of genetic material beneficial to the

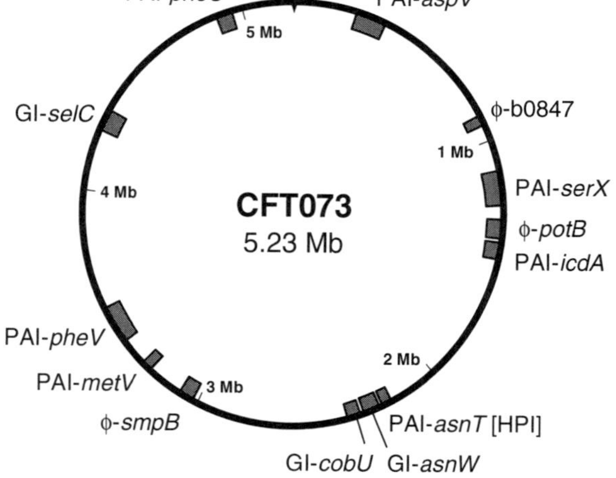


**Figure 1.** PAIs, GIs, and phage ($\phi$) of UPEC strain CFT073. Thirteen genomic insertions of >30 kb, relative to *E. coli* K-12, are shown as shaded boxes and denoted by the gene that resides most closely to the point of insertion.

recipient strain is stably maintained while DNA of lower selective value is lost to maintain overall genome size (56, 67).

Fitness islands generate a competitive advantage for UPEC in the unique niche of the urinary tract. These fitness islands code for factors such as toxins that may damage host tissues and release nutrients for utilization by the bacterium, factors to attenuate the host immune response, antimicrobial resistance genes, protein secretion systems, and adhesive organelles to protect against the flow of urine. Furthermore, the acquisition of multiple, nonidentical adhesins and iron acquisition systems provides UPEC with functional redundancy and niche specificity, permitting adaptation to a variety of environmental conditions both outside and within a host (e.g., survival in the intestinal tract versus the urinary tract; and once within the urinary tract, conferring apparent organ specificity [preferential colonization of the bladder or kidney]). This increase in flexibility correlates with increased fitness. Examples of genes that increase the fitness of the microorganism and are frequently encoded on GIs include toxins, fimbriae (pili), iron acquisition and transport systems, protein secretion systems, and those involved in the avoidance of host defense mechanisms.

## TOXINS

Acquisition of toxin genes provides an excellent example of how the addition of as little as a single gene may increase bacterial fitness. Several well-known toxins have been identified in UPEC, including hemolysin (*hlyA*) (79), cytotoxic necrotizing factor 1 (*cnf*1) (12), secreted autotransporter toxin (*sat*) (27, 28), and the autotransporter toxins *picU* and *vat* (formerly known as *tsh*) (34, 71, 72). The potent exotoxin α-hemolysin (Hly), the most extensively characterized member of the RTX (repeats in toxin) family of proteins, produces large clear zones of β-hemolysis on erythrocyte-containing agar plates. Hemolysin, encoded by the *hlyCABD* operon, is more frequently associated with UPEC strains (49% of pyelonephritis strains; 40% of cystitis strains) than ABU strains (20% of strains) or fecal *E. coli* strains (12% of strains) (40). Furthermore, hemolysin induces extensive shedding of the transitional epithelial cells that comprise the murine urothelium within 24 hours of infection (95).

Hemolysin is often genetically linked to another UPEC-associated toxin, *cnf*1 (12). Although some *E. coli* strains express *hly* alone, *cnf*1 is rarely detected in the absence of *hly* (7, 12, 20). Indeed, both *hly* and *cnf*1 genes are present on $PAI_{J96}$II-like GIs. Independent mutations in *cnf*1 and *hly* have been shown to attenuate UPEC (78, 107). UPEC strains CFT073 and 536 have acquired PAIs containing the *hly* genes but not the *cnf*1 gene, with the latter strain containing two copies of the *hly* operon on separate GIs (PAI-CFT073-*pheV*, PAI $I_{536}$, and PAI $II_{536}$, respectively). Conversely, PAIs containing *hly* and *cnf*1 are present in UPEC isolates J96 and UTI89 ($PAI_{J96}$II and $PAI_{UTI89}$II, respectively). Table 1 outlines the GIs identified in UPEC strains CFT073, 536, UTI89, and J96. BLAST analysis reveals that UPEC strain F11 and ABU isolate 83972 contain both the *hly* operon and *cnf*1, suggesting that these strains also contain a $PAI_{J96}$II-like GI. The presence of the *hly* and *cnf*1 genes in *E. coli*

**Table 1.** Features of UPEC strains

| Strain (reference[s]) | Serotype, clinical syndrome | Genome size (bp) | No. of CDS | GIs/PAIs[a] | GI/PAI size (kb) | Associated tRNA gene | Virulence or fitness factors |
|---|---|---|---|---|---|---|---|
| CFT073 (58, 106) | O6:K2:H1 Pyelonephritis Urosepsis | 5,231,428 | 5,379 | PAI-CFT073-*aspV* | 100 | *aspV* | *fbpABCD*, *cdiA*, *picU*, *tosA* |
| | | | | *ϕ*-CFT073-b0847 | 33 | | |
| | | | | PAI-CFT073-*serX* | 113 | *serX* | *mchBCDEF*, *sfa*/*foc*, *iroBCDEN*, antigen 43 |
| | | | | *ϕ*-CFT073-*potB* | 44 | | |
| | | | | PAI-CFT073-*icdA* | 54 | | *sitABCD* |
| | | | | PAI-CFT073-*asnT* (HPI) | 32 | *asnT* | *ybt* (*fyuA*)[b] |
| | | | | GI-CFT073-*asnW* | 54 | *asnW* | |
| | | | | GI-CFT073-*cobU* | 44 | | *hma*, *fbpABCD*_2 |
| | | | | *ϕ*-CFT073-*smpB* | 48 | | |
| | | | | PAI-CFT073-*metV* | 32 | *metV* | Type VI secretion system, c3405-c3410 |
| | | | | PAI-CFT073-*pheV* | 123 | *pheV* | *hlyCABD*, *pap* operon, *iha*, *sat*, *sisA*, *iutA*/*iucABCD*, antigen 43, *kpsMT* |
| | | | | PAI-CFT073-*selC* | 68 | *selC* | *sisB* |
| | | | | PAI-CFT073-*pheU* | 52 | *pheU* | *pap*_2 operon |

| | | | | | | | |
|---|---|---|---|---|---|---|---|
| 536 (9) | O6:K15:H31 Pyelonephritis | 4,938,875 | 4,747 | PAI $I_{536}$ | 76.8 | *selC* | *hlyCABD* |
| | | | | PAI $II_{536}$ | 102.3 | *leuX* | *hlyCABD*, *pap* operon, *hek* |
| | | | | PAI $III_{536}$ | 68.1 | *thrW* | *iroBCDEN*, *sfa/foc* |
| | | | | PAI $IV_{536}$ (HPI) | 30.2 | *asnT* | *ybt* (*fyuA*) |
| | | | | PAI $V_{536}$ | 79.6 | *pheV* | $kps_{K15}$ gene cluster, *pix* fimbriae |
| | | | | PAI $VI_{536}$ | 54.4 | *asnW* | colibactin, *hma* |
| UTI89 (15) | O18:K1:H7 Cystitis | 5,065,741 (plus 114,230-bp plasmid) | 5,066 (plus 145-bp plasmid) | $PAI_{UTI89}II$ | 119 | *leuX* | *hlyCABD*, *pap* operon, *cnf*1, *hek* |
| | | | | $PAI_{UTI89}III$ | 61 | *serX* | *iroBCDEN*, *sfa/foc*, antigen 43, heme/hemoglobin receptor |
| | | | | $PAI_{UTI89}IV$ (HPI) | ? | *asnT* | *ybt* (*fyuA*) |
| J96 (99a) | O4:K6:H5 Pyelonephritis | NA[c] | NA[c] | $PAI_{J96}I$ | >170 | *pheV* | *hlyCABD*, *pap* |
| | | | | $PAI_{J96}II$ | >110 | *pheU*[d] | *hlyCABD*, *cnf*1, *pap*, *hra* |

[a]Only GIs >30 kb are listed in this table.
[b]Nonsense mutations and insertion elements in the yersiniabactin (*ybt*) biosynthesis genes, *irp*1 and *irp*2, in *E. coli* CFT073 prevent detectable production of yersiniabactin. The gene encoding the outer membrane receptor for yersiniabactin, *fyuA*, appears intact.
[c]NA, not available. The genome of UPEC strain J96 has not been sequenced.
[d]Formerly *pheR*.

strain 83972 is surprising as this is an ABU isolate, and the presence of such toxin genes suggests a more damaging phenotype in vivo. The ABU strain 83972 will be discussed in greater detail later in this chapter. In addition, hemolysin is often physically linked to the P-fimbrial operon (see below), providing further evidence that lateral gene transfer has played a critical role in shaping the evolution of UPEC through the acquisition of GIs.

UPEC strain CFT073 encodes 10 autotransporters, including 3 members of the SPATE (serine protease autotransporters of *Enterobacteriaceae*) family of proteins: Sat, PicU, and Vat (Tsh). SPATE proteins are identified by the presence of three common functional domains: an extended signal sequence, a passenger domain containing an active serine protease site, and a C-terminal autotransporter domain (27). Autotransporters, as their name suggests, are capable of synthesizing, autoprocessing, and exporting an active protein from the cell via a type V secretion pathway (reviewed in reference 35). Transport through the inner membrane is mediated by the signal sequence via a Sec-dependent pathway, the signal sequence is subsequently cleaved, and the autotransporter domain forms a $\beta$-barrel, pore-like structure in the outer membrane to allow export of the protein. The passenger domain protein (typically with toxin, protease, or adhesin domains) may remain physically associated with the cell surface, may be cleaved by autoproteolytic activity or by an outer membrane protease and released into the environment, or may remain covalently associated with the cell (35).

Culture supernatants from *E. coli* CFT073 containing the Sat protein exhibited serine protease activity and elicited cytopathic effects on Vero primary kidney, HK-2 bladder, and HEp-2 cell lines (28). Although female CBA/J mice experimentally inoculated with wild-type *E. coli* CFT073 mounted a vigorous antibody response against the Sat protein, disruption of *sat* did not significantly affect levels of bladder or kidney colonization relative to wild-type *E. coli* CFT073 (27). Sat activity, however, was associated with specific renal pathology, and thus contributes to pathogenesis of UPEC during UTI.

The *sat* gene is located on PAI-CFT073-*pheV* (58) and is present in *Shigella dysenteriae* 1012, *E. coli* strain UMN026 (UPEC isolate), *E. coli* strain 83972 (ABU strain), and the probiotic *E. coli* strain Nissle 1917 (26), but is absent from the three other prototypic UPEC strains: 536, UTI89, and F11. As SPATE proteins have been traditionally associated with pathogenic strains, the role of Sat in the nonpathogenic *E. coli* strains Nissle 1917 and ABU 83972 remains to be elucidated.

The autotransporters PicU and Vat (34, 72) of UPEC strain CFT073 are upregulated at 37°C and are expressed during ascending UTI of mice. PicU is a multifunctional protein; it possesses mucinolytic, hemagglutinating, and endopeptidase activity and is able to confer serum resistance through interference with the complement cascade (72). The *picU* gene is significantly associated with pyelonephritis strains (34) and is present on the *aspV* PAI of UPEC strain CFT073 (58). Although a *picU* deletion mutant was not outcompeted during cochallenge infection with wild-type *E. coli* CFT073 (34), *picU* contributes to the virulence of *E. coli* CFT073 in the bladder (57). The SPATE protein Vat has been characterized to a lesser extent than either Sat or PicU. Unlike PicU, Vat does not exhibit serine protease activity and yet it is associated with UPEC isolates from patients with pyelonephritis or cystitis (34). The *vat* gene is located on a 7.8-kb genomic islet inserted near a *thrW* tRNA gene in UPEC strain CFT073 (71), and a similar genomic islet is present in UPEC strains 536, UTI89, and F11 but not in *E. coli* strain K-12 MG1655.

## ADHESINS OF UPEC

To successfully colonize the urinary tract and resist the hydrodynamic forces associated with the flow of urine, UPEC must first adhere to uroepithelial cells lining the bladder. Fimbriae (pili) mediate adhesion to host cells through their adhesive tip protein, with each fimbrial type binding to distinct receptors within the mammalian urinary tract. Thirteen operons encoding distinct fimbriae have been identified in UPEC strain CFT073, 10 chaperone-usher pathway fimbriae, 2 type IV pili (106) and 1 curli (4). UPEC strain UTI89 contains 10 putative fimbrial operons (15), and UPEC strain 536 has 14 fimbriae, including ETEC-like fimbriae in addition to UPEC fimbriae (9). Fimbrial expression is coordinately regulated in UPEC (97), and *E. coli* strains have been shown to express primarily one fimbrial type at any given time (66). Furthermore, UPEC exhibit reciprocal regulation of fimbriae and flagella expression within the urinary tract; that is, UPEC express either fimbriae (for adherence) or flagella (for motility) (51), but these systems are not expressed concurrently.

Multiple fimbriae have been identified in UPEC, including type 1, P, F1C/S, Dr/AFA, Auf, and F9 fimbriae. Type 1 fimbriae, present on the surface of many genera within the *Enterobacteriaceae* family, are the most ubiquitous adhesins among *E. coli* (10). Expression of type 1 fimbriae is phase variable, meaning its expression is mediated by a promoter containing an invertible element upstream of the *fimAICDFGH* (c5393-c5400) operon (47). Two recombinases that mediate phase switching of the invertible element, *fimB* and *fimE* (c5391-5392), are located immediately upstream of type 1 fimbrial structural and assembly genes. Type 1 fimbriae exhibit mannose-sensitive hemagglutination and are chromosomally located in *E. coli* strain K-12 and in all sequenced UPEC strains (CFT073, 536, UTI89, and F11) (9, 15, 106). Furthermore, type 1 fimbriae are required for full virulence of UPEC in the murine urinary tract by promoting bacterial persistence and enhancing the inflammatory response to infection (17).

P fimbriae, encoded by the *pap* (pyelonephritis-associated pili) operon, are significantly associated with UPEC strains capable of causing pyelonephritis (45). *E. coli* expressing P fimbriae adhere to exfoliated uroepithelial cells in higher numbers than *E. coli* strains from individuals with ABU (19), and unlike type 1 fimbriae, P fimbriae are able to hemagglutinate human red blood cells in the presence of mannose (52). P fimbriae were the first virulence-associated factor of UPEC (52), and the *pap* operon was acquired by UPEC through HGT of genetic material into the ancestral K-12 backbone. *E. coli* strain CFT073 contains two nonidentical copies of the *pap* operon (*pap*: c3593-c3583; and *pap_2*: c5189-c5179) that are present on individual GIs (PAI-CFT073-*pheV* and PAI-CFT073-*pheU*, respectively). UPEC strains 536, UTI89, and J96 contain a single *pap* operon, each encoded on a GI ($PAI_{536}$II, $PAI_{UTI89}$II, and $PAI_{J96}$II, respectively). At the 3′ end of the PAI-encoded *pap* operon of CFT073 (PAI-CFT073-*pheV*) (58) is *papX* (c3582), a homolog of the transcriptional regulator MrpJ in *Proteus mirabilis* (73). Overexpression of PapX resulted in a decrease in the motility of UPEC strain CFT073, and conversely a *papX* deletion mutant of *E. coli* CFT073 exhibited an increase in motility and flagellin production, indicating that UPEC are capable of regulating motility through P-fimbrial expression (94). Table 2 lists the 13 fimbrial operons identified in *E. coli* strain CFT073.

Two closely related fimbriae associated with ExPEC infections, in particular infections of the urinary tract, are S (*sfa*) and F1C (*foc*) fimbriae (69). Although sharing a high degree of sequence identity, S and F1C fimbriae possess distinct adhesins in terms of their receptor specificities (68). The S/F1C fimbriae of *E. coli* strain CFT073 (c1237-c1245) form part of the PAI-CFT073-*serX* GI, while the S fimbriae of *E. coli* strains 536 and UTI89 are located on $PAI_{536}$III and $PAI_{UTI89}$III, respectively (see Table 1). F1C fimbriae are required for biofilm formation by the probiotic *E. coli* strain Nissle 1917 (53).

**Table 2.** Predicted fimbrial gene clusters in UPEC strain CFT073

| Fimbrial gene clusters | "c" no. | Location in CFT073 genome | In *E. coli* K-12? |
|---|---|---|---|
| Type 1 fimbriae (*fim*) | c5391-c5400 | *E. coli* K-12 backbone | Yes |
| F1C fimbriae (*foc*) | c1237-c1245 | PAI-CFT073-*serX* | No |
| P fimbriae (*pap*) | c3583-c3593 | PAI-CFT073-*pheV* | No |
| P fimbriae (*pap_2*) | c5179-c5189 | PAI-CFT073-*pheU* | No |
| Auf fimbriae | c4207-c4214 | 9.4-kb genomic islet | No |
| F9 fimbriae | c1931-c1936 | 6.4-kb genomic islet | Partial[a] |
| Putative chaperone-usher fimbrial operon | c3791-c3794 | 5.0-kb genomic islet | No |
| Putative chaperone-usher fimbrial operon | c0166-c0172 | 7.2-kb region with significant sequence divergence? | Yes[b] |
| Putative chaperone-usher fimbrial operon | c2878-c2884 | 6.4-kb region with significant sequence divergence? | Yes[b] |
| Putative chaperone-usher fimbrial operon | c2635-c2638 | *E. coli* K-12 backbone | Yes[c] |
| Putative type IV pili | c2394-c2395 | 22.1-kb genomic islet | No[d] |
| Type IV pili (*ppdD*, *hofBC*) | c0125-c0127 | *E. coli* K-12 backbone | Yes |
| Curli (*csg*) | c1299-c1307 | *E. coli* K-12 backbone | Yes |

[a] Truncated in middle of c1934.
[b] Genes annotated as present in *coli*BASE but were negative by CGH and BSR analyses (58). Alignment of operon with CFT073 shows significant divergence at the nucleotide level.
[c] Present in all 14 strains by CGH and BSR analyses (58).
[d] Genes only present in 3/14 *E. coli* strains (all UPEC) by CGH and BSR analyses (58).

The Dr superfamily of adhesins, first identified in UPEC (103), includes Dr hemagglutinin, Afa-I, Afa-II, Afa-III, Afa-IV, and F1845 fimbriae (64). Dr fimbriae recognize the $Dr^a$ blood group antigen decay-accelerating factor (DAF; CD55) present on human red blood cells (63, 65). *E. coli* strain CFT073 does not encode the Dr and *afa* fimbrial operons.

A novel fimbrial gene cluster denoted Auf, encoded by *aufABCDEFG* (c4214-c4207) in *E. coli* CFT073 (Table 2), is significantly associated with UPEC isolates (11). The *auf* gene cluster is present on a 9.4-kb genomic islet in UPEC strain CFT073. DNA sequence alignments of the *auf* gene cluster from *E. coli* CFT073 with UPEC strains 536, UTI89, and F11 indicate that the *auf* operon is present in a similar genomic islet in all four sequenced UPEC strains, but absent from *E. coli* strain K-12. Previous studies in our laboratory using comparative genomic hybridization (CGH) microarray analysis found the *auf* fimbrial gene sequence was present in six of seven UPEC isolates and none of three fecal/commensal strains (58).

UpaG, a recently characterized trimeric autotransporter involved in adherence to bladder epithelial cells, cell aggregation, and biofilm formation, is frequently associated with ExPEC strains (104). UpaG is located on a 6.4-kb genomic islet (c4422-c4424) in *E. coli* strain CFT073, with similar genomic islets present in UPEC strains 536, UTI89, and F11 but absent from *E. coli* strain K-12.

The acquisition and stable integration of multiple genomic islands and islets encoding fimbrial adhesins in UPEC illustrate the significance of adhesion within the urinary tract (Table 2). The abundance of horizontally acquired, coregulated fimbrial adhesins in UPEC, each with distinct adhesive moieties, demonstrates a clear functional redundancy. This hypothesis is further supported by the absence of these adhesins (e.g., P, S/F1C, Auf, and *upaG*) in the fecal/commensal *E. coli* strain K-12. Fimbrial expression has been shown to be coordinately regulated, with deletion of type 1 and P fimbriae resulting in upregulation of F1C fimbriae (97). Constitutive expression of type 1 fimbriae by *E. coli* CFT073 resulted in an apparent downregulation of P-fimbrial expression (97). Conversely, using three different approaches, Holden and colleagues demonstrated that expression of P fimbriae represses type 1 fimbrial expression (38). The acquisition of additional organelles involved in adhesion or biofilm formation confers a competitive advantage over less well-equipped UPEC strains. That is, strains with increased flexibility in adhesion or biofilm formation have an overall fitness advantage within the urinary tract.

## IRON ACQUISITION AND TRANSPORT SYSTEMS CONTRIBUTE TO FITNESS

Iron is essential for the growth and survival of almost all bacterial species. In the host, free iron is tightly bound to the iron-binding proteins hemoglobin, ferritin, transferrin, and lactoferrin, rendering it unavailable to the bacterial cells (14). The concentration of free ferric iron in the human body is maintained at very low levels ($\approx 10^{-18}$ M). Higher levels of free ferric iron are toxic to the host and have been shown to increase the virulence of a range of bacterial species (77). In response, bacteria synthesize iron-sequestering compounds called siderophores that chelate ferric iron away from host proteins or the environment and deliver them to the bacterial cells in the solubilized ferrous form. Without these high-affinity uptake and transport systems, bacterial pathogens could not successfully establish an infection within the mammalian host.

Iron has been shown to be essential for bacteria to colonize the urinary tract (102). In the pyelonephritis isolate *E. coli* CFT073 and in the ABU isolate *E. coli* 83972, expression microarrays demonstrated that all iron acquisition systems were universally expressed in vivo in mice (96) and humans (33), indicating not only that free iron is limited in this environment, but that iron is highly sought after by UPEC.

Four genetically distinct siderophore systems have been identified in *E. coli* species: enterobactin (catecholate), salmochelin (catecholate), aerobactin (hydroxamate), and yersiniabactin (thiazoline) (36). Enterobactin (also known as enterochelin) is the primary siderophore in *E. coli*, and this system has been conserved across many *E. coli* strains, including the four sequenced UPEC isolates CFT073, 536, UTI89, and F11. The biosynthesis and transport genes for enterobactin are encoded on 20 kb of contiguous sequence located near the 13-minute mark on the backbone of *E. coli* K-12 (6). Like all iron acquisition systems, enterobactin production is tightly regulated by intracellular iron levels, with the global regulator Fur (ferric uptake regulator) playing a critical role in this. The *fur* gene, present at ~15 minutes on the *E. coli* strain K-12 (6), encodes a 17-kDa protein that transcriptionally represses enterobactin production and many other iron-related and iron-independent genes in the presence of high levels of free iron (39). The three additional siderophore systems commonly found in UPEC—salmochelin, aerobactin, and yersiniabactin—are absent from *E. coli* strain K-12 and have been horizontally acquired by UPEC on genomic islands or islets, facilitating further adaptation of these strains to the unique niche of the urinary tract.

## Salmochelin

Salmochelin, a glycosylated form of enterobactin originally identified in *Salmonella enterica* (5), has also been identified in ExPEC isolates and is associated with virulence in UPEC (86). The salmochelin operon (*iroBCDEN*) is absent from *E. coli* strain K-12 and is located on a GI inserted near a tRNA gene in all three fully sequenced UPEC strains: CFT073 (PAI-CFT073-*serX*), 536 (PAI $III_{536}$; *thrW*), and UTI89 ($PAI_{UTI89}III$; *serX*); and on a GI of ~60 kb in UPEC strain F11. The outer membrane receptor for salmochelin, IroN, has been associated with virulence of UPEC within the urinary tract (86) and is implicated in the invasion of urothelial cells by ExPEC (22). An isogenic *iroN* mutant was able to colonize the bladder, kidneys, and urine of mice to levels comparable to wild type during independent challenge. It was only during a competitive infection (cochallenge) that the *iroN* mutant was outcompeted by its wild-type/parental strain (86), indicating that salmochelin contributes to the overall fitness of UPEC but is not essential for colonization.

## Aerobactin

The locus encoding the hydroxamate siderophore aerobactin consists of genes for biosynthesis (*iucABCD*; iron uptake chelate) and import (*iutA*; iron uptake transport) of aerobactin. Aerobactin has been associated with ExPEC infections (septicemia, pyelonephritis, and symptomatic and asymptomatic lower UTI) (13). Johnson and colleagues studied a group of *E. coli* isolates responsible for bacteremic UTIs and found that 78% of the urosepsis isolates contained the aerobactin system, with 21% of these strains possessing plasmid-encoded aerobactin (42). In *E. coli* CFT073, the aerobactin system is encoded on the largest GI in this strain (123 kb), along with hemolysin, P fimbriae, and Sat (PAI-CFT073-*pheV*) (58). Previous work from our laboratory revealed that the aerobactin gene cluster was present in 1/7 UPEC and in 1/3 fecal *E. coli* strains (not *E. coli* K-12) using CGH microarrays and in only 1/14 sequenced *E. coli* strains (EPEC isolate E110019) using BLAST score ratio (BSR) analysis (58). However, despite a demonstrated association between aerobactin and ExPEC strains, this gene cluster is not found in UPEC strains *E. coli* 536, UTI89, and F11. Aerobactin can also be plasmid encoded in *E. coli* (105, 109).

## Yersiniabactin

Yersiniabactin, encoded on the high-pathogenicity island (HPI), was first identified in highly pathogenic *Yersinia* spp. (*Y. pestis, Y. entercolitica*, and *Y. pseudotuberculosis*) (14), but has since been classified as "promiscuous" due to its widespread distribution among different species and genera of the *Enterobacteriaceae* family (3, 92). Yersiniabactin is a siderophore with a very high affinity for ferric iron ($K_D$ [equilibrium dissociation constant] $= 4 \times 10^{-36}$) (74). Yersiniabactin is one of the most well-studied PAIs of gram-negative bacteria and is strongly associated with ExPEC strains from phylogenetic groups B2 and D (90). The yersiniabactin HPI is a 35- to 43-kb PAI that is inserted at an *asnT* tRNA locus in *Yersinia* spp. (14, 90). Similarly, HPI is inserted near an *asnT* tRNA gene in UPEC strains CFT073 (106), 536 (9), and UTI89 (15), EAEC strain 17-2, and the probiotic *E. coli* strain DSM6601 (91). Preferential integration of HPI into the vicinity of the *asnT* tRNA gene may represent a possible mechanism for widespread transfer within the *Enterobacteriaceae* using a well-conserved sequence for homologous recombination. Yersiniabactin has been described as an iron uptake island (14).

## Heme/Hemoglobin

Heme may serve as a rich source of iron. Heme acquisition is also mediated by specific binding and import systems. *chuA*, the heme and hemoglobin utilization gene first identified in EHEC strain O157:H7 EDL933 (101), is present on an approximately 8.1-kb genomic islet in UPEC strains CFT073, 536, and UTI89 that is absent from *E. coli* K-12. ChuA is highly conserved among the phylogenetic lineages B2 and D, and the *chuA* gene is used as a target sequence in a PCR-based phylogrouping technique (16). The recently identified heme acquisition protein, Hma, functions independently of ChuA to mediate TonB-dependent hemin uptake by UPEC strain CFT073 (32), and *hma* is located on the 44-kb GI-CFT073-*cobU* in *E. coli* CFT073 (58). Homologs of *hma* exist in all sequenced UPEC strains (536, F11, and UTI89) (32), and based on DNA sequence alignments performed using *coli*BASE, a similar GI appears to be present in *E. coli* strains 536 and UTI89.

### Iha and IreA

The UPEC virulence factor Iha (IrgA homolog adhesin; c3610), originally identified on a GI in *E. coli* O157:H7 (100), is present on the *pheV* PAI of *E. coli* strain CFT073. Iha is an iron-regulated adhesin that has been associated with UPEC (41), with one study identifying the *iha* gene in 55% of urosepsis isolates (43). IreA (iron-responsive element; c5174), identified in ExPEC strain CP9 (O4:K54:H5) following upregulation in human urine (3.6-fold) and blood (6.6-fold) relative to LB broth, significantly contributes to the ability of UPEC strain CP9 to colonize the bladder of experimentally infected mice (84). Studies from our laboratory have shown that both *iha* and *ireA* are upregulated in the urine of experimentally infected mice (18.9- and 23.3-fold, respectively) (96) and in human urine ex vivo (5.87- and 7.81-fold, respectively) (2) relative to growth in LB broth in vitro. The *ireA* gene is located on the PAI-CFT073-*pheU* GI of UPEC strain CFT073 (58), and the IreA protein has shown promising results as a vaccine candidate against UTI (1). Mice vaccinated with IreA and subsequently challenged with *E. coli* strain CFT073 showed significant protection in the bladder compared to mice vaccinated with cholera toxin alone (1) (see below).

### Iron Transport

In addition to ferric iron acquisition, UPEC strains encode ferrous iron transport systems such as the *sitABCD* operon. The *sitABCD* system, also present in *S. flexneri* (83), *S. enterica* (110), and APEC strain $\chi$7122 (88), mediates the transport of iron and manganese and confers resistance to hydrogen peroxide. In *E. coli* CFT073, the *sitABCD* gene cluster is encoded on a 54-kb GI (PAI-CFT073-*icdA*) along with many bacteriophage-related genes (58). DNA sequence alignments reveal that the *sit* iron transport system is found in *E. coli* strains 536, UTI89, and F11. We identified the *sit* iron transport system in six of seven arbitrarily selected UPEC strains and only one of three fecal/commensal strains (not in *E. coli* K-12) using CGH (58). APEC strains represent a subgroup of ExPEC isolates and, thus, share many virulence and fitness factors. The *sitABCD* gene cluster has been identified on a plasmid in APEC strain $\chi$7122 (88) and is inserted into the chromosome on a GI in APEC strain O1 (44).

In addition to the systems described, numerous putative iron transport loci are present in UPEC strains, including the recently identified *fit* system (70). The *fit* iron transport system (*fitABCDER*) is present on a 7.2-kb genomic islet in UPEC strains CFT073, 536, UTI89, and F11 and APEC strain O1, and is not present in *E. coli* strain K-12. Studies in our laboratory using CGH microarrays found that the *fit* iron transport system was present in seven of seven UPEC and none of three fecal/commensal isolates (58).

### Iron Acquisition and Vaccine Development

Iron is essential for bacterial survival. Genes involved in iron acquisition are highly conserved in UPEC, and functional redundancy is common (9, 15, 106). Possession of multiple siderophores and iron sequestration and transport systems increases the flexibility and overall fitness of a UPEC strain. Examination of coincident urinary and rectal *E. coli* isolates from patients with recurrent UTIs from a quantitative metabolomic perspective revealed that expression of yersiniabactin and salmochelin is strongly associated with urinary *E. coli* isolates (36).

Given the importance of iron in vivo and the functional redundancy seen in UPEC, iron acquisition systems are a significant target for vaccine studies. A study by Alteri and colleagues used a range of approaches, including bioinformatic, genomic, transcriptional, and proteomic methods, to identify potential vaccine candidates against *E. coli* UTIs (1). Six outer membrane iron receptors were identified: ChuA, Hma, Iha, IreA, IroN, and IutA; and these antigens alone were tested in a mouse model of ascending UTI. Three antigens—Hma, IutA, and IreA—provided significant protection from experimental infection, and each of these genes is encoded on a GI in UPEC strain CFT073 (GI-CFT073-*cobU*, PAI-CFT073-*pheV*, and PAI-CFT073-*pheU*, respectively) (58). Russo and colleagues demonstrated that subcutaneous immunization with denatured IroN generated a significant IroN immunoglobulin G-specific response in serum, and protected against kidney infection in a mouse model of ascending UTI (87).

## AVOIDANCE OF HOST DEFENSE MECHANISMS

Our laboratory has recently identified two nonidentical genes, *sisA* and *sisB*, in UPEC strain CFT073 that are capable of suppressing the host inflammatory response during UTI (59). *sisA* and *sisB* are homologs of the *S. flexneri* SHI-2 PAI gene *shiA* (62), and these genes are also present on GIs: *sisA* on PAI-CFT073-*pheV* and *sisB* on GI-CFT073-*selC*. Deletion of both copies of the *shiA* homolog resulted in a hyperinflammatory response in the kidneys of experimentally inoculated CBA/J mice, and this phenotype could be restored to wild-type levels by complementation in *trans* with either *sisA* or *sisB*. Interestingly, homologs of *sisA* or *sisB* are not present in the other prototypic UPEC strains, whereas the ABU *E. coli* strain 83972 and the probiotic *E. coli* strain Nissle 1917 both contain a *sisA* homolog. Phylogenetic analysis of the *sisA* and *sisB* sequences revealed that these genes tended to group into two distinct clades—*sisA* with ExPEC strains and *sisB* with diarrheagenic *E. coli* strains—highlighting the amazing degree of genome flexibility of UPEC strains, and the significant role HGT has played in the evolution of this pathotype of *E. coli*.

## WHAT WE CAN LEARN FROM DECONSTRUCTED PATHOGENS

Several well-studied *E. coli* strains, ABU strain 83972 (33) and *E. coli* Nissle 1917 (26), may hold the key to understanding their pathogenic *E. coli* relatives. These strains have been proposed as "deconstructed pathogens"—strains that have evolved from pathogenic ancestors and no longer express virulence-associated genes—rather than commensal *E. coli* strains that acquired virulence and fitness factors horizontally over time. These strains appear to have an advantage over more pathogenic ExPEC isolates as they have retained genes that increase their overall fitness, while acquiring mutations or deletions that render true virulence factors and toxins nonfunctional, and they are able to persist in the host without triggering a damaging inflammatory response.

### ABU Strain 83972

ABU strains are capable of establishing long-term infection of the urinary tract, lasting from months to years, without producing a symptomatic infection. ABU has been described as an asymptomatic carrier state, similar to commensalism (48), with the strains well adapted to growth in the human urinary tract. The prototypic ABU strain, *E. coli* 83972, was isolated from a young girl in the 1970s who had been colonized with this strain for more than 3 years without exhibiting symptoms (33).

ABU strain 83972 emerged as a deconstructed pathogen, originating from a pyelonephritogenic *E. coli* strain that subsequently lost the ability to produce many of the virulence factors associated with UPEC isolates. ABU strain 83972 does not produce functional type 1, P (49), or F1C fimbriae (82), an adaptation that may facilitate persistence within the urinary tract without triggering an immune response. The majority of ABU strains are unable to adhere to human epithelial cells, although a subset of ABU strains were recently identified that were strongly adherent and yet failed to elicit a corresponding proinflammatory response (60). Interestingly, ABU strain 83972 has lost many of the typical UPEC virulence genes and yet has retained systems contributing to fitness, particularly those involved in iron acquisition and transport (aerobactin, salmochelin, yersiniabactin, heme/hemoglobin, and *sitABCD*) (33). *E. coli* 83972 has been successfully used as a prophylactic to treat patients with recurrent UTIs, patients with neurogenic bladders, spinal cord injury patients, and catheterized individuals (reviewed in reference 48).

### *E. coli* Nissle 1917

*E. coli* Nissle 1917 (O6:K5:H1) is a nonpathogenic commensal *E. coli* strain (26) that has been used to successfully treat inflammatory bowel disease (93). *E. coli* Nissle 1917 expresses type 1 and F1C fimbriae, two microcins, and a full complement of iron uptake systems, including enterobactin, aerobactin, salmochelin, yersiniabactin, and the *chu* heme acquisition locus (26). Four GIs have been identified in *E. coli* Nissle 1917 (26). GI $\text{I}_{\text{Nissle 1917}}$ contains two microcins, F1C fimbriae, and salmochelin and is almost identical at the nucleotide level to the PAI-CFT073-*serX* GI (58). GI $\text{II}_{\text{Nissle 1917}}$ contains *sat*, aerobactin, *iha*, and antigen 43, and this GI shares some features with the *pheV* PAI of UPEC strain CFT073. GI $\text{III}_{\text{Nissle 1917}}$ is located at an *argW* tRNA gene, and GI $\text{IV}_{\text{Nissle 1917}}$, inserted at an *asnT* tRNA gene, contains the yersiniabactin biosynthesis system (HPI of *E. coli* CFT073; PAI-CFT073-*asnT*). Similarities between the genomes of *E. coli* Nissle 1917 and UPEC strain CFT073 have been observed in several studies (26, 99), with *E. coli* Nissle 1917 containing many of the virulence-associated factors (iron acquisition systems, type 1 and F1C fimbriae) of *E. coli* CFT073, but missing pathogenicity-related genes (e.g., hemolysin and P fimbriae) required to transition into the urinary tract. *E. coli* Nissle 1917 has been used effectively as a probiotic in the treatment of inflammatory bowel disease (50, 93).

## CONCLUSION

It is clear that the evolution of *E. coli* is an ongoing process. At the time that a commensal strain of *E. coli* first appeared, there had already been hundreds of HGT events and deletion events that separated this species from what we recognized as *Salmonella* spp. This can be likened to the proposed evolution of EHEC strains outlined by Whittam (108). During the development of pathogenic strains that include the six pathotypes of diarrheagenic *E. coli* and three pathotypes of ExPEC, larger segments of DNA, termed genomic or pathogenicity islands, were acquired horizontally and conferred a competitive advantage to these strains in new niches. Some of these newly acquired genes can be considered outright virulence factors, as deletion of such genes attenuates the strains in their respective hosts or in animal models of these diseases. However, while many of the genes described throughout this chapter contribute to the overall fitness of these strains (here focused on UPEC), they are not absolutely required for colonization. It would be fair to state that these are not true virulence factors. However, the cumulative beneficial effect of acquiring multiple fitness factors on genomic islands or islets permits UPEC strains to thrive in the human urinary tract, while the ancestral *E. coli* strain K-12 cannot colonize this niche. While it is perhaps artificial to differentiate between either genomic islands (>30 kb) and islets (<30 kb) or virulence factors and fitness factors, the accumulation of these genes and the deletion of others have adapted our current pathotypes so that they now colonize specific niches, avoid the immune response, and damage the host during infection.

## REFERENCES

1. **Alteri, C. J., E. C. Hagan, K. E. Sivick, S. N. Smith, and H. L. Mobley.** 2009. Mucosal immunization with iron receptor antigens protects against urinary tract infection. *PLoS Pathog.* **5:**e1000586.

2. **Alteri, C. J., and H. L. Mobley.** 2007. Quantitative profile of the uropathogenic *Escherichia coli* outer membrane proteome during growth in human urine. *Infect. Immun.* **75:**2679–2688.
3. **Bach, S., A. de Almeida, and E. Carniel.** 2000. The *Yersinia* high-pathogenicity island is present in different members of the family *Enterobacteriaceae. FEMS Microbiol. Lett.* **183:**289–294.
4. **Barnhart, M. M., and M. R. Chapman.** 2006. Curli biogenesis and function. *Annu. Rev. Microbiol.* **60:**131–147.
5. **Baumler, A. J., T. L. Norris, T. Lasco, W. Voight, R. Reissbrodt, W. Rabsch, and F. Heffron.** 1998. IroN, a novel outer membrane siderophore receptor characteristic of *Salmonella enterica. J. Bacteriol.* **180:**1446–1453.
6. **Berlyn, M. K.** 1998. Linkage map of *Escherichia coli* K-12, edition 10: the traditional map. *Microbiol. Mol. Biol. Rev.* **62:**814–984.
7. **Blanco, J., M. Blanco, M. P. Alonso, J. E. Blanco, E. A. Gonzalez, and J. I. Garabal.** 1992. Characteristics of haemolytic *Escherichia coli* with particular reference to production of cytotoxic necrotizing factor type 1 (CNF1). *Res. Microbiol.* **143:**869–878.
8. **Blum, G., M. Ott, A. Lischewski, A. Ritter, H. Imrich, H. Tschape, and J. Hacker.** 1994. Excision of large DNA regions termed pathogenicity islands from tRNA-specific loci in the chromosome of an *Escherichia coli* wild-type pathogen. *Infect. Immun.* **62:** 606–614.
9. **Brzuszkiewicz, E., H. Bruggemann, H. Liesegang, M. Emmerth, T. Olschlager, G. Nagy, K. Albermann, C. Wagner, C. Buchrieser, L. Emody, G. Gottschalk, J. Hacker, and U. Dobrindt.** 2006. How to become a uropathogen: comparative genomic analysis of extraintestinal pathogenic *Escherichia coli* strains. *Proc. Natl. Acad. Sci. USA* **103:**12879–12884.
10. **Buchanan, K., S. Falkow, R. A. Hull, and S. I. Hull.** 1985. Frequency among *Enterobacteriaceae* of the DNA sequences encoding type 1 pili. *J. Bacteriol.* **162:**799–803.
11. **Buckles, E. L., F. K. Bahrani-Mougeot, A. Molina, C. V. Lockatell, D. E. Johnson, C. B. Drachenberg, V. Burland, F. R. Blattner, and M. S. Donnenberg.** 2004. Identification and characterization of a novel uropathogenic *Escherichia coli*-associated fimbrial gene cluster. *Infect. Immun.* **72:**3890–3901.
12. **Caprioli, A., V. Falbo, F. M. Ruggeri, L. Baldassarri, R. Bisicchia, G. Ippolito, E. Romoli, and G. Donelli.** 1987. Cytotoxic necrotizing factor production by hemolytic strains of *Escherichia coli* causing extraintestinal infections. *J. Clin. Microbiol.* **25:**146–149.
13. **Carbonetti, N. H., S. Boonchai, S. H. Parry, V. Väisänen-Rhen, T. K. Korhonen, and P. H. Williams.** 1986. Aerobactin-mediated iron uptake by *Escherichia coli* isolates from human extraintestinal infections. *Infect. Immun.* **51:**966–968.
14. **Carniel, E.** 2001. The *Yersinia* high-pathogenicity island: an iron-uptake island. *Microbes Infect.* **3:**561–569.
15. **Chen, S. L., C. S. Hung, J. Xu, C. S. Reigstad, V. Magrini, A. Sabo, D. Blasiar, T. Bieri, R. R. Meyer, P. Ozersky, J. R. Armstrong, R. S. Fulton, J. P. Latreille, J. Spieth, T. M. Hooton, E. R. Mardis, S. J. Hultgren, and J. I. Gordon.** 2006. Identification of genes subject to positive selection in uropathogenic strains of *Escherichia coli*: a comparative genomics approach. *Proc. Natl. Acad. Sci. USA* **103:**5977–5982.
16. **Clermont, O., S. Bonacorsi, and E. Bingen.** 2000. Rapid and simple determination of the *Escherichia coli* phylogenetic group. *Appl. Environ. Microbiol.* **66:**4555–4558.
17. **Connell, I., W. Agace, P. Klemm, M. Schembri, S. Marild, and C. Svanborg.** 1996. Type 1 fimbrial expression enhances *Escherichia coli* virulence for the urinary tract. *Proc. Natl. Acad. Sci. USA* **93:**9827–9832.
18. **Dobrindt, U.** 2005. (Patho-)Genomics of *Escherichia coli. Int. J. Med. Microbiol.* **295:** 357–371.
19. **Eden, C. S., L. A. Hanson, U. Jodal, U. Lindberg, and A. S. Akerlund.** 1976. Variable adherence to normal human urinary-tract epithelial cells of *Escherichia coli* strains associated with various forms of urinary-tract infection. *Lancet* **1:**490–492.

20. **Falbo, V., M. Famiglietti, and A. Caprioli.** 1992. Gene block encoding production of cytotoxic necrotizing factor 1 and hemolysin in *Escherichia coli* isolates from extra-intestinal infections. *Infect. Immun.* **60:**2182–2187.
21. **Falkow, S.** 1996. The evolution of pathogenicity in *Escherichia coli, Shigella* and *Salmonella*, p. 2723–2729. *In* F. C. Neidhardt (ed.), Escherichia coli *and* Salmonella: *Cellular and Molecular Biology.* ASM Press, Washington, DC.
22. **Feldmann, F., L. J. Sorsa, K. Hildinger, and S. Schubert.** 2007. The salmochelin siderophore receptor IroN contributes to invasion of urothelial cells by extraintestinal pathogenic *Escherichia coli in vitro. Infect. Immun.* **75:**3183–3187.
23. **Foxman, B.** 2003. Epidemiology of urinary tract infections: incidence, morbidity, and economic costs. *Dis. Mon.* **49:**53–70.
24. **Gransden, W. R., S. J. Eykyn, I. Phillips, and B. Rowe.** 1990. Bacteremia due to *Escherichia coli*: a study of 861 episodes. *Rev. Infect. Dis.* **12:**1008–1018.
25. **Groisman, E. A., and H. Ochman.** 1996. Pathogenicity islands: bacterial evolution in quantum leaps. *Cell* **87:**791–794.
26. **Grozdanov, L., C. Raasch, J. Schulze, U. Sonnenborn, G. Gottschalk, J. Hacker, and U. Dobrindt.** 2004. Analysis of the genome structure of the nonpathogenic probiotic *Escherichia coli* strain Nissle 1917. *J. Bacteriol.* **186:**5432–5441.
27. **Guyer, D. M., I. R. Henderson, J. P. Nataro, and H. L. Mobley.** 2000. Identification of Sat, an autotransporter toxin produced by uropathogenic *Escherichia coli. Mol. Microbiol.* **38:**53–66.
28. **Guyer, D. M., S. Radulovic, F. E. Jones, and H. L. Mobley.** 2002. Sat, the secreted autotransporter toxin of uropathogenic *Escherichia coli*, is a vacuolating cytotoxin for bladder and kidney epithelial cells. *Infect. Immun.* **70:**4539–4546.
29. **Hacker, J., G. Blum-Oehler, I. Muhldorfer, and H. Tschape.** 1997. Pathogenicity islands of virulent bacteria: structure, function and impact on microbial evolution. *Mol. Microbiol.* **23:**1089–1097.
30. **Hacker, J., and E. Carniel.** 2001. Ecological fitness, genomic islands and bacterial pathogenicity. A Darwinian view of the evolution of microbes. *EMBO Rep.* **2:**376–381.
31. **Hacker, J., and J. B. Kaper.** 2000. Pathogenicity islands and the evolution of microbes. *Annu. Rev. Microbiol.* **54:**641–679.
32. **Hagan, E. C., and H. L. Mobley.** 2009. Haem acquisition is facilitated by a novel receptor Hma and required by uropathogenic *Escherichia coli* for kidney infection. *Mol. Microbiol.* **71:**79–91.
33. **Hancock, V., A. S. Seshasayee, D. W. Ussery, N. M. Luscombe, and P. Klemm.** 2008. Transcriptomics and adaptive genomics of the asymptomatic bacteriuria *Escherichia coli* strain 83972. *Mol. Genet. Genomics* **279:**523–534.
34. **Heimer, S. R., D. A. Rasko, C. V. Lockatell, D. E. Johnson, and H. L. Mobley.** 2004. Autotransporter genes *pic* and *tsh* are associated with *Escherichia coli* strains that cause acute pyelonephritis and are expressed during urinary tract infection. *Infect. Immun.* **72:** 593–597.
35. **Henderson, I. R., F. Navarro-Garcia, M. Desvaux, R. C. Fernandez, and D. Ala'Aldeen.** 2004. Type V protein secretion pathway: the autotransporter story. *Microbiol. Mol. Biol. Rev.* **68:**692–744.
36. **Henderson, J. P., J. R. Crowley, J. S. Pinkner, J. N. Walker, P. Tsukayama, W. E. Stamm, T. M. Hooton, and S. J. Hultgren.** 2009. Quantitative metabolomics reveals an epigenetic blueprint for iron acquisition in uropathogenic *Escherichia coli. PLoS Pathog.* **5:**e1000305.
37. **Hentschel, U., and J. Hacker.** 2001. Pathogenicity islands: the tip of the iceberg. *Microbes Infect.* **3:**545–548.
38. **Holden, N. J., M. Totsika, E. Mahler, A. J. Roe, K. Catherwood, K. Lindner, U. Dobrindt, and D. L. Gally.** 2006. Demonstration of regulatory cross-talk between P fimbriae and type 1 fimbriae in uropathogenic *Escherichia coli. Microbiology* **152:**1143–1153.

39. **Hunt, M. D., G. S. Pettis, and M. A. McIntosh.** 1994. Promoter and operator determinants for *fur*-mediated iron regulation in the bidirectional *fepA-fes* control region of the *Escherichia coli* enterobactin gene system. *J. Bacteriol.* **176:**3944–3955.
40. **Johnson, J. R.** 1991. Virulence factors in *Escherichia coli* urinary tract infection. *Clin. Microbiol. Rev.* **4:**80–128.
41. **Johnson, J. R., S. Jelacic, L. M. Schoening, C. Clabots, N. Shaikh, H. L. Mobley, and P. I. Tarr.** 2005. The IrgA homologue adhesin Iha is an *Escherichia coli* virulence factor in murine urinary tract infection. *Infect. Immun.* **73:**965–971.
42. **Johnson, J. R., S. L. Moseley, P. L. Roberts, and W. E. Stamm.** 1988. Aerobactin and other virulence factor genes among strains of *Escherichia coli* causing urosepsis: association with patient characteristics. *Infect. Immun.* **56:**405–412.
43. **Johnson, J. R., T. A. Russo, P. I. Tarr, U. Carlino, S. S. Bilge, J. C. Vary, Jr., and A. L. Stell.** 2000. Molecular epidemiological and phylogenetic associations of two novel putative virulence genes, *iha* and $iroN_{E.\ coli}$, among *Escherichia coli* isolates from patients with urosepsis. *Infect. Immun.* **68:**3040–3047.
44. **Johnson, T. J., S. Kariyawasam, Y. Wannemuehler, P. Mangiamele, S. J. Johnson, C. Doetkott, J. A. Skyberg, A. M. Lynne, J. R. Johnson, and L. K. Nolan.** 2007. The genome sequence of avian pathogenic *Escherichia coli* strain O1:K1:H7 shares strong similarities with human extraintestinal pathogenic *E. coli* genomes. *J. Bacteriol.* **189:**3228–3236.
45. **Kallenius, G., R. Mollby, S. B. Svenson, I. Helin, H. Hultberg, B. Cedergren, and J. Winberg.** 1981. Occurrence of P-fimbriated *Escherichia coli* in urinary tract infections. *Lancet* **2:**1369–1372.
46. **Kaper, J. B., J. P. Nataro, and H. L. Mobley.** 2004. Pathogenic *Escherichia coli*. *Nat. Rev. Microbiol.* **2:**123–140.
47. **Klemm, P.** 1986. Two regulatory *fim* genes, *fimB* and *fimE*, control the phase variation of type 1 fimbriae in *Escherichia coli*. *EMBO J.* **5:**1389–1393.
48. **Klemm, P., V. Hancock, and M. A. Schembri.** 2007. Mellowing out: adaptation to commensalism by *Escherichia coli* asymptomatic bacteriuria strain 83972. *Infect. Immun.* **75:**3688–3695.
49. **Klemm, P., V. Roos, G. C. Ulett, C. Svanborg, and M. A. Schembri.** 2006. Molecular characterization of the *Escherichia coli* asymptomatic bacteriuria strain 83972: the taming of a pathogen. *Infect. Immun.* **74:**781–785.
50. **Kruis, W., P. Fric, J. Pokrotnieks, M. Lukas, B. Fixa, M. Kascak, M. A. Kamm, J. Weismueller, C. Beglinger, M. Stolte, C. Wolff, and J. Schulze.** 2004. Maintaining remission of ulcerative colitis with the probiotic *Escherichia coli* Nissle 1917 is as effective as with standard mesalazine. *Gut* **53:**1617–1623.
51. **Lane, M. C., C. J. Alteri, S. N. Smith, and H. L. Mobley.** 2007. Expression of flagella is coincident with uropathogenic *Escherichia coli* ascension to the upper urinary tract. *Proc. Natl. Acad. Sci. USA* **104:**16669–16674.
52. **Lane, M. C., and H. L. Mobley.** 2007. Role of P-fimbrial-mediated adherence in pyelonephritis and persistence of uropathogenic *Escherichia coli* (UPEC) in the mammalian kidney. *Kidney Int.* **72:**19–25.
53. **Lasaro, M. A., N. Salinger, J. Zhang, Y. Wang, Z. Zhong, M. Goulian, and J. Zhu.** 2009. F1C fimbriae play an important role in biofilm formation and intestinal colonization by the *Escherichia coli* commensal strain Nissle 1917. *Appl. Environ. Microbiol.* **75:**246–251.
54. **Lawrence, J. G.** 2005. Horizontal and vertical gene transfer: the life history of pathogens. *Contrib. Microbiol.* **12:**255–271.
55. **Lawrence, J. G., and H. Ochman.** 1998. Molecular archaeology of the *Escherichia coli* genome. *Proc. Natl. Acad. Sci. USA* **95:**9413–9417.
56. **Lawrence, J. G., and J. R. Roth.** 1999. Genomic flux: genome evolution by gene loss and acquisition, p. 263–290. *In* R. L. Charlebois (ed.), *Organization of the Prokaryotic Genome.* ASM Press, Washington, DC.

57. **Lloyd, A. L., T. A. Henderson, P. D. Vigil, and H. L. Mobley.** 2009. Genomic islands of uropathogenic *Escherichia coli* contribute to virulence. *J. Bacteriol.* **191:**3469–3481.
58. **Lloyd, A. L., D. A. Rasko, and H. L. Mobley.** 2007. Defining genomic islands and uropathogen-specific genes in uropathogenic *Escherichia coli. J. Bacteriol.* **189:**3532–3546.
59. **Lloyd, A. L., S. N. Smith, K. A. Eaton, and H. L. Mobley.** 2009. Uropathogenic *Escherichia coli* suppresses the host inflammatory response via pathogenicity island genes *sisA* and *sisB*. *Infect. Immun.* **77:**5322–5333.
60. **Mabbett, A. N., G. C. Ulett, R. E. Watts, J. J. Tree, M. Totsika, C. L. Ong, J. M. Wood, W. Monaghan, D. F. Looke, G. R. Nimmo, C. Svanborg, and M. A. Schembri.** 2009. Virulence properties of asymptomatic bacteriuria *Escherichia coli. Int. J. Med. Microbiol.* **299:**53–63.
61. **Mobley, H. L., D. M. Green, A. L. Trifillis, D. E. Johnson, G. R. Chippendale, C. V. Lockatell, B. D. Jones, and J. W. Warren.** 1990. Pyelonephritogenic *Escherichia coli* and killing of cultured human renal proximal tubular epithelial cells: role of hemolysin in some strains. *Infect. Immun.* **58:**1281–1289.
62. **Moss, J. E., T. J. Cardozo, A. Zychlinsky, and E. A. Groisman.** 1999. The *selC*-associated SHI-2 pathogenicity island of *Shigella flexneri. Mol. Microbiol.* **33:**74–83.
63. **Nowicki, B., J. P. Barrish, T. Korhonen, R. A. Hull, and S. I. Hull.** 1987. Molecular cloning of the *Escherichia coli* O75X adhesin. *Infect. Immun.* **55:**3168–3173.
64. **Nowicki, B., A. Labigne, S. Moseley, R. Hull, S. Hull, and J. Moulds.** 1990. The Dr hemagglutinin, afimbrial adhesins AFA-I and AFA-III, and F1845 fimbriae of uropathogenic and diarrhea-associated *Escherichia coli* belong to a family of hemagglutinins with Dr receptor recognition. *Infect. Immun.* **58:**279–281.
65. **Nowicki, B., J. Moulds, R. Hull, and S. Hull.** 1988. A hemagglutinin of uropathogenic *Escherichia coli* recognizes the Dr blood group antigen. *Infect. Immun.* **56:**1057–1060.
66. **Nowicki, B., M. Rhen, V. Väisänen-Rhen, A. Pere, and T. K. Korhonen.** 1984. Immunofluorescence study of fimbrial phase variation in *Escherichia coli* KS71. *J. Bacteriol.* **160:**691–695.
67. **Ochman, H., J. G. Lawrence, and E. A. Groisman.** 2000. Lateral gene transfer and the nature of bacterial innovation. *Nature* **405:**299–304.
68. **Ott, M., J. Hacker, T. Schmoll, T. Jarchau, T. K. Korhonen, and W. Goebel.** 1986. Analysis of the genetic determinants coding for the S-fimbrial adhesin (*sfa*) in different *Escherichia coli* strains causing meningitis or urinary tract infections. *Infect. Immun.* **54:** 646–653.
69. **Ott, M., H. Hoschutzky, K. Jann, I. Van Die, and J. Hacker.** 1988. Gene clusters for S fimbrial adhesin (*sfa*) and F1C fimbriae (*foc*) of *Escherichia coli*: comparative aspects of structure and function. *J. Bacteriol.* **170:**3983–3990.
70. **Ouyang, Z., and R. Isaacson.** 2006. Identification and characterization of a novel ABC iron transport system, *fit*, in *Escherichia coli. Infect. Immun.* **74:**6949–6956.
71. **Parham, N. J., S. J. Pollard, M. Desvaux, A. Scott-Tucker, C. Liu, A. Fivian, and I. R. Henderson.** 2005. Distribution of the serine protease autotransporters of the *Enterobacteriaceae* among extraintestinal clinical isolates of *Escherichia coli. J. Clin. Microbiol.* **43:**4076–4082.
72. **Parham, N. J., U. Srinivasan, M. Desvaux, B. Foxman, C. F. Marrs, and I. R. Henderson.** 2004. PicU, a second serine protease autotransporter of uropathogenic *Escherichia coli. FEMS Microbiol. Lett.* **230:**73–83.
73. **Pearson, M. M., and H. L. Mobley.** 2008. Repression of motility during fimbrial expression: identification of 14 *mrpJ* gene paralogues in *Proteus mirabilis. Mol. Microbiol.* **69:**548–558.
74. **Perry, R. D., P. B. Balbo, H. A. Jones, J. D. Fetherston, and E. DeMoll.** 1999. Yersiniabactin from *Yersinia pestis*: biochemical characterization of the siderophore and its role in iron transport and regulation. *Microbiology* **145**(Pt. 5)**:**1181–1190.
75. **Preston, G. M., B. Haubold, and P. B. Rainey.** 1998. Bacterial genomics and adaptation to life on plants: implications for the evolution of pathogenicity and symbiosis. *Curr. Opin. Microbiol.* **1:**589–597.

76. **Rasko, D. A., M. J. Rosovitz, G. S. Myers, E. F. Mongodin, W. F. Fricke, P. Gajer, J. Crabtree, M. Sebaihia, N. R. Thomson, R. Chaudhuri, I. R. Henderson, V. Sperandio, and J. Ravel.** 2008. The pangenome structure of *Escherichia coli*: comparative genomic analysis of *E. coli* commensal and pathogenic isolates. *J. Bacteriol.* **190:**6881–6893.
77. **Raymond, K. N., E. A. Dertz, and S. S. Kim.** 2003. Enterobactin: an archetype for microbial iron transport. *Proc. Natl. Acad. Sci. USA* **100:**3584–3588.
78. **Rippere-Lampe, K. E., A. D. O'Brien, R. Conran, and H. A. Lockman.** 2001. Mutation of the gene encoding cytotoxic necrotizing factor type 1 (*cnf1*) attenuates the virulence of uropathogenic *Escherichia coli*. *Infect. Immun.* **69:**3954–3964.
79. **Robinson, G. L.** 1951. The haemolysin of *Bacterium coli*. *J. Gen. Microbiol.* **5:**788–792.
80. **Rode, C. K., L. J. Melkerson-Watson, A. T. Johnson, and C. A. Bloch.** 1999. Type-specific contributions to chromosome size differences in *Escherichia coli*. *Infect. Immun.* **67:**230–236.
81. **Ronald, A.** 2002. The etiology of urinary tract infection: traditional and emerging pathogens. *Am. J. Med.* **113**(Suppl. 1A)**:**14S–19S.
82. **Roos, V., M. A. Schembri, G. C. Ulett, and P. Klemm.** 2006. Asymptomatic bacteriuria *Escherichia coli* strain 83972 carries mutations in the *foc* locus and is unable to express F1C fimbriae. *Microbiology* **152:**1799–1806.
83. **Runyen-Janecky, L. J., and S. M. Payne.** 2002. Identification of chromosomal *Shigella flexneri* genes induced by the eukaryotic intracellular environment. *Infect. Immun.* **70:** 4379–4388.
84. **Russo, T. A., U. B. Carlino, and J. R. Johnson.** 2001. Identification of a new iron-regulated virulence gene, *ireA*, in an extraintestinal pathogenic isolate of *Escherichia coli*. *Infect. Immun.* **69:**6209–6216.
85. **Russo, T. A., and J. R. Johnson.** 2000. Proposal for a new inclusive designation for extraintestinal pathogenic isolates of *Escherichia coli*: ExPEC. *J. Infect. Dis.* **181:**1753–1754.
86. **Russo, T. A., C. D. McFadden, U. B. Carlino-MacDonald, J. M. Beanan, T. J. Barnard, and J. R. Johnson.** 2002. IroN functions as a siderophore receptor and is a urovirulence factor in an extraintestinal pathogenic isolate of *Escherichia coli*. *Infect. Immun.* **70:**7156–7160.
87. **Russo, T. A., C. D. McFadden, U. B. Carlino-MacDonald, J. M. Beanan, R. Olson, and G. E. Wilding.** 2003. The siderophore receptor IroN of extraintestinal pathogenic *Escherichia coli* is a potential vaccine candidate. *Infect. Immun.* **71:**7164–7169.
88. **Sabri, M., S. Leveille, and C. M. Dozois.** 2006. A SitABCD homologue from an avian pathogenic *Escherichia coli* strain mediates transport of iron and manganese and resistance to hydrogen peroxide. *Microbiology* **152:**745–758.
89. **Schmidt, H., and M. Hensel.** 2004. Pathogenicity islands in bacterial pathogenesis. *Clin. Microbiol. Rev.* **17:**14–56.
90. **Schubert, S., P. Darlu, O. Clermont, A. Wieser, G. Magistro, C. Hoffmann, K. Weinert, O. Tenaillon, I. Matic, and E. Denamur.** 2009. Role of intraspecies recombination in the spread of pathogenicity islands within the *Escherichia coli* species. *PLoS Pathog.* **5:**e1000257.
91. **Schubert, S., A. Rakin, D. Fischer, J. Sorsa, and J. Heesemann.** 1999. Characterization of the integration site of *Yersinia* high-pathogenicity island in *Escherichia coli*. *FEMS Microbiol. Lett.* **179:**409–414.
92. **Schubert, S., A. Rakin, H. Karch, E. Carniel, and J. Heesemann.** 1998. Prevalence of the "high-pathogenicity island" of *Yersinia* species among *Escherichia coli* strains that are pathogenic to humans. *Infect. Immun.* **66:**480–485.
93. **Schultz, M.** 2008. Clinical use of *E. coli* Nissle 1917 in inflammatory bowel disease. *Inflamm. Bowel Dis.* **14:**1012–1018.
94. **Simms, A. N., and H. L. Mobley.** 2008. PapX, a P fimbrial operon-encoded inhibitor of motility in uropathogenic *Escherichia coli*. *Infect. Immun.* **76:**4833–4841.

95. **Smith, Y. C., S. B. Rasmussen, K. K. Grande, R. M. Conran, and A. D. O'Brien.** 2008. Hemolysin of uropathogenic *Escherichia coli* evokes extensive shedding of the uroepithelium and hemorrhage in bladder tissue within the first 24 hours after intraurethral inoculation of mice. *Infect. Immun.* **76:**2978–2990.
96. **Snyder, J. A., B. J. Haugen, E. L. Buckles, C. V. Lockatell, D. E. Johnson, M. S. Donnenberg, R. A. Welch, and H. L. Mobley.** 2004. Transcriptome of uropathogenic *Escherichia coli* during urinary tract infection. *Infect. Immun.* **72:**6373–6381.
97. **Snyder, J. A., B. J. Haugen, C. V. Lockatell, N. Maroncle, E. C. Hagan, D. E. Johnson, R. A. Welch, and H. L. Mobley.** 2005. Coordinate expression of fimbriae in uropathogenic *Escherichia coli*. *Infect. Immun.* **73:**7588–7596.
98. **Stamm, W. E., and T. M. Hooton.** 1993. Management of urinary tract infections in adults. *N. Engl. J. Med.* **329:**1328–1334.
99. **Sun, J., F. Gunzer, A. M. Westendorf, J. Buer, M. Scharfe, M. Jarek, F. Gossling, H. Blocker, and A. P. Zeng.** 2005. Genomic peculiarity of coding sequences and metabolic potential of probiotic *Escherichia coli* strain Nissle 1917 inferred from raw genome data. *J. Biotechnol.* **117:**147–161.
99a. **Swenson, D. L., N. O. Bukanov, D. E. Berg, and R. A. Welch.** 1996. Two pathogenicity islands in uropathogenic *Escherichia coli* J96: cosmid cloning and sample sequencing. *Infect. Immun.* **64:**3736–3743.
100. **Tarr, P. I., S. S. Bilge, J. C. Vary, Jr., S. Jelacic, R. L. Habeeb, T. R. Ward, M. R. Baylor, and T. E. Besser.** 2000. Iha: a novel *Escherichia coli* O157:H7 adherence-conferring molecule encoded on a recently acquired chromosomal island of conserved structure. *Infect. Immun.* **68:**1400–1407.
101. **Torres, A. G., and S. M. Payne.** 1997. Haem iron-transport system in enterohaemorrhagic *Escherichia coli* O157:H7. *Mol. Microbiol.* **23:**825–833.
102. **Torres, A. G., P. Redford, R. A. Welch, and S. M. Payne.** 2001. TonB-dependent systems of uropathogenic *Escherichia coli*: aerobactin and heme transport and TonB are required for virulence in the mouse. *Infect. Immun.* **69:**6179–6185.
103. **Väisänen-Rhen, V.** 1984. Fimbria-like hemagglutinin of *Escherichia coli* O75 strains. *Infect. Immun.* **46:**401–407.
104. **Valle, J., A. N. Mabbett, G. C. Ulett, A. Toledo-Arana, K. Wecker, M. Totsika, M. A. Schembri, J. M. Ghigo, and C. Beloin.** 2008. UpaG, a new member of the trimeric autotransporter family of adhesins in uropathogenic *Escherichia coli*. *J. Bacteriol.* **190:**4147–4161.
105. **Warner, P. J., P. H. Williams, A. Bindereif, and J. B. Neilands.** 1981. ColV plasmid-specific aerobactin synthesis by invasive strains of *Escherichia coli*. *Infect. Immun.* **33:**540–545.
106. **Welch, R. A., V. Burland, G. Plunkett III, P. Redford, P. Roesch, D. Rasko, E. L. Buckles, S. R. Liou, A. Boutin, J. Hackett, D. Stroud, G. F. Mayhew, D. J. Rose, S. Zhou, D. C. Schwartz, N. T. Perna, H. L. Mobley, M. S. Donnenberg, and F. R. Blattner.** 2002. Extensive mosaic structure revealed by the complete genome sequence of uropathogenic *Escherichia coli*. *Proc. Natl. Acad. Sci. USA* **99:**17020–17024.
107. **Welch, R. A., E. P. Dellinger, B. Minshew, and S. Falkow.** 1981. Haemolysin contributes to virulence of extra-intestinal *E. coli* infections. *Nature* **294:**665–667.
108. **Wick, L. M., W. Qi, D. W. Lacher, and T. S. Whittam.** 2005. Evolution of genomic content in the stepwise emergence of *Escherichia coli* O157:H7. *J. Bacteriol.* **187:**1783–1791.
109. **Williams, P. H.** 1979. Novel iron uptake system specified by ColV plasmids: an important component in the virulence of invasive strains of *Escherichia coli*. *Infect. Immun.* **26:**925–932.
110. **Zhou, D., W. D. Hardt, and J. E. Galan.** 1999. *Salmonella typhimurium* encodes a putative iron transport system within the centisome 63 pathogenicity island. *Infect. Immun.* **67:**1974–1981.

*Population Genetics of Bacteria: a Tribute to Thomas S. Whittam*
Edited by S. T. Walk and P. C. H. Feng


*Chapter 11*

# Genomic and Virulence Heterogeneity of Enteroaggregative *Escherichia coli*

***Edward G. Dudley and David A. Rasko***

Enteroaggregative *Escherichia coli* (EAEC) are defined by their pattern of adherence to HEp-2 cells growing in monolayer, which is often described as "stacked-brick adherence," "aggregative adherence" (AA), or a "honeycomb" structure" (Fig. 1A, B, and C). While first recognized as a causative agent of pediatric diarrhea in developing countries, this pathovar is now a prevalent cause of disease in children of developed countries, travelers to and adults residing in developing areas, and adults with HIV infection (1, 25, 32). Disease symptoms caused by EAEC are watery diarrhea that may contain blood and/or mucus and intestinal inflammation, characterized by the presence of fecal lactoferrin and proinflammatory cytokines such as interleukin-8 (IL-8) (20).

The known virulence factors of EAEC and their role in pathogenesis have been the subject of several reviews (12, 20, 30), and are summarized in Table 1. The current model, built predominantly through studying the EAEC strain 042, suggests that EAEC use fimbriae, collectively referred to as aggregative adherence fimbriae (AAFs) (Fig. 1D), to adhere to the epithelium of the terminal ileum and colon (4, 6, 13, 34). Although AAFs are necessary and sufficient for adherence in vitro and ex vivo, other proteins such as Shf and Air may also contribute (19, 38). Colonization is further mediated by a secreted protein called Pic (protein involved in colonization), which degrades mucin (21), and by the antiaggregative protein (Aap), which may aid the release of bacteria from mature biofilms and promote colonization of additional sites within the intestinal tract (35). Recognition of EAEC flagellar and putative tip fimbriae proteins by the host innate immune system results in an inflammatory response (20). Cytotoxicity and fluid accumulation are mediated by the secreted protease Pet (plasmid-encoded toxin) and the endotoxin East1, respectively, and possibly some secreted effectors of a type VI secretion system, although their

***Edward G. Dudley*** • Department of Food Science, Pennsylvania State University, University Park, PA 16802. ***David A. Rasko*** • Department of Microbiology and Immunology, University of Maryland School of Medicine, Baltimore, MD 21201.

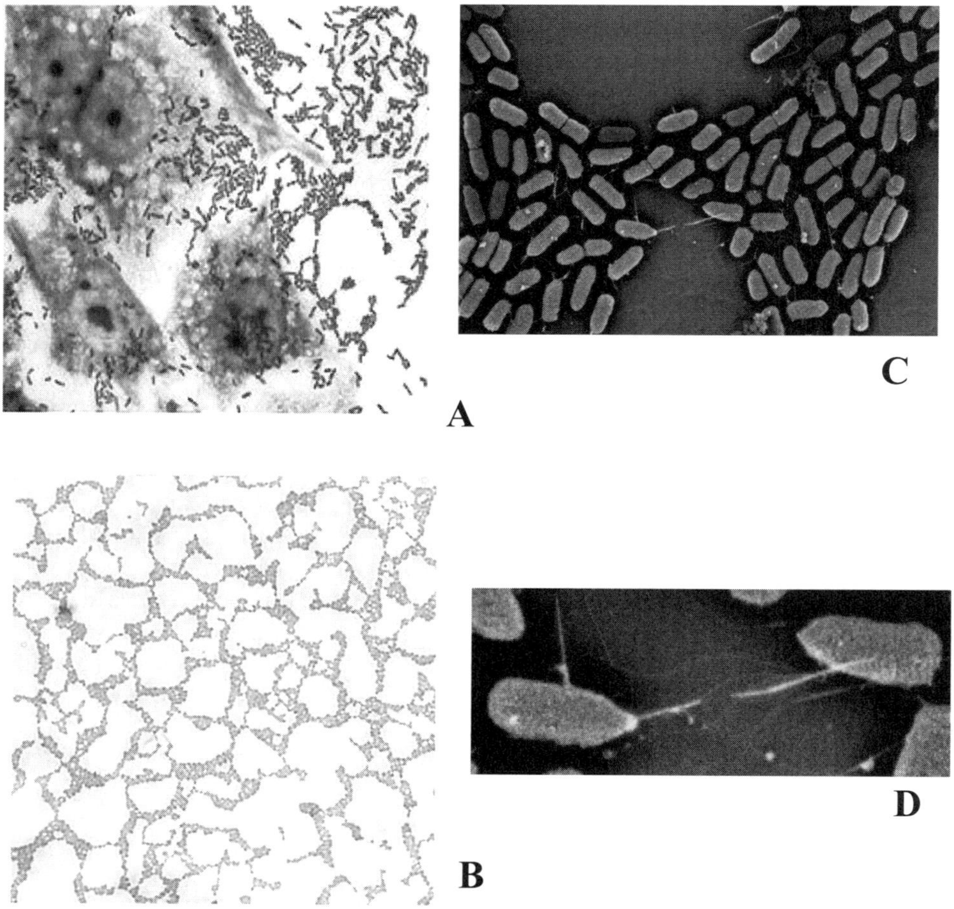


**Figure 1.** The prototypical strain EAEC 042 adheres to cells and surfaces in a stacked-brick morphology. (A) Adherence to HEp-2 cells; note the adherence both to the cells and to the surrounding glass substratum. (B) The characteristic "honeycomb" structure formed when an EAEC isolate adheres to glass or plastic. (C) Transmission electron micrograph showing details of the honeycomb structure. (D) Higher magnification of a portion of panel C showing the AAF/II extending from the surface of the bacterium. Figures 1A and B reprinted from reference 15 with permission.

precise roles in pathogenesis remain unclear (20). A large number of virulence genes are under the control of the transcriptional regulator AggR, including the *aggR* gene itself (16).

This chapter will not focus on EAEC pathogenesis, which has been recently reviewed (12, 20). Instead, given Tom Whittam's passion for the study of *E. coli* evolution, we will examine what is understood concerning the diversity of this highly heterogeneous pathovar, recent insights from whole and partial genome sequences, and how these data may eventually be used to further

**Table 1.** Virulence genes of EAEC

| Gene name(s) | Plasmid or chromosome | Proposed function | Under AggR control? | Reference(s) |
|---|---|---|---|---|
| *agg, aaf, aaf3, hdc* | Plasmid | Aggregative adherence | Yes | 4, 6, 13, 34 |
| *aggR* | Plasmid | Master regulator of virulence | Yes | 16 |
| *aap* | Plasmid | Antiaggregation; shielding fimbriae from surface charge | Yes | 35 |
| *aatPABCD* | Plasmid | Translocation of Aap across outer membrane | Yes | 35 |
| *shf* | Plasmid | Adherence | Yes | 19 |
| *pet* | Plasmid | Cytotoxin | No | |
| *astA* | Plasmid | Heat-stable enterotoxin | No | 12 |
| *aai* locus | Chromosome | Type VI secretion system | Yes | 16 |
| *pic* | Chromosome | Mucin catabolism; intestinal colonization | No | 21 |
| *air* | Chromosome | Adherence | No | 38 |
| *setBA* | Chromosome | Fluid secretion; $AB_5$ enterotoxin | No | 20 |

clarify what factors make EAEC pathogens. In the more than 2 decades since its initial description, it is still not apparent why some EAEC strains are pathogenic to humans while others appear to be nothing more than commensals. We will argue that a more thorough understanding of EAEC phylogeny using traditional methods and comparative genomics of a large number of strains is required to clarify this puzzle. However, in order to fully appreciate why this has been such a difficult problem, we must return to the past and understand how this pathovar was first defined.

## HISTORY

### The Discovery that EAEC Are Causative Agents of Pediatric Diarrhea

In 1979, Cravioto et al. (11) introduced a technique that today is referred to as the HEp-2 adherence assay. This method is used for characterizing mannose-resistant adherence patterns of *E. coli* strains isolated from clinical cases and from healthy individuals. At this point in history, determination of the somatic antigen and enterotoxin production was used to classify clinical isolates into categories such as enterotoxigenic *E. coli* (ETEC), enteropathogenic *E. coli* (EPEC), and "non-EPEC, non-ETEC." In what has subsequently become a classic paper with 458 citations (according to the ISI Web of Knowledge as of January 2010), the authors state in their discussion that "The assay can be used for the selection and study of *E. coli* strains possessing an adhesive factor different from those already described." This statement foreshadowed the identification of new pathovars, and the stage was set for the discovery of EAEC.

The term "EAEC" was originally used to mean "enteroadherent *E. coli*" and was used to define bacteria that adhered to HEp-2 cells, including strains that today would be defined as EPEC (28). However, in 1987, Nataro et al. (31) screened 516 *E. coli* isolates obtained from Chilean children, and used the term "enteroadherent-aggregative *E. coli*" to define a specific subset of non-EPEC isolates that displayed a "stacked-brick" adherence pattern on the surface of HEp-2 cells as well as the glass coverslip. Intestinal colonization by *E. coli* displaying this adherence phenotype was significantly correlated with diarrhea, and the majority of these isolates (84 out of 89) were DNA probe negative for typical markers of ETEC, EPEC, enterohemorrhagic *E. coli* (EHEC), and enteroinvasive *E. coli* (EIEC). The end goal of this study was to demonstrate the utility of adherence assays for identifying EPEC, and the casual reader may have concluded that this work serendipitously discovered an effective method of identifying a new and easily definable *E. coli* pathogen. What was to follow, though, was the reality that the adherence phenotype was a phenotypic marker for a genetically diverse set of pathogenic and nonpathogenic bacteria that would take many years of research and the help of experts in microbial evolution, like Tom Whittam, to sort out.

## Recognition of EAEC Diversity

Throughout the 1980s, nucleic acid probes were developed for *E. coli* pathovars, including ETEC, EPEC, EIEC, and EHEC (see references cited in reference 31). The first EAEC DNA probe, called the "AA probe," hybridized to a portion of a 60- to 65-MDa plasmid that was present in 89% of bacteria exhibiting AA (3). The fact that genomic DNA from 11% of isolates did not hybridize to this probe suggested early on that EAEC was a more heterogeneous group than some of the other *E. coli* pathovars. Additionally, while genetic characterization of EAEC strain 042 revealed that the plasmid encoded virulence factors including AAF, AggR, Pet, and AspU, and that AAF was necessary and sufficient for AA, distribution studies revealed that there were likely multiple alleles of such fimbriae (13, 33). Phenotypic variation of EAEC was further revealed upon discovery that not all AA-positive isolates were pathogenic in human volunteers (33). Discovering the conserved features of pathogenic isolates would be crucial for understanding which virulence determinants were critical in disease, revealing the reason for the variations in pathogenesis, and enabling the specific selection of DNA marker(s) that would accurately discriminate between virulent and avirulent strains.

## CONTRIBUTIONS OF THOMAS WHITTAM: PHYLOGENETIC ANALYSIS OF EAEC

### MLEE Investigations of EAEC Phylogeny

While the genetic basis of EAEC virulence was being investigated, Tom Whittam was applying multilocus enzyme electrophoresis (MLEE) to study this pathovar and the entire *E. coli* species. In a paper published with Luiz Trabulsi and colleagues from Brazil (7), clinical isolates of serotype O111, one of the dominant EAEC serotypes, were characterized and two electrophoretic types (ETs), ET 1 and ET 8, were determined to contain the EAEC isolates as defined by the AA probe and/or HEp-2 adherence assay. Most notably, these two ETs differed at 8/20 enzyme loci, suggesting that the EAEC pathovar may also include genetically diverse members. Additionally, ET 1 isolates of flagellar antigen type H4 were found to be AA probe negative but AA adherence positive, while ET 1 isolates that were H-nontyped were classified as EAEC by both the adherence assay and the AA probe. This confirmed the previous findings of Baudry et al. (3) that the phenotypic and genotypic methods of defining EAEC do not always correlate.

In 1999, a collaborative effort between Tom Whittam's lab and Jim Nataro's lab (14) provided the first large-scale phylogenetic determination of a worldwide collection of EAEC isolates, as well as isolates previously identified as diffusely adherent *E. coli* (DAEC). In this study (14), isolates were screened by 20-locus MLEE as well as colony hybridizations to 10 plasmid- and chromosome-encoded virulence genes. These results provided for the first time a more complete picture of EAEC diversity, and several important points came to light. First, three phylogenetic groups containing EAEC strains were identified and designated EAEC1, EAEC2, and AA/DA (Fig. 2A). The first two groups comprised almost exclusively EAEC strains (defined by the HEp-2 adherence assay), while the third group consisted of a mixture of EAEC and DAEC strains. Second, even in cases where DAEC and EAEC strains intermingled within the same phylogenetic group, these strains were clearly distinguished by the fact that DAEC did not carry EAEC virulence determinants. Thus, despite the close relatedness of EAEC and DAEC established by MLEE, the differences in their virulence gene content would suggest that the evolution of pathogenesis involved extensive horizontal gene transfer. The fact that many of the EAEC strains from different phylogenetic groups carried the same combination of virulence genes also suggested that analogous horizontal gene transfer events had occurred in distantly related lineages (i.e., convergent evolution). Third, a few strains were classified as EAEC by the HEp-2 adherence assay but were DNA probe negative for all the screened EAEC virulence factors and clustered phylogenetically outside of EAEC1, EAEC2, and AA/DA. These observations seemed at first to be trivial and expected, but two of these strains, 101-1 and C1096, were bona fide outbreak strains (9, 26). Czeczulin et al. (14) suggested that 101-1 was probe positive for *aap* and the AAF/II subunit; however, neither gene sequence is found within the 101-1 genome sequence present in GenBank. These data are consistent with the hypothesis that some patho-

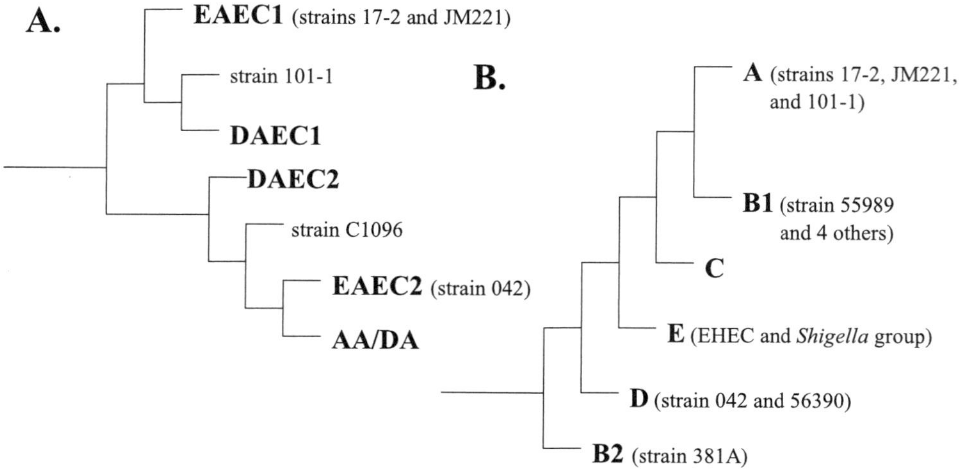


**Figure 2.** Phylogenetic analysis of EAEC isolates by two techniques. (A) MLEE analysis of EAEC and DAEC; adapted from Czeczulin et al. (14). Strains discussed in this chapter are indicated on the tree. (B) MLST using six loci; adapted from Escobar-Paramo et al. (17).

genic, atypical EAEC strains exhibit AA but are phylogenetically distinct from other AA probe-positive strains.

These results were supported by later studies, including those of Escobar-Paramo et al. (17), that analyzed the evolutionary history of different *E. coli* pathovars. In that study, multilocus sequence typing (MLST) was used to determine the phylogeny of 98 *E. coli* and *Shigella* strains and categorize them into the six major evolutionary nomenclature groups (A, B1, B2, C, D, and E) previously defined by Tom Whittam's lab (24). These strains were also screened for the presence of pathovar-associated virulence factors. This MLST analysis revealed that EAEC ($n$ = 10 AA probe-positive isolates) were phylogenetically diverse and clustered with other *E. coli* isolates of groups A, B1, B2, and D (Fig. 2B). These data provide further support for the conclusions of Czeczulin et al. (14) that EAEC may have evolved multiple times through the acquisition of a similar combination of virulence genes.

Three EAEC strains were in common between the two studies, and it is noteworthy that Escobar-Paramo et al. (17) categorized strains 17-2 and JM221 into phylogenetic group A and strain 042 into group D, while Czeczulin et al. (14) placed these two groups of strains into EAEC1 and EAEC2, respectively. This observation is interesting, as neither 17-2 nor JM221 was pathogenic in human volunteer studies (33); however, 042 caused diarrhea in three out of five volunteers. It is tempting to speculate that there is a correlation between phylogenetic grouping and pathogenesis, but more strains need to be characterized to either support or refute this speculation.

### Distribution of Virulence Genes

Prior to the paper by Czeczulin et al. (14), only a handful of verified virulence factors had been described for EAEC and the distribution of these genes was only examined by limited studies in diverse collections of EAEC strains. Czeczulin et al. identified and determined the distribution of genes encoding the AAF/I and AAF/II fimbrial subunits, AggR, Pet, Aap, AA probe, Pic (designated She in the 1999 paper), Shf, and yersiniabactin biosynthesis among a worldwide collection of EAEC isolates. It was discovered that AAF/I, AAF/II, and Pet were found in less than one-third of the isolates. This observation is suggestive that neither Pet nor a specific AAF type is required for EAEC virulence. The remaining genes were found in 50 to 80% of isolates, and it was found that the *aap* probe may be as specific as the AA probe as a DNA marker for EAEC. The observation that these two markers segregated in the same strains became clearer many years later when it was discovered that the AA probe corresponds to a gene cluster (*aatPABCD*) that is necessary for the efficient translocation of Aap through the outer membrane (35).

## CURRENT UNDERSTANDING OF EAEC DIVERSITY

The collaboration between Tom Whittam and Jim Nataro (14) provided significant insights into the diversity of EAEC strains, and in the ensuing 10 years a number of papers were published that would identify, mechanistically characterize, and study the distribution of other virulence determinants within EAEC. Some of the most significant observations concerning EAEC diversity are summarized below.

### The AA Phenotype Is Mediated by Structures Distinct from AAFs

Four distinct fimbriae, designated AAF/I, AAF/II, AAF/III, and AAF/IV, have been described to date, but with few exceptions strains only carry one type of AAF (5, 6). These fimbriae are members of a large superfamily called the Dr adhesins (6). All four of these fimbrial systems are composed of four genes, encoding the major and putative minor fimbrial subunits, as well as an usher and chaperone necessary for translocation of the fimbriae to the cell surface (Fig. 3). The AAF/I, AAF/II, and AAF/IV genes are carried on the virulence plasmid, and expression requires the transcriptional regulator AggR. AAF/III is presumably under the same regulatory control, but this has never been experimentally reported in the literature. The AAF genes are organized in an operon-like structure, except for AAF/II, where the usher and minor fimbrial subunit genes are separated from those encoding the chaperone and major fimbrial subunits by 12.9 kb (Fig. 3). The intervening region in AAF/II encodes the EAEC-specific cytotoxin Pet and the toxin East1 that is found in other *E. coli* strains as well. The gene topology of AAF/II likely explains why AAF/II-positive strains also tend to carry Pet (5, 14). Two studies (5, 6) looking at the distribution of the four AAFs in EAEC indicate that at least one of the known AAFs is found in approximately 80% of EAEC isolates.

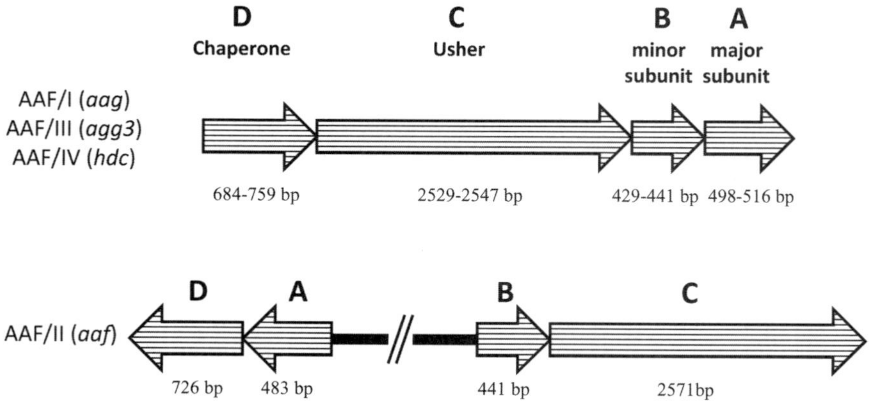


**Figure 3.** Schematic representation of the genetic organization of the AAF clusters. Arrows represent genes for the fimbriae subunits (A), putative tip adhesion (B), usher (C), and chaperone (D). The organization of genes for AAF/I, AAF/III, and AAF/IV is similar, and the range of sizes in nucleotides for the three homologs are indicated below the arrows. The four AAF/II genes are organized in two different clusters, and these are separated by 12.9 kb (indicated by break lines between *aafDA* and *aafBC*). The size of each AAF/II gene in nucleotides is indicated.

AAFs are not the only extracellular structures capable of mediating AA. For example, Monteiro-Neto et al. (29) described a 60-kDa outer membrane protein from EAEC strain 236 that was sufficient to confer the AA phenotype when expressed in a nonadherent *E. coli* strain. This protein was largely found in strains of serotype O111:H12 that displayed AA, but absent in other EAEC, EPEC, or EHEC strains screened or in O111:H12 serotypes that were weakly adherent. Additionally, a type IV pilus of an outbreak strain was found to contribute to the AA phenotype on HT-29 cells and to biofilm formation in a continuous microfermentor, but not significantly to AA on glass during static incubation (15). This suggests that AA is sometimes mediated by multiple extracellular structures in the same bacteria. Lastly, EPEC of serogroup O125ac: H6 (2) demonstrate AA on cultured cells, suggesting that other *E. coli* pathovars exhibit this adherence pattern. Collectively, these data indicate that AA should not be used as the only criterion for defining EAEC.

## The AggR Regulon Defines "Typical EAEC" That Are Associated with Virulence

The fact that diverse populations of *E. coli* exhibit AA may explain in part why some studies that define EAEC by the adherence pattern alone fail to associate this organism with disease (10). While the AA probe (see "Recognition of EAEC Diversity," above) seems to be more specific for pathogenic EAEC, the gene sequence targeted by this probe did not have a known function when it was first developed and therefore it was uncertain that this probe was a true marker of virulence. Sequencing of the virulence plasmid of EAEC strain 042 demonstrated that the AA probe was contained entirely within a gene encoding a TolC-like outer membrane protein designated AatA (35). The gene encoding this protein is the second of a cluster designated *aatPABCD*, and was proposed to encode a membrane-spanning complex for the translocation of Aap to the surface of the bacterial cell. As Aap serves a role in dispersal of EAEC aggregates during infection, this indicates that the AA probe is a true marker of virulence, and provides a stronger rationale for its use in detection of virulent EAEC isolates.

The discovery that AggR positively regulates transcription of the plasmid-encoded *aatPABCD* and *aap* (35), other virulence genes such as the AAFs (34, 39), and a large number of chromosomal genes that are part of a type VI secretion system (16) strongly suggested that the AggR regulon comprises a unique suite of virulence genes in EAEC. Indeed, microarray analysis categorized a collection of EAEC strains into two groups, and genes of the AggR regulon were found at a statistically higher frequency within "group 1" strains (27). Group 1 is now defined as "typical EAEC" (30), and it is hypothesized that these isolates cause disease with a greater frequency than "atypical EAEC" (aEAEC) that exhibit AA but lack genes of the AggR regulon. This, however, does not suggest that aEAEC are avirulent; quite to the contrary, aEAEC strains have been implicated in at least two published outbreak reports (9, 26). Both of these outbreak aEAEC strains lacked *aggR* and all members of the AggR regulon that were screened except for AAF/IV, which is found in 101-1 (6). It will be interesting to determine by future comparative genomic analysis of a larger collection of known pathogenic aEAEC strains whether this group of pathogens comprises phylogenetically related or heterogeneous organisms.

## The AAFs Are Structurally and Functionally Different

While the AAFs are necessary and sufficient for the AA phenotype and are routinely found in typical EAEC isolates, it is not clear whether these fimbriae are structurally or functionally interchangeable. Amino acid sequence comparisons of the AAF fimbrial subunits indicate that these peptides are not merely slight variants of one another, but are distantly related (Fig. 4A). Interestingly, there is greater conservation among the deduced amino acid sequences of the usher, chaperone, and putative tip adhesions (Fig. 4, panels B, C, and D, respectively) than among the major subunits (Fig. 4, panel A). Not unexpectedly, the sequence divergence of the fimbrial subunits manifests as phenotypic differences, including structural morphology and binding specificity. Electron microscopy has been used to examine the AAFs, and comparisons determined that both AAF/II and AAF/III fimbriae form 3- to 5-nm rigid fimbriae while AAF/I fimbriae form 2- to 3-nm flexible structures that bundle (4, 13). AAF/IV fimbriae have not been visualized by electron microscopy to date, although phylogenetic (6) and bioinformatic (Fig. 4A) analysis indicates that the protein is quite distinct from the other three AAFs. Despite the known differences between AAFs, it is fascinating that the AAF/I usher/chaperone system can translocate AAF/II fimbriae to the cell surface (13).

| A. | I | II | III | IV |
|---|---|---|---|---|
| AggA | | | | |
| AafA | 25 | | | |
| Agg3A | 23 | 18 | | |
| HdcA | NS | NS | NS | |

| B. | I | II | III | IV |
|---|---|---|---|---|
| AggB | | | | |
| AafB | 57 | | | |
| Agg3B | 60 | 59 | | |
| HdcB | 52 | 48 | 52 | |

| C. | I | II | III | IV |
|---|---|---|---|---|
| AggC | | | | |
| AafC | 68 | | | |
| Agg3C | 68 | 67 | | |
| HdcC | 67 | 68 | 65 | |

| D. | I | II | III | IV |
|---|---|---|---|---|
| AggD | | | | |
| AafD | 64 | | | |
| Agg3D | 66 | 71 | | |
| HdcD | 63 | 63 | 65 | |

**Figure 4.** Percent identity of deduced amino acid sequences of *aafA, aafB, aafC*, and *aafD* homologs encoded within AAF/I (genes with *agg* designation), AAF/II (*aaf*), AAF/III (*agg3*), and AAF/IV (*hdc*) clusters, as determined using BLAST with default parameters. Gene names are given on the left side of each chart, and roman numerals above indicate corresponding genes from AAF/I through AAF/IV clusters. Predicted signal sequences were removed prior to alignments. NS, no significant similarity (defined as Expect value of <0.05) was found in pairwise comparison.

Additionally, the hemagglutination properties of the AAFs differ. AAF/I agglutinates human, bovine, sheep, rabbit, and guinea pig erythrocytes, whereas the other AAFs agglutinate only a subset of these (6). A recent study demonstrated that AAF/II bound to extracellular matrix proteins including fibronectin, laminin, and collagen IV, but not to collagen I (18). Given the differences between the AAFs, it seems plausible that binding specificity also differs between them, and that this could theoretically affect colonization levels and/or the site of adherence along the human gastrointestinal tract. Whether the different AAF types affect the severity or course of disease remains to be investigated.

It is well understood that EAEC-mediated diarrhea induces the production of proinflammatory cytokines such as IL-8 and IL-1$\beta$. As with other enteric pathogens, EAEC induces cytokine secretion via Toll-like receptor-5 recognition of the structural subunit of the bacterial flagella (FliC). Harrington et al. (22) noted that IL-8 secretion is also stimulated by AafB, the proposed tip adhesion of AAF/II. As with the major subunits, there is significant amino acid sequence variability between AafB homologs and other AAFs (Fig. 4). Another question unanswered to date, then, is whether various AafB alleles stimulate IL-8 and other proinflammatory cytokines to the same levels, and whether this may affect the severity of disease.

## EAEC Encode a Diverse Set of Autotransporters

Serine protease autotransporters of *Enterobacteriaceae* (SPATEs) represent a large class of proteins that mediate translocation of the N-terminal serine protease domain across the bacterial outer membrane via a C-terminal $\beta$-barrel domain (23). These proteins are further categorized into two groups, where class I designates SPATEs that are cytotoxic to epithelial cells and class II includes those that have other functions (5). EAEC strain 042 secretes two SPATEs, designated Pet (class I) and Pic (class II). Pet is internalized into cells where it mediates toxic effects through cleavage of proteins such as fodrin (12), whereas Pic is not cytotoxic, but is believed to assist in intestinal colonization through the cleavage of mucin (21). As mentioned above, Pet is not widely found in clinical EAEC isolates, so it has been assumed that most EAEC strains mediate cytotoxicity through a mechanism independent of autotransporters. However, a recent study (5) discovered that 85% of EAEC strains screened carried genes for Pet, Sat, SigA, and/or EspP, which are class I SPATEs that were previously characterized in uropathogenic *E. coli, Shigella*, and Shiga toxin-producing *E. coli*. This study, therefore, suggests that EAEC strains lacking Pet may remain cytotoxic, which further complicates efforts to identify a definitive mechanism of EAEC pathogenesis. It was already well established that most EAEC isolates encode *pic* (14); however, this study also revealed that approximately 40% of EAEC strains carry *sepA*. This SPATE is better characterized in *Shigella flexneri*, where, similar to Pic in EAEC, it is proposed to promote intestinal colonization through the cleavage of mucin. Unlike other virulence factors in EAEC 042, *pet* and *pic* are not under the control of AggR; however, whether other SPATEs are regulated by AggR will be a question for future investigation. Even if the SPATEs are not part of the AggR regulon, a collection of autotransporters may be involved in virulence.

## Genomic Studies Will Continue To Provide Insights into EAEC Diversity

To date, two complete (strains 042 and 55989) and one shotgun-assembled (strain 101-1) EAEC genome sequences are available, providing important data for comparative studies between typical and atypical EAEC, as well as EAEC from different phylogenetic groups. In addition, a shotgun sequence for strain JM221 was recently generated (D. Rasko, unpublished results), permitting characterization and comparison of a known pathogenic strain (042) to a nonpathogenic strain (JM221). Many insights into EAEC comparative genomics were discussed in the recent paper on the EAEC 042 genome (8) and will not be reiterated here. However, the next section of this chapter will discuss additional information derived from the comparison of these genomes.

## GENOMICS OF EAEC

As mentioned above, there are now two complete genomes from typical EAEC strains (042 and 55989), another genome from a typical EAEC strain is in progress (JM221), and one draft genome from an atypical EAEC strain is available (101-1). The addition of each of these genomes to the comparative scheme allows further insight into the genetic structure of the EAEC pathovar as a whole. All of the EAEC genomes belong to distinct serogroups that are widely distributed geographically and temporally (Table 2). This is one of the few *E. coli* pathovars that have this type of representation over time, space, and serogroup, further extending the previous assertion that EAEC is a diverse pathovar. In addition to the diversity observed in the classification phenotypes, there is also diversity in the virulence factors that each genome contains (Table 1). Much of the genetic study of EAEC has focused on the virulence of the pathovar, and rightly so; however, it is becoming clear that the examination of the complete genomic repertoire of the isolates is required to further understand the level of variation within this species and these pathovars.

Genomic comparisons to identify conserved and unique genes among the EAEC have revealed that, much like the other *E. coli* species, there is a significant core genome and an ever-expanding pan-genome (37, 40). When the EAEC genomes and the *E. coli* commensal isolate, HS, are compared based on gene content, it is clear that the EAEC are a diverse group of pathogens, with >300 unique genes identified in strain 55989 and >500 in strains 101-1 and 042 (Table 3 and Fig. 5). While limited sequence data are currently available for JM221, it appears that it will follow a similar trend in carrying a significant number of unique genes. One such example observed when comparing the atypical EAEC isolate 101-1 with all the sequenced genomes was the identification of a bundle-forming pilus (*bfp*)-like gene cluster (Fig. 6). While this cluster was identified bioinformatically, antisera to the Bfp pilus obtained from J. Kaper did not recognize any surface structure on 101-1 (Rasko, unpublished). The role of this putative pilus cluster in virulence has not been verified and requires further investigation; however, it does highlight the potential for genetic exchange among the pathovars. One aspect yet to be described in EAEC is the exact mechanism by which it is virulent to the human host. Further detailed genomic studies may aid in resolving this unknown, but significant functional characterization is going to be required to understand these mechanisms.

The previous typing methods have capitalized on the presence of constituents of the core genome across the species. Now with the advent of the next-generation sequencing technologies we will be able to take greater advantage of large-scale sequencing projects in much the same way that MLST or MLEE typing did in the past. This does not mean that previous typing schemes do not hold true, but rather that the foundation from the work of Tom Whittam and others has provided the evolutionary framework of *E. coli* that we can now build upon. While genomics holds great promise for the identification of additional virulence factors, diagnostics, and typing methods, it is clear that it represents the beginning of detailed studies and not the end.

**Table 2.** Characteristics of sequenced *E. coli* isolates

| Pathovar | Strain | Serotype | Place of origin | Virulence factors | Genome status | GenBank accession no. | Genome size | Plasmid size (accession no.) |
|---|---|---|---|---|---|---|---|---|
| Laboratory-adapted | K-12 | | United States | | Complete | U00096.2 | 4,639,675 bp | None present |
| Commensal | HS | O9 | United States | | Complete | CP000802 | 4,643,538 bp | None present |
| EAEC | 042 | O44:H18 | Peru | $AggR^+$, $AAF\text{-}II^+$, pAAF plasmid $probe^+$, $Pet^+$, $Pic^+$ | Complete | N554766 | 5,241,977 bp | 113,346 bp (FN554767) |
| EAEC | 55989 | O140:H4 | Africa | $AggR^+$, AAF-III fimbriae, pAA plasmid $probe^+$, $Sat^+$, $SigA^+$, $Pic^+$ | Complete | CU928145 | 5,154,862 bp | 77,482 bp (CU928159) |
| EAEC | 101-1 | Untypeable:H10 | Japan | Heat-stable enterotoxin production, lacks pAAF plasmid | Draft (70 contigs) | AAMK00000000 | >5 Mbp | Unknown[b] |
| EAEC | JM221 | O92:H33 | Mexico | $AggR^+$, $AAF\text{-}I^+$, pEAA2 plasmid, $Sat^+$, $Pic^+$ | In progress (>300 contigs) | NA[a] | >5 Mbp | Unknown[b] |

[a]NA, not applicable.
[b]Unknown: these draft sequences contain contigs that appear to be plasmid in origin; however, no plasmids have been closed from these projects.

**Table 3.** Conserved and unique genes contained within EAEC strains and commensal *E. coli* strain HS

| Strain | No. of conserved genes in strain[a] | No. of unique genes in strain[b] |
|---|---|---|
| *E. coli* HS | 3,181 | 330 |
| *E. coli* 042 | 3,155 | 623 |
| *E. coli* 55989 | 3,251 | 362 |
| *E. coli* 101-1 | 3,228 | 528 |

[a]Conserved genes are those that have a BLAST score ratio of ≥0.8 in all isolates.
[b]Unique genes are identified as having a BLAST score ratio of <0.4 in all other tested isolates.

## CONCLUSIONS/FUTURE DIRECTIONS

Over 2 decades of research has revealed large genetic variations between *E. coli* isolates that are classified as EAEC, as well as identifying features that certain isolates have in common. It is clear that some of this observed diversity comes from originally defining this pathovar based upon an adherence pattern. Even though new gene markers can be used to identify whether an isolate is categorized as typical or atypical EAEC, which may provide insight into its pathogenic potential, it is still not clear what is necessary and sufficient for virulence. After all, typical EAEC strains JM221 and 17-2 were found to be nonpathogenic in clinical trials, and typical EAEC can be isolated from the stools of healthy individuals, indicating that virulence is much more complex than simply the presence or absence of the AggR regulon.

Certainly, the largest hurdle facing this issue is the lack of a defining, highly conserved toxin or an animal model that accurately reflects the disease seen in humans. Additionally, except for strain 042, many other "pathogenic" EAEC isolates reported in the literature are assumed to be virulent because they were the only known pathogen isolated from individuals with diarrhea; however, it cannot be ruled out that these isolates were associated with but not respon-

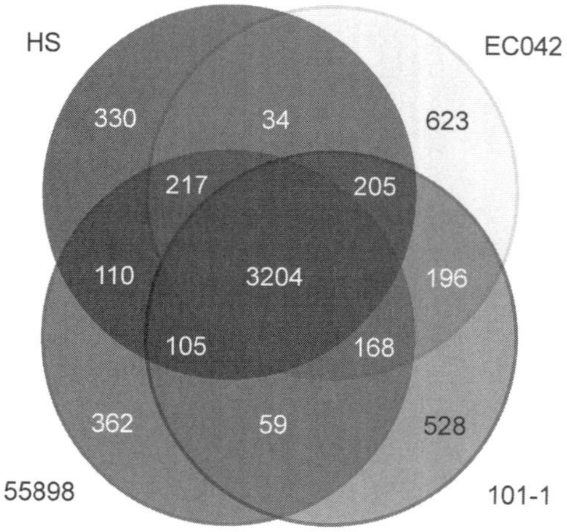


**Figure 5.** Venn diagram of gene distribution among the EAEC, showing genomic comparisons of the complete and draft EAEC genomes to that of the commensal strain *E. coli* HS using the BLAST score ratio analysis (36). Shared genes are listed in the overlapping regions of the circles, and the core genome calculated on only these four genomes is larger than when more diverse genotypes are included (37, 40).

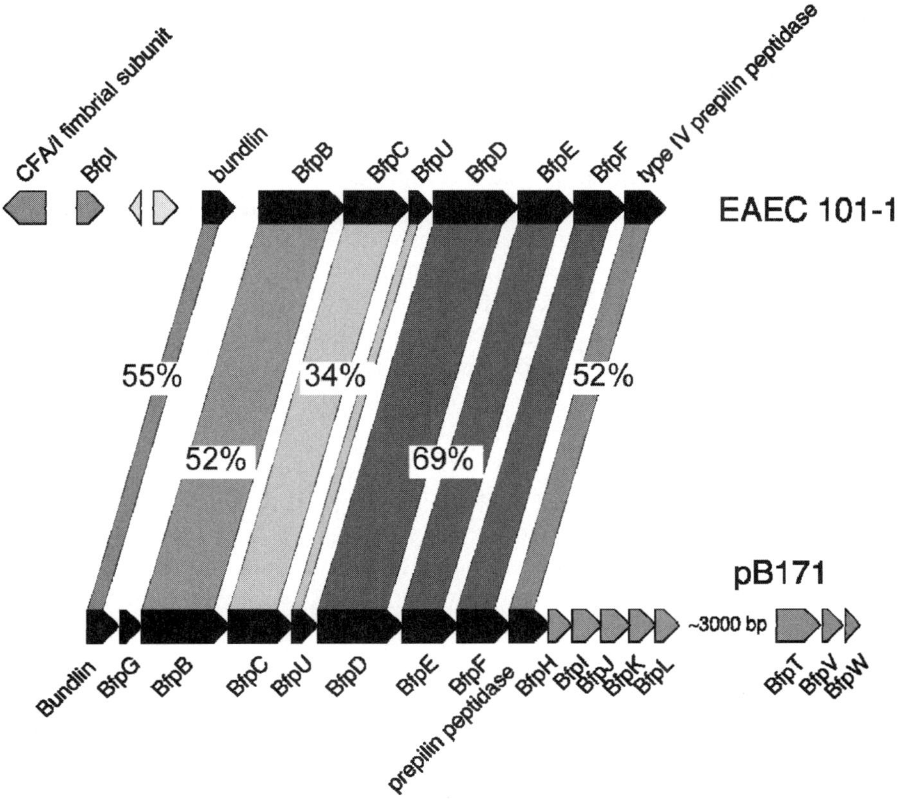


**Figure 6.** Bioinformatic identification of a bundle-forming pilus (*bfp*)-like gene cluster in *E. coli* 101-1. The cluster was not present in any of the other EAEC strains, but does share some similarity with the gene cluster encoded on the EPEC isolate B171 plasmid, pB171. Similarity between the two gene clusters is amino acid identity and is listed for each gene or group of genes. Attempts to identify a functional pilus in 101-1 with Bfp-reactive sera were unsuccessful.

sible for the symptoms. The lack of multiple proven pathogenic and nonpathogenic isolates makes this issue very difficult to address. Disease mediated by typical EAEC may also be multifactorial, requiring not only bacterial factors but host predisposition and the composition of the rest of the gastrointestinal microbiota. Therefore, much needs to be done before we can accurately predict the virulence potential of an EAEC isolate.

One possibility that we suggest and that has not been explored fully is whether additional phylogenetic studies could lead us in that direction. As noted above, it is interesting that the known pathogenic EAEC (042) falls into subgroup D (Fig. 2B), while the nonpathogenic (JM221, 17-2) strains and other nonpathogenic *E. coli* such as strains K-12, HS, and C fall into group A. If we assume for a minute that our supposition above is wrong, and that most clinical isolates of EAEC from other studies are pathogenic to humans, it would be worthwhile classifying these isolates to see whether one could statistically correlate phylogenetic group with virulence. Additionally, the development of an MLST method for categorizing isolates within a defined subgroup may help uncover whether pathogenic isolates fall into specific sequence types. While EAEC was not a large focus of the Whittam lab, Tom's efforts were seminal to the first characterization of a worldwide collection of isolates. He did what he did best by showing us the heterogeneity of this pathovar, and left it to the rest of us to figure out what unites the virulent strains. Dobzhansky is famous for the essay "Nothing in Biology Makes Sense Except in the Light of Evolution," and it would be a fitting tribute if the tools for studying *E. coli* evolution popularized by Tom and his lab group proved essential for answering this question.

**Acknowledgments.** The Dudley laboratory is funded through USDA Milk Safety grants 2008-34163-19283 and 2009-34163-20132, USDA Agriculture and Food Research Initiative grant 2009-03611, and through startup funds from the Penn State College of Agricultural Sciences and Department of Food Science. The Rasko laboratory is funded through startup funds from the state of Maryland. We thank Ian Henderson (University of Birmingham) for helpful discussions concerning EAEC phylogeny and Brad Harris (Johns Hopkins School of Medicine) for taking the scanning electron microscopy pictures.

## REFERENCES

1. **Amar, C., C. East, J. Gray, M. Iturriza-Gomara, E. Maclure, and J. McLauchlin.** 2007. Detection by PCR of eight groups of enteric pathogens in 4,627 faecal samples: re-examination of the English case-control Infectious Intestinal Disease Study (1993–1996). *Eur. J. Clin. Microbiol. Infect. Dis.* **26:**311–323.
2. **Barros, S. F., C. M. Abe, S. P. D. Rocha, R. M. Ruiz, L. Beutin, L. R. Trabulsi, and W. P. Elias.** 2008. *Escherichia coli* O125ac:H6 encompasses atypical enteropathogenic *E. coli* strains that display the aggregative adherence pattern. *J. Clin. Microbiol.* **46:** 4052–4055.
3. **Baudry, B., S. J. Savarino, P. Vial, J. B. Kaper, and M. L. Myron.** 1990. A sensitive and specific DNA probe to identify enteroaggregative *Escherichia coli*, a recently discovered diarrheal pathogen. *J. Infect. Dis.* **161:**1249–1251.
4. **Bernier, C., P. Gounon, and C. Le Bouguenec.** 2002. Identification of an aggregative adhesion fimbria (AAF) type III-encoding operon in enteroaggregative *Escherichia coli* as a sensitive probe for detecting the AAF-encoding operon family. *Infect. Immun.* **70:** 4302–4311.
5. **Boisen, N., F. Ruiz-Perez, F. Scheutz, K. A. Krogfelt, and J. P. Nataro.** 2009. High prevalence of serine protease autotransporter cytotoxins among strains of enteroaggregative *Escherichia coli*. *Am. J. Trop. Med. Hyg.* **80:**294–301.
6. **Boisen, N., C. Struve, F. Scheutz, K. A. Krogfelt, and J. P. Nataro.** 2008. New adhesin of enteroaggregative *Escherichia coli* related to the Afa/Dr/AAF family. *Infect. Immun.* **76:**3281–3292.
7. **Campos, L. C., T. S. Whittam, T. A. Gomes, J. R. Andrade, and L. R. Trabulsi.** 1994. *Escherichia coli* serogroup O111 includes several clones of diarrheagenic strains with different virulence properties. *Infect. Immun.* **62:**3282–3288.
8. **Chaudhuri, R. R., M. Sebaihia, J. L. Hobman, M. A. Webber, D. L. Leyton, M. D. Goldberg, A. F. Cunningham, A. Scott-Tucker, P. R. Ferguson, C. M. Thomas, G. Frankel, C. M. Tang, E. G. Dudley, I. S. Roberts, D. A. Rasko, M. J. Pallen, J. Parkhill, J. P. Nataro, N. R. Thomson, and I. R. Henderson.** 2010. Complete genome

sequence and comparative metabolic profiling of the prototypical enteroaggregative *Escherichia coli* strain 042. *PLoS One* **5**:e8801.

9. **Cobeljić, M., B. Miljković-Selimovic, D. Paunović-Todosijević, Z. Velicković, Z. Lepsanović, N. Zec, D. Savić, R. Ilić, S. Konstantinović, B. Jovanović, and V. Kostić.** 1996. Enteroaggregative *Escherichia coli* associated with an outbreak of diarrhoea in a neonatal nursery ward. *Epidemiol. Infect.* **117**:11–16.
10. **Cohen, M. B., J. P. Nataro, D. I. Bernstein, J. Hawkins, N. Roberts, and M. A. Staat.** 2005. Prevalence of diarrheagenic *Escherichia coli* in acute childhood enteritis: a prospective controlled study. *J. Pediatr.* **146**:54–61.
11. **Cravioto, A., R. Gross, S. Scotland, and B. Rowe.** 1979. An adhesive factor found in strains of *Escherichia coli* belonging to the traditional infantile enteropathogenic serotypes. *Curr. Microbiol.* **3**:95–99.
12. **Croxen, M. A., and B. B. Finlay.** 2010. Molecular mechanisms of *Escherichia coli* pathogenicity. *Nat. Rev. Microbiol.* **8**:26–38.
13. **Czeczulin, J. R., S. Balepur, S. Hicks, A. Phillips, R. Hall, M. H. Kothary, F. Navarro- Garcia, and J. P. Nataro.** 1997. Aggregative adherence fimbria II, a second fimbrial antigen mediating aggregative adherence in enteroaggregative *Escherichia coli. Infect. Immun.* **65**:4135–4145.
14. **Czeczulin, J. R., T. S. Whittam, I. R. Henderson, F. Navarro-Garcia, and J. P. Nataro.** 1999. Phylogenetic analysis of enteroaggregative and diffusely adherent *Escherichia coli. Infect. Immun.* **67**:2692–2699.
15. **Dudley, E. G., C. Abe, J.-M. Ghigo, P. Latour-Lambert, J. C. Hormazabal, and J. P. Nataro.** 2006. An IncI1 plasmid contributes to the adherence of the atypical enteroaggregative *Escherichia coli* strain C1096 to cultured cells and abiotic surfaces. *Infect. Immun.* **74**:2102–2114.
16. **Dudley, E. G., N. R. Thomson, J. Parkhill, N. P. Morin, and J. P. Nataro.** 2006. Proteomic and microarray characterization of the AggR regulon identifies a *pheU* pathogenicity island in enteroaggregative *Escherichia coli. Mol. Microbiol.* **61**:1267–1282.
17. **Escobar-Paramo, P., O. Clermont, A. B. Blanc-Potard, H. Bui, C. Le Bouguenec, and E. Denamur.** 2004. A specific genetic background is required for acquisition and expression of virulence factors in *Escherichia coli. Mol. Biol. Evol.* **21**:1085–1094.
18. **Farfan, M. J., K. G. Inman, and J. P. Nataro.** 2008. The major pilin subunit of the AAF/II fimbriae from enteroaggregative *Escherichia coli* mediates binding to extracellular matrix proteins. *Infect. Immun.* **76**:4378–4384.
19. **Fujiyama, R., J. Nishi, N. Imuta, K. Tokuda, K. Manago, and Y. Kawano.** 2008. The *shf* gene of a *Shigella flexneri* homologue on the virulent plasmid pAA2 of enteroaggregative *Escherichia coli* 042 is required for firm biofilm formation. *Curr. Microbiol.* **56**:474–480.
20. **Harrington, S. M., E. G. Dudley, and J. P. Nataro.** 2006. Pathogenesis of enteroaggregative *Escherichia coli* infection. *FEMS Microbiol. Lett.* **254**:12–18.
21. **Harrington, S. M., J. Sheikh, I. R. Henderson, F. Ruiz-Perez, P. S. Cohen, and J. P. Nataro.** 2009. The Pic protease of enteroaggregative *Escherichia coli* promotes intestinal colonization and growth in the presence of mucin. *Infect. Immun.* **77**:2465–2473.
22. **Harrington, S. M., M. C. Strauman, C. M. Abe, and J. P. Nataro.** 2005. Aggregative adherence fimbriae contribute to the inflammatory response of epithelial cells infected with enteroaggregative *Escherichia coli. Cell. Microbiol.* **7**:1565–1578.
23. **Henderson, I. R., F. Navarro-Garcia, M. Desvaux, R. C. Fernandez, and D. Ala'Aldeen.** 2004. Type V protein secretion pathway: the autotransporter story. *Microbiol. Mol. Biol. Rev.* **68**:692–744.
24. **Herzer, P. J., S. Inouye, M. Inouye, and T. S. Whittam.** 1990. Phylogenetic distribution of branched RNA-linked multicopy single-stranded DNA among natural isolates of *Escherichia coli. J. Bacteriol.* **172**:6175–6181.
25. **Huang, D. B., J. P. Nataro, H. L. DuPont, P. P. Kamat, A. D. Mhatre, P. C. Okhuysen, and T. Chiang.** 2006. Enteroaggregative *Escherichia coli* is a cause of acute diarrheal illness: a meta-analysis. *Clin. Infect. Dis.* **43**:556–563.

26. **Itoh, Y., I. Nagano, M. Kunishima, and T. Ezaki.** 1997. Laboratory investigation of enteroaggregative *Escherichia coli* O untypeable:H10 associated with a massive outbreak of gastrointestinal illness. *J. Clin. Microbiol.* **35:**2546–2550.
27. **Jenkins, C., C. Ijperen, E. G. Dudley, H. Chart, G. A. Willshaw, T. Cheasty, H. R. Smith, and J. P. Nataro.** 2005. Use of a microarray to assess the distribution of plasmid and chromosomal virulence genes in strains of enteroaggregative *Escherichia coli*. *FEMS Microbiol. Lett.* **253:**119–124.
28. **Mathewson, J. J., H. L. DuPont, D. R. Morgan, S. A. Thornton, and C. D. Ericsson.** 1983. Enteroadherent *Escherichia coli* associated with travellers' diarrhoea. *Lancet* **321:** 1048. (Letter.)
29. **Monteiro-Neto, V., S. Y. Bando, C. A. Moreira-Filho, and J. A. Girón.** 2003. Characterization of an outer membrane protein associated with haemagglutination and adhesive properties of enteroaggregative *Escherichia coli* O111:H12. *Cell. Microbiol.* **5:** 533–547.
30. **Nataro, J. P.** 2005. Enteroaggregative *Escherichia coli* pathogenesis. *Curr. Opin. Gastroenterol.* **21:**4–8.
31. **Nataro, J. P., J. B. Kaper, R. Robins-Browne, V. Prado, P. Vial, and M. M. Levine.** 1987. Patterns of adherence of diarrheagenic *Escherichia coli* to HEp-2 cells. *Pediatr. Infect. Dis. J.* **6:**829–831.
32. **Nataro, J. P., V. Mai, J. Johnson, W. C. Blackwelder, R. Heimer, S. Tirrell, S. C. Edberg, C. R. Braden, J. Glenn Morris, Jr., and J. M. Hirshon.** 2006. Diarrheagenic *Escherichia coli* infection in Baltimore, Maryland, and New Haven, Connecticut. *Clin. Infect. Dis.* **43:**402–407.
33. **Nataro, J. P., D. Yikang, S. Cookson, A. Cravioto, S. J. Savarino, L. D. Guers, M. L. Myron, and C. O. Tacket.** 1995. Heterogeneity of enteroaggregative "*Escherichia coli*" virulence demonstrated in volunteers. *J. Infect. Dis.* **171:**465–468.
34. **Nataro, J. P., D. Yikang, D. Yingkang, and K. Walker.** 1994. AggR, a transcriptional activator of aggregative adherence fimbria I expression in enteroaggregative *Escherichia coli*. *J. Bacteriol.* **176:**4691–4699.
35. **Nishi, J., J. Sheikh, K. Mizuguchi, B. Luisi, V. Burland, A. Boutin, D. J. Rose, F. R. Blattner, and J. P. Nataro.** 2003. The export of coat protein from enteroaggregative *Escherichia coli* by a specific ATP-binding cassette transporter system. *J. Biol. Chem.* **278:**45680–45689.
36. **Rasko, D. A., G. S. Myers, and J. Ravel.** 2005. Visualization of comparative genomic analyses by BLAST score ratio. *BMC Bioinformatics* **6:**2.
37. **Rasko, D. A., M. J. Rosovitz, G. S. A. Myers, E. F. Mongodin, W. F. Fricke, P. Gajer, J. Crabtree, V. Sperandio, and J. Ravel.** 2008. The pan-genome structure of *Escherichia coli*: comparative genomic analysis of *E. coli* commensal and pathogenic isolates. *J. Bacteriol.* **190:**6881–6893.
38. **Sheikh, J., E. G. Dudley, B. Sui, B. Tamboura, A. Suleman, and J. P. Nataro.** 2006. EilA, a HilA-like regulator in enteroaggregative *Escherichia coli*. *Mol. Microbiol.* **61:** 338–350.
39. **Sheikh, J., S. Hicks, M. Dall'Agnol, A. D. Phillips, and J. P. Nataro.** 2001. Roles for Fis and YafK in biofilm formation by enteroaggregative *Escherichia coli*. *Mol. Microbiol.* **41:**983–997.
40. **Touchon, M., C. Hoede, O. Tenaillon, V. Barbe, S. Baeriswyl, P. Bidet, E. Bingen, S. Bonacorsi, C. Bouchier, O. Bouvet, A. Calteau, H. Chiapello, O. Clermont, S. Cruveiller, A. Danchin, M. Diard, C. Dossat, M. E. Karoui, E. Frapy, L. Garry, J. M. Ghigo, A. M. Gilles, J. Johnson, C. Le Bouguénec, M. Lescat, S. Mangenot, V. Martinez-Jéhanne, I. Matic, X. Nassif, S. Oztas, M. A. Petit, C. Pichon, Z. Rouy, C. S. Ruf, D. Schneider, J. Tourret, B. Vacherie, D. Vallenet, C. Médigue, E. P. C. Rocha, and E. Denamur.** 2009. Organized genome dynamics in the *Escherichia coli* species results in highly diverse adaptive paths. *PLoS Genet.* **5:**e1000344.

*Population Genetics of Bacteria: a Tribute to Thomas S. Whittam*
Edited by S. T. Walk and P. C. H. Feng


*Chapter 12*

# Shiga Toxin-Producing *Escherichia coli*

***Louise D. Teel, Angela R. Melton-Celsa, and Alison D. O'Brien***

## INTRODUCTION

Pathogenic *Escherichia coli* strains have assembled an armamentarium of surface, soluble, and secreted factors that define their virulence potential for humans (reviewed in reference 51). Numerous among these traits are bacterial toxins that have profound effects on eukaryotic cell operations. Examples of these poisons include the heat-labile (LT) and -stable toxins (ST) of the enterotoxigenic *E. coli* (ETEC), *Shigella* enterotoxin 1 (ShET1) of enteroaggregative *E. coli* (EAEC), and *Shigella* enterotoxin 2 (ShET2) of enteroinvasive *E. coli* (EIEC). Each of the latter toxins contributes to diarrheal disease by targeting various cell signaling molecules and provoking ion secretion from those affected cells. In contrast, the Shiga toxins (Stx) Stx1 and Stx2, produced by enterohemorrhagic and other Shiga toxin-producing *E. coli* (EHEC and STEC, respectively), are protein synthesis-inhibiting toxins that contribute to the development of hemorrhagic colitis and the most life-threatening manifestation of STEC infection, hemolytic uremic syndrome (HUS). HUS is a triad of symptoms that includes acute renal failure, thrombocytopenia, and microangiopathic hemolytic anemia.

The bacterial toxins listed above have deleterious effects when elaborated by *E. coli* in the intestines of infected people. However, the potential benefit, in evolutionary terms, of toxin production to the fitness of toxigenic strains is not fully understood. Undoubtedly there is a role for toxins that induce secretory diarrhea in dissemination of bacteria into the environment. Additionally, production of LT or Stx2 enhances colonization of the host by *E. coli* in a mouse model of ETEC infection, or in mice and cattle infected with Stx2-producing strains, respectively (4, 63, 94). Improved adherence in the gut by these strains may be attributable to toxin-mediated upregulation of ancillary adherence fac-

***Louise D. Teel, Angela R. Melton-Celsa, and Alison D. O'Brien*** • Department of Microbiology and Immunology, Uniformed Services University of the Health Sciences, Bethesda, MD 20814.

tors such as fibronectin in the case of ETEC (4) and nucleolin during infection with STEC (94).

Beyond the role of toxin production in establishment and perpetuation of human infection with pathogenic *E. coli*, there is evidence to suggest that elaboration of Stx specifically promotes survival of STEC in the environment. For example, Lainhart et al. (57) have shown that *E. coli* lysogenized with bacteriophages that encode Stxs have an advantage over strains that do not elaborate Stx when these bacterial strains are cocultured in the presence of the protozoan *Tetrahymena*. *Tetrahymena* are found in soil, water, and the rumen of cattle, which is the natural reservoir of STEC. Furthermore, STEC can kill *Tetrahymena*, an interaction that may explain one selective advantage for bacteria to produce Stxs. In this chapter we will discuss the emergence and impact of STEC in human disease, the biology of the Stx family, and approaches to diagnosis, treatment, and prevention of infection with STEC.

## EMERGENCE AND EPIDEMIOLOGY OF STEC

*E. coli* O157:H7 was first recognized as a food-borne pathogen in 1982 during two sequential outbreaks of hemorrhagic colitis that occurred in Oregon and Michigan after individuals ingested undercooked contaminated hamburger patties at different branches of the same fast-food chain (92). The risk associated with undercooked hamburger was reinforced during an extensive *E. coli* O157:H7 outbreak linked to a fast-food chain in the western United States in 1993 (19, 37). Among the 732 individuals affected in that outbreak, 195 were hospitalized, 55 developed HUS, and 4 died (37). That the oral 50% infectious dose of *E. coli* O157:H7 for humans is low was first documented in that episode (38), and it was later demonstrated to be <50 organisms (114). Investigation of these outbreak strains confirmed that these *E. coli* O157 strains were lysogenized with bacteriophages that encode Stx1, a toxin that is virtually identical to the Stx expressed by *Shigella dysenteriae* type 1, and a serologically distinct but functionally similar Stx2 (71). The Stxs are discussed further below.

The occurrence of the seminal STEC outbreaks described above highlighted the need to implement laboratory methods to readily detect *E. coli* O157:H7. Detection strategies were developed based on the O157 and H7 serotypes and the fact that O157:H7 strains ferment sorbitol slowly whereas other *E. coli* strains rapidly ferment sorbitol (25). Nonetheless, the sporadic nature and apparent low frequency of O157:H7 infections compared to those attributable to other enteric pathogens, plus the expense of additional tests, initially discouraged diagnostic labs from uniformly implementing screening for O157:H7 in stool samples (15). However, infections with O157 STEC (with either the H7 flagellar type or nonmotile strains) became reportable to public health authorities in the United States in 1995, a requirement that provided added impetus for laboratories to include diagnostic testing for O157 strains (120). Between 1998 and 2007, the Centers for Disease Control and Prevention (CDC) documented 424 outbreaks of O157-mediated disease (39). Among those, a likely source for the acquisition of the O157 strain was identified in 341 outbreaks. Sixty-eight percent were associated with food, 18% with water, 8% with animal or environmental exposure, and 6% with person-to-person spread (39). Outbreaks of O157:H7-mediated disease have been associated with diverse food and environmental sources, such as apple cider, salami, roast beef, swimming water, well water, sprouts, lettuce, and spinach (7, 18, 38, 42, 64, 114). The primary reservoir for STEC is ruminants, particularly cattle, but other livestock and wild animals can also transmit STEC. The large multistate O157:H7 outbreak associated with prewashed, bagged spinach that occurred in the late summer of 2006 and caused three deaths was attributed to contamination of plants from irrigation water contaminated by feces from feral pigs or cattle (64).

It is estimated that at least 100,000 cases of STEC-associated disease occur annually in the United States and that about 70% of those are attributable to *E. coli* O157:H7 strains (89). Similarly, O157 (either H7 or $H^-$) is the most common disease-associated STEC serotype in Argentina and elsewhere in South America, as well as in Japan, Ireland, the United Kingdom, and throughout Western Europe (90), whereas the O111 serotype predominates in Australia (93). It is likely that the incidence of non-O157 STEC infections has been underestimated because they are more difficult to distinguish from normal intestinal flora than O157 strains that generally do not ferment sorbitol. The difficulty in identifying non-O157 STEC is that there are many different non-O157 serotypes; and, therefore, the best way to identify them is by screening for the presence of Stx, a measure that has been slow to be adopted by clinical laboratories (120). Nonetheless, the importance of non-O157:H7 STEC strains as human pathogens has become clear (reviewed in references 8 and 48). Between 1990 and 2007, the CDC documented 22 outbreaks of disease associated with non-O157 strains in the United States, of which the most common non-O157 agents were O111 (45%), O26 (25%), and O121 (20%), while the remaining 10% belonged to other serotypes (39). Although the non-O157:H7 STEC strains are less frequently associated with outbreaks than O157 STEC, they do occur in sporadic diarrheal disease and HUS, as was demonstrated in a prospective study undertaken in Germany and Austria (33). Similarly, in a prospective cohort study of the etiology of diarrhea in a children's hospital in Seattle, Washington, non-O157 STEC were identified in stools from children with diarrhea with nearly the same incidence as *Shigella* species (0.7 versus 0.9%, respectively) (55). Indeed, 51% of the STEC strains from stools of ill patients in Connecticut between 2000 and 2005 were reported to be non-O157:H7 isolates (20). These studies suggest that when appropriate laboratory methods are utilized for isolation and identification of non-O157 STEC they are detected in significant numbers.

## CHARACTERIZATION OF STEC STRAINS BY ENZYME PROFILES AND BY GENETICS

Early STEC isolates were named "enterohemorrhagic" *E. coli* to connote the clinical association of the newly emergent O157 and certain O26 isolates with hemorrhagic colitis and to distinguish them from the nontoxigenic enteropathogenic *E. coli* (EPEC) (59). Today we recognize that in addition to Stx production, the EHEC also contain a large pathogenicity island known as the locus of enterocyte effacement, or LEE (discussed below), and are a subset of the larger and extremely diverse group of STEC. Thomas Whittam plotted the relatedness of numerous serotypes of STEC through comparison of multilocus enzyme electrophoresis (MLEE) profiles, and these efforts led to a grouping of common STEC strains into the categories EHEC 1, EHEC 2, STEC 1, and STEC 2 (129, 130; see also http://www.shigatox.net/stec/cgi-bin/clonal). The EHEC 1 group consists of the O157:H7 and nonmotile O157 strains, while the EHEC 2 group contains the O26 and O111 (motile and nonmotile) serotypes. Common to the EHEC groups are the synthesis of one or more Stxs, a chromosomal pathogenicity island known as the LEE that is highly homologous to the LEE of EPEC, and an approximately 90-kb plasmid that encodes an enterohemolysin and other potential virulence genes. Despite these unifying characteristics, great diversity among the EHECs has been described (76). The significant genetic heterogeneity of the chromosomes and the virulence plasmids sequenced from representative EHEC strains suggests that parallel evolution and horizontal gene acquisition occurred in these varied host backgrounds to yield common virulence profiles (76). The LEE-negative STEC include STEC group 1, which is composed of diverse serotypes, including O91, O113, O146, and others that appear to have a clonal relatedness and share the H21 flagellar type. These groups contain numerous strains that have been linked with serious disease in humans, an association that has prompted the search for additional or alternative virulence factors. One of the best studied of the STEC 1 strains is the O113:H21 HUS isolate 98NK2. Strain 98NK2 and some other non-O157:H7 strains (82, 83) encode the $AB_5$ subtilase cytotoxin. The target of the subtilase, which is a potent protease, is the endoplasmic reticulum chaperone BiP (81). The STEC 2 group is also LEE negative and is composed mostly of O103 isolates. Not only did the development of the MLEE profiles allow the placement of STEC into four different groups, it also helped in the construction of a model for the evolution of the O157 STEC from a precursor of a serotype O55 EPEC (129). That model for the derivation of toxigenic O157 has been further supported by genetic methods (26, 67).

Although early characterization of STEC strains was made possible by Whittam's MLEE method, genetic techniques now allow comparison of strains at the nucleotide level. Considerable genetic variability within the O157 and O26 serotypes has been described (67, 134). Genotyping based on single nucleotide polymorphisms in 83 genetic loci from hundreds of O157 isolates revealed 39 distinct genotypes that were then grouped into 9 proposed lineages or clades. Assessment of the occurrence of severe disease associated with representative strains within these clades revealed elevated virulence potentials with certain ones, particularly clade 8, which includes the spinach outbreak strain (67). A genomic comparison of some of the most prevalent EHEC serotypes—O157, O26, O111, and O103—was recently made (76). In that study, the authors found that the genomes of EHEC are larger than those of other *E.*

*coli* strains due to the acquisition of additional genetic elements, such as prophages, integrative elements, and plasmids. However, sequence differences among the various genetic elements suggest that they were acquired at different times and indicate an independent evolution for these EHEC strains (76). That the LEE was constructed from all three types of genetic elements was also proposed.

## LEE AND OTHER POTENTIAL VIRULENCE DETERMINANTS

The approximately 43-kb locus encodes the proteins that promote intimate adherence of LEE$^+$ bacteria in the gut. Proteins encoded within the LEE include the adhesin intimin, its transmembrane receptor (Tir), numerous Esps (*E. coli* secreted proteins), and Escs (*E. coli* secretion proteins that form the type III secretion system apparatus) (32, 50). In addition to the LEE pathogenicity island, O157 strains as well as many non-O157 strains carry a large plasmid. The large plasmid of O157 strains encodes the enterohemolysin, a metalloprotease called StcE that specifically cleaves C1 esterase inhibitor (58), a type II secretion system, and a protein called ToxB that has some homology to the *Clostridium difficile* toxin B (65). Lathem and colleagues proposed that StcE may contribute to tissue damage in patients with *E. coli* O157:H7 infections by promoting proinflammatory responses, thrombotic abnormalities, and intestinal edema (58). Moreover, Grys et al. from that same group found that StcE contributes to intimate adherence of EHEC to host cells by cleavage of glycoproteins from the cell surface (40). In fact, Sheng et al. reported that an *E. coli* O157:H7 cured of pO157 persists less well in rectally challenged cattle than does the wild-type strain (104). However, Tzipori et al. did not find a role for pO157 in *E. coli* O157:H7 colonization of piglets (117). Other potential virulence determinants identified for the STEC include an autoagglutinating adhesin (Saa) and a novel autotransporter named Sab that has a role in adherence and biofilm formation (41, 84).

## PROPERTIES OF SHIGA TOXINS

Stxs are the virulence determinants of STEC that render these *E. coli* strains uniquely capable of causing HUS (reviewed in reference 71). The Stxs are members of a family of toxins that share genetic, structural, and functional features (reviewed in references 70, 71, and 96). The holotoxin has a 1A:5B noncovalently associated subunit structure. The B pentamer (7.7 kDa per monomer) is responsible for binding to a eukaryotic glycolipid receptor that is usually globotriaosylceramide ($Gb_3$). When the A subunit (32 kDa) is nicked with trypsin and reduced, an $A_1$ portion of ~28 kDa and an $A_2$ peptide of 4 kDa are generated. The $A_1$ polypeptide contains the *N*-glycosidase activity that causes depurination of a critical residue in the 28S rRNA of 60S ribosomes and, hence, inhibition of protein synthesis. This mode of action is identical to that of the plant lectin ricin. Cell death evoked by members of the Stx family occurs by apoptosis (reviewed in reference 112) that follows inhibition of protein synthesis. The $A_2$ peptide is required to noncovalently bind the $A_1$ to the B pentamer.

Stxs are cytotoxic for Vero cells (African green monkey kidney epithelium). The exquisite sensitivity of Vero cells to this family of toxins led to their early designation based on that phenotype as verocytotoxins or verotoxins. That name has endured in the scientific literature, particularly outside of the United States; however, the international STEC research community has agreed that the term "Stx" more accurately represents the structural relatedness and functional activity of the toxins, and therefore consensus has grown to adopt the Stx nomenclature uniformly internationally (7th International Symposium on Shiga Toxin [Verotoxin]-Producing *Escherichia coli* Infections [VTEC] 2009 meeting general discussion). Stxs are also enterotoxic for rabbits and lethal for mice, rabbits, and other animals (75). The current model (46, 85, 97) of the predominant pathway by which Stx intoxicates sensitive cells is as follows: (i) the B pentamer of holotoxin binds to $Gb_3$ within lipid rafts; (ii) the entire receptor-holotoxin complex is endocytosed; (iii) the complex moves by retrograde transport to the Golgi and then to the endoplasmic reticulum; and (iv) the $A_1$ subunit is released into the cytoplasm, where it targets the ribosome. Our hypothesis is that the A subunit may be nicked by proteases from intestinal mucus prior to step i above.

The Stx family contains two main types, Stx1 and Stx2, that are not neutralized by the heterologous antisera. A single STEC isolate may produce Stx1, Stx2 (or subtypes thereof), or combinations of these toxins (reviewed in reference 71). Many *E. coli* O157:H7 strains isolated in the United States make both Stx1 and Stx2. However, there is evidence that the epidemiology of toxin types has shifted over time. Zhang et al. reported that O26 strains isolated in Germany and the Czech Republic prior to 1996 typically expressed only Stx1, while later isolates expressed only Stx2, or less frequently Stx1 and Stx2 (134). A similar shift from Stx1 to Stx2 expression has been noted among O157 strains in the United States. Of over 4,000 strains isolated between 1999 and 2008, the proportion that express only Stx2 increased from 11% in 1999 to 57% in 2008 (39). Additionally, a shift has been observed in the proportion of O157 strains that produce the Stx2 subtype Stx2c, from about 4% in 1996 to about 30% in 2005 (36).

What is not clear is the relative contribution of each of these toxin types to glomerular lesion development as well as other manifestations of HUS and intestinal damage noted in humans. Stx1 of EHEC differs from the Stx of *S. dysenteriae* type 1 by only one amino acid in the A polypeptide. To date, only two subtypes of Stx1 have been described, Stx1c and Stx1d (71), and these toxins appear to occur infrequently. The crystal structures of the Stx and Stx2 holotoxins have been determined (28, 29) (Fig. 1). These models indicate that the B pentamer of Stx encircles the helical carboxy terminus of the Stx A subunit (77), but the position of the C-terminal portion of the $A_2$ peptide, which most likely protrudes from the bottom of the pentamer, has not been resolved.

Several related subtypes of Stx2 have been described (reviewed in reference 71). The subtypes are denoted as Stx2a, Stx2b, Stx2c, Stx2d, Stx2d-

**Figure 1.** Ribbon diagram of the Stx2 crystal structure. The A subunit is depicted in black while the five B subunits are shown in alternating light and dark gray. This image (Protein Data Bank ID: 1R4P) was generated with Protein Workshop 3.6.

activatable, Stx2e, Stx2f, and Stx2g. These toxins are homologous to Stx2 and most are cross-neutralizable with anti-Stx2 antibody, but they differ in amino acid sequence, biological activity, preferred cellular receptor (Stx2e that causes edema disease of swine uses globotetraosylceramide ($Gb_4$) rather than $Gb_3$ as a functional receptor), relative cytotoxicity for Vero and HeLa cells, and/or the host range of strains that produce them. Although the pathogenicity of various STEC strains that synthesize different types of Stx2 cannot be compared directly because the strains are not isogenic, we have found that an O91 strain that produces Stx2 is not virulent in the streptomycin-treated mouse model for STEC infection, whereas O91 strains that produce Stx2d-activatable are highly virulent in those mice (69). Indeed, STEC that make Stx2d-activatable are lethal for orally challenged streptomycin-treated CD-1 mice at very low doses, but STEC that synthesize Stx2 or the variants Stx2c or Stx2e have an oral 50% lethal dose ($LD_{50}$) of $10^{10}$ CFU/CD-1 mouse or greater (reviewed in reference 71). The lower $LD_{50}$ for mice of STEC that produce Stx2d-activatable versus Stx2 correlates with the capacity of Stx2d-activatable to become more cytotoxic after treatment with elastase derived from murine or human intestinal mucus (reviewed in reference 71). Activation is a process in which the C-terminal two amino acids of the $A_2$ from Stx2d-activatable (69) are cleaved by elastase to produce a molecule more toxic for Vero cells and mice. This highly potent toxin is made by about a quarter of all the Whittam STEC 1 group strains (128) that we tested (our unpublished data) and 4/5 of the Stx2-variant-producing non-O157 strains associated with sporadic cases (45) or outbreaks of diarrheal disease in Montana (21). Furthermore, Bielaszewska et al. (13) found that the presence of Stx2d-activatable in patient isolates (all non-O157 strains that were also *eae* negative) correlated with the occurrence of bloody diarrhea and HUS in those patients.

Stx1 and Stx2 have indistinguishable enzymatic activities in cell-free systems but differ in toxicity for mice (reviewed in reference 71). The $LD_{50}$ of purified Stx1 for parenterally inoculated, adult CD-1 mice is ~125 ng, compared to ~1 ng for Stx2 (reviewed in reference 71). Furthermore, Stx2, but not Stx1, is responsible for the severe necrotic renal tubular lesions and death of streptomycin-treated mice fed an EHEC strain that makes both Stx1 and Stx2 (121). This difference in toxicity is also evident when human glomerular endothelial cells are treated with purified Stx1 or Stx2; Stx2 is about 1,000-fold more toxic (62). Additionally, epidemiological data suggest a difference in Stx1 and Stx2 toxicity when delivered by *E. coli*; i.e., $stx_2$-containing EHEC O157:H7 strains are more frequently associated with HUS than strains with $stx_1$ (79, 101), a correlation also noted with non-O157 STEC strains (48). Nevertheless, strains producing only Stx1 are isolated from cases of bloody diarrhea and HUS, although certainly less frequently than Stx2-producing isolates (11, 66).

## ROLE OF BACTERIOPHAGES IN Stx PRODUCTION

Although transduction of *stx* genes into *E. coli* via bacteriophages was crucial to the emergence of STEC, the biology of these toxin-converting phages also contributes significantly to the degree of toxin expression and hence the virulence exhibited by STEC (30, 135). Not all *stx* alleles studied thus far are associated with inducible bacteriophages, and as such, they are not upregulated in conditions that induce phage (111). However, investigation of the *stx* gene-flanking regions in numerous STEC strains, as well as in *S. dysenteriae* type 1, reveals that the toxin genes are universally associated with sequences that share at least some homology with lambdoid phages (119). This sequence link to lambdoid phages indicates that *stx* loci were likely acquired by horizontal gene transfer even in *S. dysenteriae*. Studies with the prototypic Stx1 phage H-19B and Stx2 phage 933W isolated from clinical strains *E. coli* O26 strain H19 and *E. coli* O157 strain EDL933, respectively (74, 107), show that phage genetic elements strongly influence toxin expression (116, 123, 124). By virtue of their location within the late phage gene region, *stx* genes are subject to the same phage regulatory control that represses expression of the host cell lysis and phage assembly genes to maintain the lysogenic state of lambda phages (88). The CI phage repressor prevents transcription of the genes required for the phage lytic cycle by binding to specific sites within the operator region for the left and the right late gene promoters $p_L$ and $p_R$ (Fig. 2), such that transcription of the *N* antiterminator gene is repressed and lysogeny is maintained. In response to bacterial host cell DNA damage (which can be induced by UV irradiation, mitomycin C, or antibiotic exposure), the CI repressor undergoes

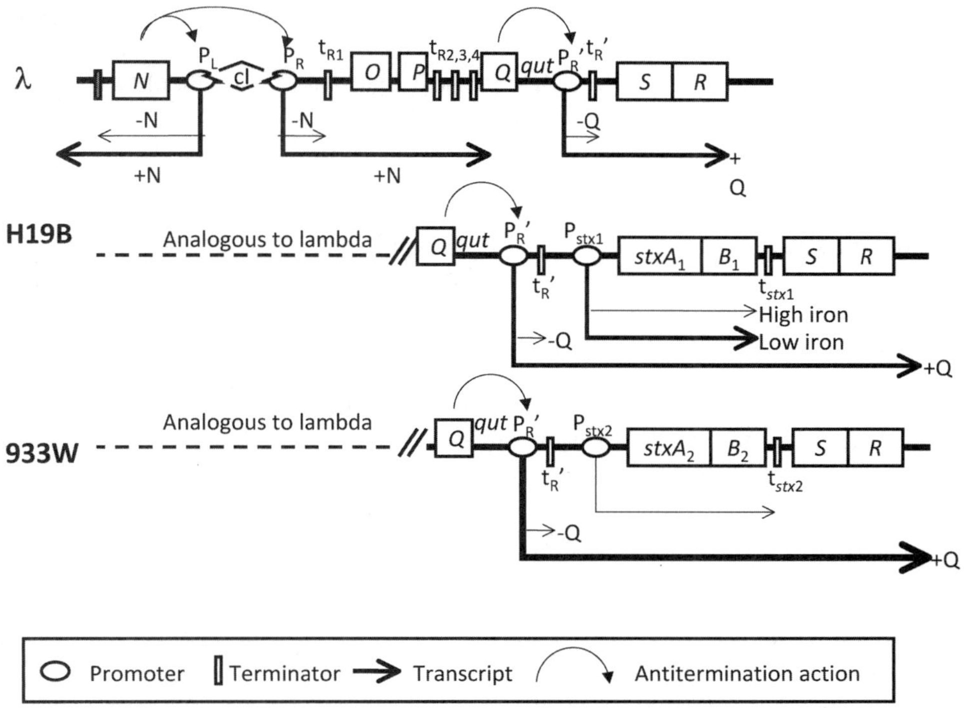


**Figure 2.** Stx expression is under the control of multiple genetic elements. Expression of *stx* and the lambdoid late genes is repressed during lysogeny by the binding of the phage repressor CI to sites between the early left and right promoter regions ($p_L$ and $p_R$). Upon phage induction, the CI repressor is cleaved and the N antiterminator is transcribed from $p_R$. N facilitates transcription through numerous terminator sequences ($t_{R1-4}$), such that the Q antiterminator is expressed. When Q binds *qut*, the $p_R'$ promoter is active and transcription proceeds through the *stx* operon and on through the *S* (holin) and *R* (endolysin) genes further downstream. Thus Q antitermination and transcription originating at $p_R'$ contributes most strongly to Stx1 and Stx2 production upon phage induction. Stx1 expression is also partially mediated by an iron-repressed (Fur) promoter, $p_{stx1}$. Very low levels of Stx1 are produced when the bacteria are grown in high iron conditions in the absence of phage induction, and in low iron conditions Stx1 expression from the $p_{stx1}$ promoter is enhanced. Low levels of transcription originating at $p_{stx2}$ occur in the absence of phage induction independent of iron concentration. Arrows denote the length of transcripts with and without antitermination. The relative strength of each promoter is indicated by the thickness of the transcript arrows. Based on diagrams by Wagner et al. (122–124) and reprinted with permission of the publisher.

autocleavage, mediated by association with the bacterial RecA protease, and transcription of the *N* antiterminator gene proceeds. N binds and modifies the RNA polymerase to increase processivity and to facilitate transcription from $p_R$ through several intervening termination sequences downstream through the *Q* antiterminator gene. Q most strongly influences the expression of both Stx1 and Stx2 following phage induction by facilitating transcription from the secondary right promoter $p_R'$ through the toxin operons. Phage-independent (native) promoters have also been identified immediately upstream of both the $stx_1$ and $stx_2$ operons (17, 110). The $stx_1$ native promoter $p_{stx1}$ has a Fur-binding site that allows for repression under high iron conditions, a phenotype that is also seen for *stx* expression in *S. dysenteriae* type 1 (17). Thus there are three promoters upstream of the $stx_1$ genes in phage H-19B that influence toxin production: the phage $p_R$ that responds to phage induction, the $p_R'$ phage promoter on which Q acts, and the Fur-regulated toxin gene promoter. The *Q* gene and $p_R'$ are not required for toxin expression under phage induction in H-19B, but this is in contrast to the situation in 933W and another Stx2 phage, Φ1:361. In these Stx2 phages, toxin production does not proceed upon phage induction unless the Q antiterminator is transcribed and acts at $p_R'$. The native promoter $p_{stx2}$ is not necessary for Stx2 expression (116, 124). It is also possible to observe spontaneous phage induction that results in Stx expression under conditions favorable for bacterial growth. Rates of spontaneous induction differ between the prototype phages, with 1 induction in 20,000 lysogens per cell generation of H-19B, compared to 1 in 7,000 per cell generation of 933W. Spontaneous induction is hypothesized to be dependent upon the basal levels of RecA within the bacterial host cell (60).

In addition to the important role that the phage plays in toxin production through its complicated, multilayered genetic regulation, the phage likely also has a role in toxin release from the bacteria as phage production lyses the bacteria. Downstream of the *stx* genes is a cassette of lambda phage lysis genes. Genes *S* and *R* encode a holin and an endolysin, respectively (reviewed in reference 125). The holin permeabilizes the host bacterial inner membrane, and the endolysin degrades the cell wall. The bacteriophage holins have been described as "clocks" (125) because the kinetics of holin expression and action are coordinated with phage induction and replication such that the timing of lysis does not precede the assembly of viable phage progeny. The duration of phage replication prior to lysis determines the burst size. In STEC deficient in holin, bacteriophage DNA and Stx accumulate within the bacterium, but the cell does not lyse (123). Variations in transcription of holin and endolysin among Stx1 and Stx2 phages may provide an explanation for the observation that Stx1 tends to remain cell associated as compared to Stx2, which is usually found in the supernatant.

It is also clear that bacteriophages that encode Stxs display a considerable amount of mosaicism, evidence of liberal recombination among other Stx-bearing and lambdoid phages (47). Indeed, the *E. coli* O157 (Sakai strain) chromosome contains 18 prophages, many of which are defective. Nonetheless, this pool of bacteriophage DNA within such strains allows for recombination and complementation among coinfecting bacteriophages and defective ones. The opportunity for the synthesis of new phages that encode Stx genes and that have the capacity to be mobilized into new hosts is inevitable (6).

## DIAGNOSTIC CONSIDERATIONS

Rapid diagnosis of infection with STEC is essential for proper treatment of the infected individual, both for initiation of appropriate supportive steps such as parenteral volume expansion (3) and for avoidance of potentially damaging antibiotic therapy (131). Because of the important role of O157:H7 STEC in disease in the United States, the CDC has recommended that clinical laboratories screen all stools from patients with community-acquired diarrhea for O157 strains using selective/differential medium, regardless of patient age, the season, or the presence of frank blood in the stool (35). Suitable selective/differential media for O157 strains include sorbitol MacConkey agar (SMAC), SMAC supplemented with cefixime and tellurite to repress the growth of other gastrointestinal organisms that are slow sorbitol fermentors, and CHROMagar O157 (Becton Dickinson, Cockeysville, MD). CHROMagar O157 was shown in one study to be somewhat more sensitive for isolation of O157 strains than SMAC (22). Once potential O157 strains are isolated, they can be confirmed by agglutination in O157 antisera or latex reagent. Such positive strains can then be promptly referred to a regional public health laboratory for further genetic epidemiologic profiling (35). It should be noted that in Germany, O157 nonmotile strains that readily ferment sorbitol are prevalent, and thus are not detectable by the strategy outlined above.

The diversity among non-O157 STEC makes their detection more difficult than O157 strains that exhibit the typical sorbitol-negative phenotype. Therefore, the unifying characteristic of Stx production is the most inclusive trait for which to screen for STEC. A variety of commercial products based on enzyme-linked immunoassays (Premier by Meridian Bioscience, Cincinnati, OH; ProSpecT Shiga Toxin Detection, Remel, Lenexa, KS), immunochromatography (Duopath, Merck, Darmstadt, Germany), or optical immunoassays (OIA Shigatox, Inverness Medical, Louisville, CO) for toxin detection are available. Although these kits are approved for direct toxin detection in fecal samples, their use as stand-alone diagnostics without culture confirmation is not recommended (56). Virtually all of the commercial toxin detection kit manufacturers also recommend a screen for toxin after enrichment of the facultative fecal flora in a broth culture subjected to overnight incubation. Suggested enrichment broth media include gram-negative broth or MacConkey broth. An acid enrichment procedure has also been suggested to enhance recovery of STEC from broth (43). Amplification of STEC in enrichment broth further increases the sensitivity of toxin detection as well as isolation of O157 when the enrichment broth is subcultured onto SMAC or other suitable screening media.

Few rapid screening strategies have been outlined for the isolation of non-O157 strains. Nonetheless, a relatively consistent trait among STEC of diverse serotypes besides Stx production is the elaboration of enterohemolysin, the action of which is detectable on medium that contains washed sheep erythrocytes (10). Although the enterohemolytic trait is found in only about 86% of non-O157 strains characterized in the United States (16), it can be used to screen primary cultures for non-O157 STEC, followed by an immunologic procedure for Stx detection, and serotype determination to identify putative non-O157 STEC strains or sorbitol-fermenting O157:NM strains (12).

To promote a rapid identification and response to outbreaks of STEC disease, clinical laboratories in the United States are urged by the CDC to refer any STEC strains that are isolated to regional public health facilities where such strains can be more thoroughly characterized. Numerous strategies are available for isolate characterization. These methods include real-time PCR to identify the primary virulence traits such as toxin type and the presence of intimin and/or enterohemolysin genes. Other protocols are designed to distinguish toxin subtypes with PCR amplification of key sequences followed by restriction fragment length polymorphism analysis (9). The drawback of the PCR followed by restriction fragment length polymorphism method for toxin subtyping is that single nucleotide polymorphisms may alter the restriction patterns and confuse the outcome. A PCR-based method for toxin gene subtyping that relies on primers specific for unique sequences within each toxin subtype gene is currently under development (F. Scheutz, S. Persson, R. Tozzoli, and S. Morabito, unpublished). Although the subtyping of toxin genes is not generally done in primary isolation of STEC, there is a clinical advantage to doing so, because production of certain subtypes such as Stx2d-activatable or Stx2c (13, 45) has been shown to correlate with more serious disease. In addition to PCR-based strategies, a high-throughput Southern blot protocol developed by Schuetz (100) has also been suggested by the CDC for use by public health laboratories in the United States.

One of the most useful epidemiologic tools for detection of common source outbreaks of STEC is pulsed-field gel electrophoresis (PFGE) for molecular profiling of *E. coli* and other enteric pathogens. DNA fingerprinting based on PFGE restriction patterns obtained by a standardized protocol (91) can readily be compared and the data shared electronically to detect similar profiles among widely distributed isolates through the CDC-based PulseNet program. This PFGE strategy has been key to the identification of multistate outbreaks in the United States and the detection of common food sources. Several state and regional public health laboratories, the Food and Drug Administration, and the U.S. Department of Agriculture participate with the CDC in the PulseNet program. In addition, PulseNet is currently expanding to link international participants as well. Additionally, this surveillance network is incorporating multilocus variable-number tandem repeat analysis and multilocus sequence typing to refine strain typing and overcome some of the limitations of PFGE alone (http://www.cdc.gov/pulsenet/).

## INTERVENTIONS FOR STEC-MEDIATED DISEASE

There are no approved medical interventions to directly treat STEC-mediated disease, so the medical response to STEC infection is largely supportive. However, Ake et al. reported success in reducing renal injury in children with O157:H7 infection with parenteral volume expansion early in the course of disease (3). The fact that early treatment worked best in the Ake study is illustrative of much of the literature regarding treatments for STEC-mediated disease—the earlier the intervention, the better for the patient—a conclusion that points to the need for rapid diagnostics (stressed above). A number of recent reviews discuss potential treatments or interventions for HUS or STEC infections in humans as well as attempts to reduce O157:H7 colonization in cattle (14, 34, 95, 99, 102). In this section we will briefly review the current status of these options.

The treatment for STEC-mediated disease is complicated by the fact that the toxin genes usually reside on an antibiotic-inducible phage, as discussed above. Several in vitro studies demonstrate that treatment of STEC cultures with various antibiotics results in an increase in the amount of toxin produced and/or released into the medium. Certain animal models suggest that such a phenomenon may exist in vivo: injection of mitomycin C, a potent phage inducer, into streptomycin-treated mice enhances the virulence of O157:H$^-$ strain E32511/HSC in those mice as compared to untreated mice (31). A recent study of gnotobiotic pigs infected with O157 strain 86-24 showed that treatment with ciprofloxacin, an antibiotic that increases Stx2 production by 86-24 in vitro, resulted in lower colonization levels but more severe neurological symptoms compared to control animals (133). In contrast, when 86-24-infected pigs were treated with azithromycin, an antibiotic that did not enhance in vitro Stx2 production, fewer animals showed neurological symptoms and died (133). Because many antibiotics induce toxin production by STEC in vitro and increase pathogenicity of STEC in vivo, a high level of concern exists that antibiotic treatment of patients infected with STEC would lead to enhanced toxin levels in the intestine, a result that would likely lead to greater disease. There is contradictory evidence in humans about the question of whether antibiotic treatment is harmful or not (see reference 80 for a review). However, a study published in the *New England Journal of Medicine* clearly indicated that the use of sulfa-containing drugs and perhaps $\beta$-lactam antibiotics increased the risk of HUS (131). Therefore, antibiotic treatment for STEC infections is contraindicated in the United States. Part of the difficulty in reaching a consensus about antibiotic treatment for STEC disease is that the studies conducted to date have had a small number of patients, used different antibiotics, and treated patients at different times in the course of disease. A randomized, controlled, prospective trial would be needed to answer the question about whether antibiotic treatment is beneficial or harmful to those infected with STEC. However, because of the contradictory and varied results from different studies, it would be difficult to design such a study. Moreover, such a study might be considered unethical. In contrast to the recommended practice in the United States, clinicians in Japan routinely used antibiotics to treat O157:H7 infections in a large outbreak centered in Sakai in 1996. They reported some reduction in risk of HUS with the use of fosfomycin in the first 2 days of illness (44). However, another group performed a retrospective analysis that compared fluoroquinolone, fosfomycin, and fosfomycin/cefotaxime treatment that suggested a lower HUS rate only in the group that received early fluoroquinolone treatment (105). In the latter study the number of patients was small (15 or fewer in each group), a fact that makes it difficult to make strong conclusions. In addition, the rate of HUS in the large Japanese outbreak was low overall, and most patients were treated with drugs, two facts that make it difficult to conclude whether antibiotics were indeed the cause of the low overall rate of HUS or whether the Sakai strain has reduced virulence compared to other O157:H7 strains. Ultimately, the use of an antibiotic that eliminates STEC without increasing toxin production may be helpful in patients who are diagnosed early; however, extreme caution should be taken in treating patients infected with STEC, as the potential to increase the chance of HUS should be avoided.

One approach to try to neutralize the Stx that leads to HUS is to absorb the toxin as it is elaborated in the gut. The first toxin absorbent tested was Synsorb Pk, a synthetic analog of $Gb_3$. Although Synsorb Pk binds both Stx1 and Stx2, it did not reduce the severity of diarrhea-associated HUS in a clinical trial (115). The reason that Synsorb Pk was ineffective as a treatment for HUS is most likely that it was given too late in the time course of the disease; i.e., the patients already had HUS and the damage had already been done. Even so, a number of toxin-binding agents have been developed in the hope that such agents, if they bind tightly enough to the Stxs, may succeed where Synsorb Pk did not, or that they may be used as a prophylactic treatment. These Stx binders include Starfish, Daisy, and Super Twig (see references 34, 95, and 99 for reviews). More recently, a new type of soluble toxin-binding agent was developed: a tetravalent peptide that binds to Stx2 and neutralizes its toxicity in vitro for Vero and Caco-2 cells and in vivo in a protein calorie-deficient mouse model and a rabbit ileal loop model (73, 126).

In a similar approach, Paton's group developed an oligosaccharide $Gb_3$ mimic that terminates in the same linkage found on $Gb_3$: Gal$\alpha$[1-4]Gal$\beta$[1-4]. However, they express the $Gb_3$ mimic from an *E. coli* C600 strain such that the strain becomes a probiotic. They were able to demonstrate that the *E. coli* strain that expressed the oligosaccharide binds Stx1 and Stx2c and protects streptomycin-treated mice from death due to a STEC strain that produces Stx2 (86). However, it is likely that the use of the Gal$\alpha$[1-4]Gal$\beta$[1-4]-loaded *E. coli* will have the same difficulty in humans that Synsorb Pk did: it will only neutralize toxin that is present in the intestines and will be unable to neutralize systemic toxin. Additionally, since C600 is widely used as an indicator strain for phage, the danger that the treatment strain could pick up the toxin phage exists.

Several groups have described an Stx2-binding activity in human serum that is not antibody. The component in the serum that binds to Stx2 (but not Stx1) was eventually shown to be serum amyloid P component (SAP) (53). However, Armstrong et al. showed that the SAP levels were not different in HUS or non-HUS patients with antibodies to O157 or in patients without HUS who did not have anti-O157 antibody (5), a result that suggests that native levels of SAP do not prevent HUS, although it does not appear that the patients were separated into groups by the toxin type (Stx1 or Stx2) of the infecting organism. However, when SAP was administered intraperitoneally to mice challenged subcutaneously with 2 $LD_{50}$s of Stx2, the mice were protected. Conversely, in another study, SAP injected intravenously into mice challenged intraperitoneally with 10 $LD_{50}$s of Stx2 did not protect (54). Additionally, the SAP appeared to reduce the capacity of a monoclonal antibody to the Stx2 B subunit (TMA-15) to bind to Stx2 and neutralize the toxin in vivo, a result that suggests that these two molecules compete for Stx2 binding.

The studies described above indicate that it will not be sufficient to try to neutralize toxin in the intestine of individuals already ill with STEC. Our approach, and that of a few other groups, has been to develop monoclonal antibodies that neutralize the Stx(s) and can be delivered intravenously (reviewed in reference 118). Monoclonal antibodies directed against Stx1 or Stx2 were developed in the 1980s by our group and subsequently were made into human/mouse chimeras to facilitate their use in humans (24). Those chimeric antibodies had the capacity to protect mice from injection with Stx1 or infection by non-O157 strain B2F1 (O91:H21, Stx2d-activatable producer) even when the antibody was administered after injection or infection (24). These antibodies have completed safety trials in humans alone and in combination (23; our unpublished results). Other humanized or human-specific anti-Stx2 antibodies (5C12 and TMA-15) have also shown the capacity to protect in animal models against the lethal effects of either B2F1 or 86-24 (O157:H7, Stx2 producer) (72, 132). In addition, TMA-15 (now called urtoxazumab) has undergone safety trials in humans (61). Our group is the only group that is pursuing a combination strategy to try to neutralize both Stx1 and Stx2. We believe that such a strategy is vital, as Stx1 has the capacity to cause HUS (52) and many STEC strains produce both Stx1 and Stx2.

No FDA-approved vaccine for use in humans against STEC-mediated disease exists. However, some groups have investigated the use of vaccines in animal models, and most of those vaccines rely on the use of genetic toxoids or consist only of the A or B subunit (reviewed in references 34, 78, and 102). Our group developed a chimeric toxoid vaccine that protects against the lethal effects of both Stx1 and Stx2 (108). Another group has looked at using the O157 lipopolysaccharide as a conjugate vaccine in which *Pseudomonas aeruginosa* exotoxin A was the other component. The latter vaccine has gone through safety trials (1). Such a vaccine would be useful in the battle to eliminate O157-mediated disease, but would not be effective against non-O157 bacteria, which are responsible for about half of all STEC infections in the United States.

An alternate approach to inducing immunity in humans is to try to eliminate the bacterium from the food supply. There is a vaccine for cattle that consists of components of the type III secretion system (87). A recent test of administration of two doses of the vaccine with an altered adjuvant showed that vaccinated cattle were about 90% less likely to be colonized by O157 than the nonvaccinated animals for a period of 3 months prior to harvest (106). Although this vaccine reduced colonization by O157 strains that encode the type III secretion system, it would not prevent colonization by STEC that lack the type III secretion system. Although the majority of the most virulent STEC do have the type III secretion system, some do not, such as the highly virulent non-

O157 strains that produce Stx2d-activatable. Other animal vaccines under evaluation include one that contains intimin expressed within plant cells (49); another that contains intimin, EspA, and Tir (68); and a third with the siderophore receptor and porin (27). The latter two vaccines were shown to reduce O157 colonization in cattle (68, 113).

Several groups are investigating the use of preharvest interventions besides vaccination to reduce O157:H7 colonization in cattle, such as antibiotics, probiotics, or O157-specific bacteriophage (reviewed in reference 98). However, none of the preharvest interventions besides vaccination have been tested in a large field study or show consistent enough results to suggest their use at this time. Other groups are investigating therapies to intervene in the pathogenesis of HUS in humans. None of these strategies has been developed to the extent that they would be used as a standard of care (see reference 14 for a review).

One final intervention point for O157-mediated illness is at the level of the consumer. As people become more aware of the potential for STEC contamination of food and water, they may be able to prevent some infections, including primary and secondary infections, by using good food-handling practices and hygiene. Because up to 20% of O157 infections are due to secondary transmission (109), and because most of those secondary cases occur in situations in which the primary case is in a child less than 6 years old or in a nursery setting, logic suggests that attention must be paid to proper handwashing, safe food preparation, avoiding contact with people who have diarrhea, and staying home from school or work when ill. Indeed, modeling studies suggest that such precautions would be helpful in reducing secondary transmission (103). In addition, some clinicians have suggested that isolation of patients with O157:H7 infections would be an important strategy in the prevention of secondary cases of STEC (2, 127).

## TOM WHITTAM'S CONTRIBUTION TO OUR UNDERSTANDING OF STEC

Tom Whittam was a frequent guest lecturer on our campus. To each visit he brought his energy and obvious fascination with his studies of *E. coli* evolution. He conveyed his work in a manner that was both stimulating and convincing. An equally enthusiastic listener, Tom prompted graduate students and postdoctoral fellows to discuss their work. Tom would often suggest methods, references, and colleagues in the field who might be helpful to that student's project. Scientifically, we most appreciated his monumental effort to systematically characterize Stx-producing *E. coli* strains and compare their profiles for elucidation of the clonal relatedness among the EHEC and the diverse non-O157 STEC. He also contributed to the effort to help standardize the inconsistent nomenclature used for the STEC and Stxs. He shared our hypothesis that something was unique and interesting about the O91:H21 strain we studied and did not hesitate to share his unpublished data related to that topic. Monumental in his contribution to microbiology was the organization of the *E. coli* reference strain collection. This resource has enabled scientists to share strains, centralize data regarding such strains, and perpetuate them for future studies. Tom was an excellent ambassador for his field of science and engendered great cross talk among scientists. His personality allowed such communications to happen fluidly and collegially. All who comment on him in this text are likely to mention the breadth of his scientific contributions. We likewise revered his warmth and camaraderie. These things were manifest in his smile, his easy gait, and his typical big, unabashed wave from across the pavilion of an ASM General Meeting. His presence and stature still loom in those halls with each of us with whom he interacted.

**Acknowledgments.** We wish to acknowledge that much of our understanding of STEC has come about through research funded by the NIH grant AI20148.

## REFERENCES

1. **Ahmed, A., J. Li, Y. Shiloach, J. B. Robbins, and S. C. Szu.** 2006. Safety and immunogenicity of *Escherichia coli* O157 O-specific polysaccharide conjugate vaccine in 2–5-year-old children. *J. Infect. Dis.* **193:**515–521.
2. **Ahn, C. K., E. Klein, and P. I. Tarr.** 2008. Isolation of patients acutely infected with *Escherichia coli* O157:H7: low-tech, highly effective prevention of hemolytic uremic syndrome. *Clin. Infect. Dis.* **46:**1197–1199.
3. **Ake, J. A., S. Jelacic, M. A. Ciol, S. L. Watkins, K. F. Murray, D. L. Christie, E. J. Klein, and P. I. Tarr.** 2005. Relative nephroprotection during *Escherichia coli* O157: H7 infections: association with intravenous volume expansion. *Pediatrics* **115:**e673–e680.
4. **Allen, K. P., M. M. Randolph, and J. M. Fleckenstein.** 2006. Importance of heat-labile enterotoxin in colonization of the adult mouse small intestine by human enterotoxigenic *Escherichia coli* strains. *Infect. Immun.* **74:**869–875.
5. **Armstrong, G. D., G. L. Mulvey, P. Marcato, T. P. Griener, M. C. Kahan, G. A. Tennent, C. A. Sabin, H. Chart, and M. B. Pepys.** 2006. Human serum amyloid P component protects against *Escherichia coli* O157:H7 Shiga toxin 2 in vivo: therapeutic implications for hemolytic uremic syndrome. *J. Infect. Dis.* **193:**1120–1124.
6. **Asadulghani, M., Y. Ogura, T. Ooka, T. Itoh, A. Sawaguchi, A. Iguchi, K. Nakayama, and T. Hayashi.** 2009. The defective prophage pool of *Escherichia coli* O157: prophage-prophage interactions potentiate horizontal transfer of virulence determinants. *PLoS Pathog.* **5:**e1000408.
7. **Besser, R. E., S. M. Lett, J. T. Weber, M. P. Doyle, T. J. Barrett, J. G. Wells, and P. M. Griffin.** 1993. An outbreak of diarrhea and hemolytic uremic syndrome from *Escherichia coli* O157:H7 in fresh-pressed apple cider. *JAMA* **269:**2217–2220.
8. **Bettelheim, K. A.** 2003. Non-O157 verotoxin-producing *Escherichia coli*: a problem, paradox, and paradigm. *Exp. Biol. Med.* (Maywood) **228:**333–344.
9. **Beutin, L., A. Miko, G. Krause, K. Pries, S. Haby, K. Steege, and N. Albrecht.** 2007. Identification of human-pathogenic strains of Shiga toxin-producing *Escherichia coli* from food by a combination of serotyping and molecular typing of Shiga toxin genes. *Appl. Environ. Microbiol.* **73:**4769–4775.
10. **Beutin, L., M. A. Montenegro, I. Ørskov, F. Ørskov, J. Prada, S. Zimmermann, and R. Stephan.** 1989. Close association of verotoxin (Shiga-like toxin) production with enterohemolysin production in strains of *Escherichia coli*. *J. Clin. Microbiol.* **27:**2559–2564.
11. **Beutin, L., S. Zimmermann, and K. Gleier.** 1998. Human infections with Shiga toxin-producing *Escherichia coli* other than serogroup O157 in Germany. *Emerg. Infect. Dis.* **4:**635–639.

12. **Bielaszewska, M.** 2009. Detection of STEC/VTEC in human samples, abstr. S07.1. *In Abstracts of the 7th International Symposium on Shiga Toxin (Verocytotoxin)-Producing* Escherichia coli *Infections.* Buenos Aires, Argentina.
13. **Bielaszewska, M., A. W. Friedrich, T. Aldick, R. Schurk-Bulgrin, and H. Karch.** 2006. Shiga toxin activatable by intestinal mucus in *Escherichia coli* isolated from humans: predictor for a severe clinical outcome. *Clin. Infect. Dis.* **43:**1160–1167.
14. **Bitzan, M.** 2009. Treatment options for HUS secondary to *Escherichia coli* O157:H7. *Kidney Int.* **75:**S62–S66.
15. **Boyce, T. G., A. G. Pemberton, J. G. Wells, and P. M. Griffin.** 1995. Screening for *Escherichia coli* O157:H7—a nationwide survey of clinical laboratories. *J. Clin. Microbiol.* **33:**3275–3277.
16. **Brooks, J. T., E. G. Sowers, J. G. Wells, K. D. Greene, P. M. Griffin, R. M. Hoekstra, and N. A. Strockbine.** 2005. Non-O157 Shiga toxin-producing *Escherichia coli* infections in the United States, 1983–2002. *J. Infect. Dis.* **192:**1422–1429.
17. **Calderwood, S. B., and J. J. Mekalanos.** 1987. Iron regulation of Shiga-like toxin expression in *Escherichia coli* is mediated by the *fur* locus. *J. Bacteriol.* **169:**4759–4764.
18. **Centers for Disease Control and Prevention.** 1999. Outbreak of *Escherichia coli* O157:H7 and *Campylobacter* among attendees of the Washington County Fair—New York, 1999. *MMWR Morb. Mortal. Wkly. Rep.* **48:**803–805.
19. **Centers for Disease Control and Prevention.** 2002. Multistate outbreak of *Escherichia coli* O157:H7 infections associated with eating ground beef—United States, June–July 2002. *MMWR Morb. Mortal. Wkly. Rep.* **51:**637–639.
20. **Centers for Disease Control and Prevention.** 2007. Laboratory-confirmed non-O157 Shiga toxin-producing *Escherichia coli*—Connecticut, 2000–2005. *MMWR Morb. Mortal. Wkly. Rep.* **56:**29–31.
21. **Centers for Disease Control and Prevention.** 1995. Outbreak of acute gastroenteritis attributable to *Escherichia coli* serotype O104:H21—Helena, Montana, 1994. *MMWR Morb. Mortal. Wkly. Rep.* **44:**501–503.
22. **Church, D. L., D. Emshey, H. Semeniuk, T. Lloyd, and J. D. Pitout.** 2007. Evaluation of BBL CHROMagar O157 versus sorbitol-MacConkey medium for routine detection of *Escherichia coli* O157 in a centralized regional clinical microbiology laboratory. *J. Clin. Microbiol.* **45:**3098–3100.
23. **Dowling, T. C., P. A. Chavaillaz, D. G. Young, A. Melton-Celsa, A. O'Brien, C. Thuning-Roberson, R. Edelman, and C. O. Tacket.** 2005. Phase 1 safety and pharmacokinetic study of chimeric murine-human monoclonal antibody cαStx2 administered intravenously to healthy adult volunteers. *Antimicrob. Agents Chemother.* **49:**1808–1812.
24. **Edwards, A. C., A. R. Melton-Celsa, K. Arbuthnott, J. R. Stinson, C. K. Schmitt, H. C. Wong, and A. D. O'Brien.** 1998. Vero cell neutralization and mouse protective efficacy of humanized monoclonal antibodies against *Escherichia coli* toxins Stx1 and Stx2, p. 388–392. *In* J. B. Kaper and A. D. O'Brien (ed.), Escherichia coli *O157:H7 and Other Shiga Toxin-Producing* E. coli *Strains.* ASM Press, Washington, DC.
25. **Farmer, J. J., III, and B. R. Davis.** 1985. H7 antiserum-sorbitol fermentation medium: a single tube screening medium for detecting *Escherichia coli* O157:H7 associated with hemorrhagic colitis. *J. Clin. Microbiol.* **22:**620–625.
26. **Feng, P., K. A. Lampel, H. Karch, and T. S. Whittam.** 1998. Genotypic and phenotypic changes in the emergence of *Escherichia coli* O157:H7. *J. Infect. Dis.* **177:**1750–1753.
27. **Fox, J. T., D. U. Thomson, J. S. Drouillard, A. B. Thornton, D. T. Burkhardt, D. A. Emery, and T. G. Nagaraja.** 2009. Efficacy of *Escherichia coli* O157:H7 siderophore receptor/porin proteins-based vaccine in feedlot cattle naturally shedding *E. coli* O157. *Foodborne Pathog. Dis.* **6:**893–899.
28. **Fraser, M. E., M. M. Chernaia, Y. V. Kozlov, and M. N. James.** 1994. Crystal structure of the holotoxin from *Shigella dysenteriae* at 2.5 Å resolution. *Nat. Struct. Biol.* **1:**59–64.

29. **Fraser, M. E., M. Fujinaga, M. M. Cherney, A. R. Melton-Celsa, E. M. Twiddy, A. D. O'Brien, and M. N. James.** 2004. Structure of Shiga toxin type 2 (Stx2) from *Escherichia coli* O157:H7. *J. Biol. Chem.* **279:**27511–27517.
30. **Fuchs, S., I. Muhldorfer, A. Donohue-Rolfe, M. Kerenyi, L. Emody, R. Alexiev, P. Nenkov, and J. Hacker.** 1999. Influence of RecA on *in vivo* virulence and Shiga toxin 2 production in *Escherichia coli* pathogens. *Microb. Pathog.* **27:**13–23.
31. **Fujii, J., T. Kita, S. Yoshida, T. Takeda, H. Kobayashi, N. Tanaka, K. Ohsato, and Y. Mizuguchi.** 1994. Direct evidence of neuron impairment by oral infection with verotoxin-producing *Escherichia coli* O157:$H^-$ in mitomycin-treated mice. *Infect. Immun.* **62:**3447–3453.
32. **Gauthier, A., and B. B. Finlay.** 2003. Translocated intimin receptor and its chaperone interact with ATPase of the type III secretion apparatus of enteropathogenic *Escherichia coli*. *J. Bacteriol.* **185:**6747–6755.
33. **Gerber, A., H. Karch, F. Allerberger, H. M. Verweyen, and L. B. Zimmerhackl.** 2002. Clinical course and the role of Shiga toxin-producing *Escherichia coli* infection in the hemolytic-uremic syndrome in pediatric patients, 1997–2000, in Germany and Austria: a prospective study. *J. Infect. Dis.* **186:**493–500.
34. **Goldwater, P. N.** 2007. Treatment and prevention of enterohemorrhagic *Escherichia coli* infection and hemolytic uremic syndrome. *Expert Rev. Anti Infect. Ther.* **5:**653–663.
35. **Gould, L. H., C. Bopp, N. Strockbine, R. Atkinson, V. Baselski, B. Body, R. Carey, C. Crandall, S. Hurd, R. Kaplan, M. Neill, S. Shea, P. Somsel, M. Tobin-D'Angelo, P. M. Griffin, and P. Gerner-Smidt.** 2009. Recommendations for diagnosis of Shiga toxin-producing *Escherichia coli* infections by clinical laboratories. *MMWR Recomm. Rep.* **58:**1–14.
36. **Gould, L. H., M. M. Leeper, D. MacCannell, P. M. Griffin, and P. Gerner-Smidt.** 2009. Trends in toxin profiles of human Shiga toxin-producing *Escherichia coli* (STEC) O157 Strains, United States, 1996–2008, abstr. P02.4.9. *In Abstracts of the 7th International Symposium on Shiga Toxin (Verocytotoxin)-Producing* Escherichia coli *Infections.* Buenos Aires, Argentina.
37. **Griffin, P. M.** 1995. *Escherichia coli* O157:H7 and other enterohemorrhagic *Escherichia coli*, p. 739–761. *In* M. J. Blaser, P. D. Smith, J. I. Ravdin, H. B. Greenberg, and R. L. Guerrant (ed.), *Infections of the Gastrointestinal Tract.* Raven Press, New York, NY.
38. **Griffin, P. M.** 1998. Epidemiology of Shiga toxin-producing *Escherichia coli* infections in humans in the United States, p. 15–22. *In* J. B. Kaper and A. D. O'Brien (ed.), Escherichia coli *O157:H7 and Other Shiga Toxin-Producing* E. coli *Strains.* ASM Press, Washington, DC.
39. **Griffin, P. M., L. H. Gould, T. Ayers, and M. M. Leeper.** 2009. Epidemiology of Shiga toxin-producing *Escherichia coli* (STEC) infections in the United States, abstr. S02.2. *In Abstracts of the 7th International Symposium on Shiga Toxin (Verocytotoxin)-Producing* Escherichia coli *Infections.* Buenos Aires, Argentina.
40. **Grys, T. E., M. B. Siegel, W. W. Lathem, and R. A. Welch.** 2005. The StcE protease contributes to intimate adherence of enterohemorrhagic *Escherichia coli* O157:H7 to host cells. *Infect. Immun.* **73:**1295–1303.
41. **Herold, S., J. C. Paton, and A. W. Paton.** 2009. Sab, a novel autotransporter of locus of enterocyte effacement-negative Shiga-toxigenic *Escherichia coli* O113:H21, contributes to adherence and biofilm formation. *Infect. Immun.* **77:**3234–3243.
42. **Hilborn, E. D., J. H. Mermin, P. A. Mshar, J. L. Hadler, A. Voetsch, C. Wojtkunski, M. Swartz, R. Mshar, M. A. Lambert-Fair, J. A. Farrar, M. K. Glynn, and L. Slutsker.** 1999. A multistate outbreak of *Escherichia coli* O157:H7 infections associated with consumption of mesclun lettuce. *Arch. Intern. Med.* **159:**1758–1764.
43. **Hu, J., D. Green, J. Swoveland, M. Grant, and D. S. Boyle.** 2009. Preliminary evaluation of a procedure for improved detection of Shiga toxin-producing *Escherichia coli* in fecal specimens. *Diagn. Microbiol. Infect. Dis.* **65:**21–26.
44. **Ikeda, K., O. Ida, K. Kimoto, T. Takatorige, N. Nakanishi, and K. Tatara.** 1999. Effect of early fosfomycin treatment on prevention of hemolytic uremic syndrome accompanying *Escherichia coli* O157:H7 infection. *Clin. Nephrol.* **52:**357–362.

45. **Jelacic, J. K., T. Damrow, G. S. Chen, S. Jelacic, M. Bielaszewska, M. Ciol, H. M. Carvalho, A. R. Melton-Celsa, A. D. O'Brien, and P. I. Tarr.** 2003. Shiga toxin-producing *Escherichia coli* in Montana: bacterial genotypes and clinical profiles. *J. Infect. Dis.* **188:**719–729.
46. **Johannes, L., and W. Römer.** 2010. Shiga toxins—from cell biology to biomedical applications. *Nat. Rev. Microbiol.* **8:**105–116.
47. **Johansen, B. K., Y. Wasteson, P. E. Granum, and S. Brynestad.** 2001. Mosaic structure of Shiga-toxin-2-encoding phages isolated from *Escherichia coli* O157:H7 indicates frequent gene exchange between lambdoid phage genomes. *Microbiology* **147:**1929–1936.
48. **Johnson, K. E., C. M. Thorpe, and C. L. Sears.** 2006. The emerging clinical importance of non-O157 Shiga toxin-producing *Escherichia coli*. *Clin. Infect. Dis.* **43:**1587–1595.
49. **Judge, N. A., H. S. Mason, and A. D. O'Brien.** 2004. Plant cell-based intimin vaccine given orally to mice primed with intimin reduces time of *Escherichia coli* O157:H7 shedding in feces. *Infect. Immun.* **72:**168–175.
50. **Kaper, J. B., S. Elliott, V. Sperandio, N. T. Perna, G. F. Mayhew, and F. R. Blattner.** 1998. Attaching-and-effacing intestinal histopathology and the locus of enterocyte effacement, p. 163–182. *In* J. B. Kaper and A. D. O'Brien (ed.), Escherichia coli *O157:H7 and Other Shiga Toxin-Producing* E. coli *Strains*. ASM Press, Washington, DC.
51. **Kaper, J. B., J. P. Nataro, and H. L. T. Mobley.** 2004. Pathogenic *Escherichia coli*. *Nat. Rev. Microbiol.* **2:**123–140.
52. **Kim, Y. B., J. Okuda, C. Matsumoto, T. Morigaki, N. Asai, H. Watanabe, and M. Nishibuchi.** 1998. Isolation of an *Escherichia coli* O157:H7 strain producing Shiga toxin 1 but not Shiga toxin 2 from a patient with hemolytic uremic syndrome in Korea. *FEMS Microbiol. Lett.* **166:**43–48.
53. **Kimura, T., S. Tani, Y. Y. Matsumoto, and T. Takeda.** 2001. Serum amyloid P component is the Shiga toxin 2-neutralizing factor in human blood. *J. Biol. Chem.* **276:**41576–41579.
54. **Kimura, T., S. Tani, M. Motoki, and Y. Matsumoto.** 2003. Role of Shiga toxin 2 (Stx2)-binding protein, human serum amyloid P component (HuSAP), in Shiga toxin-producing *Escherichia coli* infections: assumption from in vitro and in vivo study using HuSAP and anti-Stx2 humanized monoclonal antibody TMA-15. *Biochem. Biophys. Res. Commun.* **305:**1057–1060.
55. **Klein, E. J., J. R. Stapp, C. R. Clausen, D. R. Boster, J. G. Wells, X. Qin, D. L. Swerdlow, and P. I. Tarr.** 2002. Shiga toxin-producing *Escherichia coli* in children with diarrhea: a prospective point-of-care study. *J. Pediatr.* **141:**172–177.
56. **Klein, E. J., J. R. Stapp, M. A. Neill, J. M. Besser, M. T. Osterholm, and P. I. Tarr.** 2004. Shiga toxin antigen detection should not replace sorbitol MacConkey agar screening of stool specimens. *J. Clin. Microbiol.* **42:**4416–4417.
57. **Lainhart, W., G. Stolfa, and G. B. Koudelka.** 2009. Shiga toxin as a bacterial defense against a eukaryotic predator, *Tetrahymena thermophila*. *J. Bacteriol.* **191:**5116–5122.
58. **Lathem, W. W., T. E. Grys, S. E. Witowski, A. G. Torres, J. B. Kaper, P. I. Tarr, and R. A. Welch.** 2002. StcE, a metalloprotease secreted by *Escherichia coli* O157:H7, specifically cleaves C1 esterase inhibitor. *Mol. Microbiol.* **45:**277–288.
59. **Levine, M. M., J. G. Xu, J. B. Kaper, H. Lior, V. Prado, B. Tall, J. Nataro, H. Karch, and K. Wachsmuth.** 1987. A DNA probe to identify enterohemorrhagic *Escherichia coli* of O157:H7 and other serotypes that cause hemorrhagic colitis and hemolytic uremic syndrome. *J. Infect. Dis.* **156:**175–182.
60. **Livny, J., and D. I. Friedman.** 2004. Characterizing spontaneous induction of Stx encoding phages using a selectable reporter system. *Mol. Microbiol.* **51:**1691–1704.
61. **López, E. L., M. M. Contrini, E. Glatstein, A. S. González, R. Santoro, D. Allende, G. Ezcurra, E. Teplitz, T. Koyama, Y. Matsumoto, H. Sato, K. Sakai, S. Hoshide, K. Komoriya, T. Morita, R. Harning, and S. Brookman.** 2010. Safety and pharmacokinetics of urtoxazumab, a humanized monoclonal antibody, against Shiga-like toxin

2 in healthy adults and in pediatric patients infected with Shiga-like toxin-producing *Escherichia coli. Antimicrob. Agents Chemother.* **54:**239–243.

62. **Louise, C. B., and T. G. Obrig.** 1995. Specific interaction of *Escherichia coli* O157:H7-derived Shiga-like toxin II with human renal endothelial cells. *J. Infect. Dis.* **172:** 1397–1401.
63. **Lowe, R. M., D. Baines, L. B. Selinger, J. E. Thomas, T. A. McAllister, and R. Sharma.** 2009. *Escherichia coli* O157:H7 strain origin, lineage, and Shiga toxin 2 expression affect colonization of cattle. *Appl. Environ. Microbiol.* **75:**5074–5081.
64. **Maki, D. G.** 2006. Don't eat the spinach—controlling foodborne infectious disease. *N. Engl. J. Med.* **355:**1952–1955.
65. **Makino, K., K. Ishii, T. Yasunaga, M. Hattori, K. Yokoyama, C. H. Yutsudo, Y. Kubota, Y. Yamaichi, T. Iida, K. Yamamoto, T. Honda, C. G. Han, E. Ohtsubo, M. Kasamatsu, T. Hayashi, S. Kuhara, and H. Shinagawa.** 1998. Complete nucleotide sequences of 93-kb and 3.3-kb plasmids of an enterohemorrhagic *Escherichia coli* O157:H7 derived from Sakai outbreak. *DNA Res.* **5:**1–9.
66. **Manning, S. D., R. T. Madera, W. Schneider, S. E. Dietrich, W. Khalife, W. Brown, T. S. Whittam, P. Somsel, and J. T. Rudrik.** 2007. Surveillance for Shiga toxin-producing *Escherichia coli*, Michigan, 2001–2005. *Emerg. Infect. Dis.* **13:**318–321.
67. **Manning, S. D., A. S. Motiwala, A. C. Springman, W. Qi, D. W. Lacher, L. M. Ouellette, J. M. Mladonicky, P. Somsel, J. T. Rudrik, S. E. Dietrich, W. Zhang, B. Swaminathan, D. Alland, and T. S. Whittam.** 2008. Variation in virulence among clades of *Escherichia coli* O157:H7 associated with disease outbreaks. *Proc. Natl. Acad. Sci. USA* **105:**4868–4873.
68. **McNeilly, T. N., M. C. Mitchell, T. Rosser, S. McAteer, J. C. Low, D. G. Smith, J. F. Huntley, A. Mahajan, and D. L. Gally.** 2010. Immunization of cattle with a combination of purified intimin-531, EspA and Tir significantly reduces shedding of *Escherichia coli* O157:H7 following oral challenge. *Vaccine* **28:**1422–1428.
69. **Melton-Celsa, A. R., J. F. Kokai-Kun, and A. D. O'Brien.** 2002. Activation of Shiga toxin type 2d (Stx2d) by elastase involves cleavage of the C-terminal two amino acids of the $A_2$ peptide in the context of the appropriate B pentamer. *Mol. Microbiol.* **43:**207–215.
70. **Melton-Celsa, A. R., and A. D. O'Brien.** 2000. Shiga toxins of *Shigella dysenteriae* and *Escherichia coli*, p. 385–406. *In* K. Aktories and I. Just (ed.), *Handbook of Experimental Pharmacology*, vol. 145. Springer-Verlag, Berlin, Germany.
71. **Melton-Celsa, A. R., M. J. Smith, and A. D. O'Brien.** 2005. Shiga toxins: potent poisons, pathogenicity determinants, and pharmacological agents. *In* R. Curtiss III (ed.), *EcoSal:* Escherichia coli *and* Salmonella, *Cellular and Molecular Biology*, 2nd ed. ASM Press, Washington, DC. [Online.]

71a. **Moreland, J. L., A. Gramada, O. V. Buzko, Q. Zhang, and P. E. Bourne.** 2005. The Molecular Biology Toolkit (MBT): a modular platform for developing molecular visualization applications. *BMC Bioinformatics* **6:**21.

72. **Mukherjee, J., K. Chios, D. Fishwild, D. Hudson, S. O'Donnell, S. M. Rich, A. Donohue-Rolfe, and S. Tzipori.** 2002. Human Stx2-specific monoclonal antibodies prevent systemic complications of *Escherichia coli* O157:H7 infection. *Infect. Immun.* **70:** 612–619.
73. **Nishikawa, K., M. Watanabe, E. Kita, K. Igai, K. Omata, M. B. Yaffe, and Y. Natori.** 2006. A multivalent peptide library approach identifies a novel Shiga toxin inhibitor that induces aberrant cellular transport of the toxin. *FASEB J.* **20:**2597–2599.
74. **O'Brien, A. D., J. W. Newland, S. F. Miller, R. K. Holmes, H. W. Smith, and S. B. Formal.** 1984. Shiga-like toxin-converting phages from *Escherichia coli* strains that cause hemorrhagic colitis or infantile diarrhea. *Science* **226:**694–696.
75. **O'Brien, A. D., V. L. Tesh, A. Donohue-Rolfe, M. P. Jackson, S. Olsnes, K. Sandvig, A. A. Lindberg, and G. T. Keusch.** 1992. Shiga toxin: biochemistry, genetics, mode of action, and role in pathogenesis, p. 66–94. *In* P. J. Sansonetti (ed.), *Pathogenesis of Shigellosis*. Springer-Verlag, Berlin, Germany.

76. **Ogura, Y., T. Ooka, A. Iguchi, H. Toh, M. Asadulghani, K. Oshima, T. Kodama, H. Abe, K. Nakayama, K. Kurokawa, T. Tobe, M. Hattori, and T. Hayashi.** 2009. Comparative genomics reveal the mechanism of the parallel evolution of O157 and non-O157 enterohemorrhagic *Escherichia coli. Proc. Natl. Acad. Sci. USA* **106:**17939–17944.
77. **Ohmura-Hoshino, M., S. T. Ho, H. Kurazono, K. Igarashi, S. Yamasaki, and Y. Takeda.** 2003. Genetic and immunological analysis of a novel variant of Shiga toxin 1 from bovine *Escherichia coli* strains and development of bead-ELISA to detect the variant toxin. *Microbiol. Immunol.* **47:**717–725.
78. **Orth, D., K. Grif, L. B. Zimmerhackl, and R. Würzner.** 2008. Prevention and treatment of enterohemorrhagic *Escherichia coli* infections in humans. *Expert Rev. Anti Infect. Ther.* **6:**101–108.
79. **Ostroff, S. M., P. I. Tarr, M. A. Neill, J. H. Lewis, N. Hargrett-Bean, and J. M. Kobayashi.** 1989. Toxin genotypes and plasmid profiles as determinants of systemic sequelae in *Escherichia coli* O157:H7 infections. *J. Infect. Dis.* **160:**994–998.
80. **Panos, G. Z., G. I. Betsi, and M. E. Falagas.** 2006. Systematic review: are antibiotics detrimental or beneficial for the treatment of patients with *Escherichia coli* O157:H7 infection? *Aliment. Pharmacol. Ther.* **24:**731–742.
81. **Paton, A. W., T. Beddoe, C. M. Thorpe, J. C. Whisstock, M. C. J. Wilce, J. Rossjohn, U. M. Talbot, and J. C. Paton.** 2006. $AB_5$ subtilase cytotoxin inactivates the endoplasmic reticulum chaperone BiP. *Nature* **443:**548–552.
82. **Paton, A. W., and J. C. Paton.** 2005. Multiplex PCR for direct detection of Shiga toxigenic *Escherichia coli* strains producing the novel subtilase cytotoxin. *J. Clin. Microbiol.* **43:**2944–2947.
83. **Paton, A. W., P. Srimanote, U. M. Talbot, H. Wang, and J. C. Paton.** 2004. A new family of potent $AB_5$ cytotoxins produced by Shiga toxigenic *Escherichia coli. J. Exp. Med.* **200:**35–46.
84. **Paton, A. W., P. Srimanote, M. C. Woodrow, and J. C. Paton.** 2001. Characterization of Saa, a novel autoagglutinating adhesin produced by locus of enterocyte effacement-negative Shiga-toxigenic *Escherichia coli* strains that are virulent for humans. *Infect. Immun.* **69:**6999–7009.
85. **Paton, J. C., and A. W. Paton.** 2006. Shiga toxin "goes retro" in human primary kidney cells. *Kidney Int.* **70:**2049–2051.
86. **Pinyon, R. A., J. C. Paton, A. W. Paton, J. A. Botten, and R. Morona.** 2004. Refinement of a therapeutic Shiga toxin-binding probiotic for human trials. *J. Infect. Dis.* **189:**1547–1555.
87. **Potter, A. A., S. Klashinsky, Y. Li, E. Frey, H. Townsend, D. Rogan, G. Erickson, S. Hinkley, T. Klopfenstein, R. A. Moxley, D. R. Smith, and B. B. Finlay.** 2004. Decreased shedding of *Escherichia coli* O157:H7 by cattle following vaccination with type III secreted proteins. *Vaccine* **22:**362–369.
88. **Ptashne, M.** 2004. *A Genetic Switch*, 3rd ed., p. 52–61. Cold Spring Harbor Laboratory Press, Cold Spring Harbor, NY.
89. **Rangel, J. M., P. H. Sparling, C. Crowe, P. M. Griffin, and D. L. Swerdlow.** 2005. Epidemiology of *Escherichia coli* O157:H7 outbreaks, United States, 1982–2002. *Emerg. Infect. Dis.* **11:**603–609.
90. **Repetto, H. A.** 2009. Overview of HUS in Argentina, abstr. S01.1. *In Abstracts of the 7th International Symposium on Shiga Toxin (Verocytotoxin)-Producing* Escherichia coli *Infections.* Buenos Aires, Argentina.
91. **Ribot, E. M., M. A. Fair, R. Gautom, D. N. Cameron, S. B. Hunter, B. Swaminathan, and T. J. Barrett.** 2006. Standardization of pulsed-field gel electrophoresis protocols for the subtyping of *Escherichia coli* O157:H7, *Salmonella*, and *Shigella* for PulseNet. *Foodborne Pathog. Dis.* **3:**59–67.
92. **Riley, L. W., R. S. Remis, S. D. Helgerson, H. B. McGee, J. G. Wells, B. R. Davis, R. J. Hebert, E. S. Olcott, L. M. Johnson, N. T. Hargrett, P. A. Blake, and M. L.**

**Cohen.** 1983. Hemorrhagic colitis associated with a rare *Escherichia coli* serotype. *N. Engl. J. Med.* **308:**681–685.

93. **Robins-Browne, R. M., E. Elliott, and P. Desmarchelier.** 1998. Shiga toxin-producing *Escherichia coli* in Australia, p. 66–72. *In* J. B. Kaper and A. D. O'Brien (ed.), Escherichia coli *O157:H7 and Other Shiga Toxin-Producing* E. coli *Strains.* ASM Press, Washington, DC.
94. **Robinson, C. M., J. F. Sinclair, M. J. Smith, and A. D. O'Brien.** 2006. Shiga toxin of enterohemorrhagic *Escherichia coli* type O157:H7 promotes intestinal colonization. *Proc. Natl. Acad. Sci. USA* **103:**9667–9672.
95. **Rogers, T. J., and J. C. Paton.** 2009. Therapeutic strategies for Shiga toxin-producing *Escherichia coli* infections. *Expert Rev. Anti Infect. Ther.* **7:**683–686.
96. **Sandvig, K.** 2001. Shiga toxins. *Toxicon* **39:**1629–1635.
97. **Sandvig, K., O. Garred, K. Prydz, J. V. Kozlov, S. H. Hansen, and B. van Deurs.** 1992. Retrograde transport of endocytosed Shiga toxin to the endoplasmic reticulum. *Nature* **358:**510–512.
98. **Sargeant, J. M., M. R. Amezcua, A. Rajic, and L. Waddell.** 2007. Pre-harvest interventions to reduce the shedding of *E. coli* O157 in the faeces of weaned domestic ruminants: a systematic review. *Zoonoses Public Health* **54:**260–277.
99. **Scheiring, J., S. P. Andreoli, and L. B. Zimmerhackl.** 2008. Treatment and outcome of Shiga-toxin-associated hemolytic uremic syndrome (HUS). *Pediatr. Nephrol.* **23:**1749–1760.
100. **Schuetz, F.** 1997. Ph.D thesis. Statens Serum Institut, Copenhagen, Denmark.
101. **Scotland, S. M., G. A. Willshaw, H. R. Smith, and B. Rowe.** 1987. Properties of strains of *Escherichia coli* belonging to serogroup O157 with special reference to production of Vero cytotoxins VT1 and VT2. *Epidemiol. Infect.* **99:**613–624.
102. **Serna, A., and E. C. Boedeker.** 2008. Pathogenesis and treatment of Shiga toxin-producing *Escherichia coli* infections. *Curr. Opin. Gastroenterol.* **24:**38–47.
103. **Seto, E. Y., J. A. Soller, and J. M. Colford, Jr.** 2007. Strategies to reduce person-to-person transmission during widespread *Escherichia coli* O157:H7 outbreak. *Emerg. Infect. Dis.* **13:**860–866.
104. **Sheng, H., J. Y. Lim, H. J. Knecht, J. Li, and C. J. Hovde.** 2006. Role of *Escherichia coli* O157:H7 virulence factors in colonization at the bovine terminal rectal mucosa. *Infect. Immun.* **74:**4685–4693.
105. **Shiomi, M., M. Togawa, K. Fujita, and R. Murata.** 1999. Effect of early oral fluoroquinolones in hemorrhagic colitis due to *Escherichia coli* O157:H7. *Pediatr. Int.* **41:**228–232.
106. **Smith, D. R., R. A. Moxley, R. E. Peterson, T. J. Klopfenstein, G. E. Erickson, G. Bretschneider, E. M. Berberov, and S. Clowser.** 2009. A two-dose regimen of a vaccine against type III secreted proteins reduced *Escherichia coli* O157:H7 colonization of the terminal rectum in beef cattle in commercial feedlots. *Foodborne Pathog. Dis.* **6:**155–161.
107. **Smith, H. W., and M. A. Linggood.** 1971. The transmissible nature of enterotoxin production in a human enteropathogenic strain of *Escherichia coli. J. Med. Microbiol.* **4:**301–305.
108. **Smith, M. J., L. D. Teel, H. M. Carvalho, A. R. Melton-Celsa, and A. D. O'Brien.** 2006. Development of a hybrid Shiga holotoxoid vaccine to elicit heterologous protection against Shiga toxins types 1 and 2. *Vaccine* **24:**4122–4129.
109. **Snedeker, K. G., D. J. Shaw, M. E. Locking, and R. J. Prescott.** 2009. Primary and secondary cases in *Escherichia coli* O157 outbreaks: a statistical analysis. *BMC Infect. Dis.* **9:**144.
110. **Sung, L. M., M. P. Jackson, A. D. O'Brien, and R. K. Holmes.** 1990. Transcription of the Shiga-like toxin type II and Shiga-like toxin type II variant operons of *Escherichia coli. J. Bacteriol.* **172:**6386–6395.
111. **Teel, L. D., A. R. Melton-Celsa, C. K. Schmitt, and A. D. O'Brien.** 2002. One of two copies of the gene for the activatable Shiga toxin type 2d in *Escherichia coli* O91:

H21 strain B2F1 is associated with an inducible bacteriophage. *Infect. Immun.* **70:** 4282–4291.

112. **Tesh, V. L.** 2010. Induction of apoptosis by Shiga toxins. *Future Microbiol.* **5**:431–453.
113. **Thomson, D. U., G. H. Loneragan, A. B. Thornton, K. F. Lechtenberg, D. A. Emery, D. T. Burkhardt, and T. G. Nagaraja.** 2009. Use of a siderophore receptor and porin proteins-based vaccine to control the burden of *Escherichia coli* O157:H7 in feedlot cattle. *Foodborne Pathog. Dis.* **6**:871–877.
114. **Tilden, J., Jr., W. Young, A. M. McNamara, C. Custer, B. Boesel, M. A. Lambert-Fair, J. Majkowski, D. Vugia, S. B. Werner, J. Hollingsworth, and J. G. Morris, Jr.** 1996. A new route of transmission for *Escherichia coli*: infection from dry fermented salami. *Am. J. Public Health* **86:**1142–1145.
115. **Trachtman, H., A. Cnaan, E. Christen, K. Gibbs, S. Zhao, D. W. Acheson, R. Weiss, F. J. Kaskel, A. Spitzer, and G. H. Hirschman.** 2003. Effect of an oral Shiga toxin-binding agent on diarrhea-associated hemolytic uremic syndrome in children: a randomized controlled trial. *JAMA* **290:**1337–1344.
116. **Tyler, J. S., M. J. Mills, and D. I. Friedman.** 2004. The operator and early promoter region of the Shiga toxin type 2-encoding bacteriophage 933W and control of toxin expression. *J. Bacteriol.* **186:**7670–7679.
117. **Tzipori, S., H. Karch, K. I. Wachsmuth, R. M. Robins-Browne, A. D. O'Brien, H. Lior, M. L. Cohen, J. Smithers, and M. M. Levine.** 1987. Role of a 60-megadalton plasmid and Shiga-like toxins in the pathogenesis of infection caused by enterohemorrhagic *Escherichia coli* O157:H7 in gnotobiotic piglets. *Infect. Immun.* **55:**3117–3125.
118. **Tzipori, S., A. Sheoran, D. Akiyoshi, A. Donohue-Rolfe, and H. Trachtman.** 2004. Antibody therapy in the management of Shiga toxin-induced hemolytic uremic syndrome. *Clin. Microbiol. Rev.* **17:**926–941.
119. **Unkmeir, A., and H. Schmidt.** 2000. Structural analysis of phage-borne *stx* genes and their flanking sequences in Shiga toxin-producing *Escherichia coli* and *Shigella dysenteriae* type 1 strains. *Infect. Immun.* **68:**4856–4864.
120. **Voetsch, A. C., F. J. Angulo, T. Rabatsky-Ehr, S. Shallow, M. Cassidy, S. M. Thomas, E. Swanson, S. M. Zansky, M. A. Hawkins, T. F. Jones, P. J. Shillam, T. J. Van Gilder, J. G. Wells, and P. M. Griffin.** 2004. Laboratory practices for stool-specimen culture for bacterial pathogens, including *Escherichia coli* O157:H7, in the FoodNet sites, 1995–2000. *Clin. Infect. Dis.* **38**(Suppl. 3)**:**S190–S197.
121. **Wadolkowski, E. A., L. M. Sung, J. A. Burris, J. E. Samuel, and A. D. O'Brien.** 1990. Acute renal tubular necrosis and death of mice orally infected with *Escherichia coli* strains that produce Shiga-like toxin type II. *Infect. Immun.* **58:**3959–3965.
122. **Wagner, P. L., D. W. Acheson, and M. K. Waldor.** 1999. Isogenic lysogens of diverse Shiga toxin 2-encoding bacteriophages produce markedly different amounts of Shiga toxin. *Infect. Immun.* **67:**6710–6714.
123. **Wagner, P. L., J. Livny, M. N. Neely, D. W. Acheson, D. I. Friedman, and M. K. Waldor.** 2002. Bacteriophage control of Shiga toxin 1 production and release by *Escherichia coli. Mol. Microbiol.* **44:**957–970.
124. **Wagner, P. L., M. N. Neely, X. Zhang, D. W. Acheson, M. K. Waldor, and D. I. Friedman.** 2001. Role for a phage promoter in Shiga toxin 2 expression from a pathogenic *Escherichia coli* strain. *J. Bacteriol.* **183:**2081–2085.
125. **Wang, I. N., D. L. Smith, and R. Young.** 2000. Holins: the protein clocks of bacteriophage infections. *Annu. Rev. Microbiol.* **54:**799–825.
126. **Watanabe-Takahashi, M., T. Sato, T. Dohi, N. Noguchi, F. Kano, M. Murata, T. Hamabata, Y. Natori, and K. Nishikawa.** 2010. An orally applicable Shiga toxin neutralizer functions in the intestine to inhibit the intracellular transport of the toxin. *Infect. Immun.* **78:**177–183.
127. **Werber, D., B. W. Mason, M. R. Evans, and R. L. Salmon.** 2008. Preventing household transmission of Shiga toxin-producing *Escherichia coli* O157 infection: promptly separating siblings might be the key. *Clin. Infect. Dis.* **46:**1189–1196.

128. **Whittam, T. S.** 1998. Evolution of *Escherichia coli* O157:H7 and other Shiga toxin-producing *E. coli* strains, p. 195–209. *In* J. B. Kaper and A. D. O'Brien (ed.), Escherichia coli *O157:H7 and Other Shiga Toxin-Producing* E. coli *Strains.* ASM Press, Washington, DC.
129. **Whittam, T. S., I. K. Wachsmuth, and R. A. Wilson.** 1988. Genetic evidence of clonal descent of *Escherichia coli* O157:H7 associated with hemorrhagic colitis and hemolytic uremic syndrome. *J. Infect. Dis.* **157:**1124–1133.
130. **Whittam, T. S., M. L. Wolfe, I. K. Wachsmuth, F. Ørskov, I. Ørskov, and R. A. Wilson.** 1993. Clonal relationships among *Escherichia coli* strains that cause hemorrhagic colitis and infantile diarrhea. *Infect. Immun.* **61:**1619–1629.
131. **Wong, C. S., S. Jelacic, R. L. Habeeb, S. L. Watkins, and P. I. Tarr.** 2000. The risk of the hemolytic-uremic syndrome after antibiotic treatment of *Escherichia coli* O157:H7 infections. *N. Engl. J. Med.* **342:**1930–1936.
132. **Yamagami, S., M. Motoki, T. Kimura, H. Izumi, T. Takeda, Y. Katsuura, and Y. Matsumoto.** 2001. Efficacy of postinfection treatment with anti-Shiga toxin (Stx) 2 humanized monoclonal antibody TMA-15 in mice lethally challenged with Stx-producing *Escherichia coli. J. Infect. Dis.* **184:**738–742.
133. **Zhang, Q., A. Donohue-Rolfe, G. Krautz-Peterson, M. Sevo, N. Parry, C. Abeijon, and S. Tzipori.** 2009. Gnotobiotic piglet infection model for evaluating the safe use of antibiotics against *Escherichia coli* O157:H7 infection. *J. Infect. Dis.* **199:**486–493.
134. **Zhang, W. L., M. Bielaszewska, A. Liesegang, H. Tschäpe, H. Schmidt, M. Bitzan, and H. Karch.** 2000. Molecular characteristics and epidemiological significance of Shiga toxin-producing *Escherichia coli* O26 strains. *J. Clin. Microbiol.* **38:**2134–2140.
135. **Zhang, X., A. D. McDaniel, L. E. Wolf, G. T. Keusch, M. K. Waldor, and D. W. Acheson.** 2000. Quinolone antibiotics induce Shiga toxin-encoding bacteriophages, toxin production, and death in mice. *J. Infect. Dis.* **181:**664–670.

*Section IV*

# THE EVOLUTIONARY ECOLOGY OF ENTEROHEMORRHAGIC *ESCHERICHIA COLI*

*Population Genetics of Bacteria: a Tribute to Thomas S. Whittam*
Edited by S. T. Walk and P. C. H. Feng


*Chapter 13*

# The Evolutionary Model of *Escherichia coli* O157:H7

***David W. Lacher***

## THE ORIGINAL STEPWISE MODEL

The first model for the evolution of *Escherichia coli* O157:H7 was proposed over a decade ago (3). This model was based on data from multilocus enzyme electrophoresis, sorbitol fermentation, $\beta$-glucuronidase activity, and the presence/absence of the genes encoding the two main antigenic forms of Shiga toxin (Stx): $stx_1$ and $stx_2$. It proposed how, through a series of stepwise events, strains of the O55:H7 serotype evolved into O157:H7 (Fig. 1). The evolutionary steps displayed by the model progress from left to right in the figure, starting with the ancestral states and ending with the derived ones. Since all known close relatives of O157:H7 possess the locus of enterocyte effacement (LEE), the model begins with an ancestral strain containing this island (i.e., a large stretch of DNA that is composed of genes arranged into operons) inserted at the *selC* tRNA locus, as found in extant O157:H7 and O55:H7 isolates. This ancestral strain also possesses the typical *E. coli* traits of being able to express $\beta$-glucuronidase (GUD$^+$) and ferment sorbitol (SOR$^+$). From this hypothetical ancestor evolved the immediate ancestor of O157:H7, although its somatic antigen (O55) remained different (A1 in Fig. 1). The next step in the model involves the acquisition of $stx_2$ via phage transduction, resulting in a Stx2-producing O55:H7 strain (A2). Then an antigenic shift from O55 to O157 occurred via lateral transfer of the *rfb* region (2) along with the independent acquisition of the enterohemorrhagic *E. coli* (EHEC) plasmid (9) to generate the hypothetical intermediate A3. A strain with this particular combination of characteristics has yet to be identified and, therefore, represents a "missing link" in the model. From A3, two distinct evolutionary lineages emerged. In one lineage, motility was lost via a 12-bp deletion in the flagellar regulatory

***David W. Lacher*** • Division of Molecular Biology, Office of Applied Research and Safety Assessment, Center for Food Safety and Applied Nutrition, U.S. Food and Drug Administration, Laurel, MD 20708.

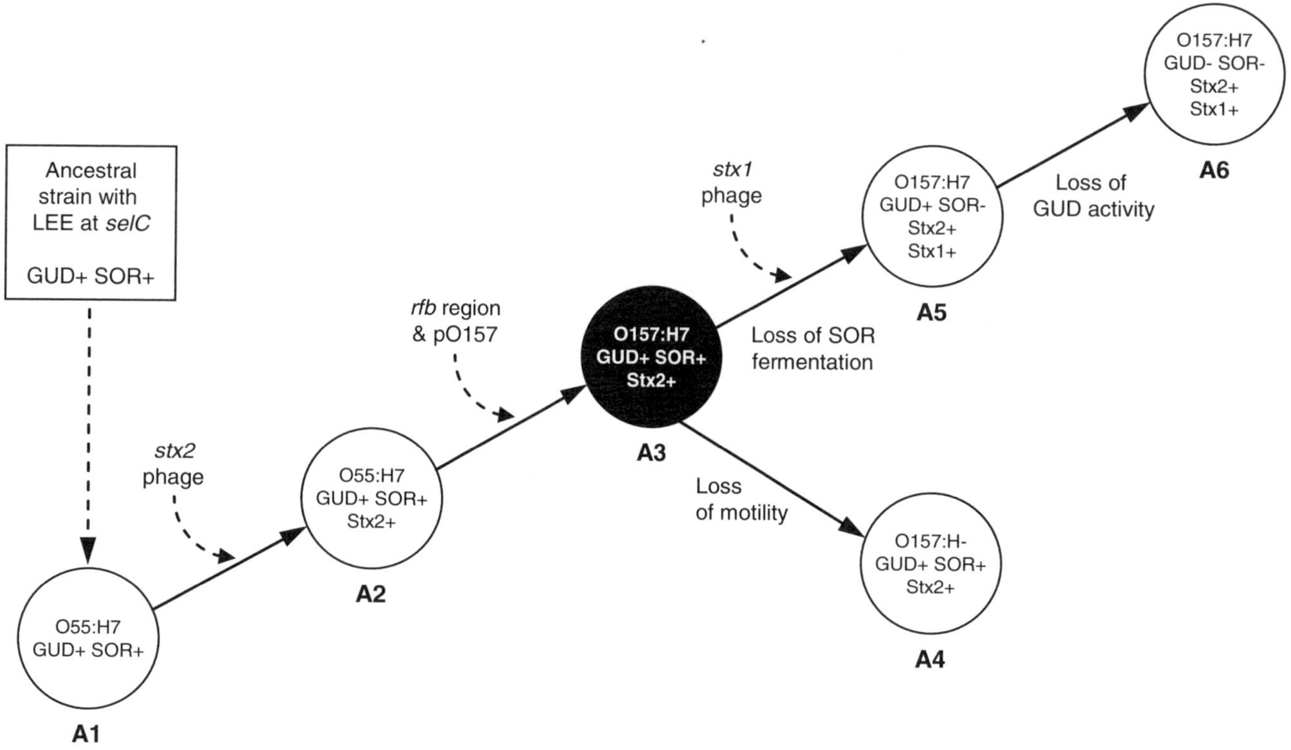


**Figure 1.** The original stepwise model for the evolution of O157:H7. This model is based on multilocus enzyme electrophoretic profiles, GUD and SOR phenotypes, and Stx production. Phenotypes of ancestors A1 to A6 are shown. A strain with the traits of ancestor A3 (black circle) has not been reported.

gene *flhC* (12), resulting in the SOR$^+$ O157:H$^-$ strains (A4) commonly referred to as the "German clone" because representative isolates were identified from an outbreak of hemolytic uremic syndrome in Germany (5–7). Along the other lineage, $stx_1$ was acquired by phage transduction and the ability to ferment sorbitol was lost, resulting in the emergence of GUD$^+$ SOR$^-$ strains that carry both Stx types (A5). In the final step, GUD activity was lost via a guanine dinucleotide insertion frameshift in *uidA* (13), producing the immediate ancestor (A6) of the typical O157:H7 clone. More details on the original stepwise model can be found in the introductory section of chapter 14.

## SNP CLADES OF O157:H7

Multilocus sequence typing (MLST) has been used to describe the population structure of numerous bacterial species. However, since only a small fraction of the genome is interrogated, typically 3 to 4 kb of concatenated sequence, MLST is limited in its ability to differentiate closely related isolates (i.e., those belonging to the same clonal group or serotype). For example, examination of the *Ec*MLST database (15) has revealed that 97% (271/280) of GUD$^-$ SOR$^-$ O157:H7 strains are the same sequence type (ST66), including the genome strains Sakai, EDL933, EC4115, and TW14359. Despite this finding, the *Ec*MLST system is still able to separate the major GUD and SOR phenotypic subtypes of the O157:H7 clonal complex, including O55:H7. In another *E. coli* MLST scheme, however, O157:H7 and O55:H7 strains are indistinguishable (20).

In order to address the limited resolution of MLST, a panel of single nucleotide polymorphisms (SNPs) was developed to subtype O157:H7 at the molecular level. Informative SNPs were identified from data generated by MLST, virulence gene sequencing (*eae, espA*, and *fimA*), comparative genome sequencing microarrays (21), and in silico comparisons of the Sakai and EDL933 genomes. Using this SNP-typing system, nine distinct clades of O157:H7 have been described (11). The genotypes identified through SNP analysis allow the stepwise model to be expanded. In the original analysis, a phylogenetic network was created using 48 parsimoniously informative SNPs. This network indicated a central group of four clades (3, 4, 5, and 7) connected by multiple paths, suggesting that either recombination or parallel mutation played a role in their diversification. This situation contrasts with that of the remaining clades (1, 2, 6, 8, and 9), which occur at the end of distinct branches.

Reanalysis of the data using all 96 SNPs described by Manning et al. (11) revealed a more detailed set of relationships among the genotypes (Fig. 2A). Much of the conflicting phylogenetic signal appears to be due to the clade 5 genotypes 21 and 22, as well as the nonclade genotype 27. Closer inspection of the strains possessing these genotypes has shown them to be mixed cultures with one component being an O157:H7 isolate and the other an unrelated *E. coli* (D. W. Lacher, unpublished data). When these three genotypes are removed from the network analysis, almost all of the conflicting signal is eliminated, resulting in a diagram that more closely resembles a bifurcating tree (Fig. 2B).

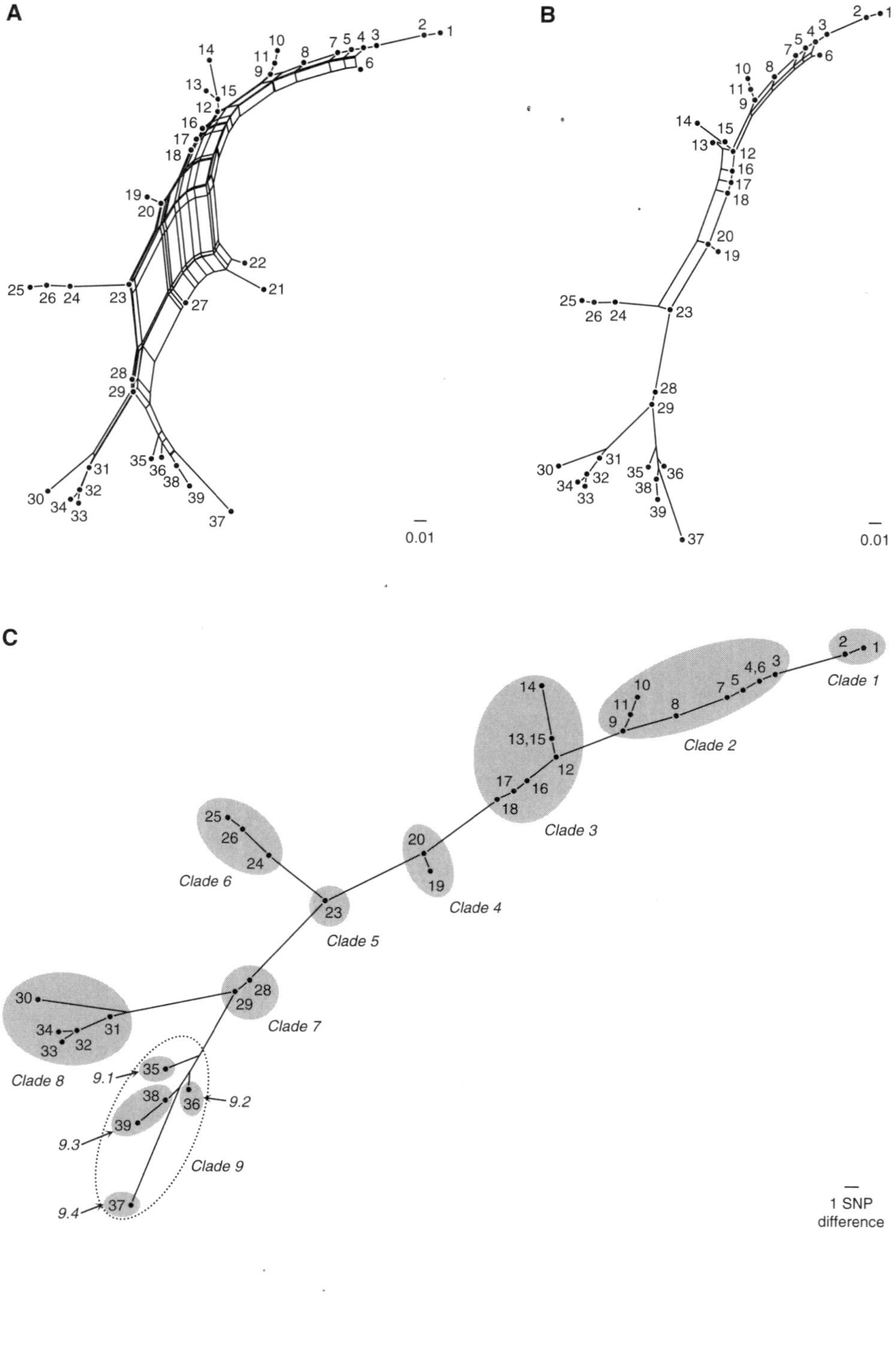
A
B
0.01
0.01
C
Clade 1
Clade 2
Clade 3
Clade 4
Clade 5
Clade 6
Clade 7
Clade 8
Clade 9
9.1
9.2
9.3
9.4
1 SNP difference

Further refinement of the data by the exclusion of seven SNP loci (three with incompatible phylogenetic signal, two with incomplete data, and two that are invariable among the 39 genotypes) results in a simple bifurcating tree on which a more detailed stepwise model can be based (Fig. 2C).

After the removal of the original clade 5 strains, genotype 23 was reclassified as the new clade 5 since it is equidistant from clades 4, 6, and 7. In addition, clade 9 has now been subdivided to reflect the identified phenotypic variants within this ancestral group. The $GUD^+$ $SOR^-$ O157:H7 strains belong to clade 9.1, while the $GUD^+$ $SOR^+$ O157:$H^-$ strains are clade 9.2. Clade 9.3 contains the O55:H7 strains, and the lone representative of clade 9.4 is ECOR 37, an $O^-$:$H^-$ member of the *E. coli* Reference Collection that has been shown to be a divergent O55:H7 isolate (4).

## Stx PROFILES

With the higher-resolution, SNP-based phylogenetic framework in place, the stepwise model can be expanded further through the examination of the Stx profiles present within the clades. In the original model, only the two main antigenic variants of Stx were described: $stx_1$ and $stx_2$. Some strains of O157:H7, however, have been shown to possess two alleles of $stx_2$, commonly referred to as $stx_2$ and $stx_{2c}$ (11, 18). To avoid confusion between the $stx_2$ allele and the $stx_2$ family, the standard $stx_2$ allele found in EDL933 and Sakai will be referred to as $stx_{2'}$, while $stx_2$ will be used as a general designation for the entire family of alleles. Eight different *stx* profiles were observed after examination of a diverse set of isolates, including the data used by Manning et al. (11) (Table 1).

Additional clade 9 representatives from a previous study (4) were reexamined using the $stx_2$ PCR-restriction fragment length polymorphism assay described by Manning et al. (11). Of the two previously reported $stx_2$-positive O55:H7 isolates, strain 5905 (the A2 representative in the original model) possesses $stx_{2c}$, while strain 3256-97 contains the $stx_{2'}$ allele. The $stx_{2c}$ allele was also found in all of the $GUD^+$ $SOR^-$ O157:H7 isolates reexamined, including the type strain G5101 (the A5 representative in the original model). As for the $stx_2$-positive $GUD^+$ $SOR^+$ O157:$H^-$ isolates, all possess the $stx_{2'}$ allele. Given the differing $stx_2$ allelic profiles of the neighboring subtypes, there is now doubt

**Figure 2.** Phylogenetic relationships among SNP genotypes. Nodes are labeled with their corresponding SNP genotype (SG). (A) Phylogenetic network generated using the neighbor-net algorithm with 96 SNPs and 39 SGs from Manning et al. (11). (B) Phylogenetic network generated using the neighbor-net algorithm with 96 SNPs and 36 SGs from Manning et al. (11). Genotypes 21, 22, and 27 were excluded from this analysis since the isolates were found to be mixed cultures. (C) Bifurcating tree generated using the neighbor-net algorithm with 89 SNPs and 36 SGs from Manning et al. (11). The nine previously described clades are marked by the shaded ellipses. SG23 has been reclassified as the new clade 5 based on the number of SNP differences between it and neighboring SGs. Clade 9 has been subdivided into its major phenotypic variants; see text for details.

**Table 1.** Stx profiles observed among the clades of O157:H7 and its close relatives

| Clade | Stx profile | | | | | | | |
|---|---|---|---|---|---|---|---|---|
| | *stx*$_1$ | *stx*$_{2'}$ | *stx*$_{2c}$ | *stx*$_1$ and *stx*$_{2'}$ | *stx*$_1$ and *stx*$_{2c}$ | *stx*$_{2'}$ and *stx*$_{2c}$ | *stx*$_1$, *stx*$_{2'}$, and *stx*$_{2c}$ | *stx* negative |
| 1 | | | | + | | | | |
| 2 | | + | | + | | | | |
| 3 | | + | | + | | | | |
| 4 | | | | | | | + | |
| 5 | | | + | | | + | | + |
| 6 | + | | + | | + | | | |
| 7 | | | + | | + | + | + | |
| 8 | | + | | | | + | | |
| 9.1 | | | + | | + | | | |
| 9.2 | | + | | | | | | + |
| 9.3 | | + | + | | | | | + |
| 9.4 | | | | | | | | + |

as to the existence of the hypothetical A3 intermediate "missing link" (see next section).

In the eight clades containing all of the GUD$^-$ SOR$^-$ isolates, the number of *stx* profiles ranged from one in clades 1 and 4 to as high as four in clade 7 (Table 1). The *stx*$_{2'}$ allele has the broadest distribution of the three Stx variants, being absent from only clades 6, 9.1, and 9.4. The distribution of *stx*$_{2c}$ is more restricted than that of *stx*$_{2'}$ in that it is completely absent from the most derived clades (1 through 3). An opposite trend is seen in the distribution of *stx*$_1$, which tends to be found in the more derived lineages (e.g., clades 1 through 4), but absent from the more basal ones (e.g., clades 9.2 through 9.4). A more detailed discussion of the Stx variants in relation to the clades can be found in chapter 17.

## AN UPDATED STEPWISE MODEL

An expanded *E. coli* O157:H7 evolution model is proposed here based on the same three assumptions as the original stepwise model: (i) that during divergence, the probability of loss of function greatly exceeds gain of function for metabolic genes; (ii) that the gain of function usually occurs via lateral transfer of genes; and (iii) that the sequence of events invoking the fewest total changes (i.e., most parsimonious) is the preferred model.

As with the original model, the revised one begins with an ancestral strain that is GUD$^+$ and SOR$^+$ with the LEE island integrated into the *selC* chromosomal locus (Fig. 3). From this strain evolved the *stx*-negative O55:H7 clone represented by A1 in the original model, now labeled as clade 9.3a. Four different lines then arose from this clade 9.3a progenitor. In one, the expression of both the somatic O and flagellar H antigens was lost, resulting in clade 9.4a, represented by strain ECOR 37. In the second, a *stx*$_2$-negative O55:H7 strain

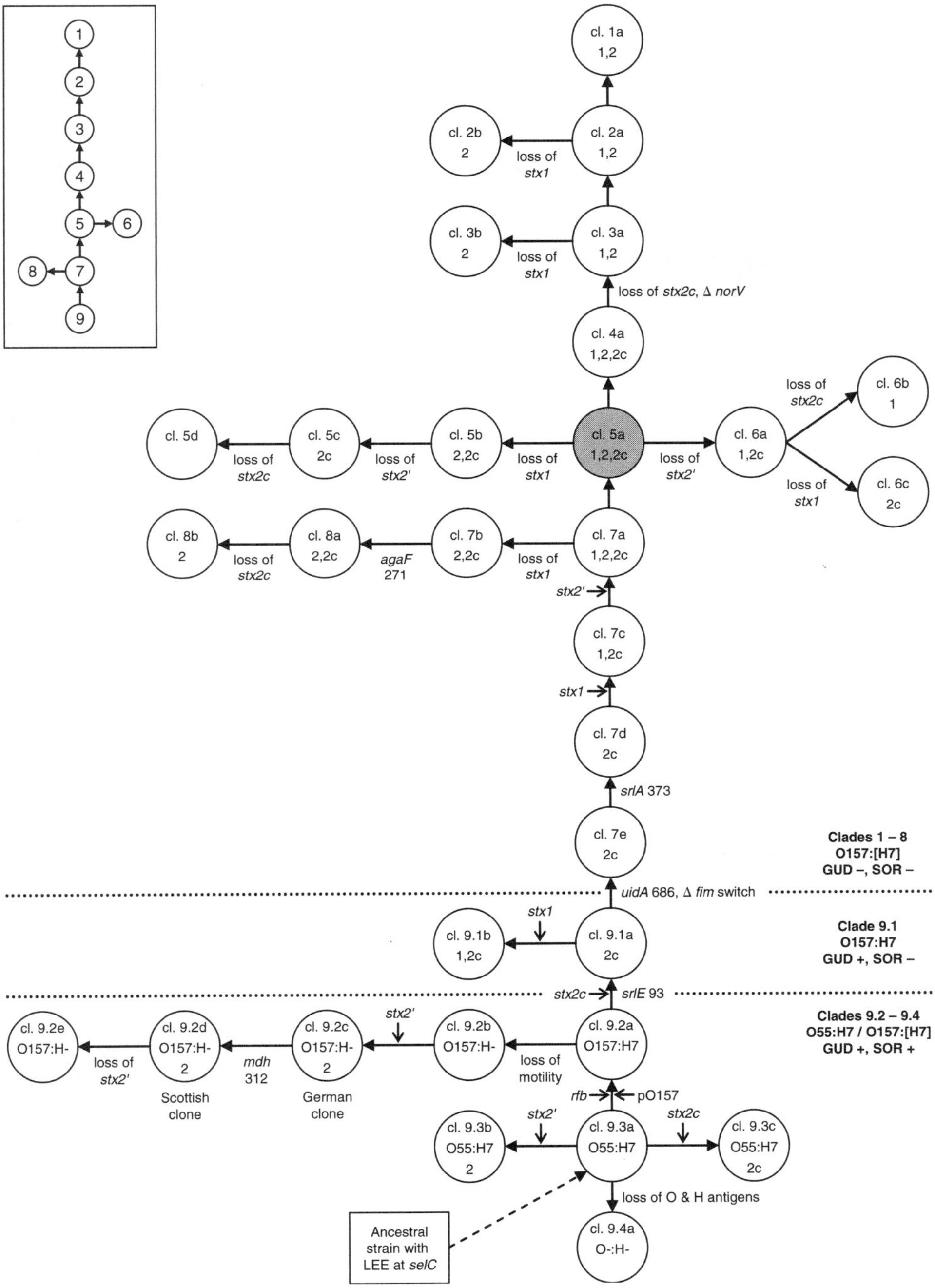


**Figure 3.** Expanded model for the stepwise evolution of O157:H7. Phenotypic and genotypic changes predicted to have occurred are indicated. A strain with the traits of clade 5a (shaded circle) has not been reported, but is predicted by the model. Abbreviations: 1, $stx_1$; 2, $stx_{2'}$; 2c, $stx_{2c}$; cl, clade. (Inset) Simplified view of clade evolution.

was converted into a $stx_{2'}$-positive one (strain 3256-97), presumably through phage transduction. A similar process occurred in a third line, except with a $stx_{2c}$-converting bacteriophage (ancestor A2 in the original model, strain 5905). In the fourth line, the antigenic shift from O55 to O157 occurred as well as the acquisition of the pO157 EHEC plasmid. This position in the model (clade 9.2a) is similar to that of A3 in the original, with the exception of clade 9.2a being *stx* negative. While a strain with the characteristics of A3 in the original model still has not been reported, a strain fitting the description of clade 9.2a has been identified. This strain, designated LSU-61, has all of the characteristics of clade 9.2a including a SNP genotype that is intermediate to the genotypes found in strains previously classified as A4 and A5. From this point, the two main lineages present in the original model (A4 and A5/A6) arose. In one lineage, as in the original model, motility was lost via a 12-bp deletion in *flhC*. The $stx_{2'}$ allele was then acquired by one of these clade 9.2b isolates, resulting in the typical SOR$^+$ "German clone" (clade 9.2c, formerly ancestor A4), which then diversified into the "Scottish clone" (clade 9.2d), as measured by an A→G transition at position 312 in the housekeeping gene *mdh* (4). These isolates have been named the "Scottish clone" because of their isolation from a patient with hemolytic uremic syndrome in Scotland in 2002, the first report of sorbitol-fermenting O157 in the United Kingdom (17) and only the second report outside of continental Europe (1). Some of the Scottish clone isolates described by Feng et al. (4) then lost $stx_{2'}$, giving rise to the clade 9.2e subtype. Atypical lineages are discussed in detail by Feng and Monday in chapter 14.

The other lineage to emerge from clade 9.2a began with the acquisition of $stx_{2c}$ and the loss of the ability to ferment sorbitol. The exact cause of the SOR$^-$ phenotype has not been determined, but comparison of the *srl* operon sequence from the Sakai O157:H7 strain to K-12 MG1655 identified two frameshift mutations, one in *srlA* and the other in *srlE*. The *srlE* mutation, a cytosine dinucleotide insertion at position 93, appears to have occurred first since GUD$^+$ SOR$^-$ O157:H7 strains possess this mutation, but not the *srlA* frameshift, which is only found in GUD$^-$ SOR$^-$ strains (see below). The GUD$^+$ SOR$^-$ O157:H7 strains that produce Stx2c are labeled as clade 9.1a (Fig. 3). A variant of this subtype then gained $stx_1$, resulting in clade 9.1b, which is represented by the ancestor A5 strain G5101 in the original model. From clade 9.1a arose the progenitor of the widespread GUD$^-$ SOR$^-$ O157:H7 clone. As mentioned previously, GUD activity was lost via a guanine dinucleotide insertion at position 686 of the *uidA* gene (13). A 16-bp deletion within the invertible genetic element (known as the *fim* switch) that is responsible for the phase variation phenotype of the type 1 fimbriae in *E. coli* (10, 16) also occurred at this step, resulting in the clade 7e subtype.

In a recent study, Shaikh et al. concluded that the *fim* switch deletion occurred after sorbitol-fermenting O157:H$^-$ diverged during the descent of the EHEC 1 clonal group (19). They, however, did not include any GUD$^+$ SOR$^-$ O157:H7 strains in their survey. Sequence analysis of additional isolates indicates that only GUD$^-$ SOR$^-$ O157:H7 and O157:NM strains possess the deletion, whereas GUD$^+$ SOR$^+$ O157:H$^-$ strains and GUD$^+$ SOR$^-$ O157:H7 contain an intact *fim* switch (refer to GenBank accession numbers HM467611, HM467612, and HM467613). These observations suggest that the deletion in the *fim* switch occurred soon after the loss of GUD expression.

From the point of clade 7e onward, all subtypes described by the new model are represented by ancestor A6 in the original model. The next change depicted by the model changes clade 7e into clade 7d. This mutation, the second sorbitol operon frameshift, involves the deletion of a single thymidine at position 373 of *srlA*. Clade 7c then arose by the acquisition of $stx_1$, followed by clade 7a with the acquisition of $stx_{2'}$. With clade 7a, the model once again bifurcates with one line losing $stx_1$ (clade 7b) and the other evolving into clade 5a. From clade 7b arose the first of the clade 8 subtypes, clade 8a. One of the hallmark phenotypes of clade 8 strains is the inability to utilize *N*-acetyl-D-galactosamine as either a carbon or nitrogen source because of a G→A transition at position 271 of *agaF* (14). The second clade 8 subtype, clade 8b, evolved from clade 8a with the loss of $stx_{2c}$. Clade 8 strains are of particular interest since it was determined that they were the cause of two O157:H7 outbreaks in the United States in 2006 (11). In addition, Manning et al. (11) report epidemiological data from 2002–2006 that indicate that the prevalence of clade 8 strains appears to be on the rise despite the overall decrease in cases of disease attributable to O157:H7 (see chapter 17 for more details).

The second line to evolve from clade 7a was clade 5a. Although a clade 5 strain possessing all three Stx variants ($stx_1$, $stx_{2'}$, and $stx_{2c}$) has not been reported, the combination of traits is predicted by the model based on the observed *stx* profiles of the neighboring isolates (Fig. 3). From clade 5a, the model splits into three distinct paths. In the first, the observed clade 5 isolates arose through the sequential loss of $stx_1$, $stx_{2'}$, and $stx_{2c}$ to give rise to clades 5b, 5c, and 5d, respectively. In the second path clade 6a evolved, losing $stx_{2'}$ in the process. Two additional clade 6 subtypes then developed from clade 6a: clade 6b by the loss of $stx_{2c}$ and clade 6c by the loss of $stx_1$. The third lineage to evolve from the clade 5a subtype is that of clade 4. All known clade 4 isolates possess $stx_1$, $stx_{2'}$, and $stx_{2c}$, so only one subtype (clade 4a) is depicted in the model. However, as seen with other clades, additional subtypes can easily result from the loss of any of the Stx variants.

In addition to the SNPs described by Manning et al. (11), the transition from clade 4 to clade 3 also involved the loss of $stx_{2c}$ as well as a 204-bp deletion in the anaerobic nitric oxide reductase gene *norV*. A simple PCR assay, developed for this revised stepwise model, demonstrated that the *norV* deletion is present only within strains belonging to clades 1, 2, and 3. This deletion, described in the genomic analysis of the 2006 spinach-associated outbreak isolate (8), may play a role in decreasing the virulence potential of strains in clades 1, 2, and 3. Kulasekara et al. hypothesize that the deletion may reduce colonization or persistence within the large intestine, while also possibly lowering Stx2 expression (8). The first step in the revised model to display the $stx_1^+$ $stx_{2'}^+$ $stx_{2c}^-$ $\Delta norV$ genotype is that of clade 3a, represented by strain EDL933. The remaining steps in the model involve either the loss of $stx_1$ as with clade

3a to 3b and with clade 2a to 2b, or additional SNP changes (clade 3 to 2 to 1). The final step in the model, clade 1a, is represented by the Sakai strain.

## THE REPEATED EXPANSION AND CONTRACTION OF *tolA*

Another insertion/deletion (indel) mutation mentioned by Kulasekara et al. in the analysis of the clade 8a spinach strain TW14359 is within the inner membrane protein-encoding gene *tolA* (8). They observed that the gene is 90 bp longer in TW14359 compared to both EDL933 and Sakai. Subsequent analysis of the *tolA* sequences from these three strains placed the indel within a tandem repeat-like region of the gene. As with *norV*, a PCR assay was developed to determine the placement of the *tolA* indel within the updated stepwise model. Sequence analysis uncovered five indel variants within the tandem repeat-like region (Fig. 4; refer to GenBank accession numbers HM467614 through HM467618). The repeats are 15 bp in length and range from a minimum of 10 copies in EDL933 and Sakai to 16 copies in TW14359. All of the repeat units encode two to four alanine residues, while lysine and glutamic acid residues are also common (Table 2). By using the revised stepwise model as a guide, the evolutionary history of this region of *tolA* can be discerned. Strains within clades 9.3 and 9.4 contain 15 repeat units, labeled A through O in Fig. 4. From this configuration, a second *tolA* allele found in clade 9.3 arose via deletion of the D and E units. Subsequently, units A, B, and C were deleted to create the allele found in clades 9.1 and 9.2. Following this deletion, units F

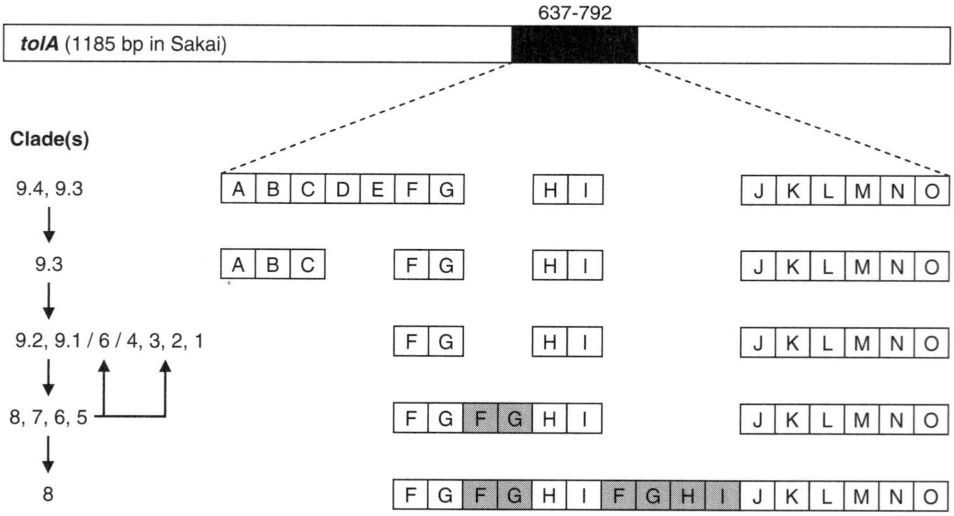


**Figure 4.** Alignment of *tolA* alleles with differing numbers of repeats. The 15-bp repeats are located between positions 637 and 792 of the Sakai allele and are labeled A through O. Repeats involved in duplication events are shaded. The proposed directionality of the observed expansion and contraction is indicated.

**Table 2.** Nucleotide and amino acid sequences of *tolA* repeats

| Repeat designation | Amino acid sequence | DNA sequence |
|---|---|---|
| A | TEAAE | ACT GAA GCT GCT GAA |
| B | KAKAE | AAA GCC AAG GCA GAA |
| C | AEKKA | GCT GAG AAG AAA GCG |
| D | AEKAA | GCT GAA AAG GCT GCA |
| F | AEKAA | GCA GAG AAA GCT GCA |
| H | AEKAA | GCA GAA AAA GCG GCT |
| I | AEKAA | GCT GAA AAG GCA GCA |
| K | AEKAA | GCA GAA AAA GCC GCC |
| N | AEKAA | GCT GAA AAA GCC GCT |
| E | ADKKA | GCT GAT AAG AAA GCG |
| G | ADKKA | GCC GAC AAA AAA GCA |
| J | ADKKA | GCT GAT AAG AAA GCA |
| L | ADKKA | GCA GAC AAA AAA GCG |
| M | AAKAA | GCT GCA AAA GCA GCA |
| O | AAKAA | GCA GCA AAA GCT GCC |

and G were duplicated in the clade 7 progenitor and retained in clades 5, 6, and 8. The clade 8 variant allele found in TW14359 then arose through the duplication of the F, G, H, and I series of repeat units. The allele found in clades 9.1 and 9.2 was restored by the deletion of the duplicated F and G repeats on at least two independent occasions—once in a clade 6 isolate and a second time in the immediate ancestor of clades 1 through 4. The observed changes in the *tolA* locus emphasize the utility of the revised model to elucidate the complex relationships among otherwise closely related isolates.

## SUMMARY

The updated stepwise evolution model presented here highlights the dynamic nature of the phage-encoded Stx variants, as each was independently acquired and lost multiple times during the diversification of O157:H7. In contrast, the losses of sorbitol fermentation, $\beta$-glucuronidase activity, type 1 fimbrial expression, and *N*-acetyl-D-galactosamine utilization, as well as a 204-bp deletion in *norV*, each occurred only once. Also, the revised model offers a more extensive and precise classification system for this important bacterial pathogen. Finally, this model reflects a collective effort on the part of multiple investigators to describe the nature, population genetic diversity, and emergence of pathogenic bacteria.

## REFERENCES

1. **Bettelheim, K. A., M. Whipp, S. P. Djordjevic, and V. Ramachandran.** 2002. First isolation outside Europe of sorbitol-fermenting verocytotoxigenic *Escherichia coli* (VTEC) belonging to O group O157. *J. Med. Microbiol.* **51:**713–714.

2. **Bilge, S. S., J. C. Vary, Jr., S. F. Dowell, and P. I. Tarr.** 1996. Role of the *Escherichia coli* O157:H7 O side chain in adherence and analysis of an *rfb* locus. *Infect. Immun.* **64:**4795–4801.
3. **Feng, P., K. A. Lampel, H. Karch, and T. S. Whittam.** 1998. Genotypic and phenotypic changes in the emergence of *Escherichia coli* O157:H7. *J. Infect. Dis.* **177:**1750–1753.
4. **Feng, P. C., S. R. Monday, D. W. Lacher, L. Allison, A. Siitonen, C. Keys, M. Eklund, H. Nagano, H. Karch, J. Keen, and T. S. Whittam.** 2007. Genetic diversity among clonal lineages within *Escherichia coli* O157:H7 stepwise evolutionary model. *Emerg. Infect. Dis.* **13:**1701–1706.
5. **Gunzer, F., H. Bohm, H. Russmann, M. Bitzan, S. Aleksic, and H. Karch.** 1992. Molecular detection of sorbitol-fermenting *Escherichia coli* O157 in patients with hemolytic-uremic syndrome. *J. Clin. Microbiol.* **30:**1807–1810.
6. **Karch, H., H. Böhm, H. Schmidt, F. Gunzer, S. Aleksic, and J. Heesemann.** 1993. Clonal structure and pathogenicity of Shiga-like toxin-producing, sorbitol-fermenting *Escherichia coli* O157:H$^-$. *J. Clin. Microbiol.* **31:**1200–1205.
7. **Karch, H., R. Wiss, H. Gloning, P. Emmrich, S. Aleksic, and J. Bockemühl.** 1990. Hemolytic-uremic syndrome in infants due to verotoxin-producing *Escherichia coli. Dtsch. Med. Wochenschr.* **115:**489–495. (In German.)
8. **Kulasekara, B. R., M. Jacobs, Y. Zhou, Z. Wu, E. Sims, C. Saenphimmachak, L. Rohmer, J. M. Ritchie, M. Radey, M. McKevitt, T. L. Freeman, H. Hayden, E. Haugen, W. Gillett, C. Fong, J. Chang, V. Beskhlebnaya, M. K. Waldor, M. Samadpour, T. S. Whittam, R. Kaul, M. Brittnacher, and S. I. Miller.** 2009. Analysis of the genome of the *Escherichia coli* O157:H7 2006 spinach-associated outbreak isolate indicates candidate genes that may enhance virulence. *Infect. Immun.* **77:**3713–3721.
9. **Levine, M. M., J. G. Xu, J. B. Kaper, H. Lior, V. Prado, B. Tall, J. Nataro, H. Karch, and K. Wachsmuth.** 1987. A DNA probe to identify enterohemorrhagic *Escherichia coli* of O157:H7 and other serotypes that cause hemorrhagic colitis and hemolytic uremic syndrome. *J. Infect. Dis.* **156:**175–182.
10. **Li, B., W. H. Koch, and T. A. Cebula.** 1997. Detection and characterization of the *fimA* gene of *Escherichia coli* O157:H7. *Mol. Cell. Probes* **11:**397–406.
11. **Manning, S. D., A. S. Motiwala, A. C. Springman, W. Qi, D. W. Lacher, L. M. Ouellette, J. M. Mladonicky, P. Somsel, J. T. Rudrik, S. E. Dietrich, W. Zhang, B. Swaminathan, D. Alland, and T. S. Whittam.** 2008. Variation in virulence among clades of *Escherichia coli* O157:H7 associated with disease outbreaks. *Proc. Natl. Acad. Sci. USA* **105:**4868–4873.
12. **Monday, S. R., S. A. Minnich, and P. C. Feng.** 2004. A 12-base-pair deletion in the flagellar master control gene *flhC* causes nonmotility of the pathogenic German sorbitol-fermenting *Escherichia coli* O157:H$^-$ strains. *J. Bacteriol.* **186:**2319–2327.
13. **Monday, S. R., T. S. Whittam, and P. C. Feng.** 2001. Genetic and evolutionary analysis of mutations in the *gusA* gene that cause the absence of beta-glucuronidase activity in *Escherichia coli* O157:H7. *J. Infect. Dis.* **184:**918–921.
14. **Mukherjee, A., M. K. Mammel, J. E. LeClerc, and T. A. Cebula.** 2008. Altered utilization of *N*-acetyl-D-galactosamine by *Escherichia coli* O157:H7 from the 2006 spinach outbreak. *J. Bacteriol.* **190:**1710–1717.
15. **Qi, W., D. W. Lacher, A. C. Bumbaugh, K. E. Hyma, L. M. Ouellette, T. M. Large, C. L. Tarr, and T. S. Whittam.** 2004. *Ec*MLST: an online database for multi locus sequence typing of pathogenic *Escherichia coli*, p. 520–521. *In Proceedings of the 2004 IEEE Computational Systems Bioinformatics Conference.*
16. **Roe, A. J., C. Currie, D. G. Smith, and D. L. Gally.** 2001. Analysis of type 1 fimbriae expression in verotoxigenic *Escherichia coli*: a comparison between serotypes O157 and O26. *Microbiology* **147:**145–152.
17. **Rosser, T., T. Dransfield, L. Allison, M. Hanson, N. Holden, J. Evans, S. Naylor, R. La Ragione, J. C. Low, and D. L. Gally.** 2008. Pathogenic potential of emergent sorbitol-fermenting *Escherichia coli* O157:NM. *Infect. Immun.* **76:**5598–5607.

18. **Schmitt, C. K., M. L. McKee, and A. D. O'Brien.** 1991. Two copies of Shiga-like toxin II-related genes common in enterohemorrhagic *Escherichia coli* strains are responsible for the antigenic heterogeneity of the O157:H$^-$ strain E32511. *Infect. Immun.* **59:**1065–1073.
19. **Shaikh, N., N. J. Holt, J. R. Johnson, and P. I. Tarr.** 2007. Fim operon variation in the emergence of enterohemorrhagic *Escherichia coli*: an evolutionary and functional analysis. *FEMS Microbiol. Lett.* **273:**58–63.
20. **Wirth, T., D. Falush, R. Lan, F. Colles, P. Mensa, L. H. Wieler, H. Karch, P. R. Reeves, M. C. Maiden, H. Ochman, and M. Achtman.** 2006. Sex and virulence in *Escherichia coli*: an evolutionary perspective. *Mol. Microbiol.* **60:**1136–1151.
21. **Zhang, W., W. Qi, T. J. Albert, A. S. Motiwala, D. Alland, E. K. Hyytia-Trees, E. M. Ribot, P. I. Fields, T. S. Whittam, and B. Swaminathan.** 2006. Probing genomic diversity and evolution of *Escherichia coli* O157 by single nucleotide polymorphisms. *Genome Res.* **16:**757–767.

*Population Genetics of Bacteria: a Tribute to Thomas S. Whittam*
Edited by S. T. Walk and P. C. H. Feng


## Chapter 14

# Evolutionary Emergence and Impact of Atypical *Escherichia coli* O157:H7 Strains

***Peter C. H. Feng and Steven R. Monday***

## INTRODUCTION

Shiga toxin-producing *Escherichia coli* (STEC) of serotype O157:H7 were first recognized as a food-borne pathogen in 1982 when they were implicated in several outbreaks caused by contaminated ground beef. This serotype is now a well-established pathogenic group known to cause many of the hemorrhagic colitis (HC) and hemolytic uremic syndrome (HUS) cases worldwide (21).

In 1988, Whittam et al. (35) used multilocus enzyme electrophoresis (MLEE) to characterize the genetic diversity of 100 strains of *E. coli* serotype O157:H7 that were implicated in HC or HUS and established that O157:H7 strains are closely related and belong in a clonal group that is distantly related to other STEC. Furthermore, they hypothesized the clonal descent of O157:H7 from ancestral cells and later tested the hypothesis by examining 1,300 *E. coli* isolates from various serotypes and pathogenic groups by MLEE (34). That study showed that the O157:H7 clone was most closely related to the O55:H7 strains that are enteropathogenic *E. coli* (EPEC) and proposed that O157:H7 emerged from O55:H7 via acquisition of additional virulence traits by horizontal transfer and recombination (34). These fundamental findings on the clonality of O157:H7 were supported by additional studies and later expanded to include other traits and by using techniques like single nucleotide polymorphism (SNP) analysis. One such genetic marker was an SNP in the *uidA* open reading frame (ORF) that encodes $\beta$-glucuronidase (GUD) in *E. coli* (5). The +93 *uidA* SNP (+93 refers to the 93rd nucleotide site from the 5′ end of the ORF) had no effect on GUD expression or function but was unique and highly conserved in the O157:H7 serotype (5). These and many other findings contributed and eventually led to the conception of an evolutionary model for the emergence of the O157:H7 clone (Fig. 1) (11).

***Peter C. H. Feng and Steven R. Monday*** • Division of Microbiology, U.S. Food and Drug Administration, College Park, MD 20740.

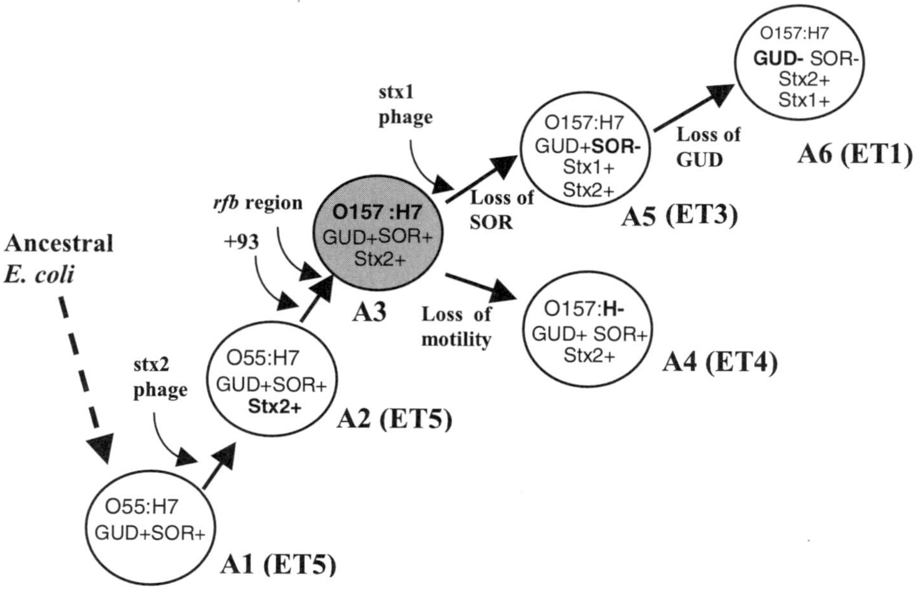


**Figure 1.** Model for the evolutionary emergence of *E. coli* O157:H7.

The model postulates that *E. coli* O157:H7 evolved from an ancestral *E. coli* that had the locus of enterocyte effacement pathogenicity island, via O55: H7 and a series of O157 clonal groups, by stepwise acquisition/loss of virulence and phenotypic traits. These clonal groups were originally predicted collectively based on phenotypes, the +93 *uidA* SNP, and electrophoretic types (ETs) determined by MLEE. Some of the clonal complex determinations were based on results derived from many isolates, while others were established from only a handful of strains, and sometimes from a single strain. Later, however, as more isolates were found, the strains in the clonal complexes were reanalyzed by multilocus sequence typing (MLST) (28) to determine their sequence types (STs), verified, and redefined as clonal groups A1 to A6 (Fig. 1). At the center of the model is a hypothetical "intermediate" (ancestor A3), postulated to have evolved from the A2 clonal complex of O55:H7 strains that are closely related but ancestral to O157:H7 (34). In the 12 years since the model has been published, the A3 intermediate has never been isolated, making it the "missing link" hypothesized to exist and from which evolved two distinct pathways. One pathway led to the emergence of the A5 clonal complex of GUD-positive O157: H7 strains, which subsequently lost GUD expression to give rise to the prototypic O157:H7 clonal complex (A6). Also from A3 emerged a divergent lineage caused in part by the loss of motility that gave rise to the A4 clonal complex composed of sorbitol-fermenting (SOR) O157 strains (SF O157).

Of all of these clonal complexes in the model, serotype O157:H7 strains from the A6 clonal complex continue to be the most important STEC pathogen, accounting for the majority of HUS infections worldwide. A6 clonal complex strains, therefore, are readily available for research and are also the most extensively studied. Prototypic A6 O157:H7 strains carry phenotypic markers, such as the absence of sorbitol fermentation and GUD activity, resistance to tellurite, and the presence of both somatic (O) 157 and flagellar (H) 7 antigens. Virulence factors most often carried by O157:H7 include Shiga toxins (Stxs) 1 and 2, encoded by phage-borne $stx_1$ and $stx_2$ genes; intimin, encoded by the γ-*eae* allele, which is found mostly in O157:H7, O55:H7, and a few other serotypes; and enterohemolysin, encoded by *ehxA* (27). In addition, the +93 *uidA* SNP mentioned earlier is also a stable genetic marker for O157:H7, and therefore these traits and markers are often used to develop diagnostic assays to test for O157:H7 in food and clinical samples. However, O157:H7 strains are not homogeneous, and atypical variants are now being isolated (6), sometimes fairly frequently (19).

Atypical variants do not exhibit one or more of the prototypic markers discussed above, and they are often undetectable by assays routinely used to test for O157:H7. A number of atypical strains have been characterized, and in some cases the genetic cause for their atypical phenotype has been determined. This chapter reviews the findings on some of these strains, examines the genetic mutations underlying their atypical phenotypes, addresses the impact they have on public health, and discusses how they fit into the evolutionary model of O157:H7.

## GUD VARIANTS

Almost all generic *E. coli* strains produce GUD, so a popular assay to identify *E. coli* utilizes the fluorogenic GUD substrate 4-methylumbelliferyl-β-D-glucuronide (9). Unlike generic *E. coli*, however, most O157:H7 strains do not exhibit GUD activity, and so this trait is a useful screening marker. Genetic studies showed that although O157:H7 strains do not exhibit activity, they do carry the *uidA* gene, which codes for GUD (12). Preliminary sequence analysis showed that O157:H7 strains had the entire *uidA* gene but contained several SNPs not present in the *uidA* gene of generic *E. coli* (10). Although none of the SNPs identified accounted for the absence of GUD activity, an SNP at the 93rd nucleotide position in the gene turned out to be highly conserved among O157:H7 strains examined and, consequently, became a very useful genetic marker in diagnostics (13) and evolutionary studies (11).

In 1995, a $GUD^+$ phenotypic variant of O157:H7 (strain G5101) was isolated from an HC patient that had all of the other prototypic O157:H7 traits (18). To examine the cause for the absence of GUD expression in O157:H7 strains, the entire *uidA* ORF, including regulatory regions, of GUD-positive and -negative O157:H7 strains was sequenced. When compared to *E. coli* K-12 (GenBank accession number D908051), the *uidA* gene of GUD-positive O157:H7 strains was found to have many SNPs but only three nonsynonymous mutations (+93, +191, and +758) that altered the amino acid sequence in the mature protein (25). However, as both strains are GUD positive, these mutations had no effect on GUD expression. The *uidA* gene sequence of the GUD-negative O157:H7 strain (ATCC 35150, GenBank AF305917) differed from that of the GUD-positive O157:H⁻ (H7) strain (493-89, GenBank AF305918) at four sites within the ORF (+686, +776, +1370, and +1671). Of these, the +1671 mutation was silent, while those at +776 and +1370 altered amino acid sequences. But more importantly, the +686 mutation in the ATCC 35150 *uidA* is a G-G dinucleotide insertion, which not only altered amino acid sequence but also caused a frameshift in the ORF and introduced a premature TGA stop codon at +711 that is identical to that which terminated translation of the functional *uidA*. Sequencing confirmed that the G-G insertion at +686 was present only in the GUD-negative O157:H7 strains. Without altering the +776 and +1370 mutations, deletion of the +686 G-G dinucleotide insertion by site-directed mutagenesis restored GUD activity in O157:H7, thus confirming that the absence of GUD expression in O157:H7 was solely due to the insertion mutation at +686 of *uidA* (25).

Some of the SNP mutations in *uidA* (e.g., +93, +758, and +776) seemed to be conserved among O157:H7 strains, so other *E. coli* strains were examined to determine the distribution of these mutations and to see how they fit into the stepwise evolutionary model (11). The +758 SNP that is not present in K-12 and had previously been thought to have appeared before the A1 clonal complex turned out not to be pertinent to the model. The +758 SNP was not only in the O55:H7 and O157:H7 strains but also in other *E. coli* strains, and therefore it is most likely the wild-type sequence and its absence in K-12 is the anomaly. The +93 SNP is likely to have occurred after the emergence of O55:H7 and before the hypothetical A3 "intermediate" (Fig. 2). In support of this hypothesis, the +93 SNP has been found in all the strains in the clonal groups that were derived from the A3 intermediate, including A5, A6, and the SF O157 strains in the divergent A4 clonal group. The +776 SNP evolved more recently and probably appeared during the emergence of the A5 and A6 clonal groups (Fig. 2). Last, the A5 clonal group strains acquired two *uidA* mutations, +1370 and the G-G insertion at +686, to give rise to the prototypic GUD-negative O157:H7 strains in A6 (25). Collectively, the occurrence and distribution of *uidA* polymorphisms are consistent with and support the proposed evolutionary model of O157:H7 (11).

Initially, the GUD-positive O157:H7 clonal group at A5 was represented solely by strain G5101, isolated in 1995 (18). More recently, however, other GUD-positive O157:H7 strains have been isolated from clinical samples in the United States. Additionally, several GUD-positive strains were implicated in outbreaks and sporadic infections of diarrhea and HC in Hokkaido, Japan (26). Analysis of the newly found U.S. and Japanese isolates showed that they were identical to G5101 in phenotypic traits, virulence factors, and genetic markers. Pulsed-field gel electrophoresis (PFGE) (4) of these strains showed that the Japanese strains were nearly identical and different from G5101 and the other U.S. strains, which also showed more genetic diversity (14). However, analysis of the same strains using MLST showed that both the U.S. and Japanese strains had ST65 genotype observed for the A5 clonal type strain G5101. Hence, de-

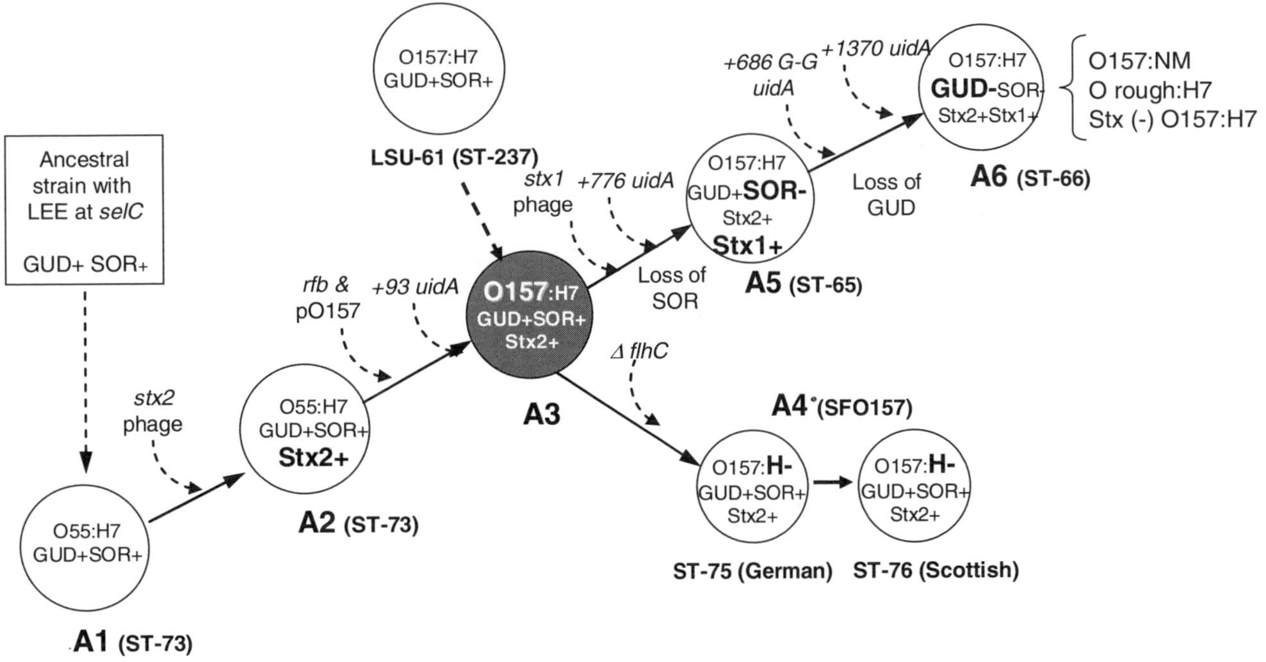


**Figure 2.** Model for the evolutionary emergence of *E. coli* O157:H7 modified to include atypical O157:H7 variants. LEE, locus of enterocyte effacement. Modified from Feng et al. (14).

spite PFGE profile differences, all these GUD-positive O157:H7 strains are in the A5 clonal group (14).

## NONMOTILE VARIANTS

The definitive identification of O157:H7 strains is based on the serological presence of both the O and the H antigens. The somatic O157 antigen is stable and easily identified. However, detection of the flagellar H7 antigen can be problematic. Flagella are part of the *E. coli* motility mechanism, which is under complex regulatory control affected by multiple physiologic and environmental variables (1). Under certain conditions, O157:H7 strains cease to produce flagella and become nonmotile, which abrogates the ability to serologically determine its H antigen type. During diagnostics, nonmotile O157 strains are passed repeatedly through motility agar medium, and if motility cannot be restored by induction, they are designated as O157:NM (NM = nonmotile). However, nonmotility in O157 strains may also arise via genetic mutations, and those strains are often designated as O157:H$^-$. However, in routine analysis, it is difficult to determine whether a nonmotile strain is O157:NM or O157:H$^-$ without additional studies; hence, strains that are "phenotypically" and "genetically" nonmotile are often grouped together as O157:NM.

The isolation rate of O157:NM is increasing worldwide, and in some countries it has surpassed that of O157:H7. Many O157:NM strains are also pathogenic and have been implicated in human infections causing both HC and HUS in Germany and the United Kingdom (3). From a diagnostic standpoint, the inability to determine H types of O157 isolates is a significant health concern. The O157 serogroup is large and, in addition to the pathogenic O157:H7 serotype, includes strains that have other H types, such as H3, H12, H16, H38, and H45. For the most part, these O157, non-H7 strains are not STEC and are not associated with disease (nonpathogenic). But NM variants of these strains have been mistakenly identified as O157:H7 by laboratories that only tested for the O157 antigen or were unable to type for H7 because the isolate was not motile. It is, therefore, critical to be able to determine the H type of an O157 isolate in order to determine if it is an O157:H7 strain.

To determine the H type and the pathotype (virulence factors) of nonmotile strains, we obtained and characterized a collection of 38 clinical O157:NM isolates from around the world. We found that all of the isolates carried the *fliC* gene, which codes for flagella, but most of the isolates failed to express a functional flagellar protein, based on Western blotting analyses with an antiflagellar monoclonal antibody (8). In addition, PCR analysis (13) of these isolates showed that most carried either $stx_1$, $stx_2$, or both, along with *γ-eae, ehxA*, and the +93 *uidA* SNP. A few other O157:NM isolates that did not have an *stx* gene, the +93 *uidA* SNP, or other classical O157:H7 virulence factors also differed phenotypically from O157:H7. A subset of isolates with and without typical O157:H7 virulence factors were subjected to multiple passages through motility agar medium to induce the production of flagella. After induction, we found that motile isolates lacking typical O157:H7 virulence factors did not react with anti-H7 antibody, suggesting that they are O157, non-H7 serotypes. In contrast, all of the motile, *stx*-bearing isolates reacted with anti-H7 antibody, confirming that they are O157:H7 isolates that did not produce a functional flagellar protein when first serotyped. These results were supported by PCR-restriction fragment length polymorphism analysis of *fliC*, which showed that the *stx*-bearing O157:NM isolates carried the *fliC* allele that encodes the H7 flagellar antigen (16). These findings caution that isolation of O157:NM isolates during routine diagnostic testing should not be disregarded, as those isolates that carry *stx* genes are likely to be simply serologic variants of O157:H7.

MLEE analysis of O157:NM isolates showed that the *stx*-bearing isolates belonged to the most common ET of O157:H7 strains (8). In contrast, the non-*stx*-bearing, non-H7 variants did not belong to this ET and were distantly related. Consistent with the MLEE results, analysis by MLST showed that the *stx*-bearing isolates were ST66, the most common genotype for O157:H7. Hence, these results are in complete agreement with the phenotypic and PCR characterizations above and support the notion that non-*stx*-bearing, non-H7 isolates do not belong to the O157:H7 clonal group. Important exceptions to this conclusion are the German SF O157 isolates that were noticeably different from typical O157:H7 isolates using MLEE and MLST. Motility could not be induced in these isolates and they failed to react with flagellar antibody in Western blot analysis, suggesting that these isolates are O157:H$^-$ and that their nonmotile phenotype is due to genetic defects (see the following section).

## SF O157—NONMOTILITY

As discussed in the previous section, we found that the majority of Shiga toxin-producing O157:NM strains are actually phenotypic variants of O157:H7 and they can sometimes be induced to express the H7 antigen (8). The exception, however, is the SF O157 German strains, which carry the H7 *fliC* allele but cannot be induced to express a functional protein. This observation suggests that their null phenotype may be due to a genetic defect. Motility in *E. coli* is regulated by a complex, three-tiered cascade of operons (class I, II, and III), containing over 40 genes (1). At the top of this hierarchy is the *flhDC* operon, which is the transcriptional activator and master regulator of flagella biosynthesis that turns on class II genes and, in turn, activates the expression of class III genes, including *fliC*. Genetic analysis of the motility genes of SF O157 (strain 493-89) and those of O157:H7 (strain 35150) showed no major genetic defects in class II and III genes. Similarly, comparison of the class I genes (*flhDC*) from strain 35150 (GenBank AE005411) with those of strain 493-89 (GenBank AY225162) showed no differences in the operon promoter region and only a single nucleotide substitution in the *flhD* ORF. However, the *flhC* sequence of 493-89 contains a 12-bp deletion that results in the absence of the amino acids phenylalanine, valine, glutamate, and serine from the native FlhC protein, and PCR analysis confirmed that the 12-bp *flhC* deletion was found only in the SF O157 but not in the other O157:NM or O157:H7 strains. *trans*-complementation of 493-89 with its own *flhC*, in which the 12-bp deletion had been repaired by site-directed mutagenesis, restored motility to 493-89 as well as serological reactivity with anti-H7 (24), thereby confirming that *ΔflhC* caused the nonmotile phenotype in SF O157. The highly conserved nature of the 12-bp *flhC* deletion in SF O157 is consistent with previous findings that these strains are in a unique clone (11). The fact that SF O157 strains carried the H7 *fliC* gene also confirms their close genetic relation to O157:H7 (11) and is consistent with the evolutionary model that postulated the divergence of SF O157 from O157:H7, in part by the loss of motility (24).

Previous MLEE analysis showed that SF O157 strains belong to the same ET, which was represented by a handful of German strains. More recently, however, SF O157 strains, which are increasingly causing sporadic infections and outbreaks of HUS, have been isolated in the Czech Republic, Finland, Scotland, and other countries (19). Molecular analysis of SF O157 strains from various European countries showed that the PFGE profiles of some SF O157 strains isolated from Finland were identical to some of the German strains that were isolated up to 10 years earlier. Likewise, some Scottish strains shared profile identity to Finnish strains that were isolated 4 years earlier. The fact that there are profile identities among strains isolated from different geographical areas and years apart suggests that these SF O157 strains may have a common origin. Despite PFGE profile identities and similarities, however, MLST analysis showed clonal differences within the SF O157 populations (14). All of the German strains, some Finnish, and a single Scottish strain had the same ST (ST75), while the remaining Scottish and Finnish strains had a distinct *mdh* allele (ST76). The original A4 clonal complex on the evolutionary model, therefore, has since been subtyped into the "German clone" (ST75) and the "Scottish clone" (ST76) (Fig. 2).

## Stx-NEGATIVE VARIANTS

The $stx_1$ and $stx_2$ genes of STEC are phage encoded and are subject to loss by phage induction. For example, Stx production can be eliminated from STEC strains by inducing the Stx-encoding phages with sublethal dosages of UV or antibiotics (22). In addition, it was shown that STEC strains can also spontaneously induce the phage and lose Stx production during routine subculture (20). The causes or the mechanisms by which *stx*-carrying phages are induced spontaneously have not been determined. This process, which apparently is not media dependent, seems to occur fairly frequently in STEC serotypes. Loss of Stx expression by strains of O157:H7 serotype is less common (17) but does occur. For example, during routine subculture of a Japanese O157:H7 isolate, two colonies were obtained that differed only in their ability to produce Stxs (7). Biochemical analyses of these strains, designated TT12A and TT12B, showed that they were identical in all traits, and serotyping confirmed both to be O157:H7 strains. PCR analysis (13) for known virulence factors determined that both carried the +93 *uidA* SNP, *γ-eae* allele, and *ehxA* and therefore were identical, except that TT12A had both $stx_1$ and $stx_2$ genes and expressed both toxins while TT12B did not. In vitro testing showed that both strains exhibited the typical attaching-and-effacing (A/E) phenotype on HeLa cell monolayers caused by intimin. However, the extent of cellular damage caused by TT12B was minimal as compared to TT12A (7). This difference may be attributed to the cytotoxic effects on HeLa cells by Stx1 and Stx2 that are being produced by TT12A and serves as further evidence that TT12B produced neither Stx. The findings that these two strains are nearly identical in traits and markers suggest that TT12B was probably derived from TT12A by the loss of Stx production. An analogous finding was reported recently, in which an Stx2-producing SF O157 strain that was isolated from patients at the time of infection gave rise to an identical strain that had lost Stx2 production in the follow-up samples taken 3 to 8 days later (23). That study also identified genetic "hot spots" in SF O157 for the insertion of the $stx_2$-encoding phages and proposed that there is active recycling of $stx_2$ phages between the Stx2-positive and -negative SF O157 strains (23).

Strains of O157:H7 that do not produce Stx are not common, but have been isolated from many sources, including bagged spinach, ground meats, and cattle (7). From a food safety perspective, the isolation of Stx-negative O157:H7 poses an interesting dilemma. For example, it is often unclear whether phage (Stx) loss had occurred as a result of the isolation procedure, in which case a toxigenic strain is likely to still be in the product and, therefore, poses a health risk. In contrast, if the loss of Stx occurs prior to product contamination, new questions are raised concerning the virulence potential of *stx*-negative O157:H7 strains. Others have raised similar concerns with regard to Stx loss and pathotype and postulated that these organisms most likely do not cause severe disease (17). Consistent with that speculation, patients who were infected with Stx-negative SF O157 strains did not develop HUS. However, Stx-negative SF O157 strains have also been isolated from HUS patients (17), but it is uncertain at what stage of infection the *stx* loss occurred. Therefore, it remains uncertain whether *stx*-negative O157:H7 strains would be capable of causing severe diseases.

Molecular subtyping analysis of XbaI-digested genomic DNA by PFGE showed diversity in profiles among the Stx-negative O157:H7 strains. Three strains isolated from meats in Japan shared profile similarity with each other, as did the two strains isolated from cattle in France, with both of these groups sharing enough similarity to cluster in the same clade (7). Strain TT12B, however, which is also Stx negative, did not cluster with those Stx-negative strains, instead clustering with the Stx-positive O157:H7 strains, including TT12A. The PFGE profiles of TT12A and TT12B differed by only two bands and shared 90% similarity, indicating these are similar strains. A Southern blot of the PFGE gel sequentially probed with $stx_1$- and $stx_2$-specific DNA probes confirmed that both *stx* genes were present only in TT12A (7).

Clonal analysis of TT12A and TT12B by MLEE confirmed that both are ET 1 and, therefore, clonal strains of O157:H7. More recent analysis by MLST concurred with the MLEE results in that both TT12A and TT12B had ST66, the most common genotype for O157:H7 strains. The findings that TT12A and TT12B have identical traits, markers, and molecular subtypes suggest that

TT12B was derived from TT12A and that TT12B represents a subset of the A6 clonal group that had lost Stx production, most likely via phage induction.

## O-ROUGH VARIANTS

The somatic (O) 157 and the flagellar (H) 7 antigens are key markers that are extensively used in diagnostics to identify the O157:H7 serotype. In 1998, an atypical O157:H7 strain that does not express the O-type 157 was isolated from a beef sample in Malaysia. This strain, designated MA6, carried $stx_2$ only but had all the typical O157:H7 virulence factors and markers including the genes required for O-antigen synthesis (15). However, it did not produce the O157 antigen (O rough) and, therefore, would be undetectable with serological assays routinely used in testing for O157:H7.

Most *E. coli* O-antigen genes involved in the biosynthesis and assembly of the four sugar nucleotides are located in the *rfb* operon, except for three genes, including *gne*, that are outside of the operon (29). The assembled O antigen is anchored to the core antigen, which comprises an inner and an outer component that are encoded by genes carried in the *waa* operon. Comparison of O-antigen gene sequences of MA6 and O157:H7 showed no major genetic defects or SNPs that may account for the absence of O157 antigen expression in MA6. One exception, however, was the *gne* gene, which codes for UDP-GalNAc 4-epimerase, an enzyme essential for the synthesis of one of the oligosaccharide subunits. The MA6 *gne* ORF was found to contain (at +385) the 1,310-bp IS*629* insertion element (GenBank GU183138) that was absent in O157:H7 (30). *trans*-complementation of MA6 with the *gne* allele from O157:H7 restored O157 antigen expression in MA6, as evidenced by serological reactivity with anti-O157, thereby confirming that IS*629* insertion within the *gne* gene caused the absence of O157 antigen expression in MA6 (30).

The recognition site of IS*629* is unknown, but since the disruption of *gne* by an IS element was deemed rare, it was speculated that the MA6 O-rough phenotype was a random mutational event. More recently, however, another O-rough O157:H7 strain (CB7326) was isolated from an HC patient in Germany (2). Like MA6, CB7326 has all of the typical O157:H7 traits and virulence factors, including O157 antigen genes, but unlike MA6, CB7326 has both *stx* genes. Interestingly, the null phenotype of CB7326 was also caused by IS*629* insertion within the *gne* locus, but at +711 instead of at +385 as in MA6 (31). The observation that IS*629* has repeatedly inserted into the *gne* locus suggests that this mutational event may be under strong selection and that similar O-rough O157:H7 strains may be emerging. In addition, MA6 was originally isolated from food, but the isolation of CB7326 from an HC patient raises additional concerns about the virulence potential of O-rough O157:H7 strains.

Molecular analysis of MA6 and CB7326 by PFGE showed that these strains had distinct profiles and only shared ~80% similarity. Neither strain was similar to any of the reference O157:H7 strains from the ATCC. However, MLST analysis showed both O-rough strains belonged to the common O157:H7 genotype (ST66) for O157:H7 and are within the A6 clonal complex. However, these O-rough O157:H7 strains represent a subset of the A6 clonal group that had lost O157 expression due to IS*629* insertion in the *gne* gene (Fig. 2).

## "MISSING LINK"?

At the center of the O157:H7 evolutionary model is a hypothetical "intermediate" (A3) or "missing link." It is hypothesized that an isolate of the A3 genotype would ferment sorbitol, have GUD activity, and carry $stx_2$. In addition, the model predicts that the present-day representatives of the A6 (O157:H7) and A4 (SF O157 strains) clonal groups shared and were derived from the common A3 ancestor.

A representative A3 intermediate isolate has, thus far, not been found. Some strains have come close to the expected phenotypic traits, but failed to meet the genetic profile. For example, strain LSU-61, isolated from deer, was identified serotypically as O157:H7. It fermented sorbitol; was GUD positive and tellurite resistant; carried *γ-eae* and *ehxA*; and had the +93 but not the +776 *uidA* SNP. However, LSU-61 did not carry either $stx_1$ or $stx_2$. Except for the absence of Stxs, which is not unexpected since these are phage-encoded genes and can be lost by induction (see "Stx-Negative Variants," above), strain LSU-61 had all of the expected traits of the A3 intermediate. Molecular analysis, however, showed that the PFGE profile of LSU-61 only had 60% similarity to strains in the A2 clonal complex that preceded A3 or those in the A4, A5, and A6 clones that were derived from A3. Furthermore, MLST analysis showed that LSU-61 had a distinct *fadD* allele, placing it in genotype ST237, which had not been encountered previously and is represented solely by the LSU-61 strain (14). These genetic findings do not fully support that LSU-61 is the A3 "missing link," but all of the data confirm that it is related to typical O157:H7 strains. The absence of the +776 *uidA* SNP suggests that LSU-61 represents an ancestral lineage of A5, but probably evolved after the A3 lineage because the representatives of A4 do not carry the +776 *uidA* SNP. The observations that LSU-61 has the O157 antigen, ferments sorbitol, and has the +93 *uidA* SNP also suggest that it emerged after A3 and its position lies somewhere in between the A3 intermediate and the A5 clonal complex. The exact position of LSU-61 and how it fits into the evolutionary model remains to be determined (Fig. 2) (14).

## CONCLUSIONS

Since the model for the evolutionary emergence of O157:H7 was published in 1998, the backbone of the model has been supported by other studies and remains essentially unchanged. Still, much of the transition process from one clonal complex to another is not well defined. Genomic comparison of O157:H7 versus *E. coli* K-12 using microarrays showed major differences in genome content, which is predicted to have occurred as a result of gene acquisition events (36). Consistent with this hypothesis, the entire *rfb* operon that codes for the O antigen along with genes adjacent to this operon were found to have cotransferred broadly into O55 and O157 (33). Subsequent recombinational events caused antigenic shifts that led to the emergence of the O55:H7 and O157:H7 lineages (33). Similarly, about 85% of the genes that were observed to be different in genomic comparison between O55:H7 and O157:H7 were found to be phage related (36), and therefore the roles of phages and other mobile genetic elements in the genetic diversification process cannot be understated. The significance of phages in the acquisition and loss of Stx traits is apparent as hot spots for the integration and excision of $stx_1$ (32) and $stx_2$ phages (23) have been identified. Importantly, as there may be active recycling of *stx*-bearing phages in some strains, the association of *stx* and clonal complexes is probably not a stable, dependable marker. Likewise, the O157:H7 genome contains many insertion elements, and some, like IS*629*, exist in multiple copies and have caused important gene disruptions. For example, IS*629* insertion in *gne* caused the null phenotype of the O-rough O157:H7 variants; however, in this instance, the insertion only caused phenotypic changes and it is uncertain whether other insertions may have led to other significant genotypic alterations.

The A1 to A6 clonal complexes continue to be defined and the examination of additional representative strains continues to expand and modify our view of O157:H7 evolution. Still, some of these clonal complexes remain poorly defined due to the limited availability of strains. For example, strain ECOR37, a marmoset *E. coli* isolate that has *γ-eae* and untypeable O and H antigens, was shown by MLEE to be closely related to the O55:H7 strains in A1. PCR analysis confirmed that ECOR37 had both the O55 *wzx* gene required for the export of O lipopolysaccharide and the H7 *fliC* gene and, therefore, is an O55:H7 strain. However, analogous to the LSU-61 and the A3 situation, ECOR37 shared only 60% similarity in PFGE profiles with the A1 strains and had distinct *mdh* and *clpX* alleles (genotype ST77) as compared to A1 clonal strains (ST73) (14). Hence, ECOR37 is speculated to represent an ancestral lineage of A1, but its position in the model is uncertain. Some of the other clonal complexes in the model, such as A4, A5, and A6, are now better characterized as more strains have been isolated and examined. The A5 clonal complex of GUD-positive O157:H7 strains isolated from different worldwide locations has identical multilocus genotypes, and although there are gene content differences among these strains (36), the complex seems to be fairly homogeneous. On the other hand, the A4 clonal group of SF O157 strains, which were also obtained from different countries, exhibited more diversity and is now subdivided into two slightly different genotypic subclones. Last, atypical variants of O157:H7 continue to emerge and to increase in isolation frequencies. Some of these, such as the O-rough:H7 strain, were once thought to be very rare, but a similar strain has now been found elsewhere. Many of these atypical O157:H7 variants have a public health impact, as they are often pathogenic and will cause illness but, due to the lack of trait marker(s), are not easily detected by assays routinely used to test for O157:H7. Although phenotypically distinct, many of the atypical O157:H7 variants were found to have identical multilocus genotypes (ST66) and belong to the A6 clonal group. Therefore, such strains represent newly identified O157:H7 variants that merely lost typical phenotypic features.

**Author's Note.** As a research microbiologist at the FDA, the focus of my (P.C.H.F.) work has always been food safety; hence, after publishing about the conserved nature of the +93 *uidA* SNP in O157:H7, my sole intention was to use it diagnostically to ensure the safety of foods. That is, until I met Tom Whittam in 1995, when a mutual friend and colleague, I. Kaye Wachsmuth, invited Tom to give a seminar at the USDA and Kaye thought that we should meet. I attended Tom's talk on clonality and evolution and found it interesting, although not too relevant to my work at the FDA on food safety. In our subsequent conversation, however, as I related the uniqueness of the +93 *uidA* SNP in O157:H7, Tom immediately recognized the potential of this marker in evolutionary studies, thus marking the beginning of our long and productive collaboration. I would have never imagined at the time that a chance meeting with Tom would take me on the exciting journey into the world of bacterial population genetics.

## REFERENCES


1. **Aldridge, P., and K. T. Hughes.** 2002. Regulation of flagellar assembly. *Curr. Opin. Microbiol.* **5:**160–165.
2. **Beutin, L., G. Krause, S. Zimmermann, S. Kaulfuss, and K. Gleier.** 2004. Characterization of Shiga toxin-producing *Escherichia coli* strains isolated from human patients in Germany over a 3-year period. *J. Clin. Microbiol.* **42:**1099–1108.
3. **Bitzan, M., K. Ludwig, M. Klemt, H. König, J. Büren, and D. E. Müller-Wiefel.** 1993. The role of *Escherichia coli* O157 infections in the classical (enteropathic) haemolytic uraemic syndrome: results of Central European, multicentre study. *Epidemiol. Infect.* **110:**183–196.
4. **Centers for Disease Control and Prevention.** 2003. *Standardized Molecular Subtyping of Foodborne Bacterial Pathogens by Pulsed-Field Gel Electrophoresis: CDC Training Manual.* CDC, Atlanta, GA.
5. **Feng, P.** 1993. Identification of *Escherichia coli* serotype O157:H7 by DNA probe specific for an allele of *uidA* gene. *Mol. Cell. Probes* **7:**151–154.
6. **Feng, P.** 1995. *Escherichia coli* serotype O157:H7: novel vehicles of infection and emergence of phenotypic variants. *Emerg. Infect. Dis.* **1:**16–21.
7. **Feng, P., M. Dey, A. Abe, and T. Takeda.** 2001. Isogenic strain of *Escherichia coli* O157:H7 that has lost both Shiga toxin 1 and 2 genes. *Clin. Diagn. Lab. Immunol.* **8:** 711–717.
8. **Feng, P., P. I. Fields, B. Swaminathan, and T. S. Whittam.** 1996. Characterization of nonmotile *Escherichia coli* O157 and other serotypes by using an anti-flagellin monoclonal antibody. *J. Clin. Microbiol.* **34:**2856–2859.
9. **Feng, P., and P. A. Hartman.** 1982. Fluorogenic assays for the immediate confirmation of *Escherichia coli*. *Appl. Environ. Microbiol.* **43:**1320–1329.
10. **Feng, P., and K. A. Lampel.** 1994. Genetic analysis of *uidA* gene expression in enterohemorrhagic *Escherichia coli* serotype O157:H7. *Microbiology* **140:**2101–2107.
11. **Feng, P., K. A. Lampel, H. Karch, and T. S. Whittam.** 1998. Genotypic and phenotypic changes in the emergence of *Escherichia coli* O157:H7. *J. Infect. Dis.* **177:**1750–1753.
12. **Feng, P., R. Lum, and G. W. Chang.** 1991. Identification of *uidA* gene sequences in $\beta$-D-glucuronidase assay negative *Escherichia coli*. *Appl. Environ. Microbiol.* **57:**320–323.
13. **Feng, P., and S. R. Monday.** 2000. Multiplex PCR for detection of trait and virulence factors in enterohemorrhagic *Escherichia coli* serotypes. *Mol. Cell. Probes* **14:**333–337.
14. **Feng, P. C., S. R. Monday, D. W. Lacher, L. Allison, A. Siitonen, C. Keys, M. Eklund, H. Nagano, H. Karch, J. Keen, and T. S. Whittam.** 2007. Genetic diversity among clonal lineages within the *Escherichia coli* O157:H7 stepwise evolutionary model. *Emerg. Infect. Dis.* **13:**1701–1706.
15. **Feng, P., R. C. Sandlin, C. H. Park, R. A. Wilson, and M. Nishibuchi.** 1998. Identification of a rough strain of *Escherichia coli* O157:H7 that produces no detectable O157 antigen. *J. Clin. Microbiol.* **36:**2339–2341.

16. **Fields, P. I., K. Blom, H. J. Hughes, L. O. Helsel, P. Feng, and B. Swaminathan.** 1997. Molecular characterization of the gene encoding H antigen in *Escherichia coli* and development of a PCR-RFLP test for the identification of *E. coli* O157:H7 and O157:NM. *J. Clin. Microbiol.* **35:**1066–1070.
17. **Friedrich, A. W., W. Zhang, M. Bielaszewska, A. Mellmann, R. Köck, A. Fruth, H. Tschäpe, and H. Karch.** 2007. Prevalence, virulence profiles, and clinical significance of Shiga toxin-negative variants of enterohemorrhagic *Escherichia coli* O157 infection in humans. *Clin. Infect. Dis.* **45:**39–45.
18. **Hayes, P. S., K. Blom, P. Feng, J. Lewis, N. A. Strockbine, and B. Swaminathan.** 1995. Isolation and characterization of a $\beta$-D-glucuronidase-producing strain of *Escherichia coli* O157:H7 in the United States. *J. Clin. Microbiol.* **33:**3347–3348.
19. **Karch, H., and M. Bielaszewska.** 2001. Sorbitol-fermenting Shiga toxin-producing *Escherichia coli* O157:H$^-$ strains: epidemiology, phenotypic and molecular characteristics, and microbiological diagnosis. *J. Clin. Microbiol.* **39:**2043–2049.
20. **Karch, H., T. Meyer, H. Rüssmann, and J. Hessemann.** 1992. Frequent loss of Shiga-like toxin genes in clinical isolates of *Escherichia coli* upon subcultivation. *Infect. Immun.* **60:**3464–3467.
21. **Karmali, M. A.** 1989. Infection by verocytotoxin-producing *Escherichia coli. Clin. Microbiol. Rev.* **2:**15–38.
22. **Matsushiro, A., K. Sato, H. Miyamoto, T. Yamamura, and T. Honda.** 1999. Induction of prophages of enterohemorrhagic *Escherichia coli* O157:H7 with norfloxacin. *J. Bacteriol.* **181:**2257–2260.
23. **Mellmann, A., S. Lu, H. Karch, J. G. Xu, D. Harmsen, M. A. Schmidt, and M. Bielaszewska.** 2008. Recycling of Shiga toxin 2 genes in sorbitol-fermenting enterohemorrhagic *Escherichia coli* O157:NM. *Appl. Environ. Microbiol.* **74:**67–72.
24. **Monday, S. R., S. A. Minnich, and P. C. Feng.** 2004. A 12-base-pair deletion in the flagellar master control gene *flhC* causes nonmotility of the pathogenic German sorbitol-fermenting *Escherichia coli* O157:H$^-$ strains. *J. Bacteriol.* **186:**2319–2327.
25. **Monday, S. R., T. S. Whittam, and P. C. Feng.** 2001. Genetic and evolutionary analysis of mutations in the *gusA* gene that cause the absence of $\beta$-glucuronidase activity in *Escherichia coli* O157:H7. *J. Infect. Dis.* **184:**918–921.
26. **Nagano, H., T. Okui, O. Fujiwara, Y. Uchiyama, N. Tamate, N. Kumada, Y. Morimoto, and S. Yano.** 2002. Clonal structure of Shiga toxin (Stx)-producing and $\beta$-D-glucuronidase-positive *Escherichia coli* O157:H7 strains isolated from outbreaks and sporadic cases in Hokkaido, Japan. *J. Med. Microbiol.* **51:**405–416.
27. **Nataro, J. P., and J. B. Kaper.** 1998. Diarrheagenic *Escherichia coli. Clin. Microbiol. Rev.* **11:**142–201.
28. **Qi, W., D. W. Lacher, A. C. Bumbaugh, K. E. Hyma, L. M. Ouellette, T. M. Large, C. L. Tarr, and T. S. Whittam.** 2004. EcMLST: an online database for multilocus sequence typing of pathogenic *Escherichia coli*, p. 520–521. *In Proceedings of the 2004 IEEE Computational Systems Bioinformatics Conference.*
29. **Raetz, C. R., and C. Whitfield.** 2002. Lipopolysaccharide endotoxins. *Annu. Rev. Biochem.* **71:**635–700.
30. **Rump, L. V., P. C. H. Feng, M. Fischer, and S. R. Monday.** 2010. Genetic analysis for the lack of expression of the O157 antigen in an O rough:H7 *Escherichia coli* strain. *Appl. Environ. Microbiol.* **76:**945–947.
31. **Rump, L. V., L. Beutin, M. Fischer, and P. C. H. Feng.** 2010. Characterization of a *gne*::IS*629* O rough:H7 *Escherichia coli* strain from a hemorrhagic colitis patient. *Appl. Environ. Microbiol.* **76:**5290–5291.
32. **Shaikh, N., and P. I. Tarr.** 2003. *Escherichia coli* O157:H7 Shiga toxin-encoding bacteriophages: integrations, excisions, truncations, and evolutionary implications. *J. Bacteriol.* **185:**3596–3605.
33. **Tarr, P. I., L. M. Schoening, Y.-L. Yea, T. R. Ward, S. Jelacic, and T. S. Whittam.** 2000. Acquisition of the *rfb-gnd* cluster in evolution of *Escherichia coli* O55 and O157. *J. Bacteriol.* **182:**6183–6191.

34. **Whittam, T. S., M. L. Wolfe, I. K. Wachsmuth, F. Ørskov, I. Ørskov, and R. A. Wilson.** 1993. Clonal relationships among *Escherichia coli* strains that cause hemorrhagic colitis and infantile diarrhea. *Infect. Immun.* **61:**1619–1629.
35. **Whittam, T. S., I. K. Wachsmuth, and R. A. Wilson.** 1988. Genetic evidence of clonal descent of *Escherichia coli* O157:H7 associated with hemorrhagic colitis and hemolytic uremic syndrome. *J. Infect. Dis.* **157:**1124–1133.
36. **Wick, L. M., W. Qi, D. W. Lacher, and T. S. Whittam.** 2005. Evolution of genomic content in the stepwise emergence of *Escherichia coli* O157:H7. *J. Bacteriol.* **187:**1783–1791.

*Population Genetics of Bacteria: a Tribute to Thomas S. Whittam*
Edited by S. T. Walk and P. C. H. Feng


*Chapter 15*

# Thomas Whittam, Shiga Toxin-Producing *Escherichia coli*, and the Clinical Consequences of Clonality

***Shana R. Leopold and Phillip I. Tarr***

## OVERVIEW

Tom Whittam did much in the past 25 years to guide our thinking about the emergence of pathogens in the *Escherichia coli* genospecies. His work chaperoned us from the confining concepts of bacterial taxonomy into broader appreciations of the interplay between phylogeny, virulence, and genome structure. In this chapter, we review the impact of Tom's contributions to the field of microbial evolution and to the evolution of our understanding of the enteric pathogenicity of *E. coli*. We two authors have different and complementary perspectives. One of us (Leopold) recently received a Ph.D. in the laboratory of the other (Tarr). So, S. R. Leopold is the scientific granddaughter of Tom Whittam. Tom served on her thesis committee, and provided outstanding ideas and guidance, and imparted his enthusiasm, even after he became ill. P. I. Tarr and Tom Whittam were long-standing collaborators (14, 18, 56, 86, 91, 93, 96). The comments in this chapter that are in the first person refer to conversations and collaborations between Tom and P. I. Tarr that began in the late 1980s.

## PATHOGENIC *E. COLI*: THE FIRST HALF CENTURY

Theodor Escherich, a German pediatrician, described the organism that belongs to the genus that now honors his name. In December 1884, he delivered a lecture in Munich on bacteriologic methods to grow and characterize the infant intestinal microbiota (33), and 7 months later he described *Bacterium coli commune* (89). His thesis on infantile bacteria (33) was published in *Fortschritte der Medizinische* (32); its English translation demonstrates the massive

***Shana R. Leopold and Phillip I. Tarr*** • Departments of Pediatrics and Molecular Microbiology, Washington University School of Medicine, St. Louis, MO 63105.

research that formed the basis for his work (34, 35). Escherich found *B. coli commune* in the urine of girls, and additionally suggested that coliform bacilli were responsible for contagious intestinal disturbances (89). Thus, elements of Tom Whittam's work were presaged by Escherich early in the last century: *B. coli commune* was introduced as an inhabitant of stool as a commensal organism, in urine as a pathogen, and as a cause of enteric illnesses. An initial question prompted by Escherich's work was whether or not a single isolate of *B. coli commune* (i.e., *E. coli*) could express multiple phenotypes (commensalism, virulence) or if it was a broader species with good and bad actors. (The term *Escherichia coli*, in lieu of *B. coli commune*, was proposed in 1919 [23], but *E. coli* was not officially adopted by the International Committee on Bacteriological Nomenclature until nearly 4 decades later [50].)

Several discoveries resolved this debate against the Dr. Jekyll/Mr. Hyde hypothesis by demonstrating diversity within this species (i.e., there were Dr. Jekylls as well as Mr. Hydes among *E. coli*). Serum isolated from a patient suffering from diarrhea agglutinated bacteria isolated from other patients during the same outbreak, but not *E. coli* isolated from healthy patients, nor other enteric pathogens (83). Therefore, it became apparent that there was a range of additional phenotypes that inferred a large, diverse pool of bacteria. Additional progress was made, also in Germany, in the 1920s: Adam demonstrated groups of *E. coli* that were biochemically distinct, which he termed "*dyspepsiekoli*" (3, 5, 6). (Adam's work was dismissed by the German pediatric establishment, which favored an etiologic role for food as a cause of this illness [46].) In 1934, Dulaney and Michelson at the Memphis General Hospital discovered a variant of *E. coli* associated with severe gastroenteritis they termed *Bacterium coli mutabile* (83). However, there was so much conservation in the biochemical profiles of *E. coli* that such characteristics did not provide sufficient discriminatory power with which to begin to address the many different kinds of organisms that could be classified within this group.

Increased understanding of its surface structures led to an appreciation of diversity among *E. coli* organisms (4, 40), and attention turned to surface structures of *E. coli* that also relate to motility and capsules. *E. coli*'s motility is based on flagella, and flagella belong to multiple different antigenic types called the H antigens. The "H" is an abbreviation for "*Hauch*," which means "breath" in German, because the gossamer film produced by motile *E. coli* on an agar plate resembles the film from warm breath on a cold window. The "O" antigens represent different sugar side chains of bacterial lipopolysaccharide. The "O" designation simply means "*ohne* [without] *Hauch*," or immunogenic substances separate from those of the flagellar (H) antigens. *E. coli* "K" (*Kapsel*) antigens were described in 1945 and represent cell envelope and polysaccharide capsule proteins, which include fimbriae (13). The O:H:K designations constitute a serotype, and by the late 1940s provided some degree of subspeciation to *E. coli* (83).

## IN DIVERSITY, THERE IS DISEASE

Immunologic traits of putatively pathogenic *E. coli* led to a greater ability to identify subsets of enteric bacteria and to link these organisms to human diseases. Varela et al., in Mexico City in the 1940s, identified a predominant bacillus from children with fulminant and fatal epidemic infantile gastroenteritis, and demonstrated that infected patients, but not controls, had the same serologic response (99). Later in that decade, Bray, in London, identified specific serogroups of *E. coli* that caused epidemic infantile diarrhea (20, 21). These organisms, which were subsequently called "enteropathogenic" *E. coli* (EPEC) (67), were soon recognized to cause diarrhea worldwide, and the chief differentiations were made using serology; serogroups O55, O111, and O127 predominated (8, 10, 26–30, 36, 39, 43, 45, 53, 58, 65, 66, 72, 74, 78, 79, 97).

EPEC strains continued to cause outbreaks and sporadic cases of infantile diarrhea through the 1980s in developed countries (19, 25, 85), and remain major causes of morbidity and mortality in children in resource-poor settings (55, 71). They have as their cardinal phenotype the ability to adhere to epithelial cells in clusters, and to induce the attaching-and-effacing lesion in the cells to which they adhere. EPEC mediate this phenotype via intimin, which was identified in 1990 (49), and which is encoded on the locus of enterocyte effacement pathogenicity island (47). This region of the EPEC chromosome was a major focus of Tom Whittam's work.

## A TOPOLOGY IS BORN

Over the past several decades, our understanding of the diversity of *E. coli* began to take form. A critical advance in this field was the development of multilocus enzyme electrophoresis (MLEE). This technique quantifies genetic variation in the form of electrophoretic mobilities of conserved housekeeping proteins encoded in the *E. coli* chromosome. Differences in the amino acid sequence of homologous proteins alter the rate at which they electrophorese across a starch gel. Therefore, comparison of a panel of proteins can be used to quantify the relatedness of different isolates. MLEE had previously been developed to study eukaryotic species (105), and in 1973 Robert Milkman used the electrophoretic mobilities of five proteins to investigate the diversity of *E. coli*. He studied 829 isolates collected from 156 samples from a range of natural sources (61). He observed that while there was much genetic variation between bacteria within a single host (i.e., an animal carrying the *E. coli*), the five proteins could be classified into relatively few allele types. This led the author to conclude that the neutral hypothesis, which states that evolutionary changes are driven by random genetic drift, was incorrect and that the selectionist hypothesis (mutations are the result of natural selection) was supported.

The diversity of the natural population of *E. coli* became more apparent as more loci and larger populations of organisms were examined (24, 70), and a later study by Whittam et al. suggested that both natural selection and genetic drift are factors in the evolution of the species (105). In 1984, MLEE was completed on the *E. coli* Reference collection (ECOR), which is a panel of 72 isolates representing diverse geographic and host origins (69). The resulting genetic analysis categorized the species into four main groups (A, B1, B2, and D) and one minor group (E), and generated a widely used phylogenetic refer-

CLONAL NATURE OF O157:H7 ESCHERICHIA COLI ASSOCIATED WITH OUTBREAKS OF HEMORRHAGIC COLITIS. T. S. Whittam[1], I. K. Wachsmuth[3], and R. A. Wilson[2]. Departments of [1]Biology and [2]Veterinary Science, Pennsylvania State Univ., University Park, PA, U.S.A. & [3]Centers for Disease Control, Atlanta, GA, U.S.A.

Strains of *Escherichia coli* collected from epidemic outbreaks and sporadic cases of hemorrhagic colitis (HC) share an uncommon set of attributes including expression of O157:H7 antigens, production of verotoxins, and an inability to rapidly ferment sorbitol. To test the hypothesis that recent outbreaks of hemorrhagic colitis are caused by organisms representing a single geograhically-widespread clone, we examined 49 O157:H7 isolates using multilocus enzyme electrophoresis to assay genetic (allelic) variation in 20 s ruc ural genes that encode soluble enzymes. By characterizing each strain by its combination of electromorphs (alleles), we found that isola es of O157:H7 from four epidemics of HC in Oregon, Michigan, North Carolina, and Washington and 32 cases of sporadic HC cases from throughout North America were identical or nearly identical in electrophoretic type (multilocus genotype). Because of the extensive genetic diversity in the *E. coli* population as a whole, these results lend strong support to the hypothesis that O157:H7 strains associated with outbreaks and sporadic cases of hemorrhagic colitis are members of a single pathogenic clone that has recently spread through North America.

GEN-1

ence that has served as the foundation for investigators in the generation since (44). By the early 1980s, Tom Whittam had become an authority on the population genetics of *E. coli*. As fate would have it, his newly gained expertise would soon revolutionize our understanding of a novel *E. coli* pathogen and a significant threat to public safety.

## THE SCIENTIST AND THE PATHOGEN: TOM WHITTAM AND THE EMERGENCE OF EHEC

In two nearly simultaneous publications in 1983, a new group of pathogenic *E. coli* strains was introduced to the medical and microbiologic communities: Karmali et al. reported that stools from children with hemolytic uremic syndrome (HUS) contained a cytotoxin, and then went on to isolate and study the *E. coli* that produced this toxin (52). These toxigenic *E. coli* strains belonged to multiple serotypes, including O157:H7. Several years prior to that discovery, such cytotoxin-producing organisms had been recovered from humans and foods, but there was no association with disease (54); these *E. coli* organisms were termed "verotoxigenic." A week after Karmali et al. published their paper, Riley and colleagues at the Centers for Disease Control described two outbreaks of painful bloody diarrhea, which were attributed to the consumption of undercooked hamburgers. Many of these patients were infected with *E. coli* O157:H7 (81). Soon thereafter, O'Brien et al. likened the toxin produced by *E. coli* O157:H7 to that produced by *Shigella dysenteriae* serotype 1 (68). It is quite likely that similar infections had occurred before these reports: an apple cider outbreak in Ontario in 1980 was probably caused by *E. coli* (90); Barnard and Kibel suggested in 1965 that HUS might be attributed to pathogenic *E. coli* (though of course without knowledge of *E. coli* O157:H7) (12); and Kaplan et al. speculated that there was an infectious cause to clustered occurrences of HUS as early as 1975 (51).

Tom Whittam immediately grasped the importance of this "new" group of diarrheagenic *E. coli* and devoted much of his early career to studying it. At the 1st International Symposium and Workshop on Verocytotoxin (Shiga-like Toxin)-Producing *Escherichia coli* (VTEC) Infections in Toronto (July 12 to 15, 1987), Tom reported on the clonality of *E. coli* O157:H7, using MLEE (109) (Fig. 1), and he published his findings soon thereafter (106, 107). (Henry Smith, another giant in our field whose death at the height of his career in 2005 also left a large void, gave a platform presentation in that same session.) During that summer, Peggie Neill and our colleagues (Carla Clausen, Bob Hickman, and Dennis Christie) at the Seattle Children's Hospital and Medical Center (as it was called at the time), in Washington, had just published our first report associating *E. coli* O157:H7 with HUS (63). We were also aware of still-emerging data demonstrating the near exclusive association between enteric

**Figure 1.** Early presentation on EHEC by Tom Whittam. (Left) Abstract book cover. (Right) Abstract of talk delivered by Tom Whittam on the afternoon of Tuesday, July 14, 1987.

infection with this serotype and HUS. Specifically, if sorbitol MacConkey agar was used to isolate *E. coli* from stool early in the disease, almost all cases of childhood HUS could be attributed to a single pathogen (95). We also soon recognized that the putative cardinal virulence trait of *E. coli* O157:H7, namely, the possession of one or more Shiga toxin genes, was not by itself sufficient to cause severe human disease. There had to be something "else" that set this "clone" apart, and that conferred singular virulence on a rare serotype.

Soon after the Toronto meeting, I called Tom at Pennsylvania State University to get some elaboration on his talk. He answered my inquiries, and explained his working hypothesis about the emergence of O157:H7 from a less virulent serotype, O55:H7. At the time, I was a trainee and had little background in bacterial genetics. Not only that, I was a physician. Despite these limitations, Tom expounded graciously and patiently on his theories of clonality. Others might have been annoyed at the need to provide any explanation to a neophyte like me. In contrast, Tom was genuinely excited that someone outside his field was interested in his work, and made sure that I understood as much as he could convey. Without the framework Tom provided to our field, we would have lacked clarity about the phylogeny of a major set of pathogens. I didn't realize it at the time, but this conversation initiated a series of fruitful scientific collaborations between the two of us that lasted until Tom's death.

I next saw Tom in Miami at the September 1988 ASM meeting, when he and I delivered back-to-back talks at the same session. He was quite interested in our O157:H7 genotype data from a geographically defined region (subsequently published [94]), and was fascinated by our proposal that organisms containing $stx_2$ were more virulent than those containing $stx_1$ and $stx_2$ (75). Tom saw the value of analyzing panels of well-provenanced strains with clinical correlations from his Pacific Northwest "field station." Also, by focusing on isolates from single points of presentation or from a well-circumscribed geographic region, we minimized interregional murkiness, and in doing so we were able to identify epidemiologically defined isolates that are often not represented in collections of convenience.

Over the next 2 decades, Tom and the people he mentored delineated the emergence scenario of *E. coli* O157:H7 (see chapters 13 and 17). He saw the terrific opportunity presented by this terrible pathogen: not only do we have a "newly" emerged pathogen in our midst, but the group of organisms from which it emerged are extant (*E. coli* O55:H7) and available for study. He extended his early MLEE work to show that *E. coli* O157:H7 and its close relative, sorbitol-fermenting *E. coli* O157:H$^-$ (41), and *E. coli* O55:H7 have similar or identical isozyme forms (102, 108). The relation to *E. coli* O55:H7 is noteworthy because it was a frequently recovered EPEC serotype in infantile diarrhea in the 1950s, and it shares many properties with *E. coli* O157:H7 (84). Indeed, O157:H7 and O55:H7 differ at only a single electromorph by MLEE, namely, 6-phosphogluconate dehydrogenase (EC 1.1.1.44), which is encoded by the *gnd* locus. Interestingly, *gnd* is adjacent to the *rfb* region in *E. coli* (9), which encodes the O antigens that obviously differ between O157 and O55 strains. The close relationship between sorbitol-fermenting *E. coli* O157:H⁻ and typical non-sorbitol-fermenting *E. coli* O157:H7 is also noteworthy, because the pathogenicity of the former group of EHEC is undisputed, thereby reinforcing the virulence primacy of the EHEC 1 clade (7, 48, 77, 100).

Tom extended his analysis of the emergence of *E. coli* O157:H7 as sequences of new loci became available. For example, mutations at the *uidA* and *gusA* loci were harmonious with his proposed descent scenario (37, 62). His advice, encouragement, and collaboration helped us define mutations in the *rfb* region (16), *gnd* (96), and the Fim operon (87) as *E. coli* O157 emerged. Tom also identified different *eae* (intimin) alleles as different classes of EPEC and Shiga toxin-producing *E. coli* (STEC) were described, and framed his work in the context of microbial evolution, epidemiology, and clinical medicine. The finding of a suite of similar (but not identical) virulence genes in a limited data set (genes encoding Shiga toxins and intimin) (59) helped to firm up our concept that phylogeny and virulence were inextricably linked among the *E. coli*: many organisms that were not pathogens had one of these loci, but only closely related *E. coli* strains with a definable set of genes encoding virulence factors could cause human disease.

Even when Tom was not directly involved in my work, he was an influence. Even though I was practicing evolutionary research without a license, Tom always offered encouragement. The most memorable example occurred in 2001. While genotyping several isolates of *E. coli* O157:H7, on a whim Nurmohammad Shaikh and I interrogated the putative insertion site of the $stx_1$ bacteriophage, using the recently released genome sequences of this pathogen (42, 76). Instead of finding a strong band across this presumably unoccupied site, we saw a very weak band in our gel (Fig. 2). After excluding each of the usual denials (nonspecific priming, contaminated reagents, mislabeled specimens, spillover from one lane to another), Nur and I reviewed the emergence scenario developed by Tom for the EHEC 1 clade, and formulated the only hypothesis that remained: something, probably a bacteriophage that did not contain *stx*, occupied this site. After conferring with Tom, we demonstrated that this was indeed the case, and our work led to even more precise refinement of the sequence of events as *E. coli* O157:H7 emerged from its less pathogenic progenitor, *E. coli* O55:H7 (14, 56, 87, 88).

## "YOU CAN SEE A LOT JUST BY OBSERVING"

With this statement (110), Yogi Berra could have been describing Tom Whittam's scientific modus operandi. Though Tom did perform hypothesis-driven experimental research, he, his collaborators, and his protégés made more major contributions by scrutinizing sequences and then formulating and proving important evolutionary hypotheses, much like mathematicians prove their postulates. Sean Reid and Tom described the parallel emergence of virulence in pathogenic *E. coli* (80). Tom also used the tightly delimited group EHEC 1 to answer other questions in evolutionary biology. For example, Nelson and Selander postulated that *gnd* "'hitchhiked" with the adjacent *rfb* region (64), and

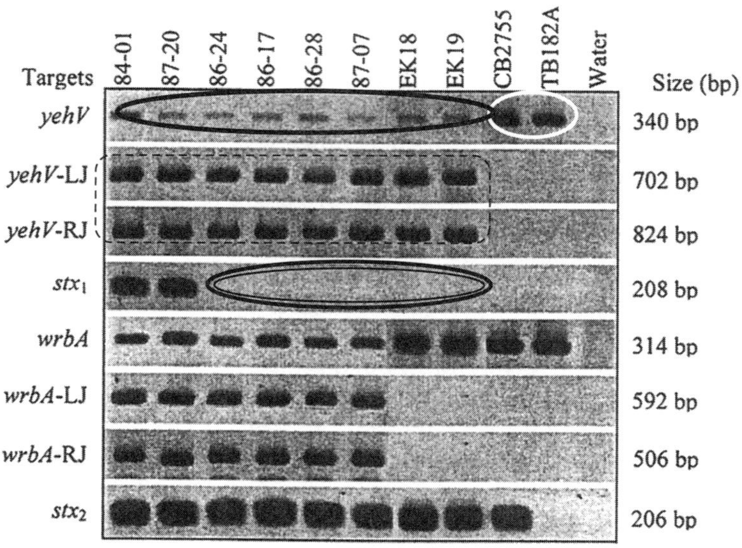


**Figure 2.** Unexplained faint band in a gel. Primers within *yehV* in *E. coli* O55:H7 produced a prominent 340-bp amplicon (white oval) as expected because this gene is not disrupted by a bacteriophage. Primers across the same site in $stx_1^-$ $stx_2^+$ as well as $stx_1^-$ $stx_2^+$ *E. coli* O157:H7 yield fainter bands (black oval), whereas we anticipated a stronger band, as in the nontoxigenic progenitor strain, *E. coli* O55:H7, amplified using the same primer set. Subsequent work demonstrated that a truncated bacteriophage occupied this site in *E. coli* O157:H7 that did not contain $stx_1$ (black dashed oval); the bacteriophage-chromosome junction amplicons are within the white dashed borders. Additional amplifications led us to deduce the precise phylogeny of emergence of the three major *E. coli* O157:H7 subgroups (14, 56, 87, 88). These data eventually led us to formulate a precise emergence of the EHEC 1 clade (see Fig. 3). Reproduced with permission from Fig. 1 in reference 88.

that this diversifying selection is responsible for the high degree of allelic variation at this locus (11, 17, 31, 103). When our group noticed that the O55:H7 *gnd* gene, which we had just cloned, failed to hybridize to *E. coli* O157:H7 strains, Tom immediately grasped the significance of this observation and prompted us to sequence *gnd* in a panel of evolutionarily instructive O157 and O55 isolates. These data made a very strong case for cotransfer of the *gnd* and the *rfb* loci (96). Also, using a "small area" pathogen collection, Tom provided phylogenetic context for putatively diarrheagenic *E. coli* in children in the Seattle area (18). Of the many STEC isolates that do not belong to the EHEC 1 clade, a single phylogenetic cluster causes most human disease, i.e., the EHEC 2 clade. This clade, too, was characterized by Tom as a closely related set of virulent *E. coli* (101, 104, 108). His findings have been confirmed, using additional loci and whole-genome sequencing (15, 73). When Tom found relatively rare pathogenic STEC that could not be assigned to EHEC 1 or 2 bins, he described them thoroughly and, as always, from an evolutionary perspective (91, 92).

The significance of these clusterings should not be underestimated: Tom's groupings of STEC that are associated with human disease illuminate why, of the many *E. coli* strains that contain *stx* genes, only a small subset have been credibly and repeatedly linked to human disease (22, 60, 100).

In recent years, Tom's work focused on the delineation of subgroups of *E. coli* O157:H7 (38, 82). He was instrumental in our formulation of multiple subgroups within this serotype (87, 88), development of the concepts of constrained radiation of *E. coli* O157:H7 subgroups, the small effective population sizes of these subsets (56) (Fig. 3), and identification of different bovine and human subsets of *E. coli* O157:H7 (14, 98). He proposed that some clones of *E. coli* O157:H7 are associated with worse human disease (57) and reported their plausibly related in vitro phenotypes (1, 2, 82). (For more on Tom's most recent work, see chapter 17.)

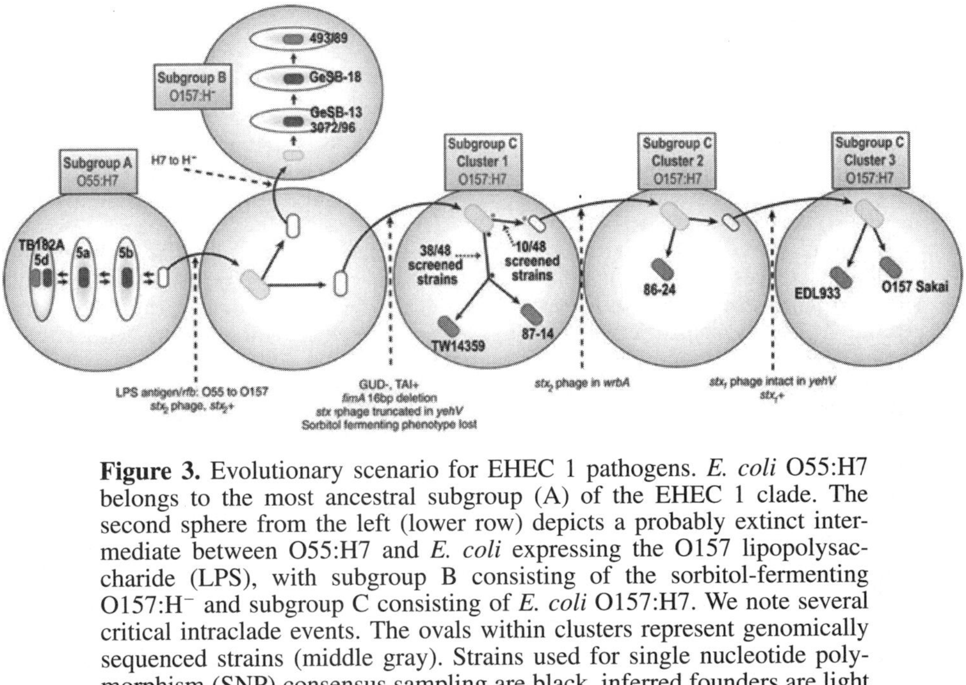


**Figure 3.** Evolutionary scenario for EHEC 1 pathogens. *E. coli* O55:H7 belongs to the most ancestral subgroup (A) of the EHEC 1 clade. The second sphere from the left (lower row) depicts a probably extinct intermediate between O55:H7 and *E. coli* expressing the O157 lipopolysaccharide (LPS), with subgroup B consisting of the sorbitol-fermenting O157:H$^-$ and subgroup C consisting of *E. coli* O157:H7. We note several critical intraclade events. The ovals within clusters represent genomically sequenced strains (middle gray). Strains used for single nucleotide polymorphism (SNP) consensus sampling are black, inferred founders are light gray, and postulated organisms that are immediate progenitors to the next cluster or subgroup are white ovals. In cluster 1, screened strains were assigned to the main branch if they had each of the three signature SNPs among the 111 shared SNPs (38/48), and the minor branch if they lacked this set of SNPs (10/48). More extensive evolutionary detail is provided in Fig. S1 in reference 56. Distances not drawn to scale. Reprinted with permission from reference 56. GUD, $\beta$-D-glucuronidase; TAI, tellurite resistance- and adherence-conferring island.

## THE LEGACY

Tom Whittam's investigations into the emergence of diarrheagenic *E. coli*, and most particularly of the O157:H7 EHEC 1 clade, stand as monumental and enduring contributions to the field. His coauthors and their diverse disciplines attest to the broadness of his influence. For example, our most recent coauthored paper, which proved the correctness of a model first proposed by Tom in the 1980s, included a microbiology graduate student (S. R. Leopold), microbial pathogenesis experts, genome biologists, a mathematician, a veterinary microbiologist, and the physician (P. I. Tarr) inspired by Tom Whittam's talk in Toronto in 1987. Tom's insights, perseverance, analytical skills, collegiality, and work ethic will inspire us for the rest of our careers, as we strive to continue his seminal work in microbial evolution and pathogenesis.

**Acknowledgments.** We thank Seth Walk and Peter Feng for helpful comments in the preparation of this manuscript, Alejandro Cravioto for sharing his knowledge of diarrheagenic *E. coli* with us, and Christine Musser for assistance with manuscript assembly and formatting. S.R.L. also thanks her thesis committee (David Haslam, Doug Berg, Elaine Mardis, Scott Hultgren, and the late Tom Whittam) for their input into her thesis dissertation, from which the historical summary in this chapter was taken.

## REFERENCES

1. **Abu-Ali, G. S., L. M. Ouellette, S. T. Henderson, D. W. Lacher, J. T. Riordan, T. S. Whittam, and S. D. Manning.** 2010. Increased adherence and expression of virulence genes in a lineage of *Escherichia coli* O157:H7 commonly associated with human infections. *PLoS One* **5:**e10167.
2. **Abu-Ali, G. S., L. M. Ouellette, S. T. Henderson, T. S. Whittam, and S. D. Manning.** 2010. Differences in adherence and virulence gene expression between two outbreak strains of enterohaemorrhagic *Escherichia coli* O157:H7. *Microbiology* **156:**408–419.
3. **Adam, A.** 1927. Dyspepsiekoli. Zur Frage der bakteriellen Ätiologie der sogenannten alimentären Intoxikation. *Jahrbuch Kinderheilkd.* **116:**8.
4. **Adam, A.** 1956. *Sauglings-Enteritis.* Georg Thieme Verlag, Stuttgart, Germany.
5. **Adam, A.** 1923. Über die Biologie der Dyspepsiecoli und ihre Beziehungen zur Pathogenese der Dyspepsie und Intoxikation. *Jahrbuch Kinderheilkd.* **101:**295.
6. **Adam, A., and C. Froboese.** 1925. Untersuchungen zur Pathologie der Durchfallserkrankungen des Säuglings. *Jahrbuch Kinderheilkd.* **39:**267.
7. **Alpers, K., D. Werber, C. Frank, J. Koch, A. W. Friedrich, H. Karch, M. an der Heiden, R. Prager, A. Fruth, M. Bielaszewska, G. Morlock, A. Heissenhuber, A. Diedler, A. Gerber, and A. Ammon.** 2009. Sorbitol-fermenting enterohaemorrhagic *Escherichia coli* O157:$H^-$ causes another outbreak of haemolytic uraemic syndrome in children. *Epidemiol. Infect.* **137:**389–395.
8. **Armstrong, J. G.** 1958. Acute gastroenteritis; a review of 821 cases treated at the Hospital for Sick Children during the year 1956. *J. Pediatr.* **53:**704–718.
9. **Bachmann, B. J.** 1990. Linkage map of *Escherichia coli* K-12, edition 8. *Microbiol. Rev.* **54:**130–197.
10. **Banerjea, A., D. N. Chatterji, and K. Pramanick.** 1957. Gastro-enteritis in different age-groups and its association with certain serological types of *Escherichia coli*. *Ann. Biochem. Exp. Med.* **17:**99.
11. **Barcak, G. J., and R. E. Wolf, Jr.** 1988. Comparative nucleotide sequence analysis of growth-rate-regulated *gnd* alleles from natural isolates of *Escherichia coli* and from *Salmonella typhimurium* LT-2. *J. Bacteriol.* **170:**372–379.

12. **Barnard, P. J., and M. Kibel.** 1965. The haemolytic-uraemic syndrome of infancy and childhood. A report of eleven cases. *Cent. Afr. J. Med.* **11:**31–34.
13. **Bergan, T. (ed.).** 1984. *Methods in Microbiology*, vol. 14. Academic Press, London, United Kingdom.
14. **Besser, T. E., N. Shaikh, N. J. Holt, P. I. Tarr, M. E. Konkel, P. Malik-Kale, C. W. Walsh, T. S. Whittam, and J. L. Bono.** 2007. Greater diversity of Shiga toxin-encoding bacteriophage insertion sites among *Escherichia coli* O157:H7 isolates from cattle than in those from humans. *Appl. Environ. Microbiol.* **73:**671–679.
15. **Bielaszewska, M., R. Prager, L. Vandivinit, A. Müsken, A. Mellmann, N. J. Holt, P. I. Tarr, H. Karch, and W. Zhang.** 2009. Detection and characterization of the fimbrial *sfp* cluster in enterohemorrhagic *Escherichia coli* O165:H25/NM isolates from humans and cattle. *Appl. Environ. Microbiol.* **75:**64–71.
16. **Bilge, S. S., J. C. Vary, Jr., S. F. Dowell, and P. I. Tarr.** 1996. Role of the *Escherichia coli* O157:H7 O side chain in adherence and analysis of an *rfb* locus. *Infect. Immun.* **64:**4795–4801.
17. **Bisercic, M., J. Y. Feutrier, and P. R. Reeves.** 1991. Nucleotide sequences of the *gnd* genes from nine natural isolates of *Escherichia coli*: evidence of intragenic recombination as a contributing factor in the evolution of the polymorphic *gnd* locus. *J. Bacteriol.* **173:**3894–3900.
18. **Bokete, T. N., T. S. Whittam, R. A. Wilson, C. R. Clausen, C. M. O'Callahan, S. L. Moseley, T. R. Fritsche, and P. I. Tarr.** 1997. Genetic and phenotypic analysis of *Escherichia coli* with enteropathogenic characteristics isolated from Seattle children. *J. Infect. Dis.* **175:**1382–1389.
19. **Boyer, K. M., N. J. Petersen, I. Farzaneh, C. P. Pattison, M. C. Hart, and J. E. Maynard.** 1975. An outbreak of gastroenteritis due to *E. coli* 0142 in a neonatal nursery. *J. Pediatr.* **86:**919–927.
20. **Bray, J.** 1973. Bray's discovery of pathogenic *Esch. coli* as a cause of infantile gastroenteritis. *Arch. Dis. Child.* **48:**923–926.
21. **Bray, J.** 1945. Isolation of antigenically homogeneous strains of *Bact. coli neapolitanum* from summer diarrhoea of infants. *J. Pathol. Bacteriol.* **57:**239.
22. **Brooks, J. T., E. G. Sowers, J. G. Wells, K. D. Greene, P. M. Griffin, R. M. Hoekstra, and N. A. Strockbine.** 2005. Non-O157 Shiga toxin-producing *Escherichia coli* infections in the United States, 1983–2002. *J. Infect. Dis.* **192:**1422–1429.
23. **Castellani, A., and A. J. Chalmers.** 1919. *Manual of Tropical Medicine*, 3rd ed. William Wood, New York, NY.
24. **Caugant, D. A., B. R. Levin, G. Lidin-Janson, T. S. Whittam, C. Svanborg Edén, and R. K. Selander.** 1983. Genetic diversity and relationships among strains of *Escherichia coli* in the intestine and those causing urinary tract infections. *Prog. Allergy* **33:**203–227.
25. **Clausen, C. R., and D. L. Christie.** 1982. Chronic diarrhea in infants caused by adherent enteropathogenic *Escherichia coli*. *J. Pediatr.* **100:**358–361.
26. **Cooper, M. L., E. W. Walters, H. M. Keller, J. M. Sutherland, and H. J. Wiseman.** 1955. Epidemic diarrhea among infants associated with the isolation of a new serotype of *Escherichia coli*: *E. coli* 0127:B8. *Pediatrics* **16:**215–227.
27. **Darrasse, H., L. Le Minor, D. Piechaud, and P. Nicolle.** 1957. Pathogenic enterobacteria from Dakar. *Bull. Soc. Pathol. Exot. Filiales* **50:**257–281. (In French.)
28. **Davydov, V. P.** 1958. Serologic characteristics of *Escherichia coli* isolated in infants with gastrointestinal diseases. *Zh. Mikrobiol. Epidemiol. Immunobiol.* **29:**56–60. (In Russian.)
29. **Derham, R. J., and M. M. Rogerson.** 1957. Gastroenteritis in infancy. *AMA J. Dis. Child.* **93:**113–121.
30. **Drimmer-Herrnheiser, H., and A. L. Olitzki.** 1951. The association of *Escherichia coli* (serotypes 0111:B4 and 055:B5) with cases of acute infantile gastroenteritis in Jerusalem. *Acta Med. Orient.* **10:**219–223.
31. **Dykhuizen, D. E., and L. Green.** 1991. Recombination in *Escherichia coli* and the definition of biological species. *J. Bacteriol.* **173:**7257–7268.

32. **Escherich, T.** 1885. Die darmbakterien des neugeborenen und säuglings. *Fortschr. Med.* **3:**515–522, 547–554.
33. **Escherich, T.** 1886. *Die Darmbakterien des Säuglings und Ihre Beziehungen zur Physiologie der Verdauung.* Fernidnand Enke, Stuttgart, Germany.
34. **Escherich, T.** 1988. The intestinal bacteria of the neonate and breast-fed infant. 1884. *Rev. Infect. Dis.* **10:**1220–1225.
35. **Escherich, T.** 1989. The intestinal bacteria of the neonate and breast-fed infant. 1885. *Rev. Infect. Dis.* **11:**352–356.
36. **Ewing, W. H., H. W. Tatum, and B. R. Davis.** 1957. The occurrence of *Escherichia coli* serotypes associated with diarrheal diseases in the United States. *Public Health Lab.* **15:**118.
37. **Feng, P., K. A. Lampel, H. Karch, and T. S. Whittam.** 1998. Genotypic and phenotypic changes in the emergence of *Escherichia coli* O157:H7. *J. Infect. Dis.* **177:**1750–1753.
38. **Feng, P. C., S. R. Monday, D. W. Lacher, L. Allison, A. Siitonen, C. Keys, M. Eklund, H. Nagano, H. Karch, J. Keen, and T. S. Whittam.** 2007. Genetic diversity among clonal lineages within *Escherichia coli* O157:H7 stepwise evolutionary model. *Emerg. Infect. Dis.* **13:**1701–1706.
39. **Fournelle, H. J., I. L. Wallace, and V. Rader.** 1958. A bacteriological and parasitological survey of enteric infections in an Alaskan Eskimo area. *Am. J. Public Health Nations Health* **48:**1489–1497.
40. **Goldschmidt, R.** 1933. Untersuchungen zur Atiologie der Durchfallserkrankungen des Säuglings. *Jahrbuch Kinderheilkd.* **139:**318.
41. **Gunzer, F., H. Böhm, H. Rüssmann, M. Bitzan, S. Aleksic, and H. Karch.** 1992. Molecular detection of sorbitol-fermenting *Escherichia coli* O157 in patients with hemolytic-uremic syndrome. *J. Clin. Microbiol.* **30:**1807–1810.
42. **Hayashi, T., K. Makino, M. Ohnishi, K. Kurokawa, K. Ishii, K. Yokoyama, C. G. Han, E. Ohtsubo, K. Nakayama, T. Murata, M. Tanaka, T. Tobe, T. Iida, H. Takami, T. Honda, C. Sasakawa, N. Ogasawara, T. Yasunaga, S. Kuhara, T. Shiba, M. Hattori, and H. Shinagawa.** 2001. Complete genome sequence of enterohemorrhagic *Escherichia coli* O157:H7 and genomic comparison with a laboratory strain K-12. *DNA Res.* **8:**11–22.
43. **Herweg, J. C., J. N. Middelkamp, and H. K. Thornton.** 1956. *Escherichia coli* diarrhea; the relationship of certain serotypes of *Escherichia coli* to sporadic and epidemic cases of infantile diarrhea. *J. Pediatr.* **49:**629–641.
44. **Herzer, P. J., S. Inouye, M. Inouye, and T. S. Whittam.** 1990. Phylogenetic distribution of branched RNA-linked multicopy single-stranded DNA among natural isolates of *Escherichia coli. J. Bacteriol.* **172:**6175–6181.
45. **Hinton, N. A., and R. R. MacGregor.** 1958. A study of infections due to pathogenic serogroups of *Escherichia coli. Can. Med. Assoc. J.* **79:**359–364.
46. **Holzel, A.** 1974. Correspondence: Bray's discovery of pathogenic *Esch. coli* as a cause of infantile gastroenteritis. *Arch. Dis. Child.* **49:**668.
47. **Jarvis, K. G., J. A. Girón, A. E. Jerse, T. K. McDaniel, M. S. Donnenberg, and J. B. Kaper.** 1995. Enteropathogenic *Escherichia coli* contains a putative type III secretion system necessary for the export of proteins involved in attaching and effacing lesion formation. *Proc. Natl. Acad. Sci. USA* **92:**7996–8000.
48. **Jenke, C., D. Harmsen, T. Weniger, J. Rothganger, E. Hyytia-Trees, M. Bielaszewska, H. Karch, and A. Mellmann.** Phylogenetic analysis of enterohemorrhagic *Escherichia coli* O157, Germany, 1987–2008. *Emerg. Infect. Dis.* **16:**610–616.
49. **Jerse, A. E., J. Yu, B. D. Tall, and J. B. Kaper.** 1990. A genetic locus of enteropathogenic *Escherichia coli* necessary for the production of attaching and effacing lesions on tissue culture cells. *Proc. Natl. Acad. Sci. USA* **87:**7839–7843.
50. **Judicial Commission of the International Committee on Bacteriological Nomenclature.** 1958. Conservation of the family name Enterobacteriaceae, of the name of the

type genus, and designation of the type species. *Intl. Bull. Bacteriol. Nomencl. Taxon.* **8:**73–74.
51. **Kaplan, B. S., R. W. Chesney, and K. N. Drummond.** 1975. Hemolytic uremic syndrome in families. *N. Engl. J. Med.* **292:**1090–1093.
52. **Karmali, M. A., B. T. Steele, M. Petric, and C. Lim.** 1983. Sporadic cases of haemolytic-uraemic syndrome associated with faecal cytotoxin and cytotoxin-producing *Escherichia coli* in stools. *Lancet* **1:**619–620.
53. **Kauffmann, F.** 1954. *Enterobacteriaceae*, 2nd ed. Ejnar Munksgaard, Copenhagen, Denmark.
54. **Konowalchuk, J., J. I. Speirs, and S. Stavric.** 1977. Vero response to a cytotoxin of *Escherichia coli. Infect. Immun.* **18:**775–779.
55. **Lanata, C. F., W. Mendoza, and R. E. Black.** 2002. *Improving Diarrhoea Estimates.* World Health Organization, Geneva, Switzerland.
56. **Leopold, S. R., V. Magrini, N. J. Holt, N. Shaikh, E. R. Mardis, J. Cagno, Y. Ogura, A. Iguchi, T. Hayashi, A. Mellmann, H. Karch, T. E. Besser, S. A. Sawyer, T. S. Whittam, and P. I. Tarr.** 2009. A precise reconstruction of the emergence and constrained radiations of *Escherichia coli* O157 portrayed by backbone concatenomic analysis. *Proc. Natl. Acad. Sci. USA* **106:**8713–8718.
57. **Manning, S. D., A. S. Motiwala, A. C. Springman, W. Qi, D. W. Lacher, L. M. Ouellette, J. M. Mladonicky, P. Somsel, J. T. Rudrik, S. E. Dietrich, W. Zhang, B. Swaminathan, D. Alland, and T. S. Whittam.** 2008. Variation in virulence among clades of *Escherichia coli* O157:H7 associated with disease outbreaks. *Proc. Natl. Acad. Sci. USA* **105:**4868–4873.
58. **Martineau, B., R. Raymond, and G. Jeliu.** 1958. Bacteriological and clinical study of gastro-enteritis with enteropathogenic *Escherichia coli* 0127:B8. *Can. Med. Assoc. J.* **79:**351–359.
59. **McGraw, E. A., J. Li, R. K. Selander, and T. S. Whittam.** 1999. Molecular evolution and mosaic structure of $\alpha$, $\beta$, and $\gamma$ intimins of pathogenic *Escherichia coli. Mol. Biol. Evol.* **16:**12–22.
60. **Mellmann, A., M. Bielaszewska, R. Köck, A. W. Friedrich, A. Fruth, B. Middendorf, D. Harmsen, M. A. Schmidt, and H. Karch.** 2008. Analysis of collection of hemolytic uremic syndrome-associated enterohemorrhagic *Escherichia coli. Emerg. Infect. Dis.* **14:**1287–1290.
61. **Milkman, R.** 1973. Electrophoretic variation in *Escherichia coli* from natural sources. *Science* **182:**1024–1026.
62. **Monday, S. R., T. S. Whittam, and P. C. Feng.** 2001. Genetic and evolutionary analysis of mutations in the *gusA* gene that cause the absence of $\beta$-glucuronidase activity in *Escherichia coli* O157:H7. *J. Infect. Dis.* **184:**918–921.
63. **Neill, M. A., P. I. Tarr, C. R. Clausen, D. L. Christie, and R. O. Hickman.** 1987. *Escherichia coli* O157:H7 as the predominant pathogen associated with the hemolytic uremic syndrome: a prospective study in the Pacific Northwest. *Pediatrics* **80:**37–40.
64. **Nelson, K., and R. K. Selander.** 1994. Intergeneric transfer and recombination of the 6-phosphogluconate dehydrogenase gene (*gnd*) in enteric bacteria. *Proc. Natl. Acad. Sci. USA* **91:**10227–10231.
65. **Neter, E.** 1955. *Escherichia coli* diarrhea: an outbreak among infants on a surgical ward. *AMA Am. J. Dis. Child.* **89:**564–566.
66. **Neter, E., R. F. Korns, and R. E. Trussel.** 1953. Association of *Escherichia coli* serogroup 0111 with two hospital outbreaks of epidemic diarrhea of the newborn infant in New York State during 1947. *Pediatrics* **12:**377–383.
67. **Neter, E., O. Westphal, O. Luderitz, R. M. Gino, and E. A. Gorzynski.** 1955. Demonstration of antibodies against enteropathogenic *Escherichia coli* in sera of children of various ages. *Pediatrics* **16:**801–808.
68. **O'Brien, A. O., T. A. Lively, M. E. Chen, S. W. Rothman, and S. B. Formal.** 1983. *Escherichia coli* O157:H7 strains associated with haemorrhagic colitis in the United States produce a *Shigella dysenteriae* 1 (SHIGA) like cytotoxin. *Lancet* **1:**702.

69. **Ochman, H., and R. K. Selander.** 1984. Standard reference strains of *Escherichia coli* from natural populations. *J. Bacteriol.* **157:**690–693.
70. **Ochman, H., T. S. Whittam, D. A. Caugant, and R. K. Selander.** 1983. Enzyme polymorphism and genetic population structure in *Escherichia coli* and *Shigella*. *J. Gen. Microbiol.* **129:**2715–2726.
71. **Ochoa, T. J., L. Ecker, F. Barletta, M. L. Mispireta, A. I. Gil, C. Contreras, M. Molina, I. Amemiya, H. Verastegui, E. R. Hall, T. G. Cleary, and C. F. Lanata.** 2009. Age-related susceptibility to infection with diarrheagenic *E. coli* in infants from peri-urban areas of Lima, Peru. *Clin. Infect. Dis.* **49:**1694–1702.
72. **Ocklitz, H. W., E. F. Schmidt, and J. Bahr.** 1957. Clinical comparison of specific intestinal infections in infants. *Z. Kinderheilkd.* **79:**290–316. (In German.)
73. **Ogura, Y., T. Ooka, A. Iguchi, H. Toh, M. Asadulghani, K. Oshima, T. Kodama, H. Abe, K. Nakayama, K. Kurokawa, T. Tobe, M. Hattori, and T. Hayashi.** 2009. Comparative genomics reveal the mechanism of the parallel evolution of O157 and non-O157 enterohemorrhagic *Escherichia coli*. *Proc. Natl. Acad. Sci. USA* **106:**17939–17944.
74. **Olarte, J., M. Ramos-Alvarez, and E. Galindo.** 1957. Isolation of Shigella, Salmonella & coli enteropathogens in rectal swabs of 802 cases of sporadic diarrhea. *Bol. Med. Hosp. Infant. Mex.* **14:**257–262; English transl. 263–268. (In Spanish.)
75. **Ostroff, S. M., P. I. Tarr, M. A. Neill, J. H. Lewis, N. Hargrett-Bean, and J. M. Kobayashi.** 1989. Toxin genotypes and plasmid profiles as determinants of systemic sequelae in *Escherichia coli* O157:H7 infections. *J. Infect. Dis.* **160:**994–998.
76. **Perna, N. T., G. Plunkett III, V. Burland, B. Mau, J. D. Glasner, D. J. Rose, G. F. Mayhew, P. S. Evans, J. Gregor, H. A. Kirkpatrick, G. Pósfai, J. Hackett, S. Klink, A. Boutin, Y. Shao, L. Miller, E. J. Grotbeck, N. W. Davis, A. Lim, E. T. Dimalanta, K. D. Potamousis, J. Apodaca, T. S. Anantharaman, J. Lin, G. Yen, D. C. Schwartz, R. A. Welch, and F. R. Blattner.** 2001. Genome sequence of enterohaemorrhagic *Escherichia coli* O157:H7. *Nature* **409:**529–533.
77. **Pollock, K. G., M. E. Locking, T. J. Beattie, H. Maxwell, I. Ramage, D. Hughes, J. Cowieson, L. Allison, M. Hanson, and J. M. Cowden.** 2010. Sorbitol-fermenting *Escherichia coli* O157, Scotland. *Emerg. Infect. Dis.* **16:**881–882.
78. **Quilligan, J. J., Jr., M. Biddle, J. W. Dedrick, M. L. Scott, and I. Burnett.** 1957. Recovery of enteropathogenic types of *Escherichia coli* from infants with diarrhea. *AMA J. Dis. Child.* **93:**420–424.
79. **Quilligan, J. J., Jr., and S. Shadomy.** 1958. Enteropathogenic *Escherichia coli* diarrhea. *AMA J. Dis. Child.* **95:**134–138.
80. **Reid, S. D., C. J. Herbelin, A. C. Bumbaugh, R. K. Selander, and T. S. Whittam.** 2000. Parallel evolution of virulence in pathogenic *Escherichia coli*. *Nature* **406:**64–67.
81. **Riley, L. W., R. S. Remis, S. D. Helgerson, H. B. McGee, J. G. Wells, B. R. Davis, R. J. Hebert, E. S. Olcott, L. M. Johnson, N. T. Hargrett, P. A. Blake, and M. L. Cohen.** 1983. Hemorrhagic colitis associated with a rare *Escherichia coli* serotype. *N. Engl. J. Med.* **308:**681–685.
82. **Riordan, J. T., S. B. Viswanath, S. D. Manning, and T. S. Whittam.** 2008. Genetic differentiation of *Escherichia coli* O157:H7 clades associated with human disease by real-time PCR. *J. Clin. Microbiol.* **46:**2070–2073.
83. **Robins-Browne, R. M.** 1987. Traditional enteropathogenic *Escherichia coli* of infantile diarrhea. *Rev. Infect. Dis.* **9:**28–53.
84. **Rodrigues, J., I. C. Scaletsky, L. C. Campos, T. A. Gomes, T. S. Whittam, and L. R. Trabulsi.** 1996. Clonal structure and virulence factors in strains of *Escherichia coli* of the classic serogroup O55. *Infect. Immun.* **64:**2680–2686.
85. **Rothbaum, R. J., R. A. Giannella, and J. C. Partin.** 1982. Diarrhea caused by adherent enteropathogenic *E. coli*. *J. Pediatr.* **101:**486.
86. **Samadpour, M., J. E. Ongerth, J. Liston, N. Tran, D. Nguyen, T. S. Whittam, R. A. Wilson, and P. I. Tarr.** 1994. Occurrence of Shiga-like toxin-producing *Escherichia coli*

in retail fresh seafood, beef, lamb, pork, and poultry from grocery stores in Seattle, Washington. *Appl. Environ. Microbiol.* **60:**1038–1040.

87. **Shaikh, N., N. J. Holt, J. R. Johnson, and P. I. Tarr.** 2007. *Fim* operon variation in the emergence of enterohemorrhagic *Escherichia coli*: an evolutionary and functional analysis. *FEMS Microbiol. Lett.* **273:**58–63.
88. **Shaikh, N., and P. I. Tarr.** 2003. *Escherichia coli* O157:H7 Shiga toxin-encoding bacteriophages: integrations, excisions, truncations, and evolutionary implications. *J. Bacteriol.* **185:**3596–3605.
89. **Shulman, S. T., H. C. Friedmann, and R. H. Sims.** 2007. Theodor Escherich: the first pediatric infectious diseases physician? *Clin. Infect. Dis.* **45:**1025–1029.
90. **Steele, B. T., N. Murphy, G. S. Arbus, and C. P. Rance.** 1982. An outbreak of hemolytic uremic syndrome associated with ingestion of fresh apple juice. *J. Pediatr.* **101:**963–965.
91. **Tarr, C. L., T. M. Large, C. L. Moeller, D. W. Lacher, P. I. Tarr, D. W. Acheson, and T. S. Whittam.** 2002. Molecular characterization of a serotype O121:H19 clone, a distinct Shiga toxin-producing clone of pathogenic *Escherichia coli*. *Infect. Immun.* **70:**6853–6859.
92. **Tarr, C. L., A. M. Nelson, L. Beutin, K. E. Olsen, and T. S. Whittam.** 2008. Molecular characterization reveals similar virulence gene content in unrelated clonal groups of *Escherichia coli* of serogroup O174 (OX3). *J. Bacteriol.* **190:**1344–1349.
93. **Tarr, P. I., C. R. Clausen, T. S. Whittam, and R. A. Wilson.** 1998. Antibiotic resistance and O antigen expression in *Escherichia coli*. *J. Infect. Dis.* **177:**1774–1775.
94. **Tarr, P. I., M. A. Neill, C. R. Clausen, J. W. Newland, R. J. Neill, and S. L. Moseley.** 1989. Genotypic variation in pathogenic *Escherichia coli* O157:H7 isolated from patients in Washington, 1984–1987. *J. Infect. Dis.* **159:**344–347.
95. **Tarr, P. I., M. A. Neill, C. R. Clausen, S. L. Watkins, D. L. Christie, and R. O. Hickman.** 1990. *Escherichia coli* O157:H7 and the hemolytic uremic syndrome: importance of early cultures in establishing the etiology. *J. Infect. Dis.* **162:**553–556.
96. **Tarr, P. I., L. M. Schoening, Y.-L. Yea, T. R. Ward, S. Jelacic, and T. S. Whittam.** 2000. Acquisition of the *rfb-gnd* cluster in evolution of *Escherichia coli* O55 and O157. *J. Bacteriol.* **182:**6183–6191.
97. **Thomas, M. E., and R. E. Charter.** 1956. Aetiology of juvenile diarrhoea; laboratory investigation of specimens referred from general practices in North-East London. *Br. Med. J.* **2:**339–341.
98. **Vanaja, S. K., A. C. Springman, T. E. Besser, T. S. Whittam, and S. D. Manning.** 2010. Differential expression of virulence and stress fitness genes between *Escherichia coli* O157:H7 strains with clinical or bovine-biased genotypes. *Appl. Environ. Microbiol.* **76:**60–68.
99. **Varela, G. A., A. Aguirre, and J. Carrillo.** 1946. *E. coli*—gomes, nueva especie aislada de un caso mortal de diarrea. *Bol. Med. Hosp. Infant. Mex.* **3:**623–627.
100. **Werber, D., S. C. Behnke, A. Fruth, R. Merle, S. Menzler, S. Glaser, L. Kreienbrock, R. Prager, H. Tschäpe, P. Roggentin, J. Bockemühl, and A. Ammon.** 2007. Shiga toxin-producing *Escherichia coli* infection in Germany: different risk factors for different age groups. *Am. J. Epidemiol.* **165:**425–434.
101. **Whittam, T. S.** 1998. Evolution of *Escherichia coli* O157:H7 and other Shiga toxin-producing *E. coli* strains, p. 195–209. *In* J. B. Kaper and A. D. O'Brien (ed.), Escherichia coli *O157:H7 and Other Shiga Toxin-Producing* E. coli *Strains*. ASM Press, Washington, DC.
102. **Whittam, T. S.** 1995. Genetic population structure and pathogenicity in enteric bacteria, p. 217–245. *In* S. Baumberg, J. P. W. Young, E. M. H. Wellington, and J. R. Saunders (ed.), *Population Genetics of Bacteria*. Cambridge University Press, Cambridge, United Kingdom.
103. **Whittam, T. S., and S. E. Ake.** 1993. Genetic polymorphisms and recombination in natural populations of *Escherichia coli*, p. 223–245. *In* N. Takahata and A. G. Clark (ed.), *Mechanisms of Molecular Evolution*. Sinauer Associates, Sunderland, MA.

104. **Whittam, T. S., and E. A. McGraw.** 1996. Clonal analysis of EPEC serogroups. *Rev. Microbiol.* **27:**7–16.
105. **Whittam, T. S., H. Ochman, and R. K. Selander.** 1983. Multilocus genetic structure in natural populations of *Escherichia coli. Proc. Natl. Acad. Sci. USA* **80:**1751–1755.
106. **Whittam, T. S., I. K. Wachsmuth, and R. A. Wilson.** 1988. Genetic evidence of clonal descent of *Escherichia coli* O157:H7 associated with hemorrhagic colitis and hemolytic uremic syndrome. *J. Infect. Dis.* **157:**1124–1133.
107. **Whittam, T. S., and R. A. Wilson.** 1988. Genetic relationships among pathogenic *Escherichia coli* of serogroup O157. *Infect. Immun.* **56:**2467–2473.
108. **Whittam, T. S., M. L. Wolfe, I. K. Wachsmuth, F. Ørskov, I. Ørskov, and R. A. Wilson.** 1993. Clonal relationships among *Escherichia coli* strains that cause hemorrhagic colitis and infantile diarrhea. *Infect. Immun.* **61:**1619–1629.
109. **Whittam, T. S., I. K. Wachsmuth, and R. A. Wilson.** 1987. Clonal nature of O157: H7 *Escherichia coli* associated with outbreaks of hemorrhagic colitis, abstr. GEN-1. *In Abstracts of the 1st International Symposium and Workshop on Verocytotoxin (Shiga-like Toxin)-Producing* Escherichia coli *(VTEC) Infections.*
110. **Wisnia, S., and D. Schlossberg.** 1997. *Wit and Wisdom of Baseball.* Publications International Limited, Lincolnwood, IL.

*Population Genetics of Bacteria: a Tribute to Thomas S. Whittam*
Edited by S. T. Walk and P. C. H. Feng


*Chapter 16*

# Sorbitol-Fermenting Enterohemorrhagic *Escherichia coli* O157:H⁻

***Helge Karch, Andrea Ammon, Phillip I. Tarr, and Martina Bielaszewska***

Enterohemorrhagic *Escherichia coli* (EHEC) is a pathogenic subgroup of Shiga toxin (Stx)-producing *E. coli* (STEC). EHEC strains have the particular ability to cause severe intestinal disease and the hemolytic uremic syndrome (HUS) (50). The prototypic EHEC, *E. coli* O157:H7, lacks the ability to produce $\beta$-D-glucuronidase and to ferment sorbitol during 24 h of incubation, the features that distinguish it from most other *E. coli* strains, including commensal and pathogenic strains (47, 82). The latter characteristic has been used as a basis for the development of culture strategies for the isolation of this organism from patients' stools and environmental samples (53). Indeed, this "low-tech" screening methodology for *E. coli* O157:H7 is more sensitive than toxin-based methodologies (18, 27, 46, 63, 80).

A relative of EHEC O157:H7, namely, sorbitol-fermenting (SF) *E. coli* O157:$H^-$ (nonmotile), is an important and fascinating human pathogen. These strains are emerging, in particular, in Europe (5, 16, 23, 25, 32, 34, 44, 54, 61, 65, 72). Tom Whittam was directly involved in our work on SF *E. coli* O157 since its inception, so we are honored to review here our work on SF *E. coli* O157:$H^-$.

## IDENTIFICATION OF SF *E. COLI* O157:$H^-$ IN GERMANY

In autumn 1988, six children from a small village in Bavaria with fewer than 5,000 residents developed HUS. All needed peritoneal dialysis after admission to a university hospital in Munich (40). This small outbreak was locally an extraordinary event. The standard bacteriological procedures on the stools

***Helge Karch and Martina Bielaszewska*** • Institute for Hygiene, University of Münster, 48149 Münster, Germany. ***Andrea Ammon*** • European Centre for Disease Prevention and Control (ECDC), 17183 Stockholm, Sweden. ***Phillip I. Tarr*** • Department of Pediatrics and Department of Molecular Microbiology, Washington University School of Medicine, St. Louis, MO 63110.

yielded negative results, which was quite surprising as the specimens were screened on sorbitol MacConkey (SMAC) agar (53) and we would have expected at least a few to be positive for *E. coli* O157:H7 (76). Therefore, the challenge of identifying the etiology of these HUS cases was given to my (H. Karch) laboratory, because we had applied a PCR and colony blot hybridization to the detection of the Stx-encoding genes (38, 39). We were quite motivated to find the infecting pathogen, because the PCR with primers MK1/MK2 (39) from the enrichment culture was positive, the SMAC agar results notwithstanding. Therefore, we performed a colony blot hybridization test. The colonies identified with this procedure agglutinated in anti-O157 serum. However, to our considerable surprise, these organisms were nonmotile. They also fermented sorbitol and produced $\beta$-D-glucuronidase (40). This was the first isolation of SF EHEC O157:H$^-$, a clone of pathogens on which my laboratory has worked for over 2 decades.

## SF EHEC O157:H$^-$ STRAINS POSE IMMENSE DIAGNOSTIC CHALLENGES

Because of their ability to ferment sorbitol after overnight incubation, SF EHEC O157:H$^-$ strains cannot be distinguished from the physiological intestinal microflora using SMAC agar (53). This medium remains the most commonly used, rapid, and appropriate way to isolate EHEC O157:H7. Moreover, in contrast to non-sorbitol-fermenting (NSF) EHEC O157:H7 strains that possess the *ter* gene cluster encoding tellurite resistance (77, 79) and can be selectively isolated on tellurite-containing media such as cefixime-tellurite (CT)-SMAC agar (87), SF EHEC O157:H$^-$ strains do not possess the *ter* genes or tellurite resistance (9). Therefore, they are tellurite susceptible and do not grow on CT-SMAC (9). Thus, because of their ability to ferment sorbitol and lack of tellurite resistance, SF EHEC O157:H$^-$ strains not only are overlooked in the procedures used to isolate NSF EHEC O157:H7, but they are suppressed by such methods.

## OUTBREAKS BY SF *E. COLI* O157:H$^-$

SF EHEC O157:H$^-$ has caused several outbreaks of HUS, the largest of which involved 38 cases (Table 1). Common features of these outbreaks were the initial increase in the number of pediatric patients with HUS and the much lower proportion of patients with uncomplicated diarrhea (2, 3). These findings indicated that a high proportion of patients infected with the pathogen would develop HUS, and indeed, our data on those with diarrhea due to SF EHEC O157:H$^-$ showed that this life-threatening outcome occurs in every second patient infected with these organisms (H. Karch, unpublished data). Therefore, the rate of progression to HUS in SF EHEC O157:H$^-$ infections (~50%) is profoundly higher than the rates of ~10% and ~15% reported in Germany (83) and the United States (78), respectively, for infections with EHEC O157:H7. This high rate of HUS development after SF EHEC O157 infection was also

**Table 1.** Outbreaks caused by SF EHEC O157:$H^-$

| Dates | Country | No. of HUS cases | No. of deaths | Source | Reference |
|---|---|---|---|---|---|
| Sep.–Nov. 1988 | Germany | 6 | 0 | Unknown | 40 |
| Dec. 1995–Mar. 1996 | Germany | 28 | 3 | Raw meat-containing sausage (teewurst)[a] | 3 |
| Mar. 2002 | Germany | 6 | 0 | Unknown | 69 |
| Oct.–Dec. 2002 | Germany | 38 | 4 | Apple juice[a] Curdled cheese[a] | 2 |
| Apr.–May 2006 | Scotland | 8 (18)[b] | 0 | Unknown | 65 |
| Apr.–June 2006 | Germany | 12 | 0 | Unknown | 70 |
| July–Aug. 2009 | Germany | 5 | 1 | Unknown | 31 |

[a]Based on case-control studies; SF EHEC O157:$H^-$ organisms were not isolated from the foods.
[b]Total number of cases.

reported in Scotland, where in almost one-half of patients infected with SF EHEC O157:$H^-$ during 2006 the infection progressed to HUS (65, 72), and in Austria, where 7 of 11 patients from whom SF EHEC O157:$H^-$ strains were isolated between 2002 and 2008 suffered from HUS (61).

## EARLY MOLECULAR INVESTIGATIONS OF SF EHEC O157:$H^-$

At the beginning of the 1990s, we sought to identify virulence factors of SF EHEC O157:$H^-$ and to determine whether or not these strains were only slight phenotypic variants of NSF EHEC O157:H7 or if they were substantially different. We discovered that SF EHEC O157:$H^-$ and NSF EHEC O157:H7 strains have a ca. 90-kb plasmid in common, along with Stx-converting phages and *eae*. Moreover, adherence to HEp-2 cells was well developed in the SF EHEC O157 strains. In pulsed-field gel electrophoresis, the SF EHEC O157:$H^-$ strains had identical or closely related XbaI digestion patterns that differed markedly from those of EHEC O157:H7 and other *E. coli* O157 strains (41). This suggested that SF EHEC O157:$H^-$ probably represented a new, distinct clone within the *E. coli* O157 serogroup (41).

## THE FIRST EVOLUTIONARY MODEL IN THE EMERGENCE OF *E. COLI* O157:H7

Because Tom Whittam had shown (85, 86) that NSF *E. coli* O157:H7 had a clonal population structure, it was of immediate interest to determine whether the SF *E. coli* O157:$H^-$ is indeed a distinct clone as it appeared from pulsed-field gel electrophoresis. Tom contacted me and asked for the prototype SF *E. coli* O157:$H^-$ strain 493-89 (41) for his *E. coli* reference collection at Pennsylvania State University. Based on the genetic relationships between EHEC O157:H7 and SF EHEC O157:$H^-$ by determination of allele profiles at 20 enzyme loci determined by multilocus enzyme electrophoresis (with Peter Feng and Keith Lampel), we proposed a new model for the evolution of *E. coli* O157 (26). The O157:H7 complex included not only the SF *E. coli* O157:H$^-$, but also, surprisingly, all *E. coli* O55:H7 isolates. The most divergent members, however, were the SF *E. coli* O157:H$^-$ strains from Germany. These isolates differed from *E. coli* O157:H7 at two enzyme loci. We formulated an evolutionary model that includes several steps in the emergence of *E. coli* O157:H7 (26). The SF EHEC O157:H$^-$ clonal lineage has been proposed to have a common *E. coli* O55:H7 ancestor with EHEC O157:H7, but to have diverged at an early stage of the evolution, retaining the ancestral ability to ferment sorbitol and to express $\beta$-D-glucuronidase activity, but losing its motility during this process. Whereas EHEC O157:H7 strains possess $stx_1$, $stx_2$, and $stx_{2c}$ genes, either alone or in combination (28, 36, 62), the SF EHEC O157:H$^-$ strains harbor only $stx_2$ genes, which later were shown by us to be phage borne (10, 55), the common chromosomal integration site of the $stx_2$-encoding phage being *yecE* (10, 55). Although the nucleotide sequences of $stx_2$ genes in five SF EHEC O157:H$^-$ strains were identical (10), the $stxA_2$ and $stxB_2$ subunit genes of the SF EHEC O157:H$^-$ strains differed from the corresponding subunits of the classical $stx_2$ of EHEC O157:H7 strain EDL933 in seven and one nucleotide, respectively, and encoded an Stx2 protein that differed from Stx2 in EDL933 in one amino acid residue in each of the subunits (10). The presence of duplicate $stx_2$ genes in SF EHEC O157:H$^-$ strains has been recently reported (10), and the occurrence of this phenomenon correlates with an increased production of Stx2 in vitro (10). The level of Stx production could be a critical factor determining the pathogenicity of EHEC, but the published data on the correlation between the amount of Stx produced and the severity of the associated disease are equivocal. Whereas in some studies the quantities of Stx produced in vitro correlate with clinical outcome (24, 57), an in vivo study did not show this correlation (21). Indeed, HUS development in the patients included in our study correlated with infections involving strains harboring a single $stx_2$ copy and producing less Stx2 in vitro (10). One explanation for this observation could be that in the latter strains, mitomycin C induced Stx2 production significantly more than in strains harboring two $stx_2$ genes (10). Thus, prophage-inducing agents present in the human colon might induce $stx_2$ genes, thereby promoting the ability of the host strains to cause HUS.

Shortly after publishing the evolutionary model of EHEC O157 (26), Tom's group demonstrated, using multilocus sequence typing, that SF EHEC O157:H$^-$ belongs to the EHEC 1 clade, which also contains EHEC O157:H7 and *E. coli* O55:H7 (67). Recent work has illuminated the timing of descent of the SF O157:H$^-$ clone. Leopold et al. (49) interrogated the backbone chromosomes of EHEC 1 clade members. They calculated that SF O157:H$^-$ and NSF O157:H7 differ in the times when they descended from a common ancestor. They also provided evidence that, unlike the O157:H7 clusters or *E. coli* O55:H7, there appear to be very few subsets of SF O157:H$^-$ (49). For more information concerning the evolutionary model of O157:H7 and the EHEC 1 clade, see chapter 13.

## WHAT WE KNOW TODAY: A COMPARISON BETWEEN EHEC O157:H7 AND SF EHEC O157:H⁻

EHEC O157:H7 and SF EHEC O157:H⁻ differ in several important aspects, including their phenotypes, the epidemiology of the infection (including the seasonality, age of the patients, and reservoirs and routes of transmission), and the risk of the progression of the infection to HUS (Table 2).

**Table 2.** Major phenotypic and epidemiological differences between EHEC O157:H7 and SF EHEC O157:H⁻

| Characteristic | EHEC O157:H7 | SF EHEC O157:H⁻ | Reference(s) |
|---|---|---|---|
| Phenotypes | | | |
| Sorbitol fermentation | − | + | 40, 47, 82 |
| β-D-Glucuronidase activity | − | + | 40, 47 |
| Motility | + | − | 40, 82 |
| Tellurite resistance | + | − | 9, 77, 79 |
| Phage type | Broad spectrum | 88, 23 | 5, 25, 51, 64 |
| Shiga toxin type | 1, 2, and 2c alone or in combinations | 2 | 36, 41, 62 |
| Cytolethal distending toxin V | Infrequent (~5%); phage type associated (PT2, 4, 8, 14, and 34) | Common (~80%) | 30, 35 |
| EHEC hemolysin | + | − | 5, 25, 43, 73 |
| EspP | + (α) | − | 11, 12, 14 |
| Sfp fimbriae | − | + | 13, 29, 58 |
| Curli fimbriae | Variable (infrequent) | Common (~80%) | 52, 72, 81 |
| Type 1 fimbriae | − | + | 71, 72, 74 |
| Iha | + | − | 77 |
| Epidemiologic characteristics | | | |
| Seasonality | Summer | Winter | 2, 3, 66 |
| Age of patients | >3 years | <3 years | 32, 78 |
| Reservoirs | Cattle and other animals | Unknown (humans?) | 17, 44, 59 |
| Transmission routes | Food, water, direct contact with animals and patients | Unknown | 4, 19, 22, 33, 44, 75, 84 |
| Outbreaks | Mostly diarrhea cases, HUS in a small subset (<5%) of symptomatic individuals | Almost exclusively HUS cases ("HUS outbreaks only") | 2, 3, 4, 19, 56, 66 |
| Risk of HUS | ~10–15% | ~50% | 65, 72, 78, 83; Karch, unpublished |

## Differences in Diagnostically Useful Phenotypes and Putative Virulence Factors

Besides the above-mentioned differences in motility, sorbitol fermentation, $\beta$-D-glucuronidase activity, and tellurite susceptibility, which have diagnostic implications, the two groups of EHEC O157 pathogens differ by the spectrum of Stx types they produce (Table 2) and by expression of several non-Stx toxins that may contribute to the pathogenesis of EHEC O157-mediated diseases (7, 45). Specifically, cytolethal distending toxin V (35), which causes an irreversible injury to microvascular endothelial cells (8) that are the major targets affected during HUS (7, 78), is produced by the majority of SF EHEC O157:H$^-$ strains (35), but only by a small subset of EHEC O157:H7, restricted to strains belonging to particular phage types (30) (Table 2). In contrast, EHEC hemolysin, another toxin with endothelium-injuring capacity (1), is regularly expressed by EHEC O157:H7 (73), but usually not by SF EHEC O157:H$^-$ (25, 43), though the latter organisms possess the structural gene encoding this toxin (14). The highly proteolytically active serine protease EspP$\alpha$, which cleaves human coagulation factor V (12) and might therefore hinder hemostasis, is expressed by EHEC O157:H7 strains (11, 12), but the respective gene, which is encoded on a large plasmid of EHEC O157:H7 (12, 15), is absent from the large plasmid of SF EHEC O157:H$^-$ organisms (14).

Although both EHEC O157:H7 and SF EHEC O157:H$^-$ possess a type III secretion system that is responsible for the intimate attachment and formation of attaching-and-effacing lesions (37, 41), other adhesins might contribute to the significantly higher adherence of SF EHEC O157:H$^-$ than EHEC O157:H7 to intestinal epithelial cells in vitro that has been recently reported (72). Rosser et al. (72) demonstrated that curli fimbriae are the main factors responsible for the increased adherence of SF EHEC O157:H$^-$ to Caco-2 cells. Notably, curli fimbriae were expressed by almost 80% of 66 SF EHEC O157:H$^-$ strains tested (72), whereas expression of these structures by EHEC O157:H7 is variable and rather low (52, 81). Another adhesin type that is regularly produced by SF EHEC O157:H$^-$ but not by EHEC O157:H7 is Sfp fimbriae (13, 58) (Fig. 1). A recent study from our laboratory demonstrated that Sfp fimbriae are expressed under oxygen-reduced conditions and in a medium simulating the colonic environment. The expression of Sfp fimbriae in SF *E. coli* O157:H$^-$ correlated with increased adherence of the strains to the human intestinal cells including Caco-2 and HCT-8 lines (58). The expression of Sfp and curli fimbriae at 37°C and under conditions that resemble the natural site of infection suggests that these structures may contribute to the adherence of the organisms to the human intestinal epithelium during infection. Moreover, SF EHEC O157:H$^-$ strains regularly express type 1 fimbriae (72), but EHEC O157:H7 strains are unable to produce these fimbriae because of a 16-bp deletion in the *fimA* promoter (71) and a mutation in the FimH adhesin (74). On the other hand, another adhesin that has been shown to mediate adherence of EHEC O157:H7 to epithelial (nonintestinal) cells, namely, the IrgA homolog adhesin (Iha) (77), is absent from SF EHEC O157:H$^-$ (Table 2).

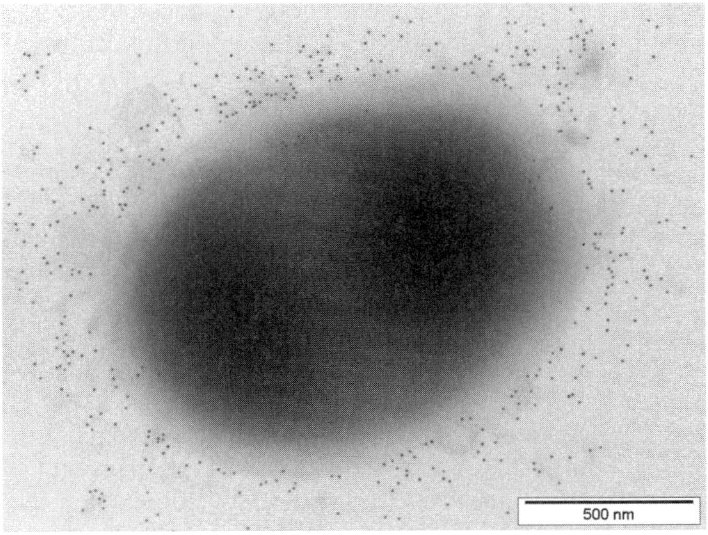


**Figure 1.** Electron microscopic demonstration of Sfp fimbriae in an SF EHEC O157:H$^-$ strain using immunogold staining with anti-Sfp antibody. Photomicrograph by Lilo Greune, Institute of Infectiology, University of Münster, Germany.

## Differences in Epidemiology of Infections

The most important presently known differences in the epidemiology of the two EHEC O157 pathogens are that (i) SF EHEC O157:H$^-$ infections occur most frequently during winter months (2, 3), whereas EHEC O157:H7 infections peak in summer (66); and (ii) SF EHEC O157:H$^-$ infections mainly affect children under 3 years of age (32), whereas EHEC O157:H7 infections are more common in children older than 3 years (32, 78). These features of the infections caused by the two EHEC O157 pathogens point to differences in the nature of the reservoirs and vehicles for transmission. Although it is well established that cattle and other domestic and wild animals are major reservoirs for EHEC O157:H7, and consumption of contaminated foods and water and direct contact with the animals are major transmission routes (4, 17, 19, 22, 33, 56, 59, 66, 75), SF EHEC O157:H$^-$ has only occasionally been isolated from animals, including cattle (6, 48, 60, 61) and a pony (68). The rarity of these organisms in animals is noteworthy because immunomagnetic bead technology (20, 42) was employed, which would have found them had they been present. The low rate of isolation of these pathogens from animals raises the intriguing possibility that humans might be the main reservoirs (44), as is the case for classical enteropathogenic *E. coli* (EPEC), enteroinvasive *E. coli* (EIEC), *Shigella* spp., and *Salmonella enterica* serovar Typhi. Vehicles and routes of transmission of SF EHEC O157:H$^-$ infection are currently poorly understood, but the extended onset of the infection within small geographic areas (space clusters) supports the involvement of a continuing source, rather than a widely disseminated contaminated food product. Although foods (raw sausage, apple juice, curdled cheese) have been implicated as possible sources in outbreaks caused by SF EHEC O157:H$^-$ based on case-control studies, the outbreak strains were never isolated from food or environmental samples (2, 3). Thus, vehicles and routes of transmission of SF EHEC O157:H$^-$ infection are mostly unknown, but need to be identified so as to control human infections in the future.

In contrast to outbreaks caused by EHEC O157:H7, where most infected individuals develop diarrhea but not HUS (4, 19, 56, 66), in outbreaks of infection with SF EHEC O157:H$^-$ in Germany (2, 3) there has been a predominance of patients with HUS and a much lower proportion of patients with diarrhea ("HUS outbreaks only") (Table 2). The reason for the apparently higher virulence of SF EHEC O157:H$^-$ as compared to EHEC O157:H7, as suggested by the high rate of HUS development in outbreaks (2, 3, 65) and sporadic cases of SF EHEC O157:H$^-$ infection (61), is not completely understood. However, it is hypothesized that one reason might be a more efficient adherence of SF EHEC O157:H$^-$ to the intestinal mucosa and thus an increased delivery of Stxs and other virulence factors into the bloodstream, which in turn is more likely to lead to HUS (72).

## SUMMARY

SF *E. coli* O157:H$^-$ strains are a very interesting, and so far mostly geographically restricted, type of pathogenic EHEC. This is fortunate, as they appear to be hypervirulent. In general populations, it is very difficult to identify these pathogens in infected patients using standard techniques. A variety of strategies other than toxin testing has been proposed, but it is not clear that such algorithms will increase the yield of detection. Moreover, many infected patients do not even have diarrhea, let alone bloody diarrhea, so the opportunity to perform community-based diagnosis before HUS ensues is extremely limited. We continue our quest to develop strategic detection methods for, and to find the source(s) of, SF *E. coli* O157:H$^-$. However, we will forever miss the scientific companionship of Tom Whittam, whose insight, knowledge, and collegiality inspired our careers and our work.

## REFERENCES

1. **Aldick, T., M. Bielaszewska, W. Zhang, J. Brockmeyer, H. Schmidt, A. W. Friedrich, K. S. Kim, M. A. Schmidt, and H. Karch.** 2007. Hemolysin from Shiga toxin-negative *Escherichia coli* O26 strains injures microvascular endothelium. *Microbes Infect.* **9:**282–290.
2. **Alpers, K., D. Werber, C. Frank, J. Koch, A. W. Friedrich, H. Karch, M. an der Heiden, R. Prager, A. Fruth, M. Bielaszewska, G. Morlock, A. Heissenhuber, A. Diedler, A. Gerber, and A. Ammon.** 2009. Sorbitol-fermenting enterohaemorrhagic *Escherichia coli* O157:H$^-$ causes another outbreak of haemolytic uraemic syndrome in children. *Epidemiol. Infect.* **137:**389–395.

3. **Ammon, A., L. R. Petersen, and H. Karch.** 1999. A large outbreak of hemolytic uremic syndrome caused by an unusual sorbitol-fermenting strain of *E. coli* O157:H$^-$. *J. Infect. Dis.* **179:**1274–1277.
4. **Bell, B. P., M. Goldoft, P. M. Griffin, M. A. Davis, D. C. Gordon, P. I. Tarr, C. A. Bartleson, J. H. Lewis, T. J. Barrett, and J. G. Wells.** 1994. A multistate outbreak of *Escherichia coli* O157:H7-associated bloody diarrhea and hemolytic uremic syndrome from hamburgers. The Washington experience. *JAMA* **272:**1349–1353.
5. **Bielaszewska, M., H. Schmidt, M. A. Karmali, R. Khakhria, J. Janda, K. Bláhová, and H. Karch.** 1998. Isolation and characterization of sorbitol-fermenting Shiga toxin (verocytotoxin)-producing *Escherichia coli* O157:H$^-$ strains in the Czech Republic. *J. Clin. Microbiol.* **36:**2135–2137.
6. **Bielaszewska, M., H. Schmidt, A. Liesegang, R. Prager, W. Rabsch, H. Tschäpe, A. Cízek, J. Janda, K. Bláhová, and H. Karch.** 2000. Cattle can be a reservoir of sorbitol-fermenting Shiga toxin-producing *Escherichia coli* O157:H$^-$ strains and a source of human diseases. *J. Clin. Microbiol.* **38:**3470–3473.
7. **Bielaszewska, M., and H. Karch.** 2005. Consequences of enterohaemorrhagic *Escherichia coli* infection for the vascular endothelium. *Thromb. Haemost.* **94:**312–318.
8. **Bielaszewska, M., B. Sinha, T. Kuczius, and H. Karch.** 2005. Cytolethal distending toxin from Shiga toxin-producing *Escherichia coli* O157 causes irreversible G2/M arrest, inhibition of proliferation, and death of human endothelial cells. *Infect. Immun.* **73:** 552–562.
9. **Bielaszewska, M., P. I. Tarr, H. Karch, W. Zhang, and W. Mathys.** 2005. Phenotypic and molecular analysis of tellurite resistance among enterohemorrhagic *Escherichia coli* O157:H7 and sorbitol-fermenting O157:NM clinical isolates. *J. Clin. Microbiol.* **43:**452–454.
10. **Bielaszewska, M., R. Prager, W. Zhang, A. W. Friedrich, A. Mellmann, H. Tschäpe, and H. Karch.** 2006. Chromosomal dynamism in progeny of outbreak-related sorbitol-fermenting enterohemorrhagic *Escherichia coli* O157:NM. *Appl. Environ. Microbiol.* **72:** 1900–1909.
11. **Brockmeyer, J., M. Bielaszewska, A. Fruth, M.-L. Bonn, A. Mellmann, H.-U. Humpf, and H. Karch.** 2007. Subtypes of the plasmid-encoded serine protease EspP in Shiga toxin-producing *Escherichia coli*: distribution, secretion and proteolytic activity. *Appl. Environ. Microbiol.* **73:**6351–6359.
12. **Brunder, W., H. Schmidt, and H. Karch.** 1997. EspP, a novel extracellular serine protease of enterohaemorrhagic *Escherichia coli* O157:H7 cleaves human coagulation factor V. *Mol. Microbiol.* **24:**767–778.
13. **Brunder, W., A. Salam Khan, J. Hacker, and H. Karch.** 2001. A novel fimbrial gene cluster encoded by the large plasmid of sorbitol-fermenting enterohemorrhagic *Escherichia coli* O157:H$^-$. *Infect. Immun.* **69:**4447–4457.
14. **Brunder, W., H. Karch, and H. Schmidt.** 2006. Complete sequence of the large virulence plasmid pSFO157 of the sorbitol-fermenting enterohemorrhagic *Escherichia coli* O157:H$^-$ strain 3072/96. *Int. J. Med. Microbiol.* **296:**467–474.
15. **Burland, V., Y. Shao, N. T. Perna, G. Plunkett, H. J. Sofia, and F. R. Blattner.** 1998. The complete DNA sequence and analysis of the large virulence plasmid of *Escherichia coli* O157:H7. *Nucleic Acids Res.* **26:**4196–4204.
16. **Buvens, G., D. Piérard, S. Hachimi-Idrissi, and S. Lauwers.** 2009. First sorbitol-fermenting Verocytotoxin-producing *Escherichia coli* O157:H$^-$ isolated in Belgium. *Acta Clin. Belg.* **64:**59–64.
17. **Caprioli, A., S. Morabito, H. Brugère, and E. Oswald.** 2005. Enterohaemorrhagic *Escherichia coli*: emerging issues on virulence and modes of transmission. *Vet. Res.* **36:** 289–311.
18. **Carroll, K. C., K. Adamson, K. Korgenski, A. Croft, R. Hankemeier, J. Daly, and C. H. Park.** 2003. Comparison of a commercial reversed passive latex agglutination assay to an enzyme immunoassay for the detection of Shiga toxin-producing *Escherichia coli*. *Eur. J. Clin. Microbiol. Infect. Dis.* **22:**689–692.

19. **Centers for Disease Control and Prevention.** 2006. Ongoing multistate outbreak of *Escherichia coli* serotype O157:H7 infections associated with consumption of fresh spinach—United States, September 2006. *MMWR Morb. Mortal. Wkly. Rep.* **55:**1045–1046.
20. **Chapman, P. A., D. J. Wright, and C. A. Siddons.** 1994. A comparison of immunomagnetic separation and direct culture for the isolation of verocytotoxin-producing *Escherichia coli* O157 from bovine faeces. *J. Med. Microbiol.* **40:**424–427.
21. **Cornick, N. A., S. Jelacic, M. A. Ciol, and P. I. Tarr.** 2002. *Escherichia coli* O157: H7 infections: discordance between filterable fecal Shiga toxin and disease outcome. *J. Infect. Dis.* **186:**57–63.
22. **Crump, J. A., A. C. Sulka, A. J. Langer, C. Schaben, A. S. Crielly, R. Gage, M. Baysinger, M. Moll, G. Withers, D. M. Toney, S. B. Hunter, R. M. Hoekstra, S. K. Wong, P. M. Griffin, and T. J. Van Gilder.** 2002. An outbreak of *Escherichia coli* O157:H7 infections among visitors to a dairy farm. *N. Engl. J. Med.* **347:**555–560.
23. **Editorial Team.** 2006. *E. coli* O157 infections in the UK. *Euro Surveill.* **11:**E060601.2.
24. **Eklund, M., K. Leino, and A. Siitonen.** 2002. Clinical *Escherichia coli* strains carrying *stx* genes: *stx* variants and *stx*-positive virulence profiles. *J. Clin. Microbiol.* **40:**4585–4593.
25. **Eklund, M., M. Bielaszewska, U. Nakari, H. Karch, and A. Siitonen.** 2006. Molecular and phenotypic profiling of sorbitol-fermenting *Escherichia coli* O157:$H^-$ human isolates from Finland. *Clin. Microbiol. Infect.* **12:**634–641.
26. **Feng, P., K. A. Lampel, H. Karch, and T. S. Whittam.** 1998. Genotypic and phenotypic changes in the emergence of *Escherichia coli* O157:H7. *J. Infect. Dis.* **177:**1750–1753.
27. **Fey, P. D., R. S. Wickert, M. E. Rupp, T. J. Safranek, and S. H. Hinrichs.** 2000. Prevalence of non-O157:H7 Shiga toxin-producing *Escherichia coli* in diarrheal stool samples from Nebraska. *Emerg. Infect. Dis.* **6:**530–533.
28. **Friedrich, A. W., M. Bielaszewska, W.-L. Zhang, M. Pulz, T. Kuczius, A. Ammon, and H. Karch.** 2002. *Escherichia coli* harboring Shiga toxin 2 gene variants: frequency and association with clinical symptoms. *J. Infect. Dis.* **185:**74–84.
29. **Friedrich, A. W., K. V. Nierhoff, M. Bielaszewska, A. Mellmann, and H. Karch.** 2004. Phylogeny, clinical associations, and diagnostic utility of the pilin subunit gene (*sfpA*) of sorbitol-fermenting, enterohemorrhagic *Escherichia coli* O157:$H^-$. *J. Clin. Microbiol.* **42:**4697–4701.
30. **Friedrich, A. W., S. Lu, M. Bielaszewska, R. Prager, P. Bruns, J. G. Xu, H. Tschäpe, and H. Karch.** 2006. Cytolethal distending toxin in *Escherichia coli* O157:H7: spectrum of conservation, structure, and endothelial toxicity. *J. Clin. Microbiol.* **44:**1844–1846.
31. **Fruth, A., S. Nielsen, R. Prager, A. Graff, C. Frank, M. Lüttgehetmann, A. Spode, and D. Müller-Wiefel.** 2010. A localised outbreak of hemolytic uremic syndrome caused by sorbitol-fermenting enterohemorrhagic *Escherichia coli* O157:$H^-$ in Germany, 2009, abstr. MSV05, p. 213. *In Abstracts of the 3rd Common Congress of the German Society for Hygiene and Microbiology* (*DGHM*) *and the Society for General and Applied Microbiology* (*VAAM*), Hannover, Germany, 28 to 31 March 2010.
32. **Gerber, A., H. Karch, F. Allerberger, H. M. Verweyen, and L. B. Zimmerhackl.** 2002. Clinical course and the role of Shiga toxin-producing *Escherichia coli* infection in the hemolytic-uremic syndrome in pediatric patients, 1997–2000, in Germany and Austria: a prospective study. *J. Infect. Dis.* **186:**493–500.
33. **Holme, R.** 2003. Drinking water contamination in Walkerton, Ontario: positive resolutions from a tragic event. *Water Sci. Technol.* **47:**1–6.
34. **Jakubczak, A., J. Szych, and K. Januszkiewicz.** 2008. Characterization of first sorbitol-fermenting Shiga toxin-producing *Escherichia coli* O157:$H^-$ strain isolated in Poland. *Med. Dosw. Mikrobiol.* **60:**173–181. (In Polish.)
35. **Janka, A., M. Bielaszewska, U. Dobrindt, L. Greune, M. A. Schmidt, and H. Karch.** 2003. The cytolethal distending toxin (*cdt*) gene cluster in enterohemorrhagic *Escherichia coli* O157:$H^-$ and O157:H7: characterization and evolutionary considerations. *Infect. Immun.* **71:**3634–3638.

36. **Jelacic, J. K., T. Damrow, G. S. Chen, S. Jelacic, M. Bielaszewska, M. Ciol, H. M. Carvalho, A. R. Melton-Celsa, A. D. O'Brien, and P. I. Tarr.** 2003. Shiga toxin-producing *Escherichia coli* in Montana: bacterial genotypes and clinical profiles. *J. Infect. Dis.* **188:**719–729.
37. **Kaper, J. B., J. P. Nataro, and H. L. Mobley.** 2004. Pathogenic *Escherichia coli. Nat. Rev. Microbiol.* **2:**123–140.
38. **Karch, H., and T. Meyer.** 1989. Evaluation of oligonucleotide probes for identification of Shiga-like-toxin-producing *Escherichia coli. J. Clin. Microbiol.* **27:**1180–1186.
39. **Karch, H., and T. Meyer.** 1989. Single primer pair for amplifying segments of distinct Shiga-like-toxin genes by polymerase chain reaction. *J. Clin. Microbiol.* **27:**2751–2757.
40. **Karch, H., R. Wiss, H. Gloning, P. Emmrich, S. Aleksic, and J. Bockemühl.** 1990. Hemolytic-uremic syndrome in infants due to verotoxin-producing *Escherichia coli. Dtsch. Med. Wochenschr.* **115:**485–495. (In German.)
41. **Karch, H., H. Böhm, H. Schmidt, F. Gunzer, S. Aleksic, and J. Heesemann.** 1993. Clonal structure and pathogenicity of Shiga-like toxin-producing, sorbitol-fermenting *Escherichia coli* O157:H$^-$. *J. Clin. Microbiol.* **31:**1200–1205.
42. **Karch, H., C. Janetzki-Mittmann, S. Aleksic, and M. Datz.** 1996. Isolation of enterohemorrhagic *Escherichia coli* O157 strains from patients with hemolytic-uremic syndrome by using immunomagnetic separation, DNA-based methods, and direct culture. *J. Clin. Microbiol.* **34:**516–519.
43. **Karch, H., and M. Bielaszewska.** 2001. Sorbitol-fermenting Shiga toxin-producing *Escherichia coli* O157:H$^-$ strains: epidemiology, phenotypic and molecular characteristics, and microbiological diagnosis. *J. Clin. Microbiol.* **39:**2043–2049.
44. **Karch, H., P. I. Tarr, and M. Bielaszewska.** 2005. Enterohaemorrhagic *Escherichia coli* in human medicine. *Int. J. Med. Microbiol.* **295:**405–418.
45. **Karch, H., A. W. Friedrich, A. Gerber, L. B. Zimmerhackl, M. A. Schmidt, and M. Bielaszewska.** 2006. New aspects in the pathogenesis of enteropathic hemolytic uremic syndrome. *Semin. Thromb. Hemost.* **32:**105–112.
46. **Klein, E. J., J. R. Stapp, C. R. Clausen, D. R. Boster, J. G. Wells, X. Qin, D. L. Swerdlow, and P. I. Tarr.** 2002. Shiga toxin-producing *Escherichia coli* in children with diarrhea: a prospective point-of-care study. *J. Pediatr.* **141:**172–177.
47. **Krishnan, C., V. A. Fitzgerald, S. J. Dakin, and R. J. Behme.** 1987. Laboratory investigation of outbreak of hemorrhagic colitis caused by *Escherichia coli* O157:H7. *J. Clin. Microbiol.* **25:**1043–1047.
48. **Lee, J. H., and S. Choi.** 2006. Isolation and characteristics of sorbitol-fermenting *Escherichia coli* O157 strains from cattle. *Microbes Infect.* **8:**2021–2026.
49. **Leopold, S. R., V. Magrini, N. J. Holt, N. Shaikh, E. R. Mardis, J. Cagno, Y. Ogura, A. Iguchi, T. Hayashi, A. Mellmann, H. Karch, T. E. Besser, S. A. Sawyer, T. S. Whittam, and P. I. Tarr.** 2009. A precise reconstruction of the emergence and constrained radiations of *Escherichia coli* O157 portrayed by backbone concatenomic analysis. *Proc. Natl. Acad. Sci. USA* **106:**8713–8718.
50. **Levine, M. M., J. G. Xu, J. B. Kaper, H. Lior, V. Prado, B. Tal, J. Nataro, H. Karch, and K. Wachsmuth.** 1987. A DNA probe to identify enterohemorrhagic *Escherichia coli* of O157:H7 and other serotypes that cause hemorrhagic colitis and hemolytic uremic syndrome. *J. Infect. Dis.* **156:**175–182.
51. **Liesegang, A., U. Sachse, R. Prager, H. Claus, H. Steinrück, S. Aleksic, W. Rabsch, W. Voigt, A. Fruth, H. Karch, J. Bockemühl, and H. Tschäpe.** 2000. Clonal diversity of Shiga toxin-producing *Escherichia coli* O157:H7/H$^-$ in Germany—a ten-year study. *Int. J. Med. Microbiol.* **290:**269–278.
52. **Low, A. S., N. Holden, T. Rosser, A. J. Roe, C. Constantinidou, J. L. Hobman, D. G. Smith, J. C. Low, and D. L. Gally.** 2006. Analysis of fimbrial gene clusters and their expression in enterohaemorrhagic *Escherichia coli* O157:H7. *Environ. Microbiol.* **8:** 1033–1047.
53. **March, S. B., and S. Ratnam.** 1986. Sorbitol-MacConkey medium for detection of *Escherichia coli* O157:H7 associated with hemorrhagic colitis. *J. Clin. Microbiol.* **23:** 869–872.

54. **Mellmann, A., M. Bielaszewska, R. Köck, A. W. Friedrich, A. Fruth, B. Middendorf, D. Harmsen, M. A. Schmidt, and H. Karch.** 2008. Analysis of collection of hemolytic uremic syndrome-associated enterohemorrhagic *Escherichia coli*. *Emerg. Infect. Dis.* **14:**1287–1290.
55. **Mellmann, A., S. Lu, H. Karch, J. Xu, D. Harmsen, M. A. Schmidt, and M. Bielaszewska.** 2008. Recycling of Shiga toxin 2 genes in sorbitol-fermenting enterohemorrhagic *Escherichia coli* O157:NM. *Appl. Environ. Microbiol.* **74:**67–72.
56. **Michino, H., K. Araki, S. Minami, S. Takaya, N. Sakai, M. Miyazaki, A. Ono, and H. Yanagawa.** 1999. Massive outbreak of *Escherichia coli* O157:H7 infection in schoolchildren in Sakai City, Japan, associated with consumption of white radish sprouts. *Am. J. Epidemiol.* **150:**787–796.
57. **Muniesa, M., M. De Simon, G. Prats, D. Ferrere, H. Panella, and J. Jofre.** 2003. Shiga toxin 2-converting bacteriophages associated with clonal variability in *Escherichia coli* O157:H7 strains of human origin isolated from a single outbreak. *Infect. Immun.* **71:**4554–4562.
58. **Müsken, A., M. Bielaszewska, L. Greune, C. H. Schweppe, J. Müthing, H. Schmidt, M. A. Schmidt, H. Karch, and W. Zhang.** 2008. Anaerobic conditions promote expression of Sfp fimbriae and adherence of sorbitol-fermenting enterohemorrhagic *Escherichia coli* O157:NM to human intestinal epithelial cells. *Appl. Environ. Microbiol.* **74:**1087–1093.
59. **Naylor, S. W., D. L. Gally, and J. C. Low.** 2005. Enterohaemorrhagic *E. coli* in veterinary medicine. *Int. J. Med. Microbiol.* **295:**419–441.
60. **Orth, D., K. Grif, M. P. Dierich, and R. Würzner.** 2006. Sorbitol-fermenting Shiga toxin-producing *Escherichia coli* O157: indications for an animal reservoir. *Epidemiol. Infect.* **134:**719–723.
61. **Orth, D., K. Grif, L. B. Zimmerhackl, and R. Würzner.** 2009. Sorbitol-fermenting Shiga toxin-producing *Escherichia coli* O157 in Austria. *Wien. Klin. Wochenschr.* **121:** 108–112.
62. **Ostroff, S. M., P. I. Tarr, M. A. Neill, J. H. Lewis, N. Hargrett-Bean, and J. M. Kobayashi.** 1989. Toxin genotypes and plasmid profiles as determinants of systemic sequelae in *Escherichia coli* O157:H7 infections. *J. Infect. Dis.* **160:**994–998.
63. **Park, C. H., H. J. Kim, D. L. Hixon, and A. Bubert.** 2003. Evaluation of the duopath verotoxin test for detection of Shiga toxins in cultures of human stools. *J. Clin. Microbiol.* **41:**2650–2653.
64. **Pearce, M. C., M. E. Chase-Topping, I. J. McKendrick, D. J. Mellor, M. E. Locking, L. Allison, H. E. Ternent, L. Matthews, H. I. Knight, A. W. Smith, B. A. Synge, W. Reilly, J. C. Low, S. W. Reid, G. J. Gunn, and M. E. Woolhouse.** 2009. Temporal and spatial patterns of bovine *Escherichia coli* O157 prevalence and comparison of temporal changes in the patterns of phage types associated with bovine shedding and human *E. coli* O157 cases in Scotland between 1998–2000 and 2002–2004. *BMC Microbiol.* **9:**276.
65. **Pollock, K. G., M. E. Locking, T. J. Beattie, H. Maxwell, I. Ramage, D. Hughes, J. Cowieson, L. Allison, M. Hanson, and J. M. Cowden.** 2010. Sorbitol-fermenting *Escherichia coli* O157, Scotland. *Emerg. Infect. Dis.* **16:**881–882.
66. **Rangel, J. M., P. H. Sparling, C. Crowe, P. M. Griffin, and D. L. Swerdlow.** 2005. Epidemiology of *Escherichia coli* O157:H7 outbreaks, United States, 1982–2002. *Emerg. Infect. Dis.* **11:**603–609.
67. **Reid, S. D., C. J. Herbelin, A. C. Bumbaugh, R. K. Selander, and T. S. Whittam.** 2000. Parallel evolution of virulence in pathogenic *Escherichia coli*. *Nature* **406:**64–67.
68. **Robert Koch-Institut.** 1999. Fallbericht: Inkomplettes HUS bei EHEC-Enteritis, übertragen durch Minipferde. *Epidemiol. Bull.* **37:**276.
69. **Robert Koch-Institut.** 2002. Häufung Sorbitol-fermentierender *E. coli* O157:$H^-$ in mehreren Bundesländern. *Epidemiol. Bull.* **15:**123.
70. **Robert Koch-Institut.** 2006. Hämolytisch-urämisches Syndrom: Hinweis auf eine Häufung in Norddeutschland. *Epidemiol. Bull.* **24:**190.

71. **Roe, A. J., C. Currie, D. G. Smith, and D. L. Gally.** 2001. Analysis of type 1 fimbriae expression in verotoxigenic *Escherichia coli*: a comparison between serotypes O157 and O26. *Microbiology* **147:**145–152.
72. **Rosser, T., T. Dransfield, L. Allison, M. Hanson, N. Holden, J. Evans, S. Naylor, R. La Ragione, J. C. Low, and D. J. Gally.** 2008. Pathogenic potential of emergent sorbitol-fermenting *Escherichia coli* O157:NM. *Infect. Immun.* **76:**5598–5607.
73. **Schmidt, H., L. Beutin, and H. Karch.** 1995. Molecular analysis of the plasmid-encoded hemolysin of *Escherichia coli* O157:H7 strain EDL 933. *Infect. Immun.* **63:** 1055–1061.
74. **Shaikh, N., N. J. Holt, J. R. Johnson, and P. I. Tarr.** 2007. *Fim* operon variation in the emergence of enterohemorrhagic *Escherichia coli*: an evolutionary and functional analysis. *FEMS Microbiol. Lett.* **273:**58–63.
75. **Swerdlow, D. L., B. A. Woodruff, R. C. Brady, P. M. Griffin, S. Tippen, H. D. J. Donnell, E. Geldreich, B. J. Payne, A. J. Meyer, and J. G. Wells.** 1992. A waterborne outbreak in Missouri of *Escherichia coli* O157:H7 associated with bloody diarrhea and death. *Ann. Intern. Med.* **117:**812–819.
76. **Tarr, P. I., M. A. Neill, C. R. Clausen, S. L. Watkins, D. L. Christie, and R. O. Hickman.** 1990. *Escherichia coli* O157:H7 and the hemolytic uremic syndrome: importance of early cultures in establishing the etiology. *J. Infect. Dis.* **162:**553–556.
77. **Tarr, P. I., S. S. Bilge, J. C. Vary, S. Jelacic, R. L. Habeeb, T. R. Ward, M. R. Baylor, and T. E. Besser.** 2000. Iha: a novel *Escherichia coli* O157:H7 adherence-conferring molecule encoded on a recently acquired chromosomal island of conserved structure. *Infect. Immun.* **68:**1400–1407.
78. **Tarr, P. I., C. A. Gordon, and W. L. Chandler.** 2005. Shiga toxin-producing *Escherichia coli* and the haemolytic uraemic syndrome. *Lancet* **365:**1073–1086.
79. **Taylor, D. E., M. Rooker, M. Keelan, L.-K. Ng, I. Martin, N. T. Perna, N. T. Burland, and F. R. Blattner.** 2002. Genome variability of O islands encoding tellurite resistance in enterohemorrhagic *Escherichia coli* O157:H7 isolates. *J. Bacteriol.* **184:** 4690–4698.
80. **Teel, L. D., J. A. Daly, R. C. Jerris, D. Maul, G. Svanas, A. D. O'Brien, and C. H. Park.** 2007. Rapid detection of Shiga toxin-producing *Escherichia coli* by optical immunoassay. *J. Clin. Microbiol.* **45:**3377–3380.
81. **Uhlich, G. A., J. E. Keen, and R. O. Elder.** 2001. Mutations in the *csgD* promoter associated with variations in curli expression in certain strains of *Escherichia coli* O157: H7. *Appl. Environ. Microbiol.* **67:**2367–2370.
82. **Wells, J. G., B. R. Davis, I. K. Wachsmuth, L. W. Riley, R. S. Remis, R. Sokolow, and G. K. Morris.** 1983. Laboratory investigation of hemorrhagic colitis outbreaks associated with a rare *Escherichia coli* serotype. *J. Clin. Microbiol.* **18:**512–520.
83. **Werber, D., S. C. Behnke, A. Fruth, R. Merle, S. Menzler, S. Glaser, L. Kreienbrock, R. Prager, H. Tschäpe, P. Roggentin, L. Bockemühl, and A. Ammon.** 2007. Shiga toxin-producing *Escherichia coli* infection in Germany: different risk factors for different age groups. *Am. J. Epidemiol.* **165:**425–434.
84. **Werber, D., B. W. Mason, M. R. Evans, and R. L. Salmon.** 2008. Preventing household transmission of Shiga toxin-producing *Escherichia coli* O157 infection: promptly separating siblings might be the key. *Clin. Infect. Dis.* **46:**1189–1196.
85. **Whittam, T. S., I. K. Wachsmuth, and R. A. Wilson.** 1988. Genetic evidence of clonal descent of *Escherichia coli* O157:H7 associated with hemorrhagic colitis and hemolytic uremic syndrome. *J. Infect. Dis.* **157:**1124–1133.
86. **Whittam, T. S., and R. A. Wilson.** 1988. Genetic relationships among pathogenic *Escherichia coli* of serogroup O157. *Infect. Immun.* **56:**2467–2473.
87. **Zadik, P. M., P. A. Chapman, and C. A. Siddons.** 1993. Use of tellurite for the selection of verocytotoxigenic *Escherichia coli* O157. *J. Med. Microbiol.* **39:**155–158.

*Population Genetics of Bacteria: a Tribute to Thomas S. Whittam*
Edited by S. T. Walk and P. C. H. Feng


*Chapter 17*

# Molecular Evolution of Enterohemorrhagic *Escherichia coli* and Application to Epidemiology

***Galeb S. Abu-Ali, A. Cody Springman, Lindsey M. Ouellette, Lukas Wick, Weihong Qi, Sivapriya Kalisan Vanaja, Teresa M. Bergholz, David W. Lacher, Seth T. Walk, Jillian A. Tietjen, Hans Steinsland, James T. Riordan, and Shannon D. Manning***

## INTRODUCTION

Tom Whittam was well known for his application and development of population genetic methods to study the evolution and emergence of bacterial pathogens. His pioneering research into the structure of *Escherichia coli* populations not only advanced the field of microbial evolution and genetics, but it also provided a contextual framework for investigating variation in the epidemiology and virulence of bacterial populations. While most of Tom's research focused on *E. coli*, his methodologies have been applied to the study of many other pathogens, including *Burkholderia cenocepacia* (34), *Campylobacter jejuni* (4), group B *Streptococcus* (24), *Mycobacterium tuberculosis* (12), and

***Galeb S. Abu-Ali and David W. Lacher*** • Center for Food Safety and Applied Nutrition, U.S. Food and Drug Administration, Laurel, MD 20708. ***A. Cody Springman, Lindsey M. Ouellette, and Shannon D. Manning*** • Microbiology and Molecular Genetics, Michigan State University, East Lansing, MI 48824. ***Lukas Wick*** • Biosynth AG, 9422 Staad, Switzerland. ***Weihong Qi*** • Functional Genomics Center Zurich Uni, ETH Zurich, CH-8057 Zurich, Switzerland. ***Sivapriya Kalisan Vanaja*** • Molecular Genetics and Microbiology, University of Massachusetts Medical School, Worcester, MA 01655. ***Teresa M. Bergholz*** • Department of Food Science, Cornell University, Ithaca, NY 14853. ***Seth T. Walk*** • Department of Internal Medicine, Division of Infectious Diseases, University of Michigan Health System, Ann Arbor, MI 48109. ***Jillian A. Tietjen*** • University of Michigan Genetic Counseling Department, Ann Arbor, MI 48109. ***Hans Steinsland*** • Department of Biomedicine, University of Bergen, N-5009 Bergen, Norway. ***James T. Riordan*** • Department of Cell Biology, Microbiology and Molecular Biology, University of South Florida, Tampa, FL 33063.

*Salmonella enterica* serovar Enteritidis (10). This chapter highlights the work of his laboratory involving the use of various population genetic strategies for characterizing *E. coli* pathotypes, particularly enterohemorrhagic *E. coli* (EHEC) O157:H7, while focusing on contributions to epidemiology and public health.

## MLEE

Early studies by Tom Whittam and colleagues focused on applying multilocus enzyme electrophoresis (MLEE) to pathogenic and commensal *E. coli* strains isolated from various sources to examine genetic diversity within and between populations. Among these strains, a significant proportion included uropathogenic *E. coli* (UPEC) and EHEC from both animals and humans with enteric disease. The goal of these studies was to compare the genetic relationships among human and animal strains to identify evolutionary processes that have contributed to the emergence of EHEC O157:H7 and to assess the distribution of pathogenic *E. coli* clones.

Initially, population genetic analyses of MLEE data from 100 EHEC O157:H7 strains demonstrated that strains isolated from humans with hemorrhagic colitis and hemolytic uremic syndrome (HUS) represented a well-defined clonal group (38) and were not closely related to human-derived strains of non-O157 Shiga toxin (Stx)-producing *E. coli* (STEC). Specifically, 25 multilocus enzyme profiles, or electrophoretic types (ETs), were detected in this study, with EHEC O157:H7 strains clustering together and sharing alleles on average at 90% of the 17 enzymes analyzed (38).

A follow-up study comparing 194 O157 strains from human patients and animals, namely, pigs with colibacillosis (39), identified 12 distinct flagellar antigens (i.e., H types). Additionally, 72% of the O157 strains contained K88 fimbriae, which are plasmid-encoded adhesins that aid in attachment to host tissues and are associated with virulence in swine. The expression of K88 fimbriae was not exclusive to one clonal group, but was also present in divergent genotypes (39). In all, 33 different ETs were identified that differed, on average, at 44% of the loci and clustered into five major clonal groups. One clonal group was composed of strains from porcine colibacillosis cases, while the other group was composed of strains from humans with urinary tract infections. Interestingly, these studies showed that *E. coli* strains of the O157 serogroup are genetically diverse and that clinically distinct O157:H7 strains were not closely related. Moreover, these results illustrate that *E. coli* strains of a given clonal group may be more likely to cause certain infections than others, a finding that is likely due to variation in the ability of distinct genotypes to acquire, maintain, and express foreign DNA.

To determine whether specific genotypes are associated with human enteric disease, MLEE was used to examine 1,300 *E. coli* strains representing 16 serotypes, including EHEC O157:H7 from patients with hemorrhagic colitis and HUS. Overall, 72% of the strains represented 1 of 15 different ETs (40), a less diverse subset than was previously identified among infected animals (39). Each of these 15 ETs comprised strains isolated from various geographic locations, suggesting a widespread distribution of common genotypes. Evolutionary analysis demonstrated that the serogroups are polyphyletic, where representatives of the same serogroup differed, on average, at 28% of the enzyme loci (40). Additionally, the phylogeny illustrated that EHEC O157:H7 was most closely related to enteropathogenic *E. coli* (EPEC) O55:H7, which had previously been found to be associated with infantile diarrhea. Strains of serotype O55:H7 lack Stxs and the EPEC adherence factor (EAF) plasmid that is important for adherence to intestinal epithelial cells (20). These data gave rise to the hypothesis that EHEC O157:H7 evolved from an EPEC O55:H7-like ancestor via successive acquisition of mobile DNA elements, which likely contributed to its emergence as a human pathogen (40). This evolutionary model has withstood repeated scrutiny (11) and has been reiterated in the postgenomic era (30). The study of microbial population structure was, therefore, demonstrated to be critical for monitoring the frequency and distribution of common clones associated with clinical illness, as well as evolutionary processes that may be important for pathogen emergence.

## MLST

MLEE allows for the quantification of population genetic variation at the protein level, or so-called nonsynonymous variation. DNA sequencing quantifies both nonsynonymous and synonymous variation at the nucleotide level, thereby allowing for more precise analyses of informative genetic polymorphisms. Whittam and colleagues more recently applied multilocus sequence typing (MLST) to a collection ($n = 157$) of non-O157 STEC strains recovered via surveillance from human patients with enteric disease in two geographic locations to further investigate the phylogenetic relationship between STEC strains with varying serotypes (J. Tietjen, D. Lacher, L. Ouellette, R. Mosci, Q. Phan, J. Fontana, J. Hadler, J. Andrews, J. T. Rudrik, and S. T. Manning, unpublished data). A total of 40 sequence types (STs) were identified that represented more than 10 distinct clonal groups (members of a clonal group share at least six of seven alleles). Interestingly, two lineages consisted of strains exclusive to one geographic location, suggesting geographic differences in the distribution of STEC clones. Thirteen (33%) of the 40 STs represented genotypes that have been identified previously from different countries around the world (Whittam et al., unpublished). In contrast, 77 out of 80 (96%) EHEC O157:H7 strains isolated from one population had an identical genotype, ST66, which is consistent with prior studies using MLEE (38–40). The remaining three strains were categorized as ST63, ST64, and ST65; all are closely related to ST66, but possess unique single nucleotide polymorphisms (SNPs) in one of the seven housekeeping genes (Whittam et al., unpublished).

## CGH

Gene content varies considerably between *E. coli* lineages. The evolution of O157:H7 from its O55:H7-like ancestor involved a series of independent gene acquisition and loss events. To better understand the evolution of genomic content, Whittam and colleagues used spotted oligonucleotide arrays representing the genomes of O157:H7 Sakai, O157:H7 EDL933, and K-12 MG1655 to compare representatives of the lineages that span the O157 stepwise evolutionary model (for more on the model, see chapter 13) (41).

Diversity in gene content among closely related EHEC O157 lineages far exceeded the nucleotide diversity of conserved metabolic genes (41), confirming that clonal divergence is driven by recombination, more than mutation, in *E. coli* populations (15). Although O157:H7 genome sequence data previously demonstrated that variable gene content is principally attributable to phage recombination, this comparative genomic hybridization (CGH) study revealed the successive acquisition of genomic islands, some of which are absent in ancestral lineages. For example, the tellurite resistance and adherence-conferring island (TAI) is present in extant O157:H7, but is absent in the O157:H$^-$ "German clone" (35, 41). Furthermore, the ancestral O55:H7 and O157:H$^-$ German clone share genes with K-12 MG1655 that are not found in the extant O157:H7 population (41). Collectively, these data indicate that as O157:H7 evolved into an adapted and specialized pathogen, it not only acquired novel virulence factors, but it lost those genes typically found in its less virulent progenitors.

The same CGH array platform was also used to assess the gene content in a set of human and animal clinical isolates of serotypes O26:H11, O111:H8, O118:H16, O111:H11, O153:H11, and O15:H11 (2). Using the O157:H7 genome as a reference, CGH data demonstrated that while O157:H7-specific phage genes were widely distributed among these non-O157 strains, non-O157:H7 phage genes were conserved at higher levels in the O55:H7 progenitor. Additionally, those O157:H7 genes that were vertically acquired from O55:H7 are not disseminated in the clonal complex containing the non-O157 strains (2) (for more information on clonal complexes, see chapter 3). Phylogenetic analyses of the distribution of parsimoniously informative gene loci further indicated that O111:H8 and O118:H16 were close to identical in gene composition. Conversely, O26:H11 represented a more diverse lineage (2) that due to a potentially higher frequency of lateral gene transfer (7) and increasing tendency to harbor Stx2 (25) may give rise to a new lineage of EHEC.

## SNPs

Although the prior MLEE and MLST studies increased our understanding of the association between specific intestinal *E. coli* genotypes and clinical infection, both methods rely on characterizing a few genes representing only a fraction of the *E. coli* genome. Furthermore, MLST does not have the discriminating power to investigate relatedness between closely related but distinct EHEC O157:H7 strains (27). Similar to MLEE and MLST data, the systematic analysis of SNPs is amenable to both population genetic and phylogenetic analyses, but also provides greater resolution of more closely related bacterial genotypes and offers insights into genome divergence. SNP data can also be used to examine epidemiological associations between bacterial genotypes and clinical disease. A study of *M. tuberculosis* cluster analyses based on 112 SNPs identifieed six phylogenetically distinct clusters, which were associated with the geographic origin of the strains as well as the patient's birthplace (12). SNP cluster group 1, the most ancestral cluster identified in the phylogeny, predominated among patients from the Indian subcontinent, suggesting that *M. tuberculosis* first emerged on the subcontinent and spread elsewhere (12).

The success of SNP subtyping of *M. tuberculosis* prompted the development of a similar SNP genotyping scheme for EHEC O157:H7 in the Whittam laboratory. The platform was designed to resolve genome-wide genetic variation and establish epidemiological associations by quantifying the frequency of nucleotide polymorphisms at 96 loci in 83 genes (23). These loci include 68 sites found by comparative genome sequencing microarrays (42), 15 polymorphisms found in housekeeping genes (18), 4 polymorphisms detected between two complete O157:H7 genomes (16, 28), and 9 polymorphisms in three virulence genes (*eae, espA*, and *fimA*). For every SNP locus, hairpin-shaped primers were designed by adding a 5′ tail complementary to the 3′ end of each linear primer and real-time PCR was used to identify the SNP at each locus (23). In this analysis, the diversity of EHEC O157:H7 was estimated among strains recovered from clinical sources; the SNP genotypes were organized into related groups, or clades; and associations between genotypes and disease severity were identified.

Initially, 403 EHEC O157:H7 and closely related strains from outbreaks, clusters, and sporadic cases of disease around the world were examined (23). Pairwise comparisons of the nucleotide profiles identified 39 SNP genotypes, and the number of nucleotide differences between genotypes ranged from 1 to 57, with an average of 23.1 $\pm$ 1.6 (standard deviation). The nucleotide diversity, a measure of the degree of polymorphisms, was 0.212 $\pm$ 0.199 among all 403 strains, indicating that two strains selected at random differed, on average, at 20% of the SNP loci. The minimum evolution algorithm was used to infer phylogenetic relationships among the 39 SNP genotypes. Nine significant clusters (or clades) were resolved, with the deepest node (clade 9) representing the ancestral lineage containing close relatives with wild-type *E. coli* phenotypes (e.g., GUD$^+$, $\beta$-glucuronidase positive; SOR$^+$, sorbitol positive). In contrast, the evolutionarily derived lineages (clades 1 to 8) comprised strains with contemporary O157:H7 phenotypes (e.g., GUD$^-$, SOR$^-$) (Fig. 1) (23).

Based on the SNP genotype profiles, a minimum set of 32 SNP loci was devised, which distinguishes between all 39 SNP genotypes; an additional 135 O157:H7 strains were examined using this minimum SNP set. Among all 538 strains evaluated, SNP genotypes 9 (35%) and 30 (18%) predominated. Twenty genotypes, however, were only represented by one or two strains, while six were represented by strains isolated from multiple continents. Virtually all strains could be classified into one of the eight clades identified in the minimum evolution analysis, with more than 75% belonging to clade 2 (46.6%), 3 (10.6%), 7 (7.1%), or 8 (25.6%). None of the strains examined were part of the ancestral lineage (clade 9) (23).

Subsequent analyses of the 39 SNP genotypes uncovered phylogenetically informative loci, as defined by the presence of two nucleotide differences in 2

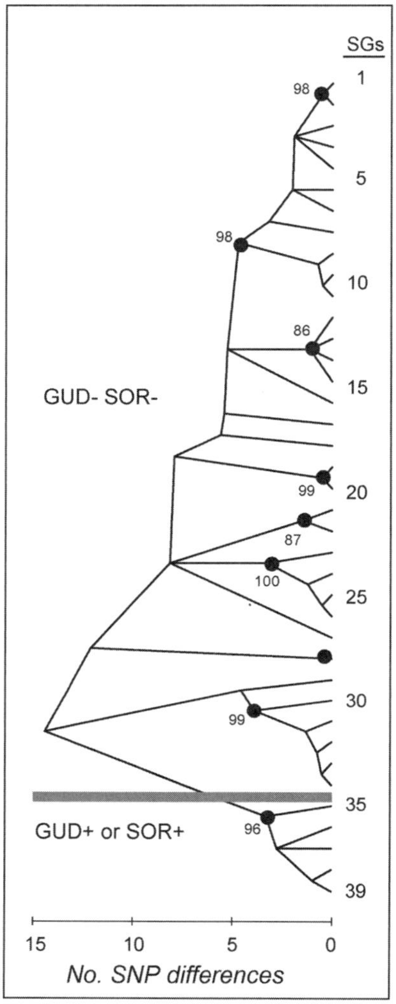


**Figure 1.** Minimum evolution tree showing the phylogenetic relationships among 39 SNP genotypes (SGs) of *E. coli* O157:H7. Numbers at the nodes are the bootstrap confidence values that mark nine distinct clusters, or clades. The gray line separates strains with ancestral traits (GUD$^+$, SOR$^+$) from those with derived traits (GUD$^-$, SOR$^-$). Illustration created by Thomas S. Whittam, 2007.

or more genotypes. Among the 96 SNP loci, 71 sites had complete data, and of these, there were 23 singletons and 48 parsimoniously informative sites. The 48 parsimoniously informative sites were used to construct a phylogenetic network to determine if the informative sites support a single phylogeny or a set of conflicting relationships. In this analysis, the 39 genotypes were resolved into 25 distinct nodes, with 10 containing 2 or more genotypes with the same profiles across all 48 loci. Rather than producing a unique bifurcating tree, the network analysis identified a central group of four clades (clades 3, 4, 5, and 7) connected by multiple paths, suggesting that recombination or recurrent mutation has contributed to the divergence of the central clades from the evolutionarily derived lineages. In contrast, clades 1, 2, 6, and 8 have no evidence of a conflicting phylogenetic signal, which is indicative of independent diversification or divergence without recombination in background polymorphisms (23). Recent data, however, contradict the evidence for recombination among the central clades, as additional investigation of those strains identified mixed stock cultures in the clade 5 strains and the strain corresponding to SNP genotype 27. Therefore, omitting those strains from the phylogenetic network analysis fails to provide evidence for recombination between the central clades (unpublished data), a finding that is consistent with a follow-up study (21). More information regarding the reanalysis of the SNP data can be found in chapter 13 of this volume.

## REAL-TIME SNP PROFILING

Because SNP genotyping was able to distinguish O157:H7 lineages, a real-time PCR method was developed to rapidly detect the presence of SNPs specific for the predominant O157:H7 clades (32). Hairpin-shaped primers targeting four SNPs (SNPs 539, 1060, 438, and 630) specific for clades 1, 2, 3, and 8 were used to examine 83 O157:H7 and 2 non-O157:H7 (control) strains. The assays were performed from both purified DNA and direct colony picks on sorbitol MacConkey agar. In the development stage, real-time PCR differentiated a set of 21 strains into the correct O157:H7 phylogenetic clades identified previously. Strains from clades 8 and 1 could be distinguished from all others using primers for SNP 539 and SNP 1060, respectively ($P < 0.001$). In the validation stage, colony picks from a blinded sample ($n = 62$) of strains from clades 2, 3, and 8 were evaluated. Of these 62 strains, 50 (81%) were correctly assigned. Within this proportion, all 20 clade 8 strains were correctly identified, as were 18 of 20 clade 3 strains and both non-O157 negative controls. Only 10 of 20 clade 2 strains were properly identified, with the remainder being categorized as members of clade 1, for which there were no representative strains chosen. As suspected, review of sequencing data demonstrated that SNP 1060 contains a different polymorphism for this subset of clade 1 strains. In all, these findings demonstrate that screening for these four SNPs can rapidly differentiate the predominant O157:H7 genotypes from colonies on sorbitol MacConkey agar, the medium commonly used for identifying EHEC O157:H7 in clinical laboratories, within 24 hours (32). If combined with *stx* detection methods, the utilization of a rapid O157:H7 genotyping system could potentially have an enormous impact on disease treatment.

## EPIDEMIOLOGICAL ASSOCIATIONS BETWEEN O157:H7 CLADES AND DISEASE SEVERITY

After strains were assigned to SNP genotypes and clades, epidemiological data for 333 outbreak and sporadic cases of EHEC O157:H7 disease in Michigan between 2001 and 2006 (22) were used to look for associations between clades and disease severity. One strain per outbreak cluster and patient was included in univariate and multivariate logistic regression analyses. Disease in younger patients (0 to 18 years of age) and cases of HUS were significantly more likely to be caused by a clade 8 strain than all of the other O157:H7 clades. In contrast, clade 7 strains were significantly less likely to be associated with bloody diarrhea and other symptoms relative to all of the other clades. Additionally, a significant increase (10% in 2002 to 46% in 2006; Mantel-Haenszel chi-square test = 32.5; df = 1; $P < 0.001$) in the frequency of disease caused by clade 8 strains was observed (23) even though the overall frequency of O157:H7 cases has decreased since 2002 (22). Despite the strong association observed between clade 8 strains and HUS, it is important to note that a flexible HUS case definition may have resulted in some case misclassifications. Nevertheless, among the 11 HUS cases identified, 8 had clade 8 infections, thereby providing justification for conducting additional large-scale studies in diverse populations to better understand the association between HUS and O157:H7 clade genotypes.

SNP genotyping was also performed on a subset of EHEC O157:H7 strains implicated in major food-borne outbreaks since 1982. Clade 1 strains, which were identified frequently among strains from Michigan, include the O157:H7 genome strain Sakai (16) from the 1996 Japanese outbreak linked to contaminated radish sprouts. Although this outbreak reportedly affected more than 8,000 people, the hospitalization and HUS rates were reported to be low (13, 17, 26). These rates, however, are questionable because of a poorly defined case definition that included abdominal pain from individuals in the epidemic area, but not culture positivity or the presence of diarrhea (P. I. Tarr, personal communication). Indeed, one study reported the HUS rate during the Sakai outbreak to be between 10 and 15% (19). Clade 3, which includes the genome strain EDL933 (28) from the first human O157:H7 outbreak linked to hamburgers from popular fast-food restaurant outlets in Michigan and Oregon (31) and the nursing home outbreak in Walla Walla, Washington (14), also contributed to relatively low rates of HUS. In contrast, clade 8 outbreaks included the O157:H7 strains implicated in the 2006 multistate outbreaks linked to contaminated spinach (9) and lettuce (8). More than 275 people were infected during these two outbreaks that resulted in remarkably high rates of hospitalization, death, and HUS (average 13%), a rate that is three times higher than the average HUS rate for 350 U.S. outbreaks (29). These high rates have led to speculation that a hypervirulent O157:H7 strain has emerged and contributes to more severe clinical disease, a hypothesis that is supported by the epidemiological study of SNP genotypes in Michigan (23). Nevertheless, it is important to note that the comparison of HUS rates among clades from different outbreaks is complicated because of differing case definitions and age distributions of affected patients; therefore, further investigation into bacterial phylogeny-ordained risk for HUS is warranted.

## VIRULENCE CHARACTERISTICS ASSOCIATED WITH O157:H7 STRAINS OF DIFFERENT CLADES

Because of the variation in disease severity observed among patients infected with diverse O157:H7 genotypes (23), additional studies have been conducted in an effort to uncover intrinsic differences important for virulence. First, the clade 8 spinach outbreak strain (TW14359) genome was sequenced and compared to the Sakai (clade 1) (16) and EDL933 (clade 3) (28) genomes. A comparison of 2,741 shared genes among the three genomes demonstrated that the Sakai and EDL933 genomes are more similar to each other in gene content and nucleotide sequence than TW14359 (23). In addition, TW14359 possesses genetic elements absent in the other genomes, including the Stx2c-encoding bacteriophage (23). By using the SNP genotyping phylogeny as a framework, it is possible to investigate virulence gene diversity, allelic variation, and gene expression differences among genotypes to identify bacterial genomic determinants of colonization, pathogenesis, and transmissibility.

## DIFFERENCES IN STRESS RESISTANCE PROPERTIES AMONG CLADES

For successful food-borne transmission, *E. coli* O157:H7 must survive the production environment on the farm; harsh conditions encountered during food process, storage, and preparation; and gastric acidity and bile salts encountered after ingestion. Assessments of variation in stress resistance properties among clades are just emerging, and the available data indicate that clade 8 strains are more resistant to stresses present during food processing as well as stresses present in the host. Chlorinated water is commonly used to wash and sanitize fresh-cut produce. A comparison of chlorine resistance across O157:H7 clades found that clade 8 strains were more resistant to chlorine compared to strains from clades 2, 4, 6, and 9 (37). A comparison of survival in a simulated gastric environment (pH 2.5) indicated that O157:H7 strains were more acid resistant than EHEC O26:H11 strains (5). When the O157:H7 survival data were analyzed by clade, clade 8 strains ($n = 4$) had a significantly ($P = 0.03$) higher survival rate compared to strains from other clades ($n = 4$). The average survival rate of clade 8 strains is two times higher than the average survival rate of strains from all other clades in this acidic environment. Additionally, after incubation in apple juice at 4°C for 24 hours, clade 8 strains had an average survival rate in the simulated gastric environment that is three times higher than the average survival rate of strains from all other clades. Further work is necessary to continue to characterize the variation in stress survival among the clades, as increased stress resistance may play a role in improved transmissibility of clade 8 strains.

## NONRANDOM DISTRIBUTION OF Stx VARIANTS

Since Stx production has been linked to virulence, the frequency among O157 clades of one or more of three Stx variants ($stx_1$, $stx_2$, and $stx_{2c}$) was determined by clade for 519 of 528 strains using SNP data (23). Although $stx_1$ was found in ~65% of the 519 O157:H7 strains tested, the distribution was not nonrandom across clades, with a high frequency of $stx_1$ in clade 2 (95.1% of all $stx_1$-positive strains are in clade 2) but not clade 8 (3.7%). In contrast, the

*$stx_2$* gene was identified in virtually all (98.5%) O157:H7 strains evaluated, occurring most frequently in clade 2 (46.8% of 519 strains) and clade 8 (25.4%) strains. Most noteworthy is the nonrandom distribution of *$stx_{2c}$*, which was concentrated in clades 4, 6, 7, and 8 (Color Plate 3), and the association between clade 8 and the presence of both *$stx_2$* and *$stx_{2c}$* relative to other *$stx_{2c}$*-positive clades ($P < 0.0001$). The epidemiological association with HUS, however, could not be explained by *$stx_{2c}$* in clade 8 strains, as only 4 of 11 HUS patients had *$stx_{2c}^{+}$* strains. Although it is possible that *$stx_{2c}$* is rapidly lost during infection, these data suggest that clade 8 strains possess alternative factors, or may differentially express virulence genes that contribute to more severe disease (23).

## VARIATION IN ADHERENCE TO EPITHELIAL CELLS

Because the O157:H7 spinach outbreak strain TW14359 differed considerably in gene content (23) and gene expression relative to the other O157:H7 genomes, the ability to adhere to bovine mammary epithelial (MAC-T) cells was also assessed relative to Sakai (3). Cell association assays, which measure both adherence and invasion, were conducted 30 and 60 minutes postinfection of MAC-T cells. After 30 minutes, there was no significant difference in the level of adherence to MAC-T cells for both strains ($P = 0.28$), but after 1 hour, TW14359 had 2.5 times higher adherence levels than Sakai ($P = 0.02$); differences were not due to variation in growth phase or starting cell densities.

Association assays were also conducted using a larger sample of strains representing clades 8 ($n = 12$) and 2 ($n = 12$), which were previously shown to represent the most abundant clades and are also divergent via phylogenetic analyses (23). After 1 hour postinfection, there were also considerable differences in the level of adherence between these two clades, with 10 of the 12 clade 8 strains having higher association levels than all 12 clade 2 strains combined (1). On average, the clade 8 strains adhered to MAC-T cells 2.31 times better than Sakai (95% confidence interval, 1.84 to 2.78), whereas the average level of adherence for clade 2 strains was only 1.03 relative to Sakai (95% confidence interval, 0.84 to 1.23). These differences were statistically significant ($P = 0.0001$) in a mixed analysis of variance model, where the biological replicate was nested within the strain effect, and suggest that these two phylogenetically distinct lineages vary in their ability to colonize the intestinal tract (1).

## DIFFERENTIAL EXPRESSION AMONG SHARED GENES

Global gene expression patterns were examined in the O157:H7 Sakai strain relative to a strain linked to a U.S. outbreak from fresh spinach (TW14359). Expression was measured following exposure of bacteria to MAC-T bovine epithelial cells using multigenome oligoarrays targeting 6,088 open reading frames from O157:H7 Sakai and EDL933, K-12 MG1655, and the pO157 plasmid (3). In all, 915 genes were differentially expressed ($P < 0.05$)

between the two strains and 206 had a change difference between 1.5- and 8.3-fold. A total of 36 of the 41 genes (e.g., *eae, tir, espA, espB*, and *sepZ*) from the locus of enterocyte effacement (LEE) pathogenicity island, the key colonization factor of attaching-and-effacing *E. coli*, were upregulated in TW14359 relative to Sakai. Similarly, the pO157-borne 12-gene *etpC-O* operon that encodes a widely conserved type II secretion system (T2SS) of gram-negative bacteria (33) was upregulated in TW14359 (1.20-fold). Although the change difference was low, the finding that all nine microarray probes for T2SS indicated upregulation in TW14359 implies a biologically meaningful observation. Two T2SS effectors with demonstrated roles in adherence, the plasmid-borne *tagA/stcE* (1.83-fold) and the chromosomal *adfO* (1.22-fold), were also upregulated in TW14359, as were the Stx2 genes ($stx_{2a}$ and $stx_{2b}$). In Sakai, microarrays revealed an increase in the expression of 14 genes involved in flagellation and chemotaxis. These data, which were confirmed by quantitative PCR, suggest discrepant regulation of virulence and motility genes in these two O157:H7 outbreak strains (3).

A similar study was performed to compare global gene expression among 12 clade 2 and 12 clade 8 strains using the same conditions; strains in each clade were matched by *stx* profile. Analyses of gene expression demonstrated remarkable differences between clades, whereas within-clade differences were marginal (1). Only four genes were differentially expressed within clade 2, including $stx_{1a}$ and $stx_{1b}$, a putative prophage repressor, and an unknown gene from the Stx1 prophage (Sp15). Fourteen genes were differentially expressed within clade 8, most of which were unknown phage genes, though expression of $stx_{2a}$ and $stx_{2b}$ was observed to be 1.559- and 1.961-fold higher, respectively, in $stx_2^+$ clade 8 strains versus $stx_2/stx_{2c}^+$ clade 8 strains. Although the number of differentially expressed genes between clades 2 and 8 varied between groups with different *stx* profiles, expression of any gene in either clade did not appear to be influenced by the *stx* profile.

In total, there was significant differential expression of 604 genes between the two clades, with 186 genes having a $>$1.5-fold increase in expression. Most noteworthy was that 29 LEE genes were upregulated in clade 8 relative to clade 2, with discrete differences between LEE operons (*LEE1* through *LEE5*). Apart from *sepL* (1.42-fold), transcription was highest in the *LEE4* genes (1.92 $\pm$ 0.13-fold), especially for the *espADB* polycistron that is stimulated by contact with the host cell. The expression of *LEE5* effectors and chaperones was slightly lower (1.715 $\pm$ 0.131-fold), followed by *LEE3* (1.66 $\pm$ 0.14-fold) and *LEE2* (1.50 $\pm$ 0.09-fold), which encode the membrane-bound type III secretion system components. *LEE1*, which encodes the Ler regulator and inner membrane proteins of the TTSS, and *orf1, rorf3, grlR*, and *orf29* were not significantly differentially expressed, though LEE regulators *rpoS, grlA*, and *gadX* were. Because microarrays can underestimate gene expression differences, LEE operon genes *ler, sepZ, escN, espA, espB, tir*, and *eae* were subjected to quantitative PCR. The observed differences in the upregulation of LEE operons in clade 8 were confirmed by quantitative PCR and were more pronounced (Fig. 2). The *toxB* gene, which has been shown to stimulate expression of LEE4

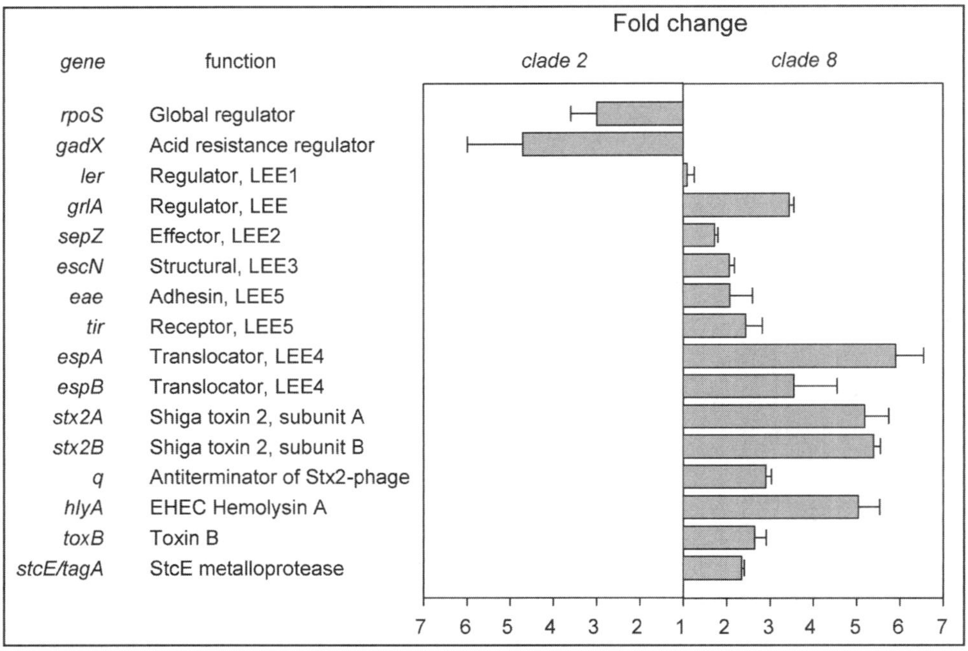


**Figure 2.** Relative differences in gene expression between clades 8 and 2 of *E. coli* O157:H7, as detected by quantitative PCR.

proteins and facilitate adhesion to epithelial cells posttranscriptionally, also was upregulated in clade 8 strains, as was the *tagA/stcE* protease, though the latter was only significant by quantitative PCR. Finally, relative to clade 2, expression of $stx_{2a}$ and $stx_{2b}$ was elevated in $stx_2^+$ clade 8 strains by 4.067- and 6.109-fold, respectively, and in $stx_2/stx_{2c}^+$ clade 8 strains by 2.606- and 3.116-fold, respectively.

A third set of transcriptome analyses were conducted to examine expression differences among bovine-associated O157:H7 strains with distinct Stx-encoding bacteriophage insertion site genotypes as determined in a prior study (6). Four strains representing the clinical genotype, or the O157:H7 type that predominated in human disease cases, were compared with four strains of the bovine-associated genotype (36), which infrequently causes human disease and predominates in cattle (6). SNP genotyping demonstrated that three of the four clinical strains belonged to clade 8, while all four bovine-associated strains were members of clade 7, a lineage previously shown to cause less severe disease (23). Similar to the previous comparisons, a number of known O157:H7 virulence genes, including LEE genes *hlyA* (enterohemolysin), *toxB*, and the pO157-encoded T2SS genes, were significantly upregulated in the clinical strains versus the bovine-associated strains. In contrast, genes essential for acid resistance such as glutamate decarboxylase system genes, acid fitness island genes, and other stress fitness-associated genes were upregulated in the bovine-associated strains relative to clinical strains. Genes involved in the nitrogen regulatory response to ammonia-limiting conditions also had increased expression in the bovine-associated strains. For 15 genes, expression differences were confirmed using quantitative PCR, and the increased expression of acid resistance genes in the bovine-associated strains was confirmed using a model stomach assay, which assesses survival in complex acidic conditions (pH 2.5) that mimic the human stomach. Interestingly, the bovine-associated strains had a sixfold-higher survival rate than the clinical strains. These results suggest that clinical strains more commonly cause human illness because of some other adaptive trait(s), and current work is focused on the possibility that enhanced expression of O157:H7 virulence factors is a key adaptive event in the pathogenesis of clade 8 strains. By contrast, bovine-associated strains might better resist adverse environmental conditions such as low pH, volatile fatty acid stress, and hyperosmolarity, thereby enabling them to survive well in cattle (36). It is currently unclear why bovine strains that are seemingly more resistant to stress and presumably more fit in conditions between hosts (transmission) are underrepresented in clinical disease (see chapter 18 for more information concerning EHEC strains that circulate among reservoir hosts). However, these gene expression data and phenotypic analyses demonstrate that the divergent clade 8 subpopulation of O157:H7 is phenotypically distinct from other lineages and has increased expression of shared virulence genes.

## SUMMARY OF STUDIES INVOLVING EHEC

Taken together, the data generated by Tom Whittam and his research team have demonstrated significant genetic variation among EHEC genotypes that contribute to human disease and that a subset of EHEC O157:H7 strains have an enhanced propensity for causing more severe disease. Assessing the widespread distribution of EHEC in the United States is important to verify these findings and to determine which genotypes frequently cause more severe clinical outcomes, like HUS. Furthermore, identifying the intrinsic differences between strains that correlate with disease severity is necessary for developing rapid diagnostic methods that could ultimately have a significant impact on both outbreak mitigation and disease treatment.

### REFERENCES

1. **Abu-Ali, G., L. M. Ouellette, S. Henderson, D. W. Lacher, J. T. Riordan, T. S. Whittam, and S. D. Manning.** 2010. Increased adherence and expression of virulence genes in a lineage of *Escherichia coli* O157:H7 commonly associated with human infections. *PLoS One* **5:**e10167.
2. **Abu-Ali, G. S., D. W. Lacher, L. M. Wick, W. Qi, and T. S. Whittam.** 2009. Genomic diversity of pathogenic *Escherichia coli* of the EHEC 2 clonal complex. *BMC Genomics* **10:**296.
3. **Abu-Ali, G. S., L. M. Ouellette, S. T. Henderson, T. S. Whittam, and S. D. Manning.** 2010. Differences in adherence and virulence gene expression between two outbreak strains of enterohaemorrhagic *Escherichia coli* O157:H7. *Microbiology* **156:**408–419.

4. **Bell, J. A., J. L. St Charles, A. J. Murphy, V. A. Rathinam, A. E. Plovanich-Jones, E. L. Stanley, J. E. Wolf, J. R. Gettings, T. S. Whittam, and L. S. Mansfield.** 2009. Multiple factors interact to produce responses resembling spectrum of human disease in *Campylobacter jejuni* infected C57BL/6 IL-$10^{-/-}$ mice. *BMC Microbiol.* **9:**57.
5. **Bergholz, T. M., and T. S. Whittam.** 2007. Variation in acid resistance among enterohaemorrhagic *Escherichia coli* in a simulated gastric environment. *J. Appl. Microbiol.* **102:**352–362.
6. **Besser, T. E., N. Shaikh, N. J. Holt, P. I. Tarr, M. E. Konkel, P. Malik-Kale, C. W. Walsh, T. S. Whittam, and J. L. Bono.** 2007. Greater diversity of Shiga toxin-encoding bacteriophage insertion sites among *Escherichia coli* O157:H7 isolates from cattle than in those from humans. *Appl. Environ. Microbiol.* **73:**671–679.
7. **Bielaszewska, M., W. Zhang, A. Mellmann, and H. Karch.** 2007. Enterohaemorrhagic *Escherichia coli* O26:H11/H$^{-}$: a human pathogen in emergence. *Berl. Munch. Tierarztl. Wochenschr.* **120:**279–287.
8. **Centers for Disease Control and Prevention.** December 14, 2006. Multistate outbreak of *E. coli* O157 infections, November–December 2006. Centers for Disease Control and Prevention. [Online.] http://www.cdc.gov/ecoli/2006/december/121406.htm.
9. **Centers for Disease Control and Prevention.** 2006. Ongoing multistate outbreak of *Escherichia coli* serotype O157:H7 infections associated with consumption of fresh spinach—United States, September 2006. *MMWR Morb. Mortal. Wkly. Rep.* **55:**1045–1046.
10. **Cho, S., T. S. Whittam, D. J. Boxrud, J. M. Bartkus, and A. M. Saeed.** 2008. Allele distribution and genetic diversity of VNTR loci in *Salmonella enterica* serotype Enteritidis isolates from different sources. *BMC Microbiol.* **8:**146.
11. **Feng, P., K. A. Lampel, H. Karch, and T. S. Whittam.** 1998. Genotypic and phenotypic changes in the emergence of *Escherichia coli* O157:H7. *J. Infect. Dis.* **177:**1750–1753.
12. **Filliol, I., A. S. Motiwala, M. Cavatore, W. Qi, M. H. Hazbon, M. Bobadilla del Valle, J. Fyfe, L. Garcia-Garcia, N. Rastogi, C. Sola, T. Zozio, M. I. Guerrero, C. I. Leon, J. Crabtree, S. Angiuoli, K. D. Eisenach, R. Durmaz, M. L. Joloba, A. Rendon, J. Sifuentes-Osornio, A. Ponce de Leon, M. D. Cave, R. Fleischmann, T. S. Whittam, and D. Alland.** 2006. Global phylogeny of *Mycobacterium tuberculosis* based on single nucleotide polymorphism (SNP) analysis: insights into tuberculosis evolution, phylogenetic accuracy of other DNA fingerprinting systems, and recommendations for a minimal standard SNP set. *J. Bacteriol.* **188:**759–772.
13. **Fukushima, H., T. Hashizume, Y. Morita, J. Tanaka, K. Azuma, Y. Mizumoto, M. Kaneno, M. Matsuura, K. Konma, and T. Kitani.** 1999. Clinical experiences in Sakai City Hospital during the massive outbreak of enterohemorrhagic *Escherichia coli* O157 infections in Sakai City, 1996. *Pediatr. Int.* **41:**213–217.
14. **Griffin, P. M., S. M. Ostroff, R. V. Tauxe, K. D. Greene, J. G. Wells, J. H. Lewis, and P. A. Blake.** 1988. Illnesses associated with *Escherichia coli* O157:H7 infections. A broad clinical spectrum. *Ann. Intern. Med.* **109:**705–712.
15. **Guttman, D. S., and D. E. Dykhuizen.** 1994. Clonal divergence in *Escherichia coli* as a result of recombination, not mutation. *Science* **266:**1380–1383.
16. **Hayashi, T., K. Makino, M. Ohnishi, K. Kurokawa, K. Ishii, K. Yokoyama, C. G. Han, E. Ohtsubo, K. Nakayama, T. Murata, M. Tanaka, T. Tobe, T. Iida, H. Takami, T. Honda, C. Sasakawa, N. Ogasawara, T. Yasunaga, S. Kuhara, T. Shiba, M. Hattori, and H. Shinagawa.** 2001. Complete genome sequence of enterohemorrhagic *Escherichia coli* O157:H7 and genomic comparison with a laboratory strain K-12. *DNA Res.* **8:**11–22.
17. **Higami, S., K. Nishimoto, T. Kawamura, T. Tsuruhara, G. Isshiki, and A. Ookita.** 1998. Retrospective analysis of the relationship between HUS incidence and antibiotics among patients with *Escherichia coli* O157 enterocolitis in the Sakai outbreak. *Kansenshogaku Zasshi* **72:**266–272. (In Japanese.)
18. **Hyma, K. E., D. W. Lacher, A. M. Nelson, A. C. Bumbaugh, J. M. Janda, N. A. Strockbine, V. B. Young, and T. S. Whittam.** 2005. Evolutionary genetics of a new

pathogenic *Escherichia* species: *Escherichia albertii* and related *Shigella boydii* strains. *J. Bacteriol.* **187:**619–628.

19. **Ikeda, K., O. Ida, K. Kimoto, T. Takatorige, N. Nakanishi, and K. Tatara.** 1999. Effect of early fosfomycin treatment on prevention of hemolytic uremic syndrome accompanying *Escherichia coli* O157:H7 infection. *Clin. Nephrol.* **52:**357–362.
20. **Kaper, J. B., J. P. Nataro, and H. L. Mobley.** 2004. Pathogenic *Escherichia coli. Nat. Rev. Microbiol.* **2:**123–140.
21. **Leopold, S. R., V. Magrini, N. J. Holt, N. Shaikh, E. R. Mardis, J. Cagno, Y. Ogura, A. Iguchi, T. Hayashi, A. Mellmann, H. Karch, T. E. Besser, S. A. Sawyer, T. S. Whittam, and P. I. Tarr.** 2009. A precise reconstruction of the emergence and constrained radiations of *Escherichia coli* O157 portrayed by backbone concatenomic analysis. *Proc. Natl. Acad. Sci. USA* **106:**8713–8718.
22. **Manning, S. D., R. T. Madera, W. Schneider, S. E. Dietrich, W. Khalife, W. Brown, T. S. Whittam, P. Somsel, and J. T. Rudrik.** 2007. Surveillance for Shiga toxin-producing *Escherichia coli*, Michigan, 2001–2005. *Emerg. Infect. Dis.* **13:**318–321.
23. **Manning, S. D., A. S. Motiwala, A. C. Springman, W. Qi, D. W. Lacher, L. M. Ouellette, J. M. Mladonicky, P. Somsel, J. T. Rudrik, S. E. Dietrich, W. Zhang, B. Swaminathan, D. Alland, and T. S. Whittam.** 2008. Variation in virulence among clades of *Escherichia coli* O157:H7 associated with disease outbreaks. *Proc. Natl. Acad. Sci. USA* **105:**4868–4873.
24. **Manning, S. D., A. C. Springman, E. Lehotzky, M. A. Lewis, T. S. Whittam, and H. D. Davies.** 2009. Multilocus sequence types associated with neonatal group B streptococcal sepsis and meningitis in Canada. *J. Clin. Microbiol.* **47:**1143–1148.
25. **Mellmann, A., M. Bielaszewska, and H. Karch.** 2009. Intrahost genome alterations in enterohemorrhagic *Escherichia coli. Gastroenterology* **136:**1925–1938.
26. **Michino, H., K. Araki, S. Minami, S. Takaya, N. Sakai, M. Miyazaki, A. Ono, and H. Yanagawa.** 1999. Massive outbreak of *Escherichia coli* O157:H7 infection in schoolchildren in Sakai City, Japan, associated with consumption of white radish sprouts. *Am. J. Epidemiol.* **150:**787–796.
27. **Noller, A. C., M. C. McEllistrem, O. C. Stine, J. G. Morris, Jr., D. J. Boxrud, B. Dixon, and L. H. Harrison.** 2003. Multilocus sequence typing reveals a lack of diversity among *Escherichia coli* O157:H7 isolates that are distinct by pulsed-field gel electrophoresis. *J. Clin. Microbiol.* **41:**675–679.
28. **Perna, N. T., G. Plunkett III, V. Burland, B. Mau, J. D. Glasner, D. J. Rose, G. F. Mayhew, P. S. Evans, J. Gregor, H. A. Kirkpatrick, G. Posfai, J. Hackett, S. Klink, A. Boutin, Y. Shao, L. Miller, E. J. Grotbeck, N. W. Davis, A. Lim, E. T. Dimalanta, K. D. Potamousis, J. Apodaca, T. S. Anantharaman, J. Lin, G. Yen, D. C. Schwartz, R. A. Welch, and F. R. Blattner.** 2001. Genome sequence of enterohaemorrhagic *Escherichia coli* O157:H7. *Nature* **409:**529–533.
29. **Rangel, J. M., P. H. Sparling, C. Crowe, P. M. Griffin, and D. L. Swerdlow.** 2005. Epidemiology of *Escherichia coli* O157:H7 outbreaks, United States, 1982–2002. *Emerg. Infect. Dis.* **11:**603–609.
30. **Reid, S. D., C. J. Herbelin, A. C. Bumbaugh, R. K. Selander, and T. S. Whittam.** 2000. Parallel evolution of virulence in pathogenic *Escherichia coli. Nature* **406:**64–67.
31. **Riley, L. W., R. S. Remis, S. D. Helgerson, H. B. McGee, J. G. Wells, B. R. Davis, R. J. Hebert, E. S. Olcott, L. M. Johnson, N. T. Hargrett, P. A. Blake, and M. L. Cohen.** 1983. Hemorrhagic colitis associated with a rare *Escherichia coli* serotype. *N. Engl. J. Med.* **308:**681–685.
32. **Riordan, J. T., S. B. Viswanath, S. D. Manning, and T. S. Whittam.** 2008. Genetic differentiation of *Escherichia coli* O157:H7 clades associated with human disease by real-time PCR. *J. Clin. Microbiol.* **46:**2070–2073.
33. **Sandkvist, M.** 2001. Type II secretion and pathogenesis. *Infect. Immun.* **69:**3523–3535.
34. **Springman, A. C., J. L. Jacobs, V. S. Somvanshi, G. W. Sundin, M. H. Mulks, T. S. Whittam, P. Viswanathan, R. L. Gray, J. J. Lipuma, and T. A. Ciche.** 2009. Genetic

diversity and multihost pathogenicity of clinical and environmental strains of *Burkholderia cenocepacia*. *Appl. Environ. Microbiol.* **75:**5250–5260.

35. **Tarr, P. I., S. S. Bilge, J. C. Vary, Jr., S. Jelacic, R. L. Habeeb, T. R. Ward, M. R. Baylor, and T. E. Besser.** 2000. Iha: a novel *Escherichia coli* O157:H7 adherence-conferring molecule encoded on a recently acquired chromosomal island of conserved structure. *Infect. Immun.* **68:**1400–1407.
36. **Vanaja, S. K., A. C. Springman, T. E. Besser, T. S. Whittam, and S. D. Manning.** 2010. Differential expression of virulence and stress fitness genes between *Escherichia coli* O157:H7 strains with clinical or bovine-biased genotypes. *Appl. Environ. Microbiol.* **76:**60–68.
37. **Wang, S., K. Deng, S. Zaremba, X. Deng, C. Lin, Q. Wang, M. L. Tortorello, and W. Zhang.** 2009. Transcriptomic response of *Escherichia coli* O157:H7 to oxidative stress. *Appl. Environ. Microbiol.* **75:**6110–6123.
38. **Whittam, T. S., I. K. Wachsmuth, and R. A. Wilson.** 1988. Genetic evidence of clonal descent of *Escherichia coli* O157:H7 associated with hemorrhagic colitis and hemolytic uremic syndrome. *J. Infect. Dis.* **157:**1124–1133.
39. **Whittam, T. S., and R. A. Wilson.** 1988. Genetic relationships among pathogenic *Escherichia coli* of serogroup O157. *Infect. Immun.* **56:**2467–2473.
40. **Whittam, T. S., M. L. Wolfe, I. K. Wachsmuth, F. Ørskov, I. Ørskov, and R. A. Wilson.** 1993. Clonal relationships among *Escherichia coli* strains that cause hemorrhagic colitis and infantile diarrhea. *Infect. Immun.* **61:**1619–1629.
41. **Wick, L. M., W. Qi, D. W. Lacher, and T. S. Whittam.** 2005. Evolution of genomic content in the stepwise emergence of *Escherichia coli* O157:H7. *J. Bacteriol.* **187:**1783–1791.
42. **Zhang, W., W. Qi, T. J. Albert, A. S. Motiwala, D. Alland, E. K. Hyytia-Trees, E. M. Ribot, P. I. Fields, T. S. Whittam, and B. Swaminathan.** 2006. Probing genomic diversity and evolution of *Escherichia coli* O157 by single nucleotide polymorphisms. *Genome Res.* **16:**757–767.

*Population Genetics of Bacteria: a Tribute to Thomas S. Whittam*
Edited by S. T. Walk and P. C. H. Feng


*Chapter 18*

# *Escherichia coli* O157:H7 in Reservoir Hosts

***Thomas E. Besser, Margaret A. Davis, and Seth T. Walk***

## OVERVIEW

Serotype O157:H7 is the charter member of the enterohemorrhagic pathogroup of *Escherichia coli* (EHEC). This serotype was first associated with disease in 1982 (111) and is now recognized as an important cause of diarrhea, bloody diarrhea, and the hemolytic uremic syndrome (HUS) globally. In the United States, >95,000 cases of O157:H7 infections are estimated to occur annually, with over 3,000 associated hospitalizations and 31 deaths per year (113a). (In this chapter, we will use "O157:H7" to indicate the non-sorbitol-fermenting, β-glucuronidase-negative clade of this species that predominates in North America despite the fact that $H^-$ variants are common.)

O157:H7 is capable of infecting humans following low dose (<100 CFU) oral exposures (132–134). The clinical presentation and the pathophysiology of infection have been topics of in-depth review (14, 130, 131, 146). O157:H7 strains may be carried asymptomatically or cause infection and result in non-bloody diarrhea, but the classic presenting complaint is bloody diarrhea (hemorrhagic colitis). A small percentage of patients with hemorrhagic colitis go on to develop HUS, characterized by microangiopathic hemolytic anemia, thrombocytopenia, and oliguric renal failure. The cardinal virulence factors of O157:H7 include the locus of enterocyte effacement (LEE) and one or more exotoxins known as Shiga toxins (Stxs). LEE encodes the cell surface protein intimin, a type III secretion system, and several effector molecules including Tir (the translocated intimin receptor), which is injected into the host intestinal epithelial cell where it translocates to the apical surface to interact tightly with intimin, contributing to the characteristic attaching-and-effacing lesion. Stxs are encoded within one or more temperate bacteriophages and are expressed when phage

***Thomas E. Besser and Margaret A. Davis*** • Department of Veterinary Microbiology and Pathology, School for Global Animal Health, Washington State University, Pullman, WA 99164. ***Seth T. Walk*** • Department of Internal Medicine, Division of Infectious Diseases, University of Michigan Health System, Ann Arbor, MI 48109.

enter the lytic cycle. Stxs contribute to human disease by binding to glycolipid receptors (globotriaosylceramide) expressed on human endothelial cells, particularly in the colon and in the kidney. Stx receptor binding mediated by the pentameric B subunits is followed by internalization of the toxin and cleavage of a specific rRNA nucleotide, resulting in impaired protein synthesis and cell death. Endothelial damage by Stxs results in local microthrombosis and interference with local blood circulation, causing colonic hemorrhage and acute renal failure. (For more information on Stxs, see chapter 12 in this volume.)

As with many other zoonotic infectious agents, research efforts targeting the diagnosis, epidemiology, microbiology, pathophysiology, treatment, and prevention of human infection with O157:H7 vastly outweigh those targeting the ecology and epidemiology of the agent in its animal reservoirs. The research on the clinical disease has provided a large body of detailed information in each of the targeted areas, but unfortunately has not resulted in effective preventive or specific treatment strategies for infected patients. Similarly, despite considerable research efforts and mandated Hazard Analysis and Critical Control Point (HACCP)-based programs within the meat and other food-producing industries, exposure of humans to O157:H7 and consequently the incidence of disease have remained unacceptably high. Therefore, the control of O157:H7 within the animal reservoir remains an important goal with considerable potential for reducing the risk of human disease. Our goal for this chapter is to identify epidemiologic and ecologic factors within the animal reservoir that are insufficiently understood but which may offer potential for control of O157:H7.

## ROLE OF O157:H7 VIRULENCE FACTORS IN ASYMPTOMATIC COLONIZATION OF ANIMAL HOSTS

Long after the key roles of LEE and Stxs in the pathogenesis of human disease were elucidated, their roles in the colonization of the animal reservoir hosts remain undefined, even though effects on animal colonization are likely the key selective force driving the evolution of this pathogen. Of the two cardinal virulence factors, the role of LEE in animal colonization is better understood, although even these studies have emphasized its role in animal models of disease rather than in asymptomatic colonization of livestock reservoir species. (i) Isogenic deletion mutants that do not express intimin are deficient in both asymptomatic colonization of cattle and colonization of mouse, rabbit, and piglet models of EHEC disease (28, 32, 72, 86). (ii) Appearance of antibody to LEE effectors correlates with clearance of O157:H7 in experimental infection of both asymptomatic colonization and animal models (51). (iii) Antibodies to intimin or Tir abrogate cellular attachment and formation of attaching-and-effacing lesions in epithelial cell cultures and in vitro organ culture systems (46, 87). (iv) Immunization with intimin abolishes disease and diminishes colonization in the rabbit and mouse animal disease models (1, 33, 65). (v) Passive antibody to intimin abolishes disease and diminishes colonization in the pig animal disease model (33). When the unique colonization site of O157:H7 at the rectoanal junction was identified (98), typical attaching-and-effacing lesions and microcolony formation were observed in vivo, demonstrating the role of LEE in bovine colonization. The limited intestinal surface area affected by these lesions in rectoanal junction-colonized cattle is entirely consistent with the lack of clinical signs typical of this host species.

The possible role of Stxs in animal colonization remains much less clear, despite recently published evidence of a potential role for Stxs in animals. This evidence includes increased adherence to epithelial cells in culture and enhanced fecal shedding in orally challenged mice (112). However, Stx-negative strains are fairly common among cattle isolates (141) and have been used in successful experimental colonization of cattle that produces similar duration and intensity of fecal shedding as Stx-positive challenges (120). Colonization of the rectoanal junction, thought to be the principal site of bovine colonization, is clearly impaired when isogenic deletion mutants of intimin, Tir, and the O157:H7 "virulence plasmid" pO157 are used, whereas $Stx^-$ strains colonized equally well as the wild type (120). However, it has been reported that shedding density (CFU/g feces) of an $Stx^+$ strain of O157:H7 exceeded that of an $Stx^-$ strain in orally challenged calves at 4 days after oral challenge (35). Efforts to detect the principal cell surface receptor of Stx, globotriaosylceramide, in cattle tissues have failed to demonstrate its presence in the bovine colonic or renal vasculature (104), but it has been detected on the apical surface of colonocytes, indicating the potential for Stx effects in the bovine intestine (61).

Recent evidence suggests that Stxs may aid in the survivorship of bacteria by defending against predators. This hypothesis developed from observations that a number of exotoxins, including a homologue of Stx, were encoded in the genome of a lambda-like bacteriophage, called APSE, that infected a facultative bacterial symbiont (*Hamiltonella defense*) of sap-feeding insects (36, 94). Pea aphids that hosted APSE-carrying *H. defense* (symbiont) were more successful at surviving endoparasitoid predation by wasps, suggesting that Stx-carrying bacteria aid in the defense of their insect host (94, 100). A similar function has been proposed for Stxs in O157:H7. Stx production by O157:H7 strain EDL933 was induced by a protozoan and bacterial predator, *Tetrahymena thermophila*. The toxin was lethal to *T. thermophila* during in vitro experiments, and head-to-head competitions of $Stx^+$ and $Stx^-$ strains showed that the toxin increased bacterial survivorship while decreasing the number of *T. thermophila* (75). These findings support the hypothesis that exotoxin-encoding bacteriophage confer resistance to predation at multiple levels (virus-microbe-host resistance or simply virus-microbe resistance). It is worth noting that soil, sediment, and water as well as the bovine alimentary tract are all environments commonly inhabited by bacterivorous protozoa. Further research is required to clarify the role that Stx plays in predation resistance and whether colonization of the bovine host is increased by Stx-carrying phage.

## EPIDEMIOLOGY OF O157:H7

O157:H7 has been isolated from a diverse set of animal host species, in which it generally is not associated with any clinical abnormalities (49). The lack of disease in animal hosts has been a consistent observation (54), although experimental challenge models, including mice, calves, rabbits, and pigs, have been developed that exhibit some of the symptoms and/or lesions of human infections (90). For example, challenge of colostrum-deprived calves at 2 days of age resulted in infection and diarrhea, presumably due to the attaching-and-effacing lesions observed in both the small and large intestines (34, 35).

Most human infections are acquired by direct animal contact or by indirect transmission via food and water contaminated with feces of colonized animals. Food vehicles of O157:H7 transmission are commonly of animal origin, including unpasteurized milk and meat products, particularly ground beef or hamburger (105, 138). Waterborne outbreaks are less frequent but typically larger (105). In recent years the infection has been increasingly associated with contaminated produce, especially lettuce and spinach that are typically consumed without cooking (123). Although up to 20% of infections during O157:H7 outbreaks are secondary infections (124), sustained transmission among susceptible humans has not been documented. Case-control studies of sporadic cases of O157:H7 generally identify similar sources of infection, including consumption of undercooked hamburgers and exposure to contaminated water (138). The presumption is that if exposure of humans to the feces of infected animal hosts could be effectively blocked, human infections would cease. This hypothesis remains to be verified. Finally, O157:H7 is surprisingly hardy and able to survive for protracted periods in environmental niches, particularly water (59, 77, 85, 139).

## ROLE OF THE RESERVOIR IN THE EPIDEMIOLOGY OF ZOONOTIC O157:H7 INFECTIONS

Many different and sometimes conflicting definitions of reservoirs exist; this chapter will use the definitions proposed by Haydon et al. (58): a reservoir is one or more epidemiologically connected populations or environments in which the pathogen can be permanently maintained and from which infection is transmitted to the target species (humans in this case). Reservoirs may include non-maintenance populations (random extinction of the agent is inevitable without reintroduction), maintenance populations (the agent may be maintained permanently), and source populations (they transmit the agent directly to humans, and may themselves be maintenance populations or just transmission links from the reservoir to humans). These distinctions are important because of their relevance to infection control.

There has been a considerable amount of research directed toward investigation of the epidemiology and ecology of O157:H7 in various animal hosts and in the environment. The literature regarding animal infections has recently been reviewed (49). O157:H7 colonization has been documented in numerous animal species, including cattle, sheep, deer, elk, goats, dogs, swine, insects, and birds (49). Human infections have been linked most frequently to cattle sources, but sheep, deer, horses, dogs, birds, and flies have also been identified as sources of human infection (49). While this evidence demonstrates multiple animal sources of human infection (transmission links), it has not been confirmed that any are maintenance reservoir hosts.

An association of O157:H7 with animal hosts, particularly cattle, was apparent from the food vehicles associated with the first outbreaks reported (53). Since that time evidence implicating cattle as an important source reservoir for human infection has extended beyond the preponderance of foods of bovine origin in outbreaks (105). Ecological studies have consistently found a geographic correlation between the incidence of human infection and cattle density in Canada (91), Scotland (62), France (57), and Sweden (71). Outbreaks associated with direct contact with ruminants and their environments at petting zoos and other animal exhibits have highlighted the importance of this transmission route (30, 39, 52, 76, 79). The parallel seasonality of cattle shedding patterns and incidence of human infections also supports a close correlation between cattle shedding and human disease (2, 56, 62, 142). Adding to the strong epidemiological evidence for the role of cattle is an abundance of genotyping studies which reported matching O157:H7 strains from human and cattle sources (7, 11, 25, 37, 69, 82, 93).

In the United States, case-control studies of sporadic cases identified direct contact with ruminants as contributing an equivalent proportion of disease as consuming pink hamburger (67, 138), and a German study of Stx-producing *E. coli* (STEC), which included O157:H7, suggests that direct ruminant contact is of higher risk for very young children than for older children and adults (140). O157:H7 transmission at petting zoos and farm visits has resulted in outbreaks affecting large numbers of children (3, 4). Other bacterial enteric zoonoses similarly show age differences with respect to the relative importance of transmission routes: the food-borne route is more important for adults and older children, while environmental exposures are more important for very young children for campylobacteriosis (47) and salmonellosis (116). Although post-harvest interventions are critical to create multiple hurdles (43) against food-borne transmission, farm-level pathogen reduction of O157:H7 in cattle is required to reduce exposure of the most susceptible and severely impacted population, very young children.

Proximity to concentrated cattle populations appears to be a risk factor for O157:H7 exposure. For example, geospatial analyses have revealed higher O157:H7 case incidence in regions of higher cattle population density (48, 71, 91). Furthermore, rural residents generally, and rural residents in regions of high cattle density specifically, have been shown to have higher average serologic responses to O157 and/or Stx1 (107, 143). Farmworkers have also been demonstrated to carry O157:H7 asymptomatically (122, 143). However, it is not clear that farmworkers suffer higher O157:H7 infection or diarrheal incidence than other occupations, and some data indicate they may experience less diarrhea on average (13).

The seemingly contrasting observations that (i) some people become infected with O157:H7 after visiting farms or petting zoos and (ii) individuals who live and work on farms rarely develop bloody diarrhea or HUS support the hypothesis that human immunity plays an important role in mitigating the onset and progression of O157:H7 disease. A screening of Canadian residents found that dairy farmers and their families had a higher frequency of antibodies toward O157 lipopolysaccharide (LPS) compared to urban families, suggesting that individuals who reside on farms develop an immune response upon repeated exposure (107). Consistent with these results, more farm-resident children (1 to 4 years of age) in Wisconsin had circulating antibodies toward O157 LPS than non-farm-resident children (13). Also, with regard to young children, it was shown that maternal antibodies against O157 LPS cross the placental barrier to the developing fetus (103). These data support the hypothesis that passive maternal immunization as well as repeated immune system exposure to O157:H7 antigens early in life have a significant protective effect. It is interesting to consider the possibility that the present-day incidence of bloody diarrhea and HUS as a result of O157:H7 infection is a result of an underexposure of much of the human population to the pathogen's animal and environmental reservoirs. A similar hypothesis (the hygiene hypothesis) has been proposed to explain the recent rise in frequency of a number of autoimmune and allergic diseases in developed versus developing countries around the world (50).

## EVOLUTION OF O157:H7 AND THE ANIMAL RESERVOIR

It has been estimated that half of all *E. coli* bacteria exist outside warm-blooded animal hosts at any given time (113), and for O157:H7 it seems reasonable that of those existing within warm-blooded animal hosts, vastly more of the population exists in wild and domestic animal than human hosts at any given time. The U.S. cattle herd alone is estimated at over 95 million animals, which seasonally vary in O157:H7 fecal shedding prevalence from near zero in the winter to 10 to 15% or higher in the summer, and contaminated feces from infected cattle are readily available in the environment to infect new hosts. In contrast, fewer than 80,000 people are estimated to become infected with O157:H7 annually. It is possible that during its evolution, there was a selective advantage to cause human diarrheal disease to enhance transmission of this agent. However, the contemporary epidemiology of O157:H7 in asymptomatic animal reservoirs suggests a very well-adapted agent that can sustain colonization of numerous animal hosts without induction of disease. Due to the severe human disease caused by O157:H7, it is easy to lose track of the probable dominant role of the reservoir in shaping the evolution of this agent (and probably most other zoonotic enteric pathogens as well). An improved understanding of bacterial factors involved in O157:H7 animal colonization and transmission among animal hosts is required to truly understand the evolution of this pathogen, and that understanding, in turn, may suggest novel control measures that can be applied to the animal reservoir in order to reduce the risk of human exposure and disease.

One indication of the importance of the reservoir in the evolution of O157:H7 is the differing genetic strain composition of cattle isolates compared to clinical isolates. Chapman and Siddons, using pulsed-field gel electrophoresis (PFGE) and phage typing, observed a pattern of segregation of isolate types in which over half of the strain types observed in cattle were not represented among human clinical isolates, although nearly all clinical isolate genotypes were observed within the cattle isolate set (23). This finding was the first suggestion of a broader bacterial diversity in the reservoir than in human disease. Kim et al. subsequently used octamer-based genomic scanning to define two lineages, one of which was overrepresented in cattle isolates and the other in human clinical isolates (70). They also showed that at least part of the divergence of the two lineages resulted from phage-related sequence differences. Subsequently, Yang et al. developed a simpler PCR-based method to identify lineage I and lineage II isolates (144). Shaikh and Tarr used another strain-typing method based on specific Stx-encoding bacteriophage chromosomal insertion sites; this method identified two large subsets of O157:H7 genotypes from human infections (118). Whereas previous studies had implicated bacteriophages in the genomic plasticity of O157:H7, Shaikh and Tarr's work showed that the Stx-encoding phages were specifically involved in at least some of these modifications. Subsequent testing of a larger set of isolates confirmed the predominance (~95%) of these two strain types among clinical isolates but their lesser (~50%) representation among isolates from cattle. The other ~50% of cattle isolates had distinct and diverse patterns of Stx-encoding bacteriophage insertions (16). Lejeune et al. noticed a similar pattern in the PCR detection of $stx_2$-*Q* regions of Stx2-encoding phages in cattle and human isolates, in which the *Q933* allele was detected in ~90% of clinical isolates but <50% of cattle isolates (76). A single nucleotide polymorphism (SNP) is similarly overrepresented in human clinical isolates compared to cattle isolates. A *tir* 255T→A transversion was found in >99% of clinical isolates compared to only 55% of cattle isolates (18). Subsequently, Clawson et al. identified additional chromosomal backbone SNPs that differentiated human disease and cattle isolates (26). Sharma et al. reported increased diversity of cattle isolate PFGE strain types compared to human isolates (119). Recently, application of several of these typing methods to a single set of O157:H7 isolates demonstrated a high degree of concordance in their classification into two groups: one set of genotypes commonly isolated from human infection, and another set commonly isolated from the reservoir but rarely isolated from human infection. Together, these data support the concept of a higher degree of O157:H7 genetic diversity in the animal reservoir, only some of which is represented among clinical isolates. The existence of strain subtypes within the bovine reservoir that carry all of the known cardinal virulence factors of clinical strains, yet much less frequently associated with disease, is intriguing and may reflect important differences in virulence not detected by routine virulence factor screening. (For more on O157:H7 virulence potential, see chapter 17, this volume.)

Feng et al. originally presented a hypothetical scenario for the stepwise emergence from an *E. coli* O55:H7 EPEC-like ancestor of the non-sorbitol-fermenting, $\beta$-glucuronidase-negative clade of O157:H7 that is most common in the United States (45). Among clinical isolates of O157:H7, three subgroups were discerned based on their Stx gene content and whether the Stx-encoding bacteriophages were located at the same chromosomal insertion sites as in the first two sequenced strains of O157:H7 (118). Subsequently, Manning et al. used SNP alleles at 95 "convenience" loci to further divide the clinical isolates into eight clades (83). Leopold et al. determined complete genomic sequences of *E. coli* O55:H7 and six *E. coli* O157:H7 or O157:H$^{-}$ strains to provide a comprehensive analysis of their interrelationships (80). More recently, Clawson et al. used pyrosequencing of pooled genomic DNA obtained from bovine as well as clinical cases to identify many additional SNPs. Allele typing of these novel SNP loci confirmed the existence of lineages within O157:H7 that were common in animal reservoir isolates but very rarely isolated from ill humans (26). These studies confirmed the value and necessity of analysis of strains from the animal reservoir as well as clinically affected humans in order to obtain a nonbiased estimate of the population structure of this enteric zoonotic pathogen, a lesson that should be applied to similar studies of other zoonotic agents.

The underrepresentation of some cattle O157:H7 genotypes among clinical isolate sets could be due to a lower probability of human exposure if, for example, they were less able to survive and persist in contaminated food and water vehicles that are thought to be responsible for much human infection. However, in a study of the gene expression differences between clinical and bovine-biased genetic (BBG) types of O157:H7, the BBG types showed increased expression of genes generally associated with environmental survival, including biofilms (135). It is also possible that BBG types are less virulent than clinical genotypes. A very high proportion of BBG types are positive for one or more Stxs; nearly all are positive for *eae*, a marker for the presence of the LEE; and they express these traits in vitro (16). Baker et al. tested a number of clinical and cattle isolates of O157:H7 in a gnotobiotic piglet model and showed that the cattle isolates induced less disease than the clinical isolates, a difference they attributed to lower Stx expression (10).

## LIMITATIONS OF OUR CURRENT UNDERSTANDING OF THE ANIMAL AND ENVIRONMENTAL RESERVOIRS OF *E. COLI* O157:H7

### Are Cattle the Maintenance Host of *E. coli* O157:H7?

Due to their prominence as sources of human infection, considerable emphasis has been placed on characterizing O157:H7 colonization of domestic ruminants, particularly cattle and sheep, compared to other host species (Table 1). This work has included prevalence; descriptive epidemiology of O157:H7 shedding; identifying farm management factors associated with high prevalence of animal infection; and investigating the effects of vaccines, probiotics, and other interventions (54). Research efforts concerning O157:H7 colonization of noncattle animal hosts have been limited to prevalence studies or to reports of associations with human infections (49). Most livestock that are known hosts of O157:H7 are typically reared outdoors most or all of the year. Even if reared in densely stocked pens, there is little or no restriction of exposures to wildlife

**Table 1.** PubMed citations of EHEC O157 reveal research emphasis

| Category[a] | No. of citations | Percentage of total |
|---|---|---|
| All EHEC | 8,302 | 100 |
| Cattle | 1,977 | 23.8 |
| Swine | 413 | 5.0 |
| Sheep | 306 | 3.7 |
| Birds | 235 | 2.8 |
| Companion animal species (dogs, cats, horses, camelids) | 150 | 1.8 |

[a] Animal host categories are not mutually exclusive. At least 573 of the animal host references included multiple animal species.

including diverse insects, birds, rodents, and other animals, so a nonbovine maintenance population could easily "spill over" into cattle. Interestingly, swine are readily colonized by O157:H7 experimentally, but surveillance studies typically find low or very low prevalence of natural carriage and swine-origin food products are rarely associated with human disease (19, 24, 44, 68, 92, 101, 145). Similarly, colonization readily develops in chickens (12, 128), but chicken produce is rarely linked with human O157:H7 disease. Perhaps significantly, in the United States, both poultry and swine are predominantly reared in confinement (indoor) settings with minimal exposures to wildlife. In contrast, wild boars are found to be naturally colonized with O157:H7. As a result of the paucity of data and differences in bacterial exposures and husbandry factors, the ability of any particular species to serve as a maintenance host is unknown.

O157:H7 is intermittently detectable on essentially all cattle farms (54). As described in more detail below, bovine colonization is strongly seasonal in many locations in North America, with a peak in late summer and early fall that closely corresponds to the pronounced seasonal peak of human O157:H7 infections (Fig. 1) (54). O157:H7 colonizes cattle at the rectoanal junction, where it occurs in much higher numbers relative to both other *E. coli* strains and other segments of the gastrointestinal tract (98). However, generally bovine colonization by O157:H7 is transient, typically lasting a few days to several weeks, with only rare longer infections (54).

Contact between cattle and O157:H7-derived antigen triggers humoral immune responses and the development of antibodies to several proteins involved in both colonization and disease (20, 64, 95, 96). However, O157:H7-exposed cattle are susceptible to recolonization almost immediately after clearance (29, 97). In fact, it may be that the prolonged shedding documented in cattle represents repeated recolonization rather than chronic shedding.

## Do Supershedders Exist?

Fecal shedding density of O157:H7 by colonized cattle is heterogeneous, and even in situations with high fecal shedding by cattle it is typically only a small fraction of cattle that shed the agent in high numbers. It is possible that the detection of O157:H7 by highly sensitive techniques such as immuno-

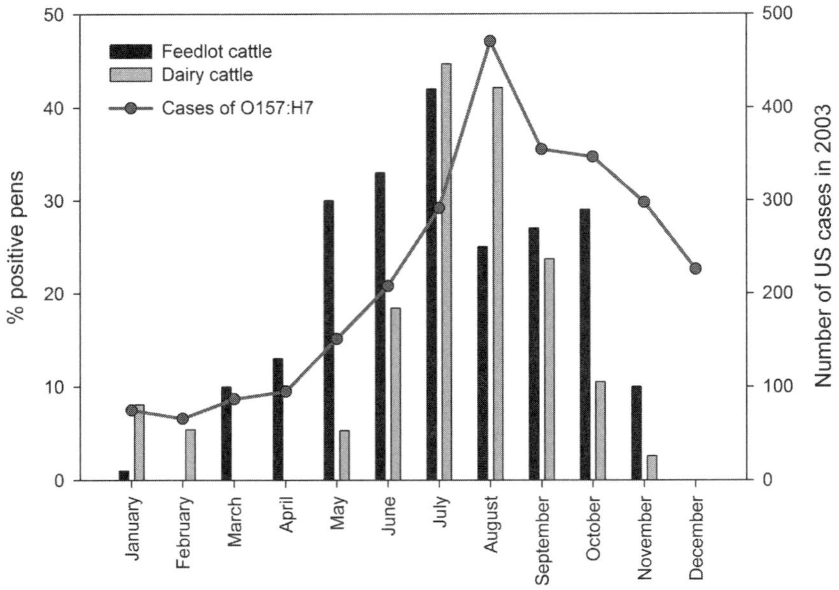


**Figure 1.** Seasonal prevalence variation of O157:H7 in dairy and beef cattle herds (126, 127) and the number of reported *E. coli* O157:H7 cases in the United States in 2003.

magnetic separations and enrichment cultures is not indicative of actual colonization; rather, the organisms are simply passing through the host gastrointestinal tract. The significance of the high-shedding animals, often termed "supershedders," in the epidemiology of a herd is clearly due to both theoretical considerations (84) and empirical demonstrations that supershedder removal from cattle pens results in a prompt decrease in shedding prevalence, and introduction of supershedders into new cohorts is associated with a prompt increase in shedding prevalence (27). However, since most data on supershedders originate from cross-sectional studies, whether supershedding is a specific characteristic of particular animals or, alternatively, a transient phase of many or most cattle is currently unknown.

## What Is the Basis of Seasonal O157:H7 Shedding by Cattle?

Seasonal variation of O157:H7 fecal shedding by cattle has been reported in geographically diverse regions, including Turkey, The Netherlands, Italy, England, Canada, the United States, and Korea (5, 17, 24, 55, 56, 60, 63, 66, 89, 102, 114, 115, 127, 136), and a similar pattern of seasonal shedding has also been reported in sheep. Several variations or interesting exceptions to the typical pattern (peak prevalence in summer and early fall months) have been reported, including wintertime prevalence peaks (2, 99, 129) or no seasonal variation (2, 21, 110). In Scotland, peak seasonal shedding in late fall and winter coincided with moving cattle from pastures to indoor housing and the presence of wild geese on winter pastures (129). In another report of a winter prevalence peak, a contravening trend toward higher shedding (CFU/g feces) in the summer was also noted (99). Typical summertime seasonal peaks in O157:H7 prevalence have been reported with variable magnitudes, ranging from no detection of *E. coli* O157:H7 during winter months (55, 73, 89, 114, 115) to >10-fold higher prevalence in summer months (127) and a number of minor variations in between (24, 38, 56). Edrington et al., summarizing data from nine studies that included grain-fed cattle on multiple farms in Mexico, the United States, and Canada that were sampled and cultured similarly, found a consistent and prominent seasonal variation (42). Recently, Williams et al. used a modeling approach to link seasonal variation in O157:H7 shedding in the bovine reservoir, contamination of beef products, and human infection (142).

An initial hypothesis to explain seasonal variation in the cattle reservoir was that summertime temperatures increased environmental replication of O157:H7 in cattle water troughs and feed sources, driving higher transmission rates (54). The increased survival of O157:H7 in cattle feces held at warmer temperatures (22 versus 3 or −10°C) may also result in increased summertime cow-to-cow transmission of O157:H7 (9). Cattle are exposed to a diversity of *E. coli* genotypes via feedstuffs, water sources, and self- and mutual grooming behaviors (31, 54). Feedstuffs are documented to be sometimes heavily contaminated with *E. coli* so as to result in $>10^8$ CFU/day (54). However, replication of the organism in feeds is generally inhibited by silage components, and no strong seasonal variation in the degree of feed contamination with generic *E. coli* is observed (54). In contrast, *E. coli* concentrations in cattle water troughs do increase during summer months, suggesting the occurrence of replication in this niche (78). Finally, water intakes of cattle increase seasonally and magnify the effects of higher *E. coli* concentrations during the summer season. Water troughs potentially maintain O157:H7 during seasons of low or negative cattle colonization, since contaminated water troughs can maintain viable O157:H7 strains with an intact ability to colonize for many months (77). Such habitats likely play a key role in O157:H7 biology on farms during prolonged wintertime periods with little or no contact with cattle.

An alternative hypothesis to account for seasonal shedding by cattle is based on observations that O157:H7 prevalence in cattle was more strongly correlated to day length than temperature, and that the use of artificial lighting was associated with altered O157:H7 shedding (42). The hypothesis is that photoperiod-associated endocrine effects, perhaps especially those with effects on immune responses, may underlie the seasonality of shedding. Subsequent efforts examining the effects of exogenously administered thyroid hormones, melatonin, and naturally present acyl-homoserine-lactone autoinducer produced interesting data and some support for the hypothesis, but no clear mechanism for potential photoperiod-associated endocrine effects has been demonstrated (40, 41).

## Does O157:H7 Have a Nonbovine Maintenance Host on Cattle Farms?

The strong seasonal reduction in cattle shedding O157:H7 is difficult to reconcile with cattle as the maintenance host of this agent, given the very

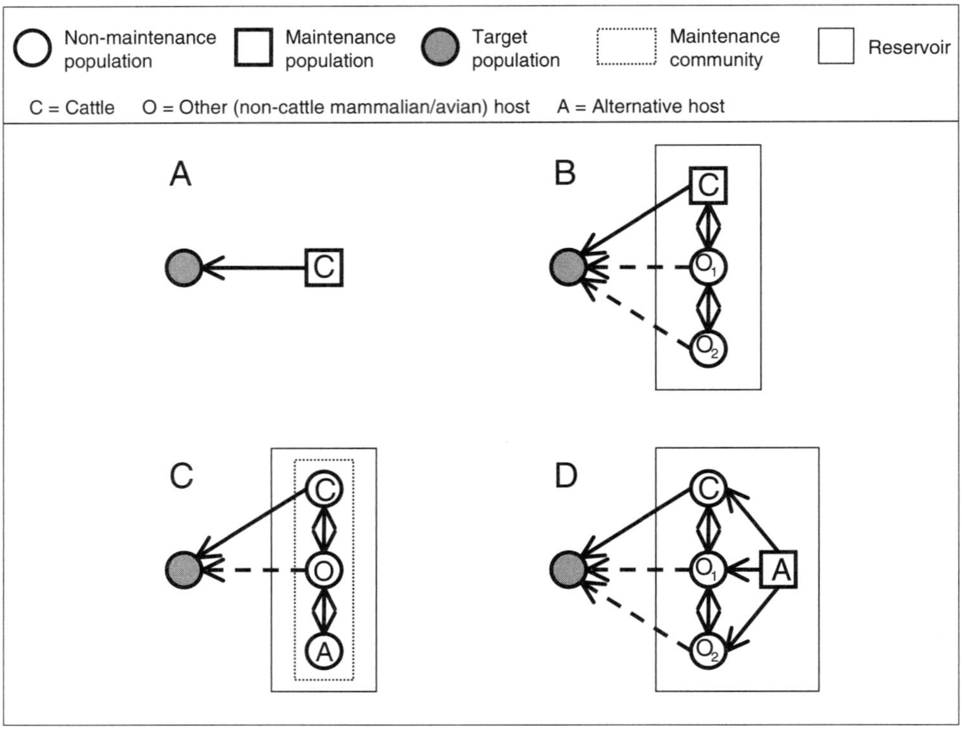


**Figure 2.** Hypothetical reservoir structures of zoonotic *E. coli* O157:H7, after Haydon et al. (58). Non-maintenance populations are species whose critical community size or transmission characteristics preclude persistence in that host alone. Maintenance populations are those capable of persistently harboring the agent without the requirement of additional hosts. The target population for *E. coli* O157:H7 in this analysis is human. Maintenance communities are assemblies of non-maintenance hosts that combined are capable of persistently maintaining the agent. The reservoir includes both maintenance and non-maintenance hosts in which the agent persists. (A) In much of the literature, cattle are described as the reservoir of *E. coli* O157:H7, implying maintenance population status. This simplification fails to take into account the dearth of cattle infection in the winter and the existence of other species (sheep, horses, birds, etc.) documented to be sources of human infection. (B) The simplified reservoir structure expanded to illustrate the existence of other host species ($O_1$, $O_2$) capable of being infected with and transmitting the agent to humans. (C) A reservoir structure illustrating the possibility that cattle and other host species, none of which alone are maintenance hosts, may collectively comprise a maintenance community. The other hosts are those observed to be sources of human infection, whereas the alternative hosts may include diverse potential cryptic hosts such as protozoa, nematodes, or coprophagous insects, non-maintenance hosts that may nevertheless enable agent persistence through seasons when C or O host infection is rare or absent. (D) An alternative possible reservoir structure, illustrating the possibility that alternative hosts are actually the maintenance population. In alternatives C and D, control measures directed at alternative hosts have significant potential for disease control.

similar management of most cattle farms throughout the annual cycle. In contrast, there is abundant evidence for the role of overall cattle farm environments as maintenance reservoirs of this agent. Specific O157:H7 PFGE types are frequently associated with individual cattle herds, and longitudinal studies have documented the persistence of predominant strains for 2 years or longer (6, 8, 15, 22, 74, 81, 106, 108, 109, 117, 121, 125, 137, 141). It's particularly interesting that this association applies not just to individual dairy farms (which generally tend to have relatively low animal population turnover) but also to large feedlots with complete population turnovers annually (79). This is consistent with the possibility that the true maintenance reservoir is more complex than just the cattle population (Fig. 2). Nevertheless, it is difficult to reject the alternative hypothesis that this pattern of persistent strains results from year-round chains of infection, providing that the sensitivity of detection is seasonally reduced. Alternatively, it is also possible that another maintenance reservoir exists on farms, such as insects, birds, or wild mammals, or even that environmental niches such as water troughs or soils are the actual maintenance reservoir of O157:H7.

## SUMMARY

There is little evidence that O157:H7 is a "professional pathogen"; rather, it is a well-adapted commensal of numerous animal hosts. It has considerable strain diversity within cattle, and only a subset of strains is strongly associated with human disease. As the agent first came to human attention due to the severity of disease, it is natural that research attention has been primarily focused on pathogenesis. However, the biggest data gaps remaining regarding this agent lie in its ecology, epidemiology, and biology in its animal and environmental reservoirs.

## REFERENCES

1. **Agin, T. S., C. Zhu, L. A. Johnson, T. E. Thate, Z. Yang, and E. C. Boedeker.** 2005. Protection against hemorrhagic colitis in an animal model by oral immunization with isogeneic rabbit enteropathogenic *Escherichia coli* attenuated by truncating intimin. *Infect. Immun.* **73:**6608–6619.
2. **Alam, M. J., and L. Zurek.** 2006. Seasonal prevalence of *Escherichia coli* O157:H7 in beef cattle feces. *J. Food. Prot.* **69:**3018–3020.
3. **Anonymous.** 2005. Outbreaks of *Escherichia coli* O157:H7 associated with petting zoos—North Carolina, Florida, and Arizona, 2004 and 2005. *MMWR Morb. Mortal. Wkly. Rep.* **54:**1277–1280.
4. **Anonymous.** 1999. VTEC O157 outbreak associated with a farm visitor centre in North Wales. *Commun. Dis. Rep. CDR Wkly.* **9:**227, 230.
5. **Aslantas, O., S. Erdogan, Z. Cantekin, I. Gulacti, and G. A. Evrendilek.** 2006. Isolation and characterization of verocytotoxin-producing *Escherichia coli* O157 from Turkish cattle. *Int. J. Food Microbiol.* **106:**338–342.
6. **Avery, S. M., E. Liebana, M. L. Hutchison, and S. Buncic.** 2004. Pulsed field gel electrophoresis of related *Escherichia coli* O157 isolates associated with beef cattle and comparison with unrelated isolates from animals, meats and humans. *Int. J. Food Microbiol.* **92:**161–169.

7. **Avery, S. M., E. Liebana, C. A. Reid, M. J. Woodward, and S. Buncic.** 2002. Combined use of two genetic fingerprinting methods, pulsed-field gel electrophoresis and ribotyping, for characterization of *Escherichia coli* O157 isolates from food animals, retail meats, and cases of human disease. *J. Clin. Microbiol.* **40:**2806–2812.
8. **Avery, S. M., A. Small, C. A. Reid, and S. Buncic.** 2002. Pulsed-field gel electrophoresis characterization of Shiga toxin-producing *Escherichia coli* O157 from hides of cattle at slaughter. *J. Food Prot.* **65:**1172–1176.
9. **Bach, S. J., K. Stanford, and T. A. McAllister.** 2005. Survival of *Escherichia coli* O157:H7 in feces from corn- and barley-fed steers. *FEMS Microbiol. Lett.* **252:**25–33.
10. **Baker, D. R., R. A. Moxley, and D. H. Francis.** 1997. Variation in virulence in the gnotobiotic pig model of O157:H7 *Escherichia coli* strains of bovine and human origin. *Adv. Exp. Med. Biol.* **412:**53–58.
11. **Barrett, T. J., H. Lior, J. H. Green, R. Khakhria, J. G. Wells, B. P. Bell, K. D. Greene, J. Lewis, and P. M. Griffin.** 1994. Laboratory investigation of a multistate food-borne outbreak of *Escherichia coli* O157:H7 by using pulsed-field gel electrophoresis and phage typing. *J. Clin. Microbiol.* **32:**3013–3017.
12. **Beery, J. T., M. P. Doyle, and J. L. Schoeni.** 1985. Colonization of chicken cecae by *Escherichia coli* associated with hemorrhagic colitis. *Appl. Environ. Microbiol.* **49:**310–315.
13. **Belongia, E. A., P. H. Chyou, R. T. Greenlee, G. Perez-Perez, W. F. Bibb, and E. O. DeVries.** 2003. Diarrhea incidence and farm-related risk factors for *Escherichia coli* O157:H7 and *Campylobacter jejuni* antibodies among rural children. *J. Infect. Dis.* **187:** 1460–1468.
14. **Besser, R. E., P. M. Griffin, and L. Slutsker.** 1999. *Escherichia coli* O157:H7 gastroenteritis and the hemolytic uremic syndrome: an emerging infectious disease. *Annu. Rev. Med.* **50:**355–367.
15. **Besser, T. E., D. D. Hancock, L. C. Pritchett, E. M. McRae, D. H. Rice, and P. I. Tarr.** 1997. Duration of detection of fecal excretion of *Escherichia coli* O157:H7 in cattle. *J. Infect. Dis.* **175:**726–729.
16. **Besser, T. E., N. Shaikh, N. J. Holt, P. I. Tarr, M. E. Konkel, P. Malik-Kale, C. W. Walsh, T. S. Whittam, and J. L. Bono.** 2007. Greater diversity of Shiga toxin-encoding bacteriophage insertion sites among *Escherichia coli* O157:H7 isolates from cattle than in those from humans. *Appl. Environ. Microbiol.* **73:**671–679.
17. **Bonardi, S., E. Maggi, A. Bottarelli, M. L. Pacciarini, A. Ansuini, G. Vellini, S. Morabito, and A. Caprioli.** 1999. Isolation of Verocytotoxin-producing *Escherichia coli* O157:H7 from cattle at slaughter in Italy. *Vet. Microbiol.* **67:**203–211.
18. **Bono, J. L., J. E. Keen, M. L. Clawson, L. M. Durso, M. P. Heaton, and W. W. Laegreid.** 2007. Association of *Escherichia coli* O157:H7 *tir* polymorphisms with human infection. *BMC Infect. Dis.* **7:**98.
19. **Booher, S. L., N. A. Cornick, and H. W. Moon.** 2002. Persistence of *Escherichia coli* O157:H7 in experimentally infected swine. *Vet. Microbiol.* **89:**69–81.
20. **Bretschneider, G., E. M. Berberov, and R. A. Moxley.** 2008. Enteric mucosal antibodies to *Escherichia coli* O157:H7 in adult cattle. *Vet. Rec.* **163:**218–219.
21. **Brichta-Harhay, D. M., M. N. Guerini, T. M. Arthur, J. M. Bosilevac, N. Kalchayanand, S. D. Shackelford, T. L. Wheeler, and M. Koohmaraie.** 2008. *Salmonella* and *Escherichia coli* O157:H7 contamination on hides and carcasses of cull cattle presented for slaughter in the United States: an evaluation of prevalence and bacterial loads by immunomagnetic separation and direct plating methods. *Appl. Environ. Microbiol.* **74:** 6289–6297.
22. **Carlson, B. A., K. K. Nightingale, G. L. Mason, J. R. Ruby, W. T. Choat, G. H. Loneragan, G. C. Smith, J. N. Sofos, and K. E. Belk.** 2009. *Escherichia coli* O157:H7 strains that persist in feedlot cattle are genetically related and demonstrate an enhanced ability to adhere to intestinal epithelial cells. *Appl. Environ. Microbiol.* **75:**5927–5937.

23. **Chapman, P. A., and C. A. Siddons.** 1994. A comparison of strains of *Escherichia coli* O157 from humans and cattle in Sheffield, United Kingdom. *J. Infect. Dis.* **170:** 251–253.
24. **Chapman, P. A., C. A. Siddons, A. T. Gerdan Malo, and M. A. Harkin.** 1997. A 1-year study of *Escherichia coli* O157 in cattle, sheep, pigs and poultry. *Epidemiol. Infect.* **119:**245–250.
25. **Cho, S., J. B. Bender, F. Diez-Gonzalez, C. P. Fossler, C. W. Hedberg, J. B. Kaneene, P. L. Ruegg, L. D. Warnick, and S. J. Wells.** 2006. Prevalence and characterization of *Escherichia coli* O157 isolates from Minnesota dairy farms and county fairs. *J. Food Prot.* **69:**252–259.
26. **Clawson, M. L., J. E. Keen, T. P. Smith, L. M. Durso, T. G. McDaneld, R. E. Mandrell, M. A. Davis, and J. L. Bono.** 2009. Phylogenetic classification of *Escherichia coli* O157:H7 strains of human and bovine origin using a novel set of nucleotide polymorphisms. *Genome Biol.* **10:**R56.
27. **Cobbold, R. N., D. D. Hancock, D. H. Rice, J. Berg, R. Stilborn, C. J. Hovde, and T. E. Besser.** 2007. Rectoanal junction colonization of feedlot cattle by *Escherichia coli* O157:H7 and its association with supershedders and excretion dynamics. *Appl. Environ. Microbiol.* **73:**1563–1568.
28. **Cornick, N. A., S. L. Booher, and H. W. Moon.** 2002. Intimin facilitates colonization by *Escherichia coli* O157:H7 in adult ruminants. *Infect. Immun.* **70:**2704–2707.
29. **Cray, W. C., Jr., and H. W. Moon.** 1995. Experimental infection of calves and adult cattle with *Escherichia coli* O157:H7. *Appl. Environ. Microbiol.* **61:**1586–1590.
30. **Crump, J. A., C. R. Braden, M. E. Dey, R. M. Hoekstra, J. M. Rickelman-Apisa, D. A. Baldwin, S. J. De Fijter, S. F. Nowicki, E. M. Koch, T. L. Bannerman, F. W. Smith, J. P. Sarisky, N. Hochberg, and P. S. Mead.** 2003. Outbreaks of *Escherichia coli* O157 infections at multiple county agricultural fairs: a hazard of mixing cattle, concession stands and children. *Epidemiol. Infect.* **131:**1055–1062.
31. **Davis, M. A., D. D. Hancock, D. H. Rice, D. R. Call, R. DiGiacomo, M. Samadpour, and T. E. Besser.** 2003. Feedstuffs as a vehicle of cattle exposure to *Escherichia coli* O157:H7 and *Salmonella enterica*. *Vet. Microbiol.* **95:**199–210.
32. **Dean-Nystrom, E. A., B. T. Bosworth, H. W. Moon, and A. D. O'Brien.** 1998. *Escherichia coli* O157:H7 requires intimin for enteropathogenicity in calves. *Infect. Immun.* **66:**4560–4563.
33. **Dean-Nystrom, E. A., L. J. Gansheroff, M. Mills, H. W. Moon, and A. D. O'Brien.** 2002. Vaccination of pregnant dams with intimin$_{O157}$ protects suckling piglets from *Escherichia coli* O157:H7 infection. *Infect. Immun.* **70:**2414–2418.
34. **Dean-Nystrom, E. A., A. R. Melton-Celsa, J. F. Pohlenz, H. W. Moon, and A. D. O'Brien.** 2003. Comparative pathogenicity of *Escherichia coli* O157 and intimin-negative non-O157 Shiga toxin-producing *E. coli* strains in neonatal pigs. *Infect. Immun.* **71:**6526–6533.
35. **Dean-Nystrom, E. A., W. C. Stoffregen, B. T. Bosworth, H. W. Moon, and J. F. Pohlenz.** 2008. Early attachment sites for Shiga-toxigenic *Escherichia coli* O157:H7 in experimentally inoculated weaned calves. *Appl. Environ. Microbiol.* **74:**6378–6384.
36. **Degnan, P. H., and N. A. Moran.** 2008. Diverse phage-encoded toxins in a protective insect endosymbiont. *Appl. Environ. Microbiol.* **74:**6782–6791.
37. **Denny, J., M. Bhat, and K. Eckmann.** 2008. Outbreak of *Escherichia coli* O157:H7 associated with raw milk consumption in the Pacific Northwest. *Foodborne Pathog. Dis.* **5:**321–328.
38. **Dunn, J. R., J. E. Keen, and R. A. Thompson.** 2004. Prevalence of Shiga-toxigenic *Escherichia coli* O157:H7 in adult dairy cattle. *J. Am. Vet. Med. Assoc.* **224:**1151–1158.
39. **Durso, L. M., K. Reynolds, N. Bauer, Jr., and J. E. Keen.** 2005. Shiga-toxigenic *Escherichia coli* O157:H7 infections among livestock exhibitors and visitors at a Texas county fair. *Vector Borne Zoonotic Dis.* **5:**193–201.
40. **Edrington, T. S., T. R. Callaway, D. M. Hallford, R. C. Anderson, and D. J. Nisbet.** 2007. Influence of exogenous triiodothyronine (T3) on fecal shedding of *Escherichia coli* O157 in cattle. *Microb. Ecol.* **53:**664–669.

41. **Edrington, T. S., T. R. Callaway, D. M. Hallford, L. Chen, R. C. Anderson, and D. J. Nisbet.** 2008. Effects of exogenous melatonin and tryptophan on fecal shedding of *E. coli* O157:H7 in cattle. *Microb. Ecol.* **55:**553–560.
42. **Edrington, T. S., T. R. Callaway, S. E. Ives, M. J. Engler, M. L. Looper, R. C. Anderson, and D. J. Nisbet.** 2006. Seasonal shedding of *Escherichia coli* O157:H7 in ruminants: a new hypothesis. *Foodborne Pathog. Dis.* **3:**413–421.
43. **Erickson, M. C., and M. P. Doyle.** 2007. Food as a vehicle for transmission of Shiga toxin-producing *Escherichia coli. J. Food Prot.* **70:**2426–2449.
44. **Farzan, A., R. M. Friendship, A. Cook, and F. Pollari.** 2010. Occurrence of *Salmonella, Campylobacter, Yersinia enterocolitica, Escherichia coli* O157 and *Listeria monocytogenes* in swine. *Zoonoses Public Health* **57:**388–396.
45. **Feng, P., K. A. Lampel, H. Karch, and T. S. Whittam.** 1998. Genotypic and phenotypic changes in the emergence of *Escherichia coli* O157:H7. *J. Infect. Dis.* **177:**1750–1753.
46. **Fitzhenry, R. J., D. J. Pickard, E. L. Hartland, S. Reece, G. Dougan, A. D. Phillips, and G. Frankel.** 2002. Intimin type influences the site of human intestinal mucosal colonisation by enterohaemorrhagic *Escherichia coli* O157:H7. *Gut* **50:**180–185.
47. **Friedman, C. R., R. M. Hoekstra, M. Samuel, R. Marcus, J. Bender, B. Shiferaw, S. Reddy, S. D. Ahuja, D. L. Helfrick, F. Hardnett, M. Carter, B. Anderson, and R. V. Tauxe.** 2004. Risk factors for sporadic *Campylobacter* infection in the United States: a case-control study in FoodNet sites. *Clin. Infect. Dis.* **38**(Suppl. 3)**:**S285–S296.
48. **Friesema, I. H., J. van de Kassteele, C. M. de Jager, A. E. Heuvelink, and W. van Pelt.** 2010. Geographical association between livestock density and human Shiga toxin-producing *Escherichia coli* O157 infections. *Epidemiol. Infect.* **8:**1–7.
49. **Garcia, A., J. G. Fox, and T. E. Besser.** 2010. Zoonotic enterohemorrhagic *Escherichia coli*: a One Health perspective. *ILAR J.* **51:**221–232.
50. **Garn, H., and H. Renz.** 2007. Epidemiological and immunological evidence for the hygiene hypothesis. *Immunobiology* **212:**441–452.
51. **Ghaem-Maghami, M., C. P. Simmons, S. Daniell, M. Pizza, D. Lewis, G. Frankel, and G. Dougan.** 2001. Intimin-specific immune responses prevent bacterial colonization by the attaching-effacing pathogen *Citrobacter rodentium. Infect. Immun.* **69:**5597–5605.
52. **Goode, B., C. O'Reilly, J. Dunn, K. Fullerton, S. Smith, G. Ghneim, J. Keen, L. Durso, M. Davies, and S. Montgomery.** 2009. Outbreak of *Escherichia coli* O157:H7 infections after petting zoo visits, North Carolina State Fair, October–November 2004. *Arch. Pediatr. Adolesc. Med.* **163:**42–48.
53. **Griffin, P. M., S. M. Ostroff, R. V. Tauxe, K. D. Greene, J. G. Wells, J. H. Lewis, and P. A. Blake.** 1988. Illnesses associated with *Escherichia coli* O157:H7 infections. A broad clinical spectrum. *Ann. Intern. Med.* **109:**705–712.
54. **Hancock, D., T. Besser, J. Lejeune, M. Davis, and D. Rice.** 2001. The control of VTEC in the animal reservoir. *Int. J. Food Microbiol.* **66:**71–78.
55. **Hancock, D. D., T. E. Besser, M. L. Kinsel, P. I. Tarr, D. H. Rice, and M. G. Paros.** 1994. The prevalence of *Escherichia coli* O157.H7 in dairy and beef cattle in Washington State. *Epidemiol. Infect.* **113:**199–207.
56. **Hancock, D. D., T. E. Besser, D. H. Rice, D. E. Herriott, and P. I. Tarr.** 1997. A longitudinal study of *Escherichia coli* O157 in fourteen cattle herds. *Epidemiol. Infect.* **118:**193–195.
57. **Haus-Cheymol, R., E. Espie, D. Che, V. Vaillant, H. de Valk, and J. C. Desenclos.** 2006. Association between indicators of cattle density and incidence of paediatric haemolytic-uraemic syndrome (HUS) in children under 15 years of age in France between 1996 and 2001: an ecological study. *Epidemiol. Infect.* **134:**712–718.
58. **Haydon, D. T., S. Cleaveland, L. H. Taylor, and M. K. Laurenson.** 2002. Identifying reservoirs of infection: a conceptual and practical challenge. *Emerg. Infect. Dis.* **8:**1468–1473.
59. **Heaton, J. C., and K. Jones.** 2008. Microbial contamination of fruit and vegetables and the behaviour of enteropathogens in the phyllosphere: a review. *J. Appl. Microbiol.* **104:**613–626.

60. **Heuvelink, A. E., B. Bleumink, F. L. van den Biggelaar, M. C. Te Giffel, R. R. Beumer, and E. de Boer.** 1998. Occurrence and survival of verocytotoxin-producing *Escherichia coli* O157 in raw cow's milk in The Netherlands. *J. Food Prot.* **61:**1597–1601.
61. **Hoey, D. E., C. Currie, R. W. Else, A. Nutikka, C. A. Lingwood, D. L. Gally, and D. G. Smith.** 2002. Expression of receptors for verotoxin 1 from *Escherichia coli* O157 on bovine intestinal epithelium. *J. Med. Microbiol.* **51:**143–149.
62. **Innocent, G. T., D. J. Mellor, S. A. McEwen, W. J. Reilly, J. Smallwood, M. E. Locking, D. J. Shaw, P. Michel, D. J. Taylor, W. B. Steele, G. J. Gunn, H. E. Ternent, M. E. Woolhouse, and S. W. Reid.** 2005. Spatial and temporal epidemiology of sporadic human cases of *Escherichia coli* O157 in Scotland, 1996–1999. *Epidemiol. Infect.* **133:** 1033–1041.
63. **Jo, M. Y., J. H. Kim, J. H. Lim, M. Y. Kang, H. B. Koh, Y. H. Park, D. Y. Yoon, J. S. Chae, S. K. Eo, and J. H. Lee.** 2004. Prevalence and characteristics of *Escherichia coli* O157 from major food animals in Korea. *Int. J. Food Microbiol.* **95:**41–49.
64. **Johnson, R. P., W. C. Cray, Jr., and S. T. Johnson.** 1996. Serum antibody responses of cattle following experimental infection with *Escherichia coli* O157:H7. *Infect. Immun.* **64:**1879–1883.
65. **Judge, N. A., H. S. Mason, and A. D. O'Brien.** 2004. Plant cell-based intimin vaccine given orally to mice primed with intimin reduces time of *Escherichia coli* O157:H7 shedding in feces. *Infect. Immun.* **72:**168–175.
66. **Karch, H., and B. Köhler.** 1999. New knowledge of the molecular biology of enterohemorrhagic *Escherichia coli* (EHEC) O157. *Gesundheitswesen* **61**(Spec. No. 1):S46–S51. (In German.)
67. **Kassenborg, H. D., C. W. Hedberg, M. Hoekstra, M. C. Evans, A. E. Chin, R. Marcus, D. J. Vugia, K. Smith, S. D. Ahuja, L. Slutsker, and P. M. Griffin.** 2004. Farm visits and undercooked hamburgers as major risk factors for sporadic *Escherichia coli* O157:H7 infection: data from a case-control study in 5 FoodNet sites. *Clin. Infect. Dis.* **38**(Suppl. 3):S271–S278.
68. **Kaufmann, M., C. Zweifel, M. Blanco, J. E. Blanco, J. Blanco, L. Beutin, and R. Stephan.** 2006. *Escherichia coli* O157 and non-O157 Shiga toxin-producing *Escherichia coli* in fecal samples of finished pigs at slaughter in Switzerland. *J. Food Prot.* **69:**260–266.
69. **Keene, W. E., E. Sazie, J. Kok, D. H. Rice, D. D. Hancock, V. K. Balan, T. Zhao, and M. P. Doyle.** 1997. An outbreak of *Escherichia coli* O157:H7 infections traced to jerky made from deer meat. *JAMA* **277:**1229–1231.
70. **Kim, J., J. Nietfeldt, and A. K. Benson.** 1999. Octamer-based genome scanning distinguishes a unique subpopulation of *Escherichia coli* O157:H7 strains in cattle. *Proc. Natl. Acad. Sci. USA* **96:**13288–13293.
71. **Kistemann, T., S. Zimmer, I. Vagsholm, and Y. Andersson.** 2004. GIS-supported investigation of human EHEC and cattle VTEC O157 infections in Sweden: geographical distribution, spatial variation and possible risk factors. *Epidemiol. Infect.* **132:**495–505.
72. **Krejany, E. O., T. H. Grant, V. Bennett-Wood, L. M. Adams, and R. M. Robins-Browne.** 2000. Contribution of plasmid-encoded fimbriae and intimin to capacity of rabbit-specific enteropathogenic *Escherichia coli* to attach to and colonize rabbit intestine. *Infect. Immun.* **68:**6472–6477.
73. **Kudva, I. T., P. G. Hatfield, and C. J. Hovde.** 1996. *Escherichia coli* O157:H7 in microbial flora of sheep. *J. Clin. Microbiol.* **34:**431–433.
74. **Lahti, E., O. Ruoho, L. Rantala, M. L. Hanninen, and T. Honkanen-Buzalski.** 2003. Longitudinal study of *Escherichia coli* O157 in a cattle finishing unit. *Appl. Environ. Microbiol.* **69:**554–561.
75. **Lainhart, W., G. Stolfa, and G. B. Koudelka.** 2009. Shiga toxin as a bacterial defense against a eukaryotic predator, *Tetrahymena thermophila*. *J. Bacteriol.* **191:**5116–5122.
76. **Lejeune, J. T., S. T. Abedon, K. Takemura, N. P. Christie, and S. Sreevatsan.** 2004. Human *Escherichia coli* O157:H7 genetic marker in isolates of bovine origin. *Emerg. Infect. Dis.* **10:**1482–1485.

77. **LeJeune, J. T., T. E. Besser, and D. D. Hancock.** 2001. Cattle water troughs as reservoirs of *Escherichia coli* O157. *Appl. Environ. Microbiol.* **67:**3053–3057.
78. **LeJeune, J. T., T. E. Besser, N. L. Merrill, D. H. Rice, and D. D. Hancock.** 2001. Livestock drinking water microbiology and the factors influencing the quality of drinking water offered to cattle. *J. Dairy Sci.* **84:**1856–1862.
79. **LeJeune, J. T., T. E. Besser, D. H. Rice, J. L. Berg, R. P. Stilborn, and D. D. Hancock.** 2004. Longitudinal study of fecal shedding of *Escherichia coli* O157:H7 in feedlot cattle: predominance and persistence of specific clonal types despite massive cattle population turnover. *Appl. Environ. Microbiol.* **70:**377–384.
80. **Leopold, S. R., V. Magrini, N. J. Holt, N. Shaikh, E. R. Mardis, J. Cagno, Y. Ogura, A. Iguchi, T. Hayashi, A. Mellmann, H. Karch, T. E. Besser, S. A. Sawyer, T. S. Whittam, and P. I. Tarr.** 2009. A precise reconstruction of the emergence and constrained radiations of *Escherichia coli* O157 portrayed by backbone concatenomic analysis. *Proc. Natl. Acad. Sci. USA* **106:**8713–8718.
81. **Liebana, E., R. P. Smith, E. Lindsay, I. McLaren, C. Cassar, F. A. Clifton-Hadley, and G. A. Paiba.** 2003. Genetic diversity among *Escherichia coli* O157:H7 isolates from bovines living on farms in England and Wales. *J. Clin. Microbiol.* **41:**3857–3860.
82. **Louie, M., S. Read, L. Louie, K. Ziebell, K. Rahn, A. Borczyk, and H. Lior.** 1999. Molecular typing methods to investigate transmission of *Escherichia coli* O157:H7 from cattle to humans. *Epidemiol. Infect.* **123:**17–24.
83. **Manning, S. D., A. S. Motiwala, A. C. Springman, W. Qi, D. W. Lacher, L. M. Ouellette, J. M. Mladonicky, P. Somsel, J. T. Rudrik, S. E. Dietrich, W. Zhang, B. Swaminathan, D. Alland, and T. S. Whittam.** 2008. Variation in virulence among clades of *Escherichia coli* O157:H7 associated with disease outbreaks. *Proc. Natl. Acad. Sci. USA* **105:**4868–4873.
84. **Matthews, L., I. J. McKendrick, H. Ternent, G. J. Gunn, B. Synge, and M. E. Woolhouse.** 2006. Super-shedding cattle and the transmission dynamics of *Escherichia coli* O157. *Epidemiol. Infect.* **134:**131–142.
85. **Maule, A.** 2000. Survival of verocytotoxigenic *Escherichia coli* O157 in soil, water and on surfaces. *Symp. Ser. Soc. Appl. Microbiol.* **2000:**71S–78S.
86. **McKee, M. L., A. R. Melton-Celsa, R. A. Moxley, D. H. Francis, and A. D. O'Brien.** 1995. Enterohemorrhagic *Escherichia coli* O157:H7 requires intimin to colonize the gnotobiotic pig intestine and to adhere to HEp-2 cells. *Infect. Immun.* **63:**3739–3744.
87. **McKee, M. L., and A. D. O'Brien.** 1996. Truncated enterohemorrhagic *Escherichia coli* (EHEC) O157:H7 intimin (EaeA) fusion proteins promote adherence of EHEC strains to HEp-2 cells. *Infect. Immun.* **64:**2225–2233.
88. **Mead, P. S., L. Slutsker, V. Dietz, L. F. McCaig, J. S. Bresee, C. Shapiro, P. M. Griffin, and R. V. Tauxe.** 1999. Food-related illness and death in the United States. *Emerg. Infect. Dis.* **5:**607–625.
89. **Mechie, S. C., P. A. Chapman, and C. A. Siddons.** 1997. A fifteen month study of *Escherichia coli* O157:H7 in a dairy herd. *Epidemiol. Infect.* **118:**17–25.
90. **Melton-Celsa, A. R., and A. D. O'Brien.** 2003. Animal models for STEC-mediated disease. *Methods Mol. Med.* **73:**291–305.
91. **Michel, P., J. B. Wilson, S. W. Martin, R. C. Clarke, S. A. McEwen, and C. L. Gyles.** 1999. Temporal and geographical distributions of reported cases of *Escherichia coli* O157:H7 infection in Ontario. *Epidemiol. Infect.* **122:**193–200.
92. **Milnes, A. S., I. Stewart, F. A. Clifton-Hadley, R. H. Davies, D. G. Newell, A. R. Sayers, T. Cheasty, C. Cassar, A. Ridley, A. J. Cook, S. J. Evans, C. J. Teale, R. P. Smith, A. McNally, M. Toszeghy, R. Futter, A. Kay, and G. A. Paiba.** 2008. Intestinal carriage of verocytotoxigenic *Escherichia coli* O157, *Salmonella*, thermophilic *Campylobacter* and *Yersinia enterocolitica*, in cattle, sheep and pigs at slaughter in Great Britain during 2003. *Epidemiol. Infect.* **136:**739–751.
93. **Mora, A., M. Blanco, J. E. Blanco, M. P. Alonso, G. Dhabi, F. Thomson-Carter, M. A. Usera, R. Bartolome, G. Prats, and J. Blanco.** 2004. Phage types and genotypes of Shiga toxin-producing *Escherichia coli* O157:H7 isolates from humans and animals

in Spain: identification and characterization of two predominating phage types (PT2 and PT8). *J. Clin. Microbiol.* **42:**4007–4015.

94. **Moran, N. A., P. H. Degnan, S. R. Santos, H. E. Dunbar, and H. Ochman.** 2005. The players in a mutualistic symbiosis: insects, bacteria, viruses, and virulence genes. *Proc. Natl. Acad. Sci. USA* **102:**16919–16926.
95. **Nart, P., N. Holden, S. P. McAteer, D. Wang, A. F. Flockhart, S. W. Naylor, J. C. Low, D. L. Gally, and J. F. Huntley.** 2008. Mucosal antibody responses of colonized cattle to *Escherichia coli* O157-secreted proteins, flagellin, outer membrane proteins and lipopolysaccharide. *FEMS Immunol. Med. Microbiol.* **52:**59–68.
96. **Nart, P., S. W. Naylor, J. F. Huntley, I. J. McKendrick, D. L. Gally, and J. C. Low.** 2008. Responses of cattle to gastrointestinal colonization by *Escherichia coli* O157: H7. *Infect. Immun.* **76:**5366–5372.
97. **Naylor, S. W., A. Flockhart, P. Nart, D. G. Smith, J. Huntley, D. L. Gally, and J. C. Low.** 2007. Shedding of *Escherichia coli* O157:H7 in calves is reduced by prior colonization with the homologous strain. *Appl. Environ. Microbiol.* **73:**3765–3767.
98. **Naylor, S. W., J. C. Low, T. E. Besser, A. Mahajan, G. J. Gunn, M. C. Pearce, I. J. McKendrick, D. G. Smith, and D. L. Gally.** 2003. Lymphoid follicle-dense mucosa at the terminal rectum is the principal site of colonization of enterohemorrhagic *Escherichia coli* O157:H7 in the bovine host. *Infect. Immun.* **71:**1505–1512.
99. **Ogden, I. D., M. MacRae, and N. J. Strachan.** 2004. Is the prevalence and shedding concentrations of *E. coli* O157 in beef cattle in Scotland seasonal? *FEMS Microbiol. Lett.* **233:**297–300.
100. **Oliver, K. M., N. A. Moran, and M. S. Hunter.** 2005. Variation in resistance to parasitism in aphids is due to symbionts not host genotype. *Proc. Natl. Acad. Sci. USA* **102:**12795–12800.
101. **Oporto, B., J. I. Esteban, G. Aduriz, R. A. Juste, and A. Hurtado.** 2008. *Escherichia coli* O157:H7 and non-O157 Shiga toxin-producing *E. coli* in healthy cattle, sheep and swine herds in Northern Spain. *Zoonoses Public Health* **55:**73–81.
102. **Paiba, G. A., J. C. Gibbens, S. J. Pascoe, J. W. Wilesmith, S. A. Kidd, C. Byrne, J. B. Ryan, R. P. Smith, M. McLaren, R. J. Futter, A. C. Kay, Y. E. Jones, S. A. Chappell, G. A. Willshaw, and T. Cheasty.** 2002. Faecal carriage of verocytotoxin-producing *Escherichia coli* O157 in cattle and sheep at slaughter in Great Britain. *Vet. Rec.* **150:**593–598.
103. **Palmeira, P., L. Yu Ito, C. Arslanian, and M. M. Carneiro-Sampaio.** 2007. Passive immunity acquisition of maternal anti-enterohemorrhagic *Escherichia coli* (EHEC) O157:H7 IgG antibodies by the newborn. *Eur. J. Pediatr.* **166:**413–419.
104. **Pruimboom-Brees, I. M., T. W. Morgan, M. R. Ackermann, E. D. Nystrom, J. E. Samuel, N. A. Cornick, and H. W. Moon.** 2000. Cattle lack vascular receptors for *Escherichia coli* O157:H7 Shiga toxins. *Proc. Natl. Acad. Sci. USA* **97:**10325–10329.
105. **Rangel, J. M., P. H. Sparling, C. Crowe, P. M. Griffin, and D. L. Swerdlow.** 2005. Epidemiology of *Escherichia coli* O157:H7 outbreaks, United States, 1982–2002. *Emerg. Infect. Dis.* **11:**603–609.
106. **Renter, D. G., J. M. Sargeant, R. D. Oberst, and M. Samadpour.** 2003. Diversity, frequency, and persistence of *Escherichia coli* O157 strains from range cattle environments. *Appl. Environ. Microbiol.* **69:**542–547.
107. **Reymond, D., R. P. Johnson, M. A. Karmali, M. Petric, M. Winkler, S. Johnson, K. Rahn, S. Renwick, J. Wilson, R. C. Clarke, and J. Spika.** 1996. Neutralizing antibodies to *Escherichia coli* Vero cytotoxin 1 and antibodies to O157 lipopolysaccharide in healthy farm family members and urban residents. *J. Clin. Microbiol.* **34:**2053–2057.
108. **Rice, D. H., K. M. McMenamin, L. C. Pritchett, D. D. Hancock, and T. E. Besser.** 1999. Genetic subtyping of *Escherichia coli* O157 isolates from 41 Pacific Northwest USA cattle farms. *Epidemiol. Infect.* **122:**479–484.
109. **Richards, H. A., D. Perez-Conesa, C. A. Doane, B. E. Gillespie, J. R. Mount, S. P. Oliver, P. Pangloli, and F. A. Draughon.** 2006. Genetic characterization of a diverse

*Escherichia coli* O157:H7 population from a variety of farm environments. *Foodborne Pathog. Dis.* **3:**259–265.

110. **Richards, M. S., J. D. Corkish, A. R. Sayers, I. M. McLaren, S. J. Evans, and C. Wray.** 1998. Studies of the presence of verocytotoxic *Escherichia coli* O157 in bovine faeces submitted for diagnostic purposes in England and Wales and on beef carcasses in abattoirs in the United Kingdom. *Epidemiol. Infect.* **120:**187–192.
111. **Riley, L. W., R. S. Remis, S. D. Helgerson, H. B. McGee, J. G. Wells, B. R. Davis, R. J. Hebert, E. S. Olcott, L. M. Johnson, N. T. Hargrett, P. A. Blake, and M. L. Cohen.** 1983. Hemorrhagic colitis associated with a rare *Escherichia coli* serotype. *N. Engl. J. Med.* **308:**681–685.
112. **Robinson, C. M., J. F. Sinclair, M. J. Smith, and A. D. O'Brien.** 2006. Shiga toxin of enterohemorrhagic *Escherichia coli* type O157:H7 promotes intestinal colonization. *Proc. Natl. Acad. Sci. USA* **103:**9667–9672.
113. **Savageau, M. A.** 1983. *Escherichia coli* habitats, cell types, and molecular mechanisms of gene control. *Am. Nat.* **122:**732–744.

113a. **Scallan, E., R. M. Hoekstra, E. F. Angulo, R. V. Tauxe, M. A. Widdowson, S. L. Roy, J. L. Jones, and P. M. Griffin.** 2011. Foodborne illness acquired in the United States—major pathogens. *Emerg. Infect. Dis.* **17:**7–15.

114. **Schouten, J. M., M. Bouwknegt, A. W. van de Giessen, K. Frankena, M. C. De Jong, and E. A. Graat.** 2004. Prevalence estimation and risk factors for *Escherichia coli* O157 on Dutch dairy farms. *Prev. Vet. Med.* **64:**49–61.
115. **Schouten, J. M., E. A. Graat, K. Frankena, A. W. van de Giessen, W. K. van der Zwaluw, and M. C. de Jong.** 2005. A longitudinal study of *Escherichia coli* O157 in cattle of a Dutch dairy farm and in the farm environment. *Vet. Microbiol.* **107:**193–204.
116. **Schutze, G. E., R. S. Kirby, E. L. Flick, R. Stefanova, K. D. Eisenach, and M. D. Cave.** 1998. Epidemiology and molecular identification of *Salmonella* infections in children. *Arch. Pediatr. Adolesc. Med.* **152:**659–664.
117. **Scott, L., P. McGee, D. Minihan, J. J. Sheridan, B. Earley, and N. Leonard.** 2006. The characterisation of *E. coli* O157:H7 isolates from cattle faeces and feedlot environment using PFGE. *Vet. Microbiol.* **114:**331–336.
118. **Shaikh, N., and P. I. Tarr.** 2003. *Escherichia coli* O157:H7 Shiga toxin-encoding bacteriophages: integrations, excisions, truncations, and evolutionary implications. *J. Bacteriol.* **185:**3596–3605.
119. **Sharma, R., K. Stanford, M. Louie, K. Munns, S. J. John, Y. Zhang, V. Gannon, L. Chui, R. Read, E. Topp, and T. McAllister.** 2009. *Escherichia coli* O157:H7 lineages in healthy beef and dairy cattle and clinical human cases in Alberta, Canada. *J. Food Prot.* **72:**601–607.
120. **Sheng, H., J. Y. Lim, H. J. Knecht, J. Li, and C. J. Hovde.** 2006. Role of *Escherichia coli* O157:H7 virulence factors in colonization at the bovine terminal rectal mucosa. *Infect. Immun.* **74:**4685–4693.
121. **Shere, J. A., K. J. Bartlett, and C. W. Kaspar.** 1998. Longitudinal study of *Escherichia coli* O157:H7 dissemination on four dairy farms in Wisconsin. *Appl. Environ. Microbiol.* **64:**1390–1399.
122. **Silvestro, L., M. Caputo, S. Blancato, L. Decastelli, A. Fioravanti, R. Tozzoli, S. Morabito, and A. Caprioli.** 2004. Asymptomatic carriage of verocytotoxin-producing *Escherichia coli* O157 in farm workers in Northern Italy. *Epidemiol. Infect.* **132:**915–919.
123. **Sivapalasingam, S., C. R. Friedman, L. Cohen, and R. V. Tauxe.** 2004. Fresh produce: a growing cause of outbreaks of foodborne illness in the United States, 1973 through 1997. *J. Food Prot.* **67:**2342–2353.
124. **Snedeker, K. G., D. J. Shaw, M. E. Locking, and R. J. Prescott.** 2009. Primary and secondary cases in *Escherichia coli* O157 outbreaks: a statistical analysis. *BMC Infect. Dis.* **9:**144.

125. **Solecki, O., M. MacRae, N. Strachan, B. A. Lindstedt, and I. Ogden.** 2009. *E. coli* O157 from sheep in northeast Scotland: prevalence, concentration shed, and molecular characterization by multilocus variable tandem repeat analysis. *Foodborne Pathog. Dis.* **6:**849–854.
126. **Stanford, K., S. J. Bach, T. H. Marx, S. Jones, J. R. Hansen, G. L. Wallins, H. Zahiroddini, and T. A. McAllister.** 2005. Monitoring *Escherichia coli* O157:H7 in inoculated and naturally colonized feedlot cattle and their environment. *J. Food Prot.* **68:**26–33.
127. **Stanford, K., D. Croy, S. J. Bach, G. L. Wallins, H. Zahiroddini, and T. A. McAllister.** 2005. Ecology of *Escherichia coli* O157:H7 in commercial dairies in southern Alberta. *J. Dairy Sci.* **88:**4441–4451.
128. **Stavric, S., B. Buchanan, and T. M. Gleeson.** 1993. Intestinal colonization of young chicks with *Escherichia coli* O157:H7 and other verotoxin-producing serotypes. *J. Appl. Bacteriol.* **74:**557–563.
129. **Synge, B. A., M. E. Chase-Topping, G. F. Hopkins, I. J. McKendrick, F. Thomson-Carter, D. Gray, S. M. Rusbridge, F. I. Munro, G. Foster, and G. J. Gunn.** 2003. Factors influencing the shedding of verocytotoxin-producing *Escherichia coli* O157 by beef suckler cows. *Epidemiol. Infect.* **130:**301–312.
130. **Tarr, P. I., C. A. Gordon, and W. L. Chandler.** 2005. Shiga-toxin-producing *Escherichia coli* and haemolytic uraemic syndrome. *Lancet* **365:**1073–1086.
131. **Tarr, P. I., M. A. Neill, D. L. Christie, and D. E. Anderson.** 1988. *Escherichia coli* O157:H7 hemorrhagic colitis. *N. Engl. J. Med.* **318:**1697.
132. **Teunis, P., K. Takumi, and K. Shinagawa.** 2004. Dose response for infection by *Escherichia coli* O157:H7 from outbreak data. *Risk Anal.* **24:**401–407.
133. **Teunis, P. F., I. D. Ogden, and N. J. Strachan.** 2008. Hierarchical dose response of *E. coli* O157:H7 from human outbreaks incorporating heterogeneity in exposure. *Epidemiol. Infect.* **136:**761–770.
134. **Tuttle, J., T. Gomez, M. P. Doyle, J. G. Wells, T. Zhao, R. V. Tauxe, and P. M. Griffin.** 1999. Lessons from a large outbreak of *Escherichia coli* O157:H7 infections: insights into the infectious dose and method of widespread contamination of hamburger patties. *Epidemiol. Infect.* **122:**185–192.
135. **Vanaja, S. K., A. C. Springman, T. E. Besser, T. S. Whittam, and S. D. Manning.** 2010. Differential expression of virulence and stress fitness genes between *Escherichia coli* O157:H7 strains with clinical or bovine-biased genotypes. *Appl. Environ. Microbiol.* **76:**60–68.
136. **Van Donkersgoed, J., T. Graham, and V. Gannon.** 1999. The prevalence of verotoxins, *Escherichia coli* O157:H7, and *Salmonella* in the feces and rumen of cattle at processing. *Can. Vet. J.* **40:**332–338.
137. **Van Donkersgoed, J., J. Berg, A. Potter, D. Hancock, T. Besser, D. Rice, J. LeJeune, and S. Klashinsky.** 2001. Environmental sources and transmission of *Escherichia coli* O157 in feedlot cattle. *Can. Vet. J.* **42:**714–720.
138. **Voetsch, A. C., M. H. Kennedy, W. E. Keene, K. E. Smith, T. Rabatsky-Ehr, S. Zansky, S. M. Thomas, J. Mohle-Boetani, P. H. Sparling, M. B. McGavern, and P. S. Mead.** 2007. Risk factors for sporadic Shiga toxin-producing *Escherichia coli* O157 infections in FoodNet sites, 1999–2000. *Epidemiol. Infect.* **135:**993–1000.
139. **Weber, D. J., and W. A. Rutala.** 2001. The emerging nosocomial pathogens *Cryptosporidium, Escherichia coli* O157:H7, *Helicobacter pylori*, and hepatitis C: epidemiology, environmental survival, efficacy of disinfection, and control measures. *Infect. Control Hosp. Epidemiol.* **22:**306–315.
140. **Werber, D., S. C. Behnke, A. Fruth, R. Merle, S. Menzler, S. Glaser, L. Kreienbrock, R. Prager, H. Tschäpe, P. Roggentin, J. Bockemühl, and A. Ammon.** 2007. Shiga toxin-producing *Escherichia coli* infection in Germany: different risk factors for different age groups. *Am. J. Epidemiol.* **165:**425–434.
141. **Wetzel, A. N., and J. T. LeJeune.** 2006. Clonal dissemination of *Escherichia coli* O157:H7 subtypes among dairy farms in northeast Ohio. *Appl. Environ. Microbiol.* **72:**2621–2626.

142. **Williams, M. S., J. L. Withee, E. D. Ebel, N. E. Bauer, W. D. Schlosser, W. T. Disney, D. R. Smith, and R. A. Moxley.** 2010. Determining relationships between the seasonal occurrence of *Escherichia coli* O157:H7 in live cattle, ground beef, and humans. *Foodborne Pathog. Dis.* **7:**1247–1254.
143. **Wilson, J. B., R. C. Clarke, S. A. Renwick, K. Rahn, R. P. Johnson, M. A. Karmali, H. Lior, D. Alves, C. L. Gyles, K. S. Sandhu, S. A. McEwen, and J. S. Spika.** 1996. Vero cytotoxigenic *Escherichia coli* infection in dairy farm families. *J. Infect. Dis.* **174:** 1021–1027.
144. **Yang, Z., J. Kovar, J. Kim, J. Nietfeldt, D. R. Smith, R. A. Moxley, M. E. Olson, P. D. Fey, and A. K. Benson.** 2004. Identification of common subpopulations of non-sorbitol-fermenting, β-glucuronidase-negative *Escherichia coli* O157:H7 from bovine production environments and human clinical samples. *Appl. Environ. Microbiol.* **70:** 6846–6854.
145. **Yeh, K. S., S. P. Chen, and J. H. Lin.** 2005. One-year (2003) nationwide pork carcass microbiological baseline data survey in Taiwan. *J. Food Prot.* **68:**458–461.
146. **Yoon, J. W., and C. J. Hovde.** 2008. All blood, no stool: enterohemorrhagic *Escherichia coli* O157:H7 infection. *J. Vet. Sci.* **9:**219–231.

# INDEX